BASIC ELECTRONICS

Suitable as a text book for undergraduate students in Engineering Colleges and Universities
in
Electrical Engineering, Electronics and Tele-communication Engineering,
Computer Science Engineering and other Engineering Branches.
Also suitable for AMIE and grad IETE syllabus and GATE, UPSC,
Institute of Metals and other Examinations.

DR. SAMBHUNATH BISWAS
B.SC. (Hons.), B.Tech., M.Tech., D.Phil.
P.R.S. LMISTE, LMSESI, (Ex.) DAAD Fellow
Professor (Ex-Head)
Departent of Electronics and Telecommunication Engineering
Bengal Engineering College (Deemed University)
Batonic Garden, Shibpur, Howrah

KHANNA BOOK PUBLISHING CO. (P) LTD.
Publisher of Engineering and Computer Books
4C/4344, Ansari Road, Darya Ganj, New Delhi-110002
Phone : 011-23244447-48 **Mobile:** +91-9910909320
E-mail : contact@khannabooks.com
Website : www.khannabooks.com

Dear Readers,

To prevent the piracy, this book is secured with HIGH SECURITY HOLOGRAM on the front title cover. In case you don't find the hologram on the front cover title, please write us to at contact@khannabooks.com or whatsapp us at +91-99109 09320 and avail special gift voucher for yourself.

Specimen of Hologram on front Cover title:

Moreover, there is a SPECIAL DISCOUNT COUPON for you with EVERY HOLOGRAM.

How to avail this SPECIAL DISCOUNT:

Step 1: Scratch the hologram
Step 2: Under the scratch area, your "coupon code" is available
Step 3: Logon to www.khannabooks.com
Step 4: Use your "coupon code" in the shopping cart and get your copy at a special discount
Step 5: Enjoy your reading!

Thanks,
Khanna Publishing House

Price: ₹ 725.00

BASIC ELECTRONICS
by **Sambhunath Biswas**

ISBN: 978-81-87522-16-4

First Edition: 2000
Reprint: 2021, 2023, **2025**

To view complete list of books, please scan the QR Code:

Published by:
Khanna Book Publishing Co. (P) Ltd.
CIN: U22110DL1998PTC095547

Visit us at: www.khannabooks.com
Write us at: contact@khannabooks.com

Printed in Bharat.

Gram : TECHNICS
E.Mail : director@becs.ac.in

Dr. S. M. Chatterjee,
B.Sc., B.Sc. Tech. (Cal.), M. Text. (Bom.),
Ph.D. (Cal.), A.M.I.E.T. (Lond.), M.I.S.T.E.,
F.A.E (Ind.), F.I.S.E., F.T.A. (Ind.), F.I.E.,
Gold Medallist, Recipients of National Unity Award,
Hind Gaurav Award & Kabi Sekhar Kalidas Roy Award.

VICE-CHANCELLOR

Phones : Off. : (033) 668-4561— 668-2916
Res. : 668-2674
Fax : (0091) (033) 668-2916

BENGAL ENGINEERING COLLEGE
(DEEMED UNIVERSITY)
P.O. BOTANIC GARDEN. HOWRAH-711 103.

Dated: 31.03.2000.

FOREWORD

During the past three decades, tremendous developments have taken place in the field of electronics and telecommunication engineering. As a result of this developments, enormous progress has been made in the fields of radio, television, computers, communications, medical electronics, fiber optics, industrial electronics, fascimile, electronic mail and internet. The subject of electronics has attracted the attention of all walks of life.

In order to meet the requirements of a good and uptodate book on basic electronics, the book written by Dr. Biswas, Professor (Ex-Head) Department of Electronics and Telecommunication Engineering is both relevant and timely. It discusses in a clear and concise way the fundamental principles and applications of basic electronics. The language is simple and easily understandable. The book has a large number of objective questions, solved problems and exercises problems. I am confident that the book will be well received and widely used.

I congratulate Dr. Biswas for his above endeavour and hope he will continue to write more such books on electronics and allied subjects in the future.

31/3/2K

(S.M. Chatterjee)

Vice-Chancellor

DEDICATION

The fond and sacred memory of

My grand father

Late Niberan Chandra Biswas

and

My Grand Mother

Late Ashalata Biswas.

About the Author

Dr. Sambunath Biswas had his schooling from Howrah Akshya Sikshayatan. He then joined the St. Xaviers College for the B.Sc. (Hons) degree in Physics from the University of Calcutta. Subsequently, he obtained the B.Tec and M.Tech degrees in Radio Physics and Electronics from the University of Calcutta in 1967 and 1968 respectively. Later on, he obtained the degree of Doctor of Philosophy from the University of Calcutta in Radio Physics and Electronics. He is also the winner of the PRS award of the University of Calcutta. He is a MOVAT medalist of the University of Calcutta. In March 1972, he joined the Department of Electronics and Telecommunication Engineering, Bengal Engineering College as a Lecturer. At present, he is working in the same department as a Professor. He was the Head of the Department of Electronics and Telecommunication Enginnering during the periods April 1988 to March 1989 and April 1995 to March 1998. He visited the Federal Republic of Germany on DAAD Fellowship during the periods December 1979 to March 1981 and also in 1992.

PREFACE

This is an age of electronics. At the dawn of the new millenium, it is no denying the fact that electronics has influenced the lifestyles of mankind in a manner never seen before. One hundred years back, our forefathers could not dream of a world connected by the magic of the internet (and World Wide Web, WWW), fascimile e-mails, mobile phones and other marvels of the present day world. The rapid developments in the field of electronics has attracted the attention of all the young and the old.

In order to understand the fundamentals of electronics, basic electronics is now taught as a compulsory subject for students of all branches of engineering. This book is planned to meet the requirements of a good and up-to-date book on basic electronics. The book discusses in a clear and concise way the fundamental principles and applications of basic electronics. The readers should find the book interesting particularly with a large number of objective questions, solved problems and exercise problems.

The author will be very happy if the book gains the appreciation of its readers. The author would be grateful for suggestion for improvement of the book.

The author is extremely grateful to the authorities of the Bengal Engineering College (Deemed University) for kind permission to publish the book.

The author gratefully acknowledges his parents Shri Sanat Kumar Biswas and Srimati Sushamarani Biswas for their heartfelt blessings and kind inspiration. The author also thanks his wife Sm. Indrani Biswas for her assistance in the preparation of the manuscript. He also places his grateful appreciation to other family members (particularly Sowvik) for their understanding support and assistance during the preparation of the text.

The author is grateful to the publishers and other staff members of M/s Khanna Book Publishing Co. (P) Ltd. for their sincere efforts in the publication of this book.

CONTENTS

CHAPTER 1
INTRODUCTION TO ELECTRONICS

CHAPTER 2
FUNDAMENTAL CONCEPTS : ENERGY BANDS IN SOLID

CHAPTER 3
SEMICONDUCTOR DIODES AND MISCELLANEOUS DEVICES

CHAPTER 4
BIPOLAR JUNCTION TRANSISTORS

CHAPTER 5
BIPOLAR TRANSISTOR BIASING

CHAPTER 6
SINGLE STAGE BJT AMPLIFIERS

CHAPTER 7 FIELD EFFECT TRANSISTORS

CHAPTER 8 POWER AMPLIFIERS

CHAPTER 9
FREQUENCY RESPONSE OF AMPLIFIERS

CHAPTER 10
FEEDBACK IN AMPLIFIERS

CHAPTER 11
OSCILLATORS AND MULTIVIBRATORS

CHAPTER 12
MODULATION AND DEMODULATION

CHAPTER 13
INTEGRATED CIRCUITS

CHAPTER 14
OPERATIONAL AMPLIFIERS

CHAPTER 15
TELEVISION

Chapter 1

INTRODUCTION TO ELECTRONICS

1.1 FUNDAMENTALS

Electronics is the branch of science which deals with the motion of electrons in solids, liquids and gases under various operating conditions. The Institution of Radio Engineers (IRE) defines the subject as the field of Science and Engineering which deals with electron devices and their utilization (Proceedings IRE, Vol. 38, 1950)

The branch of electronics engineering was originally considered as a part of electrical engineering. However, the tremendous developments during the last few decades have changed the lifestyles of human beings at the dawn of this milliennium.

The field of electronics now encompasses radio, television, computers, communications, mobile phones and Iridium, fiber optics, medical, industrial aerospace applications, facsimile, electronic mail (e mail) and Internet. All practical applications of electronics are based on the fundamental principles of electricity and magnetism. Besides, magnetism can be produced by an electric current. A simple battery can be used for supplying power to a calculater, a computer, a digital wristwatch, a television, the remote control of a colour TV, a cellular phone, a video cassette recorder or an orbiting satellite.

The study of the science and technology of electricity and electronics is very exciting. The common people have looked at the developments of electronics with astonishment and joy and have become ardent fans of it. Our forefathers could not dream of a world where a colour television displaying a live cricket match held in another continent. They could not dream of a world where all the computers are connected by Internet for almost any applications. The Internet is the World's most versatile and most easily accessible computer network that provides enormous amount of information to anyone who possesses a personal computer, a modem and an Individual internet address. The network permits millions of computer systems operating on different platforms around the globe to smoothly communicate with one another in such a fashion that the individual computer appear to be part of the network. The Internet is a conglomeration of small networks and other machines around the globe.

The internet provides opportunities for education, research, advertising, entertainment, electronic mail, access to commercial activities and other applications. Media, entertainment, commerce and education are on the verge of major transformation as the Internet becomes all pervading in global communication and interaction.

Our forefathers could not dream of a world where one can speak from one's bedroom to his friend in another continent almost with no noise (as if speaking with a friend the next room). All these have become possible with the treamendous developments is the field of electronics during the last fifty years or so. Electronics has provided mankind gifts in the form of *telephones, radio, television, facsimile, e-mail, cellular phone, Iridium, radio pagers, internet etc. Electronics has truly transformed the World into a global village.*

The developments of modern electronics can be traced to a german Physicist Heinrich Hertz who in 1887 demonstrated the effects of electromagnetic radiation through space. It has been shown that both radio waves and light waves are examples of electromagnetic radiation. Even before the demonstration of Hertz,it was proved by Michael Faraday in 1831 that a magnet in motion can generate electricity. Besides, in 1864, James Clark Maxwell predicted electromagnetic waves.

Thus, in early history, electronics owes a lot to the pioneering contributions by Heinrich Hertz, James Clark Maxwell, W. Oersted, Michael Faraday and others. Many electrical units have been named after these famous scientists. Thus Hertz is the basic unit of frequency (which indicates the number of alternations in one second). The ampere is the basic unit of current while ohm is the unit of resistance. The unit of Farad (after Michael Faraday) is the basic unit of capacitance, Henry is the basic unit of inductance, Watt is the basic unit of power while the Siemens is the basic unit of conductance. Joule is the basic unit of electrical energy and Weber is the unit of magnetic flux.

Guglielmo Marconi is credited with the transmission of first radio wave through air. However, many believe that Sir J.C. Bose did the first radio transmission experiment. Marconi is credited with wireless communication across the Atlantic is 1901. Later on in 1906, Lee De Forest invented the andion tube. The first radio broadcast was done in 1920. The first FM (frequency modulated) trammission took place in 1939 while the first FM stereo became available to listener in 1961. The first commercial TV broadcast took place in 1941 while the colour television system became available to viewers in 1953.

Perhaps, the most important revolutionary invention in the field of electronics took place with the development of transistor in 1948 by John Bardeen, Walter Brattain and William Shockley of Bell Telephone Laboratories. Since World War II, there was a need for miniaturization of electronic devices and systems. The invention of the transistor immediately made possible the miniaturisation of all types of electronic systems and equipments and resulted in more reliability, higher speed and lower cost. The following Table lists the important milestones in the field of electronics.

Table 1.1 : Important Milestones in Electronics

Year	Event
1800-1837	Volta discovers primary battery, Experiments on electricity and magnetism by Oersted, Ampere, Faraday and Henry, Ohm's law (1826) Michael Faraday (1831) discovered that a magnet in motion can generate electricity.
1838-1866	Morse perfects his telegraphy system
1845	Kirchoff's circuit laws enunciated
1876-1899	Telephony : Acoustic transducer perfected by Alexandar Graham Bell. Strowger devised automatic step by step switching.

1901	Marconi credited with first wireless communication across Atlantic.
1904-1920	Lee de Forest invented the audion (triode) based on Fleming's diode. Experiments with AM radio broadcasting, transcontinental telephone line with electronic repeaters, multiplexed carrier teleplony introduced. Armstrong perfects super heterodyne radio receiver (1918).
1923-28	Television : Mechanical image formation system demonstrated by Baird and Jenkins.
1936	Armstrong predicts FM radio
1937	Alec Reeves conceived pulse code modulation
1938-1945	World War II : Radar and microware system developed. FM used extensively for military communications.
1948-1951	Transistor invented by Shockley, Bardeen and Brattain
1953	Colour TV Standard established in USA
1955	JR Pierce proposed satellite communication
1957	First transoceanic telephone cable (36 voice channels)
1958	Development of Thyristors, first laser
1960	Maiman demonstrates first laser
1961	Integrated circuits go into commercial production
1962	Invention of semiconductor (Ga As) diode laser
1963	Solid State microware oscillator perfected by Gunn.
1965	Mariner IV transmits picture from Mars to earth
1966-1975	Wideband Communication Systems, Cable TV systems, Commercial satellite relay service, optical links using lasers and fiber optics.
1975-1988	HF power MOS devices, Gate turn off (GTO), Bipolar Power Transistors, Insulated Gate Bipolar Power Transistors (IGBT).
1990-2000	Widespread use of Fiber Optic Communication; Computers; Internet, World Wide Web, LAN, WAN, MAN, Satellite Communication, Photovoltaic modules; Radio paging, cellular telephones, e-mail, e-commerce, fax (facsimile). World reduced to a global village.

1.2 RESISTORS

Resistance is opposition to the flow of current in a circuit

The two main characteristics of a resistor are its resistance value R in ohm and the power rating P in Watts. Resistors are available in a very wide range of resistance values from a fraction of an ohm to many kilo ohm and Mega ohm (Ω). The power rating of a resistor may be as high as several hundred watts or as low as 1/10 W.

Normally, resistors which possess high values have low wattage ratings since they passess less current A typical value is 1 Ω at 1/4 W for a resistor only 2 inches long. The smaller the power rating the lower is the actual physical size of the resistor. The resistance value is, however, not related to the physical size.

Table 1.2

Electronics

- Consumer Electronics Communications Entertainment
  - Line Communications
    - Telegraphy
    - Talex
    - Telephony
    - Teleprinters
  - Wireless Communications
    - Radio Transmission
    - TV Transmission
    - Microwave Communications
    - Facsimile (FAX)
    - Fiber Optic Communications
  - Sound Systems
- Industry
  - Automatic Control Systems
  - DC Drives
  - AC Drives
  - Heating, Welding
  - Digital Computers
  - Programmable logic Controller (PLC)
  - Internet
- Defence
  - Radar ; Guided missiles ; Coded communication ; image processing
- Medical
  - X-rays
  - Electrocardiography (ECG)
  - Electroencephalography (EEG)
  - Computerized Tomography (CT)
  - Magnetic Resonance Imaging (MRI)
- Instrumentation Measurements Process Control
  - Precision Instt. Digital multimeters, (CRO) Spectrum Analyzer, Frequency Counters; PH meters Strain gauge; Load cell fiber optic; Gyroscopes and other Transducers

1.2.1 Wire wound resistors

In the wire wound resistors, a resistance wire typically made of tungsten and manganin is wrapped around on insulating core made of porcelain, cement or plain pressed paper. Typical resistance values are 1 Ω to thousand ohms. Wattage ratings of 5W to 100 W are used.

1.2.2 Carbon composition resistors

(The values range between 1 Ω to 20 M Ω). Typical resistance values are 10, 15, 18, 22, 27, 33, 39, 47, 56, 68, 82, 100 Ω and decimal multipliers of them. Thus it is possible to have resistance values readily for say 390 Ω, 3900 Ω, 39K Ω, 390K Ω, 3.9M Ω etc. The power rating is $\frac{1}{10}, \frac{1}{8}, \frac{1}{4}, \frac{1}{2}$, 1 or 2 watts. The resistor is made of finely divided carbon or graphite mixed with a powdered insulating material as a binder in the proportion needed for the desired R value. Two metal caps with leads of tinned copper wire for soldering the connections to the circuit. The resistor element is encapsulated in a plastic casing for insulation and mechanical strength.

1.2.3 Film type resistors

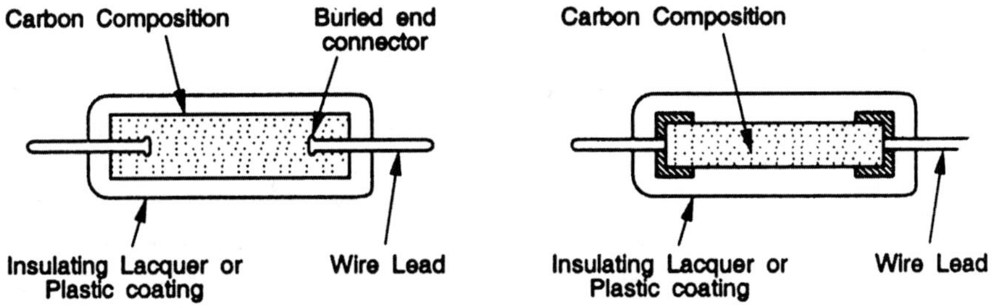

Fig 1.1 Carbon composition resistor

Film type resistors are of two types.

(a) Carbon film and (b) metal film type resistors.

Figs. 1.2 and 1.3 depict respectively the schematic diagram of carbon film and metal film resistors.

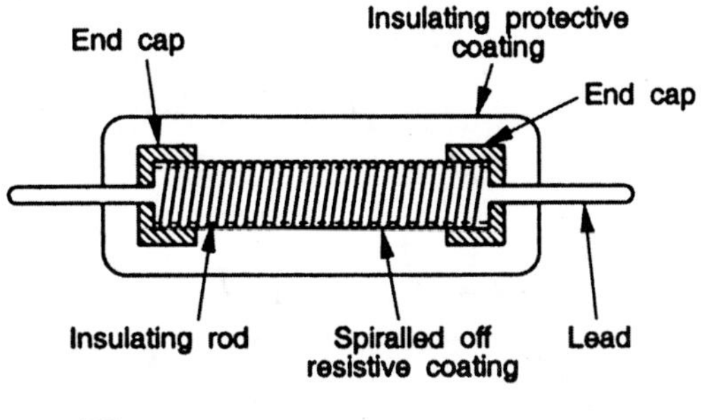

Fig. 1.2. Carbon film resistor

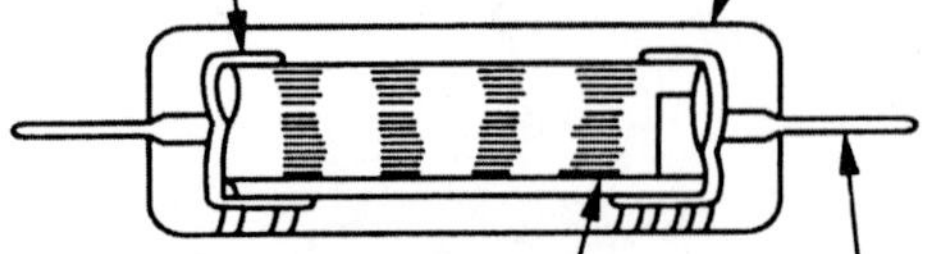

Fig. 1.3. Metal film resistor.

Carbon film resistors are fabricated by deposition of a thin layer of carbon on a ceramic substrate. Grooves are then cut over the film so that the resistive element becomes essentially a spiral. The length, width and thickness of the metal spiral determine the exact resistance value. Metal film resistors possess more precision R values. The film resistors are less influenced by temperature and ageing. They generate very little noise internally. In the case of film resistors, the diameters at the end are a bit larger compared to the body. In terms of precision resistance values, metal film are the best followed by carbon film resistors and carbon composition resistors.

1.2.4 Chip resistors or Surface mount resistors

Chip resistors are fabricated by deposition of a thick carbon film over a ceramic base. The exact resistance value is found by the composition of carbon itself, also the amount of trimming is done by carbon deposit. Surface mount resisters should never be robbed or scrapped as this will remove some of the carbon deposits and may make the resistance to change. The resistance values may be from fraction of an ohm to 1 million ohms (wattage values $\frac{1}{8}$ W to $\frac{1}{2}$ W). The resistors are temperature stable and very rugged.

Fusible resistors

It is a type of wire wound resistor that will burn open easily when the power rating is exceeded. It thus, serves the dual purpose of a fuse and a resistor.

1.2.5 Resistor colour code values

The colour codes values for resistors is given in Table 1.3 below

Table 1.3 COLOUR CODE

Colour	Value	Colour	Value
Black	0	Green	5
Brown	1	Blue	6
Red	2	Violet	7
Orange	3	Gray	8
Yellow	4	White	9

Fig. 1.4. illustrates how to read the colour values. The first digit gives the first band (Band A). The

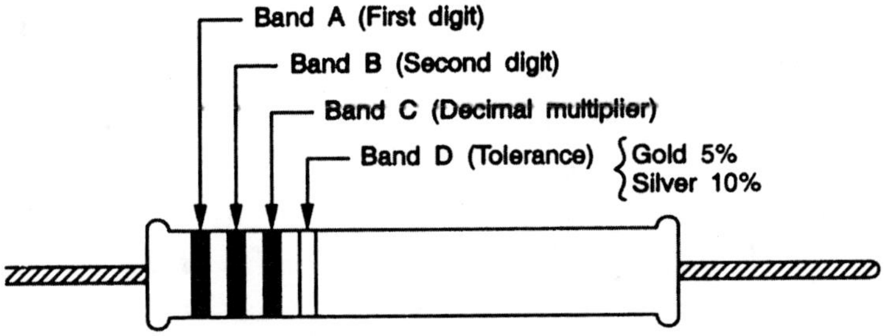

Fig. 1.4. How to read resistance values : a typical example

second digit gives the second band (Band B). The third band gives the decimal multiplier. Thus if the first three colours are red, green and orange, the value of the resistor is

$$25000 = 25\ K\Omega.$$

The fourth band gives the tolerance. Thus, if, it is silver ($\pm$ 10%), the resistance value is 25 KΩ ± 10% or the value would lie between 22·5 KΩ and 27·5 KΩ. The following table 1.4 gives the colour code for some typical resistors.

Table 1.4 : Some typical colour codes

Colour Code Value	First 4 Colour band is succession
10 Ω, ± 10%	Brown, Black, Black, Silver
120 Ω, ± 10%	Brown, Red, Brown, Silver
150 Ω, ± 10%	Brown, Green, Brown, Silver
180 Ω, ± 10%	Brown, Green, Brown, Silver
22 Ω, ± 5%	Red, Red, Orange, Gold
680 Ω, ± 10%	Blue, Gray, Yellow, Silver
3·9 Ω, ± 10%	Orange, White, Red, Gray
1 M Ω, ± 10%	Brown, Black, Green, Silver

Resistors below 10Ω

For resistance values below 10Ω, the third colour band is either gold or silver. Thus the colour combination of Orange, white and gold as the first three bands indicate a value of $39 \times 0{\cdot}1 = 3{\cdot}9\ \Omega$

For a colour combination of Orange, white and silver as the first three bands, the colour code value is $39 \times 0{\cdot}01 = 0{\cdot}39\ \Omega$. Thus the third colour band gold indicates a multiplication by $\left(\frac{1}{10}\right)$, while silver indicates, a multiplication by $\left(\frac{1}{100}\right)$

1.2.6 Five Band Code

The precision resistors are given by a five band colour code. The first three bands indicate the first three digits, the fourth band indicates the decimal multiplier and the fifth band indicates the tolerance as depicted in Fig. 1.5. In the fifth stripe, the colours brown, red, green, blue and violet indicate the following values.

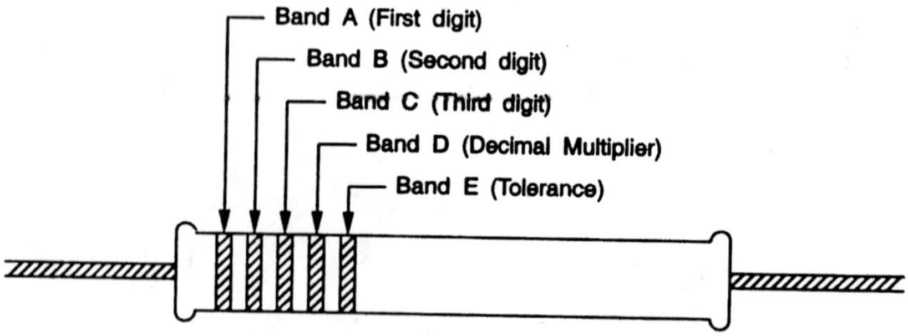

Fig. 1.5. Five band Colour coded resister.

Brown ± 1%
Red ± 2%
Green ± 0·5%
Blue ± 0·25%
Violet ± 0·1%

Mnemonics

As an aid to keep in memory the sequence of colour code values in Table 1.2, the student will remember well the following sentence (all the capital letters standing for colours)

B B ROY of Great Britain has a Very Good Wife.

EXAMPLE 1.1 *A resistor has a colour band sequence Gray, Red, Red, Gold. Find the range in which its value must be so that the manufacturer's tolerance is satisfied.*

SOLUTION With the help of the colour code table, we find

1st band	2nd band	3rd band	4th band
Gray	Red	Red	Gold
8	2	10^2	± 5%

= 8·2K Ω ± 5%

= 8200 ± 5% = 8200 ± 41

= 8159 8241 Ω

The resistance should be between 8159 and 8241Ω.

EXAMPLE 1.2 *A resistor has a colour band sequence Orange, Orange, Yellow and Silver. Find the range in which its value must be so that the manufacturer's tolerance is satisfied ?*

SOLUTION With standand colour codes, the value is

First band	Second band	Third band	Fourth band
Orange	Orange	Yellow	Silver
3	3	10^4	± 10%

The value is thus 330 K ± 10%

$= 33000 \pm 330\ \Omega$

$= 32670$ to $33330\ \Omega$

EXAMPLE 1.3 *In a precision resistor, having band colour codes, the colour band sequence is Yellow, Gray, Violet, Black and Green. Find the range within which the resistance value could be as per manufacturer's specifications.*

First band	Second band	Third band	Fourth band	Fifth band
Yellow	Gray	Violet	Black	Green
4	8	7	10°	$\pm 5\%$

SOLUTION $R = 487 \pm 0{\cdot}5\% = 487 \pm 2{\cdot}44\Omega$

$484{\cdot}56\ \Omega$ to $489{\cdot}44\ \Omega$.

Wire Wound Resistor Coding

The wire wound resistors are physically big in size and have their values printed in the insulating case. For wire wound resistors, the tolerance is $\pm 5\%$ although for precision resistors the value is $\pm 1\%$.

1.2.7 Zero ohm resistors

They are required for most printed circuit board operation where the components are inserted by automatic insertion machines (robots) rather than human hands. The actual resistance of a $\frac{1}{8}$ W zero ohm resistor is $0{\cdot}004\ \Omega$ while the value for a $\frac{1}{4}$ W zero ohm resistor is $0{\cdot}003\ \Omega$. Zero ohm resistors are available in $\frac{1}{4}$ W and $\frac{1}{8}$ W sizes. Becase the robot can handle only components such as resistors, and not wires zero ohm resistors are used. The zero ohm resistor is indicated by a single black colour band around the body of the resistor as depicted in Fig. 1.6.

Chip Resistor

For a chip resistor the following identifiable features are employed. For a chip resistor the body colour is white or off–white and the end electrodes are C-shaped. There is a dark film on one side typically, black, dark gray or green. The three digit marking is usually marked on the film. A cnip resistor is marked as 402. It's resistance value is thus $40 \times 10^2 = 4000\ \Omega$.

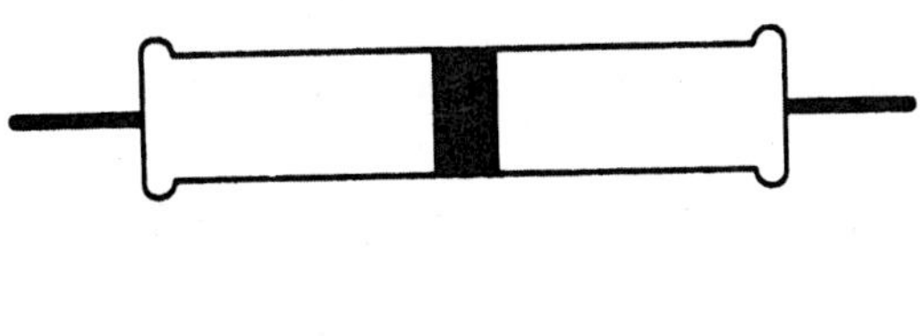

Fig. 1.6. A zero ohm resistor

Similarly, if the marking on a chip resistor has a value of 332 Ω the resistance value is 3300 Ω.

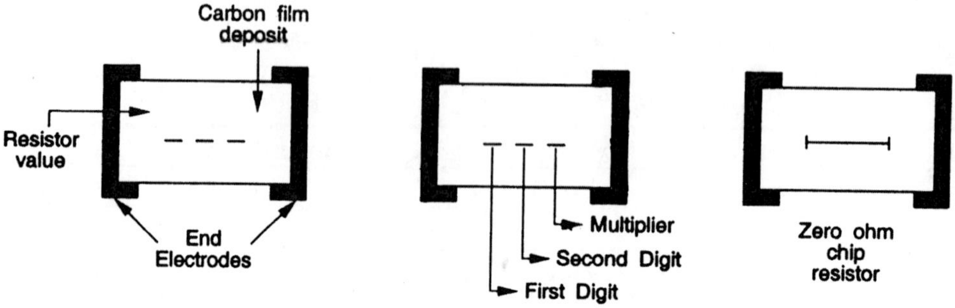

Fig. 1.7. Chip resistor colour coding.

1.2.8 Variable Resistors

A typical carbon composition type variable resistance is $\frac{3}{4}$" in diameter. It may have a thin coating on pressed paper or moulded carbon disk. Two external soldering lug terminals are connected to the two ends a and c. A metal spring wiper connects the resistor element and this acts as the middle terminal. As the control shaft is rotated, the wiper makes contact at various points and the resistance is varied. This is depicted in Fig. 1.8.

Variable carbon resistors are available in the range 1KΩ to 5 MΩ and their approximate power rating is $\frac{1}{2}$ to 2 Watts.

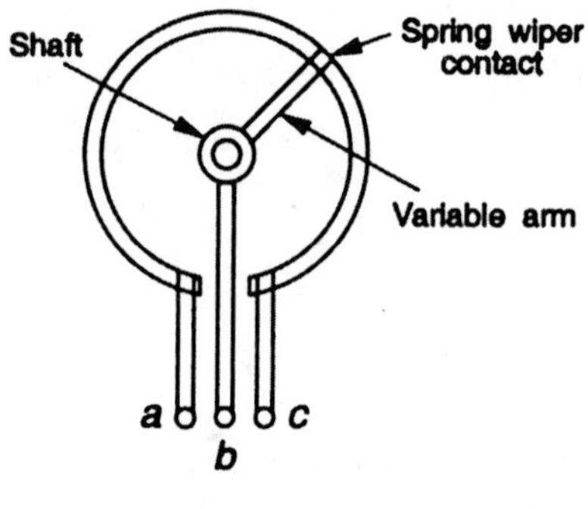

Fig. 1.8. Internal construction of a $\frac{3}{4}$" variable resistor

Decade Resistance Box

The decade resistance box is a test equipment for trying different R values in a circuit. It is a useful unit for having any one R within a wide range of values.

In the decade resistance box, there are six resistor strings, with one string for each dial switch.

The first dial connects in an R between 0 and 9 Ω. It is the unit of R × 1 dial.

The second dial is the tens or R ×10 dial and has an R of 0 to 90 Ω. The hundreds or R × 100 dial has R values from 0 to 900 Ω. The thousands or R × 1K dial has R values from 0 to 9000 Ω. Similarly the R × 10K dial has R values from 0 to 90,000 Ω. The R × 100K dial has R values of 0 to 90000 Ω. The six dial connections can be connected internally such that any value of resistance from 0 to 999,999 may be realised. As an example, when all six dials are on 3, the total value of R is R = 3 + 30 + 300 + 3000 + 30000 + 30000·ohms.

= 333,333 Ω

Rheostats and Potentiometers

A rheostat is a two terminal variable R that is connected in series with a load. It is used to vary the amount of current. The potentiometer is, however, used to tap off part of voltage connected across the two end terminals.

1.3 CAPACITANCE

By the term capacitance, we mean the ability of a dielectric to store electric charge. The higher the amount of charge stored for a given amount of voltage, the higher is the value of the capacitance. The capacitor consists of two conductors seperated by an insulator or dielectric. The voltage is applied between the conductors. The most important characteristics of a capacitor is its ability to block a steady d.c. voltage, while allowing ac signals to pass. The higher the frequency, the smaller is the opposition for a.c. voltages. Capacitors are named according to the dielectric employed. Common types are air, ceramic, mica, paper, film and electrolytic capacitors.

The dielectric constant indicates the ability of the dielectric to concentrate electric flux.

This quantity indicates the ratio of the flux in the insulator compared to the flux in the air or vaccum. Commercial capacitors may be of various types. Table 1.5 below indicates the various types of commercial capacitors.

Table 1.5 : Commercial Capacitors

Dielectric	Construction	Capacitance Range
Air	Meshed Capacitors	10-500 pF
Ceramic	Tubular	0.5-1500 pF
Ceramic	Disk	1 pF to 1 μ F
Mica	Stacked Sheets	5 to 5000 pF
Paper	Foil	0.001 to 1 μ F
Plastic Film	Foil	100pF to 100 μ F
Eletrolytic	Tantalum Aluminium	0.05 to 300 μ F 1 to 6000 μ F

For capacitors (except the electrolytic capacitors) either side can be of positive polarity. For connection of electrolytic capacitors in a circuit, the positive polarity of the charging source should be connected to the positive plate capacitor. Alternatively, the dielectric will be damaged and the capacitor may face complete destruction.

The capacitance C in farads for a parallel plate capacitor is

$$C = \varepsilon \frac{A}{d} \times 8{\cdot}85 \times 10^{-12} \text{ Farads.} \qquad \ldots(1.1)$$

where ε is the relative permittivity, A is the area in square meters of either plate, d is the distance in meters between plates. The constant $8{\cdot}85 \times 10^{-12}$ is the absolute permittivity of air or vaccum in SI.

EXAMPLE 1.4 *Find the capacitance between two plates each with area* 1m^2 *seperated by 0.25 cm or* $0{\cdot}25 \times 10^{-2}$ m *with air dielectric.*

SOLUTION Putting in the above formula,

$$C = 1 \times \frac{1}{0{\cdot}25 \times 10^{-2}} \times 8{\cdot}85 \times 10^{-12}\,\text{F}$$
$$= 3540\ \text{pF}$$

The capacitor can thus store a charge of 3540×10^{-12}C with 1 Volt.

1.3.1 Mica Capacitors

In a mica capacitor, thin mica sheets are stacked in between tinfoil sheets. Alternate sheets of tinfoils are connected together and they are brought out as one terminal for one set of plates. Similarly, the opposite terminal connects to the other set of tinplates. The unit is moulded in a bakelite enclosure. Typical dimensions of mica capacitors is $\frac{3}{4}$ in length and $\frac{1}{4}$ in thickness.

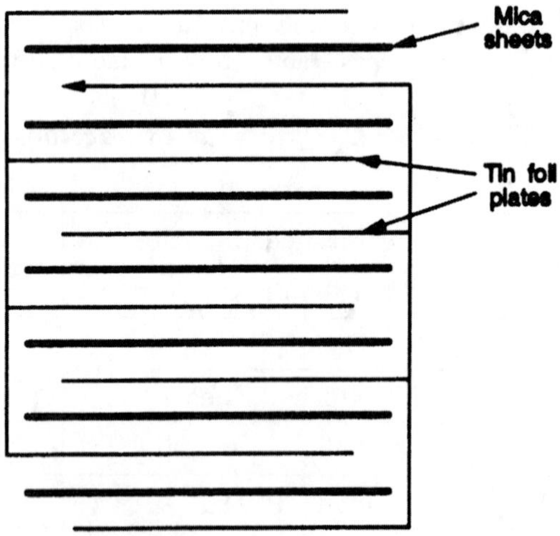

Fig. 1.9. Mica capacitor

1.3.2 Paper Capacitors

In a paper capacitor, two rolls of tinfoil conductors are seperated by a paper dielectric and rolled into a compact cylinder. The entire cylinder is enclosed in a plastic. A schematic diagram of a paper capacitor is depicted in Fig. 1.10. Paper capacitors have typically values lying between 0.001 to 1.0 μ F. The physical size for 0.05 μ F capacitor is typically 1 inch long with $\frac{3}{8}$ in diameter. A black band or a white band at one end of the capacitor indicates that the lead is connected to the **outside foil**. This lead should preferebly be connected to the ground or low potential side for taking advantage of shielding of the outer foil.

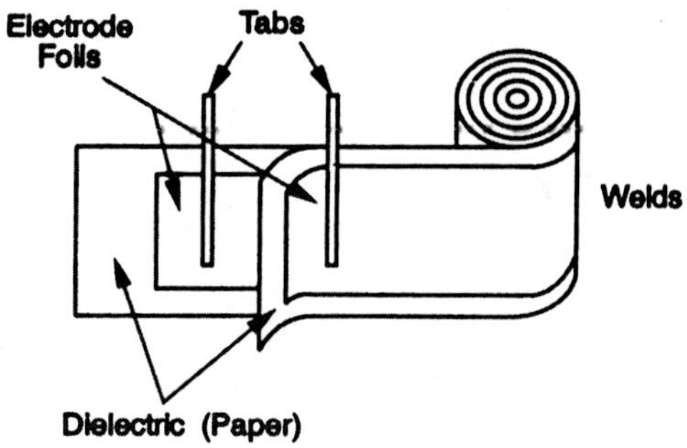

Fig. 1.10. Paper capacitor

1.3.3 Film Capacitors

In film capacitors a plastic such as polypropylene, polystrene, polycarbonate or polyethylene terepthalate (Mylar) is employed as the dielectric. Film capacitors may be of two types, the metallized type and their foil type. In the metallized type, the conductors are thin metal films, such as **aluminium or zinc** deposited (or sprayed on) a plastic film. In the foil type capacitor, thin sheets of metal foil such as aluminium or tin is used as the conductive plates. The metallized type capacitor has the advantage that it is much smaller in size for the same capacitance value. Further, the metallized type is self healing i.e. if the dielectric is punctured due to exceeding the breakdown voltage rating, the capacitor is not damaged permanently. Instead the capacitor heals itself.

Film capacitors have the important property that they are very temperature stable and are, therefore, used in circuits that require very stable capacitance values. Film capacitor values may range from 100 pF to 100 μ F.

Ceramic capacitors

The ceramic capacitors are made from earth fired under extreme heat. A very high dielectric constant ε is achieved by using titanium dioxide or a silicate. The capacitor plates are formed by depositing silver on either side of the dielectric. Ceramic capacitors are available with values of C ranging between 1 pF to 1 μ F.

1.3.4 Surface mount or chip capacitors

The surface mount or chip capacitors are constructed by placing a ceramic dielectric material between layers of conductive films. The conductive films act as the capacitor plates. Chip capacitors may have values between 0.2 pF to microfarads.

1.3.5 Variable capacitors

In the variable air capacitor (used as tuning capacitor in radio receivers), a set of fixed metal plates form together what is known as the stator. Similarly, movable plates are connected together to form the rotor.

The rotor plates rotate in or out of the fixed stator plates (without touching) and the capacitance is varied. The maximum value of capacitance is obtained when the stator and rotor plates mesh together. And the minimum value of the capacitance is obtained when the stator and rotor plates completely move out.

The temperature co-efficient of a capacitor is generally specified as parts per million (ppm) per degree Celsius, with a reference to 25°C. This a positive 500 ppm is specified as P 500 while negative 500 ppm is specified as N 500. If the temperature co-efficient is zero, it is specified as NPO.

The silver plated mica capacitors have a tolerance of ± 1%, while ceramic disk capacitors have a tolerance of ± 20%. A 32 μ F electrolytic with a tolerance of -10 percent, +50 per cent will have a capacitance value lying between 28.8 μ F to 48 μ F. It will be seen that the tolerance is less on the minus side.

The voltage rating of capacitors is specified as the maximum potential difference that can be applied across the plates without puncturing the dielectric. Typically, general purpose paper, mica and ceramic capacitors have voltage ratings between 200 to 500 V. However, ceramic capacitors with a maximum rating in the range 2-20 KV can be obtained. The electrolytic capacitors have typically ratings of 16-, 40-, 50V range. It will be seen that a higher value of capacitance can be achieved in smaller physical size, if the voltage rating is smaller. Capacitors are widely used in electronic circuits for coupling, bypassing and filtering a.c. signals.

1.3.6 Electrolytic Capacitors.

One advantage with electrolytic capacitors is that they provide the highest capacitance in the lowest physical size and lowest cost. Electrolytic capacitors may be available upto a size of 6500 μ F. The aluminium electrolytic capacitor is shown in Fig. 1.11. It will be seen that the the two aluminium electrodes are in an electrolyte of borax, phosphate or carbonate. An absorbent gauge soaks up electrolyte so as to provide to electrolysis and produce the oxide film.

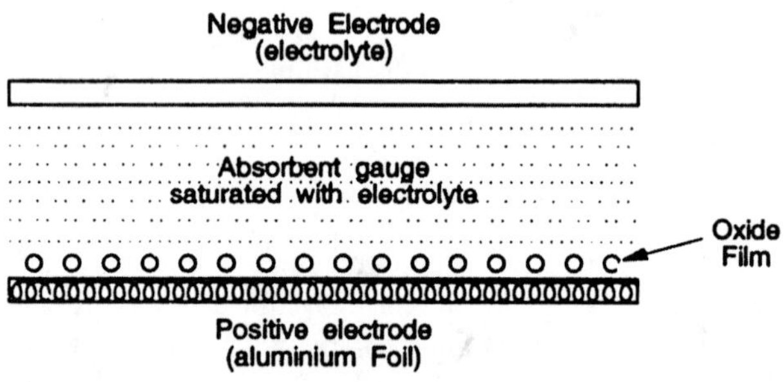

Fig. 1.11. Aluminium electrolytic capacitor construction

By the application of a d.c. voltage, the electrolytic action accumulates a thin aluminium oxide layer at the junction between the positive aluminium foil and the electrolyte. The oxide film is an insulator. The capacitance is thus formed between the positive aluminium foil and the electrolyte in the gauge seperator. The negative terminal is normally the metal can of the electrolytic capacitor which is connected to the electrolyte.

In an electrolytic capacitor, very large C values may be realised by using an extremely thin dielectric film. Also the area is increased by using very large strips of aluminium foil and gauge which is then rolled into a cylinder with very large capacitance values. Electrolytic capacitors are available with C values of 6800 μ F, 450 V rating. It will be seen that an electrolytic capacitor can give a C value 10,000 times higher compared to a paper capacitor of the same physical size.

Electrolytic capacitors are normally employed in a circuit in which there is a combination of both a d.c. and an a.c. voltage. The d.c. voltage is required to form the oxide film.

If an electrolytic capacitor is wrongly connected in a circuit such that the negative polarity is connected to the positive electrode (opposite polarity), then reverse electrolysis will form gas in the capacitor. It will become hot and it may explode. This possibility exists only in the case of electrolytic capacitors.

Two important applications of electrolytic capacitors is removal of a.c. 50 Hz or 100 Hz ripple in power supplies (100-1000 μ F) and as audio coupling capacitors in transistor amplifiers (10 to 47 μ F).

1.3.7 Tantalum Capacitors

Another type of electrolytic capacitors is the tantalum capacitors. For tantalum oxide the dielectric constant is 25 instead of 7 for aluminium oxide. Tantalum electrolytic capacitors are, however, much more costly compared to aluminium electrolytic capacitors. Tantalum capacitors give the following important characteristics namely (a) higher C value (b) lower leakage current.

1.3.8 Capacitor Coding

Capacitance values are usually specified in units of pF or μ F. If the number printed is a whole number (like 47, 68, 470) the capacitance is in pF while if the number printed is decimal fraction (like 0.22, 0.033, 0.0047), the value of the capacitance is in μ F. The manufacturer may indicate the value in different coding systems.

1.3.9 Film type capacitors

The film type capacitors are depicted as shown in Fig. 1.12.

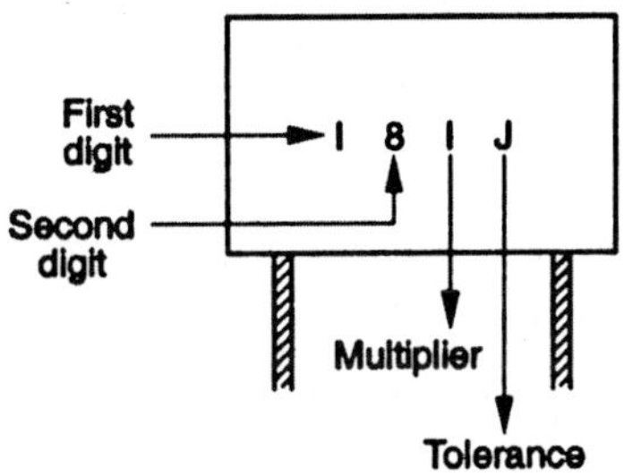

Fig. 1.12. Film capacitor coding

EXAMPLES

$181J = 18 \times 10 = 180pF \pm 5\%$

$689J = 68 \times 0{\cdot}1 = 6{\cdot}8\ pF \pm 5\%$

$689K = 68 \times 0{\cdot}1 = 6{\cdot}8\ pF \pm 10\%$

Note : *The letter R some times denotes a decimal point like* 4 R 7 = 4·7 (pF or μ F).

EXAMPLE *Find the value of the capacitance of the film capacitors in Fig. 1.13 (a) and (b)*

SOLUTION The values are respectively

3300 ± 20% in Fig. 1.13 (a)

and 15000 ± 3% in Fig. 1.13 (b)

1.3.10 Disk Ceramic Capacitors.

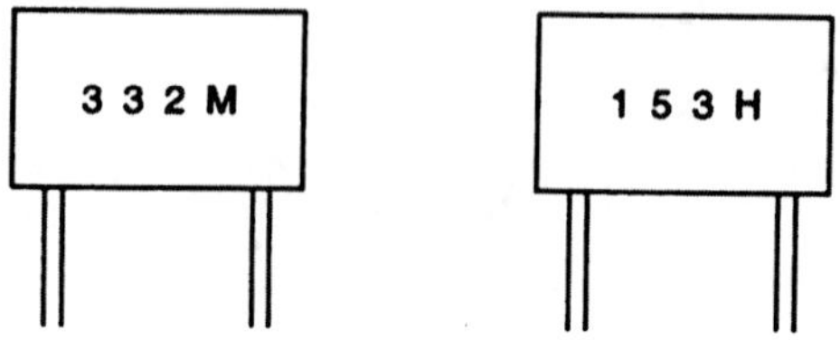

(a) (b)

Fig. 1.13. Reference example 1.4

Disk ceramic capacitors are normally employed for coupling and bypassing a.c. signals, where the allowable tolerance may be wide and/or lopsided. Fig. 1.14 depicts the Ceramic disk capacitor coding system

Typical Disk Ceramic capacitor markings

Low. Temp.	Letter Symbol	High Temp.	Numerical Symbols	Max. Cap. Change over range	Letter Symbol
+ 10° C	Z	+ 45°C	2	± 1·0%	A
− 30°C	Y	+ 65°C	4	± 1·5%	B
− 55°C	X	+ 85°C	5	± 2·1%	C
		+ 125°C	7	± 3·3%	D
				± 4·7%	E
				± 7·5%	F
				± 10%	P
				± 15%	R
				± 22%	S
				+ 22%, − 33%	T
				± 22%, − 56%	U
				± 22%, − 82%	V

Multiplier	Numerical Symbol	Tolerance Capacitance	Letter Symbol
1	0	± 5%	J
10	1	± 10%	K
100	2	± 20%	M
1000	3	+ 100%, − 0%	P
10,000	4	+ 80%, − 20%	Z

100,000	5		
—	-		
0.01	8		
0.1	9		

Capacitance value and tolerance of disk type ceramic capacitor.

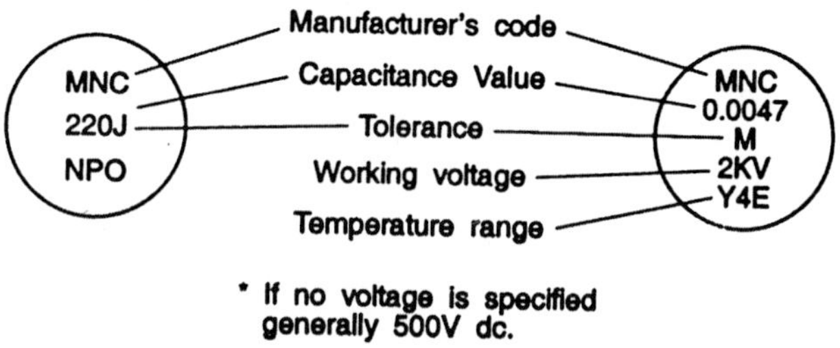

Fig. 1.14 Ceramic Disk capacitor coding system

EXAMPLE 1.5 *Find (a) the capacitance value and tolerance (b) the temperature range identification information.*

SOLUTION (a) Since the capacitance is expressed as a decimal fraction, its value is in microfarads. In this case, C = 0·068 μ F. The letter P, to the right of 0.068 indicates a capacitance tolerance of +100%, -0%. Note that the actual capacitance value can be as much as 100 percent above its coded value but only 20 percent below its coded value.

(b) The alphanumeric code Y4T, printed below the capacitance value gives further capacitor information. Referring to the problem (Y 4 T), note that Y stands for −30°C and 4 stands for +65°C respectively. The letter T indicates the maximum capacitance change over the specified temperature range (−30°C to +65°C) is +22%, −33%. If temperature changes less than the range indicated, the percent change in capacitance will be less.

Tantalum Capacitors.

Colour	Voltage rating	Capacitor Value in pF (First or Second Digit)	Multiplier
Black	4	0	—
Brown	6	1	—
Red	10	2	—
Orange	15	3	—
Yellow	20	4	10^4
Green	25	5	10^5
Blue	35	6	10^6

Violet	50	7	10^7
Gray	—	8	—
White	3	9	—

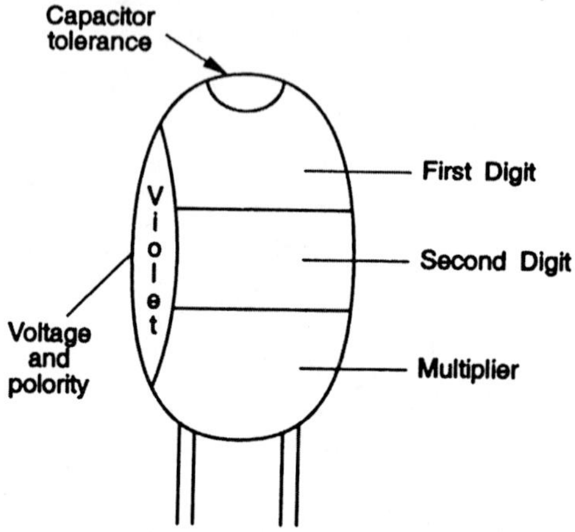

Fig. 1.15 Tantalum capacitor colour coding system

EXAMPLE 1.6 *For the tantalum capacitor shown alongside (Fig. 1.16) find the value of the capacitance C. Also determine the voltage rating and tolerance.*

SOLUTION The first two digits for blue and gray are 68 and thus the value of C is = 68×10^6 pF or 68 μF.

The violet colour at the left indicates a voltage rating of 50V. Finally, the gold dot at the top indicates a tolerance of ± 5%.

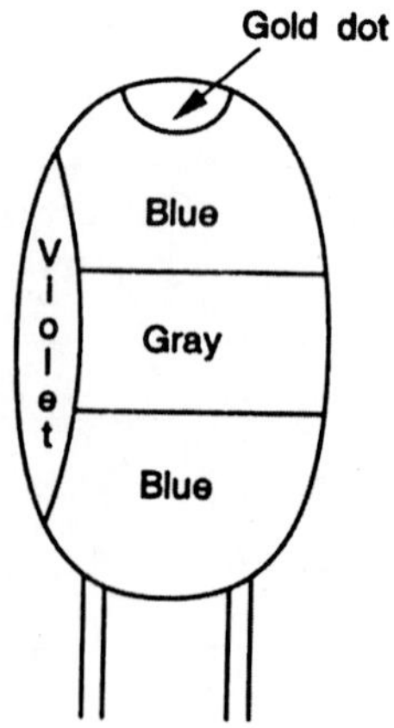

Fig. 1.16. Reference example 1.6

Variable capacitors

Besides the variable air capacitor, trimmers (also called padders) are also used. Both mica and ceramic are used as the dielectric for trimmer capacitors. (vide Fig. 1.17)

1.4 INDUCTANCE

An inductor produces an induced voltage when the current changes. A long wire possesses higher inductance value than a short wire, because as more conductor length is cut by magnetic flux, it develops more induced voltage. A coil has more inductance than the equivalent length of a straight wire because the coil concentrates the magnetic flux. Air cored coils are generally used in RF circuits because higher frequencies need less L for the required inductance. Iron cored coils are used at low frequencies, power line frequencies (50 Hz) and audio frequencies.

The inductance L of a coil of length l meter , number of turns N and area of cross section A (in square meters) is given by

$$L = \mu_r \frac{N^2 A}{l} \times 1{\cdot}26 \times 10^{-6} \text{ Henries.} \qquad ...(1.2)$$

Thus an air core coil of length 0.1m, area of cross section 2×10^{-4} m^2 and 100 turns produces an L value of 25·2 μH. If the coil is wound on an iron core (μ_r = 100), the value of L increases 100 times. An RF coil used in medium wave band may require an inductance L of 250 μ H. Iron cored inductors at 50Hz power line frequency and audio frequencies may require an L value of 1 to 25H.

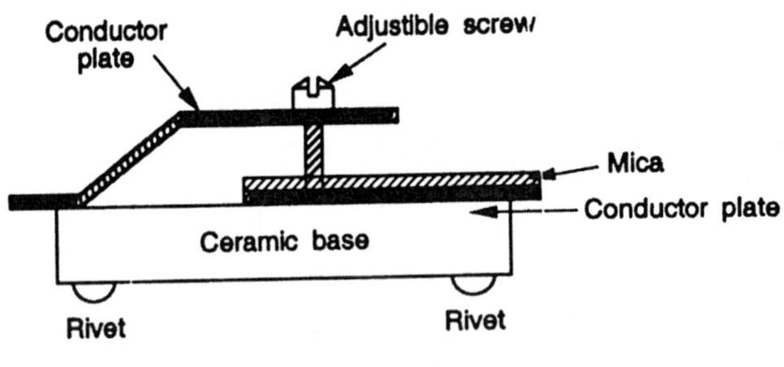

Fig. 1.17 Trimmer capacitor

1.4.1 Core Losses

The fact that a magnetic core can be hot is due to the reason that energy supplied to the coil is used in the core as heat. The two main core losses are eddy current losses and hysteresis losses. The eddy current loss is illustrated in Fig. 1.18. In iron cores, the alternating current induces voltage in the core itself. As it is a conductor, the iron core has current produced by the induced voltage. The eddy current flows in a circular path as depicted in the Fig. 1.18. The eddy current loss is used in induction heaters for heating magnetic conducting materials. The eddy current loss increases with increasing frequency.

The hysteresis loss in also another loss factor in magnetic cores. The hysteresis losses arise due to the additional power needed to reverse the magnetic field in magnetic materials when the alternating current is present. The hysteresis loss increases with frequency.

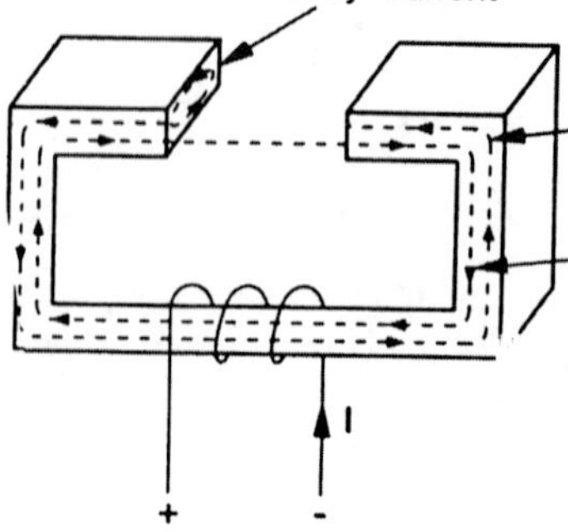

Fig. 1.18. Illustration of eddy current

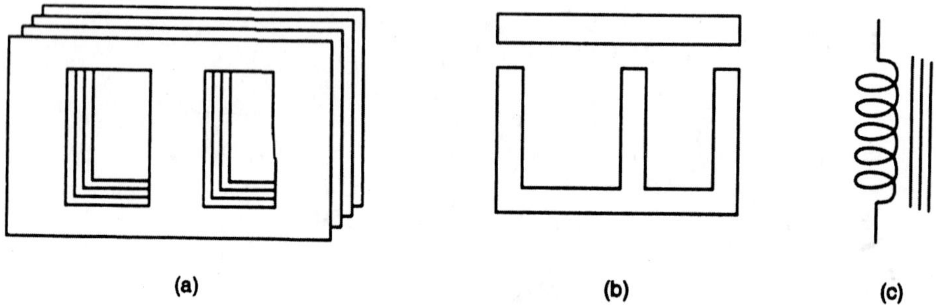

Fig. 1.19 Laminated Iron core : (a) shell type (b) *E* and *I* shaped laminations (c) Iron core

To minimize the losses at high flux densities, the core is made of laminated steel insulated from one another. This decreases the eddy currents. The steel employed reduces the hysteresis loss.

In the laminated iron core, the laminated sections are insulated from one another by a thin coating of iron oxide, silicon steel or varnish. This insulating material increases the resistance of the cross section of the core but the high flux density around the core passes unabated. Audio frequency transformers and 50 Hz power line transformers are made with laminated iron core.

At radio frequencies, the powdered iron core is used to reduce eddy currents. The individual granules are insulated and pressed into one solid form called a slug. Ferrites are ferromagnetic matrials which are synthetic ceramic materials. They have a high flux density but are insulators. They reduce the eddy current substantially.

Audio frequency chokes exhibit a high impedance to audio frequencies (upto about 20 KHz). The audio frequency chokes possess lower inductance values compared to filter chokes and are smaller in physical size. Radio frequency chokes (RFC) block radio frequencies. They possess still lower L value compared to audio frequency chokes. The permeability turned variable coil has a ferromagnetic shaft. The shaft can be moved within the coil to change the inductance values.

A transformer is fabricated by having two coils sharing the same core. The transformer has a primary winding inductance L, connected to a voltage source that produces the alternating current, while the secondary winding inductance is connected across the load resistance. The transformer transfers power from

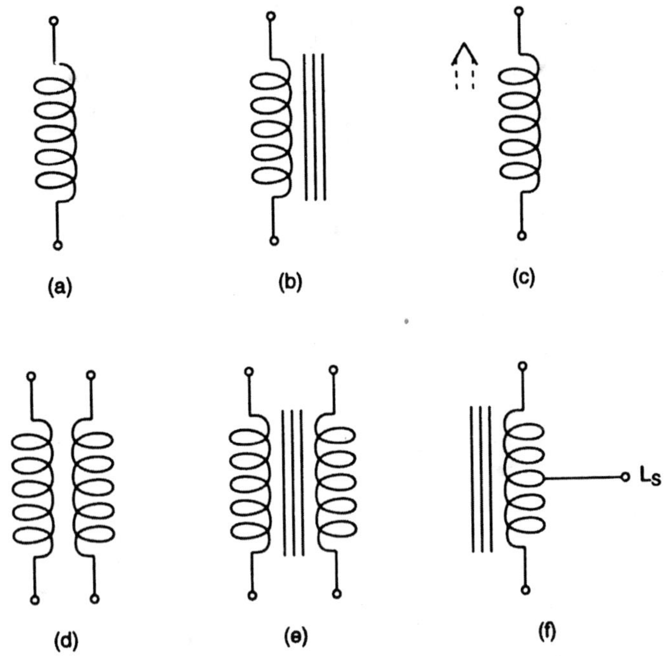

Fig. 1.20. Summary of different types of inductors (a) air cored coil (b) Iron cored (c) adjustible ferrite core (d) air core transformer (e) Iron core transformer (f) auto transformer.

the primary to the secondary, where the secondary induced voltage produces a current in the load resistance that is connected across L_2. The transformer is used to step up or step down a.c. voltage.

A transformer with an iron core is essentially a unity coupling device. The voltage ratio is thus equal to the turns ratio $\frac{V_p}{V_s} = \frac{N_p}{N_s}$ where V_p and V_s are the primary and secondary ac voltages respectively and N_p and N_s are the primary and secondary number of turns respectively.

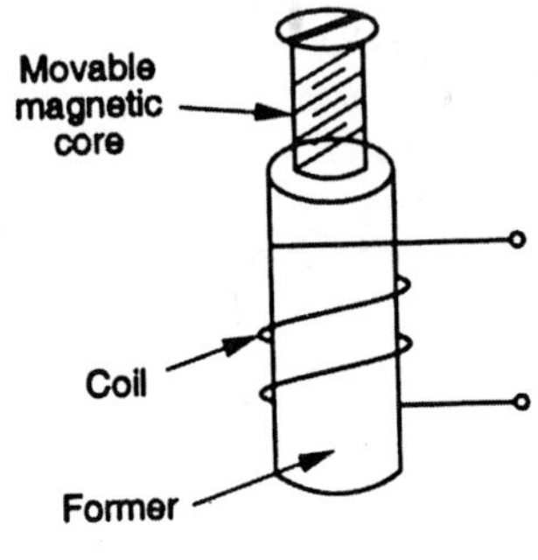

Fig. 1.21. Laminated Iron core

Assuming 100 percent efficiency for an iron core transfomer, the power supplied to the primary equals the power used in the secondary. The voltage rating of a transformer secondary is always specified under full load condition when the rated voltage is applied to the primary. The measured voltage across the unloaded secondary is normally 5 to 10 percent grater than the rated value. The current or power rating of a transformer is normally specified only for the secondary winding.

Transformers can be utilized to reflect a secondary load impedance into the primary as a new value which is either larger or smaller than its actual value. The primary impedance Z_p can be found using the formula

$$Z_p = \left(\frac{N_p}{N_s}\right)^2 Z_s \qquad ...(1.3)$$

where Z_s is the secondary impedance.

Another type of transformer, which is widely used, is an auto transformer. An auto transformer is a tapped coil used to step up or step down a.c. voltage.

1.5 ELECTRONIC COMPONENTS

An electronic system is often quite complex and consists of a large number of varied and different components. The components can be classified as shown below.

We will learn more about devices in subsequent chapter.

1.6 ENERGY SOURCES

Batteries are an important source of energy. They generate energy from the internal chemical action of the cell. The cell consists of two different electrodes which are immersed in an electrolyte. The chemical reaction between the electrolytes and the electrodes seperates the charges as ions and free electrons. The two conducting electrodes then have a potential difference which gives a voltage output from the cell. Two very common types of batteries are the alkaline cell with a voltage output of 1.5 V and the lead sulphric acid wet cell with a voltage output of 2.1 V. Six alkaline cells may be connected in series to provide 9V flat batteries commonly used in transistor radios. Six lead-sulphuric acid cells may be connected in series to yield 12V as automobile batteries.

Primary cells are cells which can not be recharged. The chemical action in the cell is not reversible. However, secondary cells can be recharged because the chemical action is reversible. The secondary cell is also called a storage cell because it can be recharged. The nominal open circuit voltage of typical cells is depicted in Table 1.6.

Table 1.6 Different types of cells

Cell Name	Type	Numinal O.C. Voltage.
Carbon-Zinc	Primary	1.5
Zinc Chloride	Primary	1.5
MnO_2 (alkaline)	Primary	1.5
Lead Acid	Secondary	2.1
Nickel-cadmium	Secondary	1.25

Nickel-iron (Edison Cell)	Secondary	1.2
Silver-cadmium	Secondary	1.1
Lithium	Primary	3.0

1.6.1 Carbon Zinc Dry Cell

The construction of the carbon-zinc dry cell and the alkaline cell is depicted in Fig. 1.22 below.

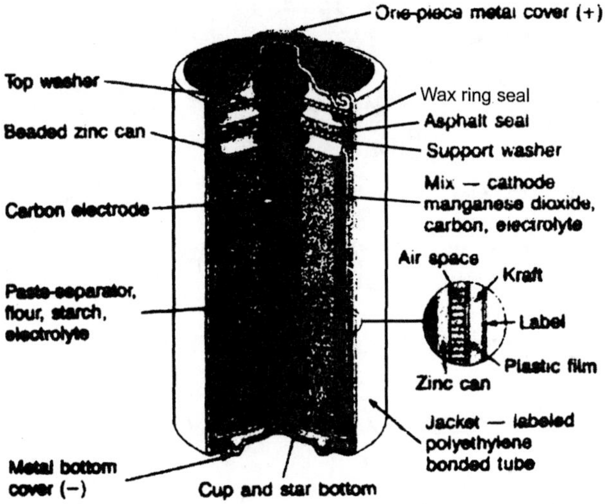

Fig. 1.22. Construction of carbon-zinc dry cell

The Carbon-zinc dry cell (also known as Leclanche' cell) provides an output voltage between 1.4 and 1.6V (with 1.5V as the nominal value). A typical D size $2\frac{1}{4}$" length 3.18 in^3 volume cell can provide upto 150mA.

In the carbon zinc dry cell, the electrolyte is a combination of ammonium chloride and zinc chloride dissolved in water. A zinc anode and a manganese dioxide cathode is placed in the wet electrolyte. For rounded cell shape, the carbon rod is at the centre which acts as the current collector. The zinc rod is positive while MnO_2, the electrolyte and the zinc is negative.

In the actual operation, polarization takes place, and ammonia releases hydrogen gas which collects around the zinc anode. Also the manganese dioxide releases oxygen that combines with hydrogen to form water. The manganese dioxide is the depolarizer. For improvement of conductivity and retaining moisture, powdered carbon is added to the depolarizer. Carbon zinc cells operate very well at a temperature of about 70°F. Higher temperatures will provide higher output. However, for a temperature above 125°F, there will be rapid deteroriation of the cell.

1.6.2 Alkaline Cell

The alkaline cell has many applications for its high efficiency with high discharge rates. It also possesses a low internal resistance. Its low input resistance is due to dense cathode material, large surface area between the anode and the electrolyte and high conductivity of the electrolyte.

In the alkaline cell, the electrolyte is potassium hydroxide and the anode is powdered zinc while the cathode is manganese dioxide. Hydroxide compounds are alkaline with negative hydroxyl ions while the acid electrolyte is positive hydrogen (H) ions. The voltage output of alkaline cell is 1.5 V.

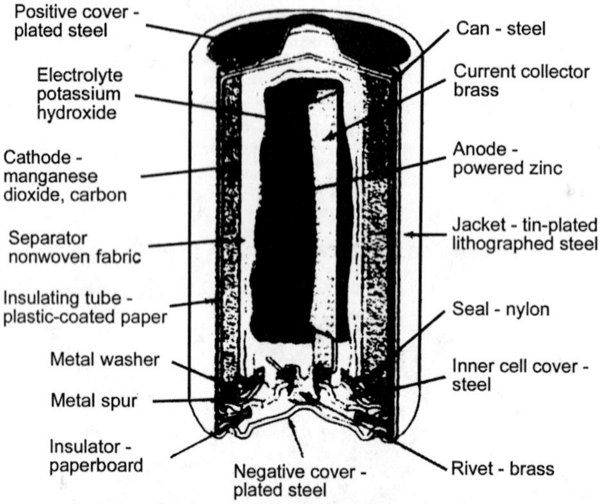

Fig. 1.23. Cutway view of an alkaline cell

1.7 INTERNAL IMPEDANCE OF A SOURCE

Let us consider the circuit of Fig. 1.24 where a battery is feeding load R_L

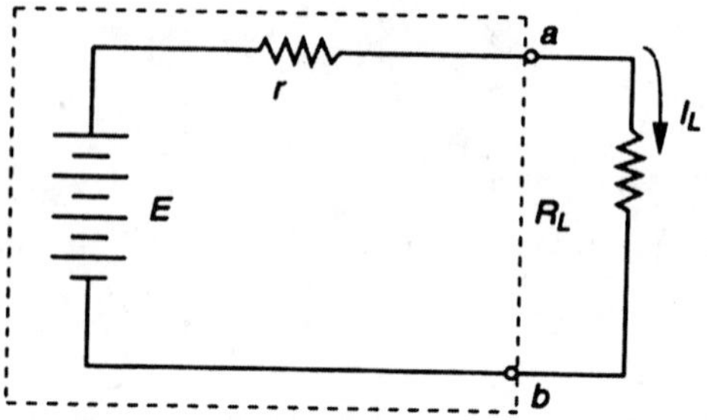

Fig. 1.24. Battery source feeding load R_L

In the Fig. 1.24, E is the battery voltage and r is the internal resistance of the battery. The terminals a and b are the available battery terminals. A load resistance of magnitude R_L is connected across ab.

Let us consider an example of $E = 6$V and $r = 0.1$ ohms. If now the load resistance is varied let us find the current I_L flowing through the load and also the output voltage V_{ab}. The current I is given by

$$I = \frac{E}{R_L + r} \quad \text{...(1.4)}$$

and the voltage

$$V_{ab} = I.\ R_L = \frac{ER_L}{r + R_L} = \frac{E}{\left(1 + \frac{r}{R_L}\right)} \quad \text{...(1.5)}$$

When $R_L = 10\text{K}\ \Omega$ in the above example,

$$I_L = \frac{6V}{10K\,\Omega + 0{\cdot}1} = 6mA$$

and $V_{ab} \cong 6V$

If now R_L is changed to 1K Ω

$$I_L = \frac{6V}{1K\,\Omega + 0{\cdot}1} = 5{\cdot}999mA$$

and $V_{ab} = 5{\cdot}999V$

If now R_L is changed to have values of 100Ω and 10Ω and 1Ω respectively, the corresponding load current and output voltage values are listed in table.

RL	IL	Vab
10K Ω	6m A	6V
1K Ω	5.999 mA	5.999 V
100	5994 mA	5.999 V
10	594 mA	5.94 V
1	5.4 5 A	5.45 V
0.1	30 A	3 V

It will be seen that with increasing load current, the value of the output voltage decreases. And when the value of R_L is of the order of r, the output voltage falls directly. This is also depicted in the graph showing the voltage across points ab plotted against I_L.

The practical voltage source can be depicted as shown in Fig. 1.26

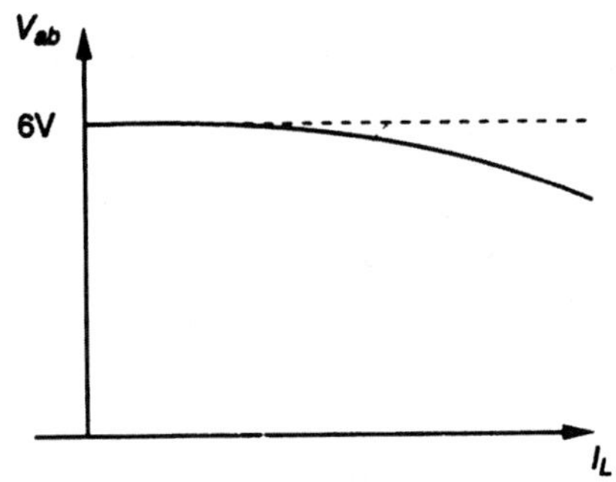

Fig. 1.25 Variation of output voltage of a voltage source with increasing load current.

A practical d.c. voltage source can be depicted as a battery in series with an internal resistance r. An ideal d.c. voltage source has zero resistance. In a practical case, the d.c. voltage source has a small but finite value. Similarly a practical a.c. voltage source can be represented as an ideal a.c. voltage source in series with an impedance Z_c representing the internal impedance of the voltage source. A typical example of a d.c. voltage source in series with a variable load is depicted in Fig. 1.27. In the circuit the values of the current and voltage V_{ab} corresponding to load resistances 10Ω, 5Ω and 1Ω are respectively as follows :

RL	IL	Vab
10Ω	0.272 A	2.72 V
5Ω	0.5 A	2.5 V
1Ω	1.5 A	1.5 V

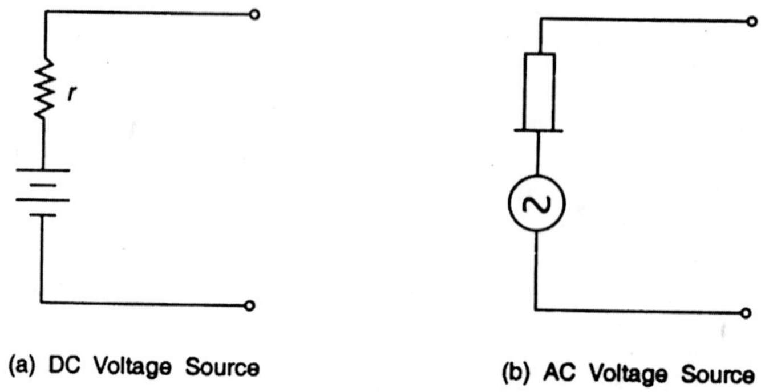

Fig. 1.26. (a) DC voltage source (b) AC voltage source

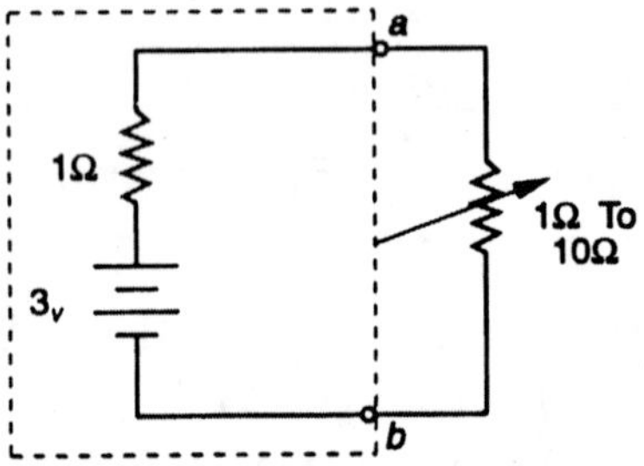

Fig. 1.27. Voltage source with variable load

with respect to the maximum value, the percentage variation in terminal voltage is

$$= \frac{2{\cdot}72\text{V} - 1{\cdot}5\text{V}}{2{\cdot}72\text{V}} \times 100 = \frac{1{\cdot}22}{2{\cdot}72} = 44{\cdot}9\%.$$

In the circuit shown, for a load impedance of 10KΩ

$$V_{T_1} = \frac{V_s}{Z_s + Z_{L_1}} \times Z_{L_1} \qquad \text{...(1.6)}$$

$$= \frac{1}{50 + 10{,}000} \times 10{,}000 = 0{\cdot}995\text{V}$$

When the load impedance is 50KΩ,

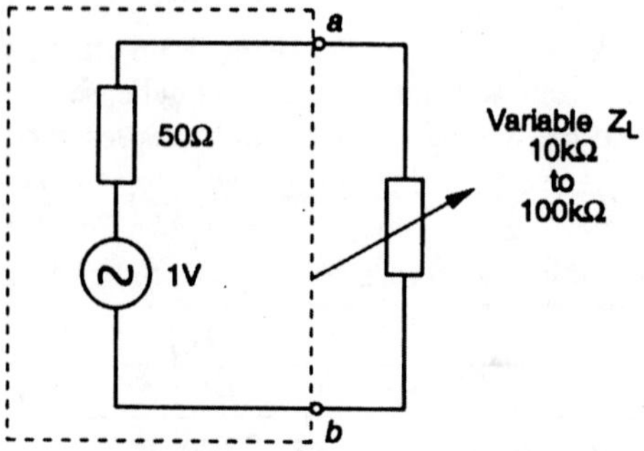

Fig. 1.28. Example of voltage source with variable load

$$V_{T_2} = \frac{V_s}{Z + Z_{L_1}} \times Z_{L_1} = \frac{1}{50 + 50 \times 10^3} \times 50 \times 10^3 = 0{\cdot}999V \qquad ...(1.7)$$

And when the load impedance is 100KΩ,

$$V_{T_3} = \frac{1}{50 + 100 \times 10^3} \times 100 \times 10^3$$

$$= 0{\cdot}9995\text{V}$$

The percentage variation in terminal voltage for the load impedance variation between 10KΩ and 100 KΩ is

$$= \frac{0{\cdot}9995 - 0{\cdot}995}{0{\cdot}9995} \times 100\% = 0{\cdot}45\%.$$

It should be evident from the above examples, that the internal impedance value of the voltage source is not alone responsible for determining whether a given voltage source would function as a good voltage source or not. Rather, the relative value $\frac{Z_s}{Z_L}\left(\text{or } \frac{r}{R_L}\right)$ determines whether a given voltage source would function as a good voltage source. The smaller the value of $\frac{Z_s}{Z_L}\left(\text{or } \frac{r}{R_L}\right)$, the better is the source as a constant voltage source.

1.8 CURRENT SOURCE

Just like a constant voltage source, there may be a constant current source—a source that supplies a constant current to a load even if the impedance varies. In the ideal case, the current supplied by a constant current source should remain constant, no matter what the load impedance is. The symbolic representations of an ideal current source and that of a practical current source are depicted in Fig. 1.29.

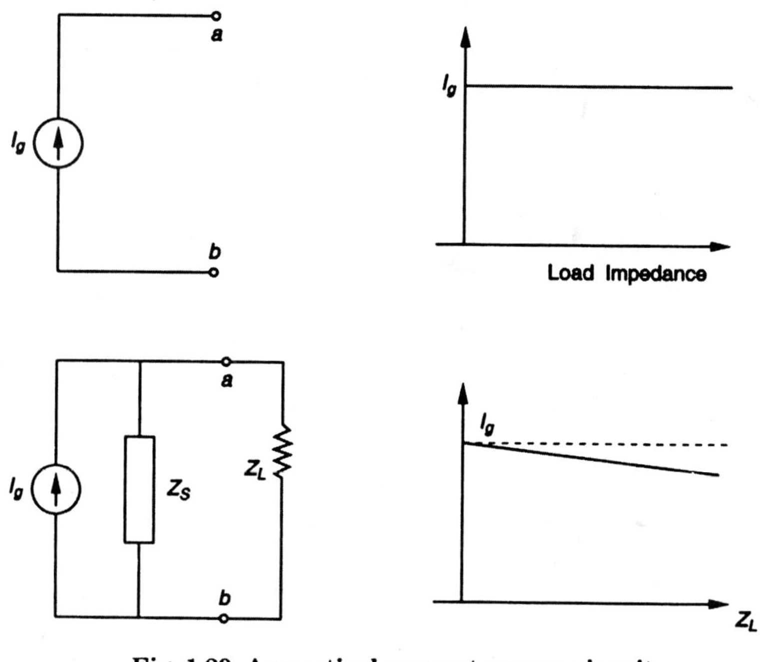

Fig. 1.29. A practical current source circuit

Equivalence between voltage source and current source

A voltage source can be represented by its equivalent current source circuit. Similar a current source can be represented by its equivalent voltage source circuit. This is illustrated by Fig. 1.30.

In the equivalent voltage source representation of Fig. 1.29 (b), the current I_{L_1} flowing through the load is

$$I_{L_1} = \frac{V_s}{Z_L + Z_{s_1}} \quad \text{... (1.8)}$$

Similary from the equivalent current source reresentation circuit of Fig. 1.30 (c),

$$I_{L_2} = \frac{I_s Z_s}{Z_L \; + Z_{s_2}} \quad \text{... (1.9)}$$

Let us take an example.

In the circuit of Fig. 1.30 (a), the equivalent current source representation Fig. 1.30 (b) may be found by first calculating the short circuit current I_{sc}.

(a) Voltage source equivalent circuit

(b) Current source equivalent circuit

In the Fig. of 1.31 (a) $I_{sc} = \dfrac{1\text{V}}{0{\cdot}5\Omega} = 2$ Amps

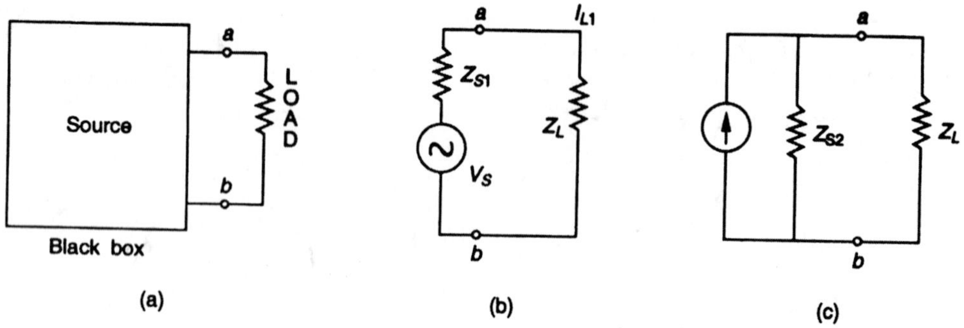

Fig. 1.30. (a) Source with external terminals a and b, (b) Voltage source representation and (c) Current source representation

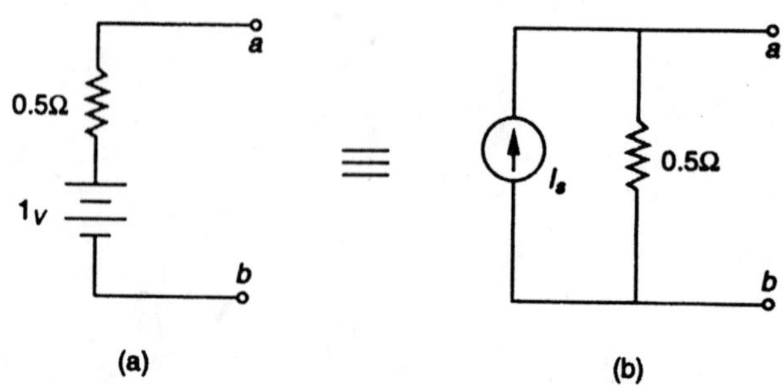

Fig. 1.31. (a) Voltage source equivalent circuit and (b) Current source equivalent circuit

Thus in Fig. 1.30 (b), I_{sc} = 2 Amps. and R_{s_2} = 0·5Ω

Similarly the following equivalent circuits can be drawn. (Fig. 1.32a)

EXAMPLE 1.7 *Find the a.c. equivalent circuit of an amplifier using a transistor. Find the output voltage V out.*

SOLUTION For the circuit of Fig. 1.33,

$$\text{Input current } i_s = \frac{6 \text{ mV}}{1200\Omega} = 5 \ \mu\text{A}$$

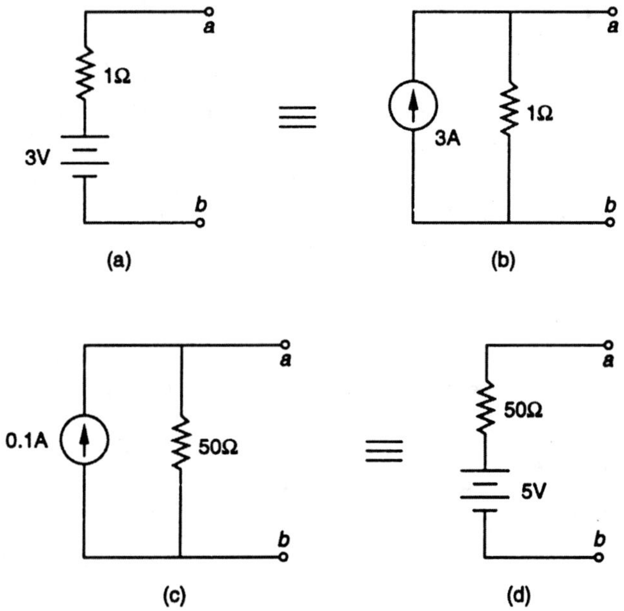

Fig. 1.32. Voltage source and current source equivalent circuits.

The current source $i_2 = 100 \ i_s = 100 \times 5\mu A = 500 \ \mu A$

The current divides into two branches

$$i_L = i_2 \times \frac{25 \text{ k}\Omega}{25 \text{ k}\Omega + 1 \text{ k}\Omega}$$

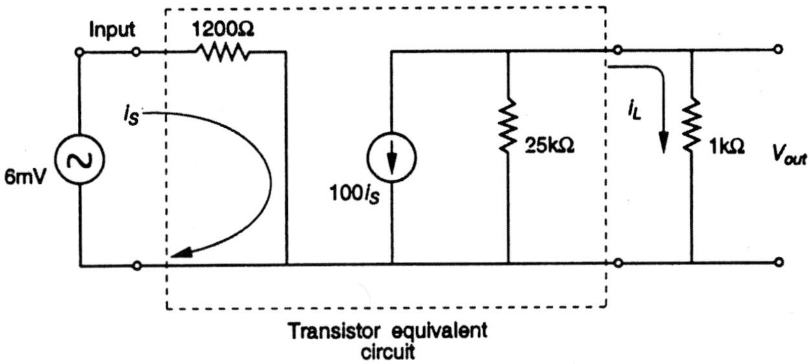

Fig. 1.33 A.C. equivalent circuit of a transistor amplifier.

$$= 500\ \mu A \times \frac{25}{26} = 480\ \mu A$$

Thus the output voltage is $v_{out} = 480\ \mu A \times 1\ k\Omega = 0{\cdot}48V$

1.9 UNITS

Physical quantities are defined in kind and magnitude. The standard value of each kind of physical quantity is the unit. The number of times the unit occurs is the number of the measured quantity. In the fields of engineering and science, two kinds of units are employed — fundamental units and derived units. In mechanics, the fundamental units are measures of length, mass and time. The length, mass and time units are also known as primary fundamental units since length, mass and time are fundamental to most other physical units. Many other units in electrical, thermal and optical disciplines are represented by fundamental units. Other units which can be expressed in terms of the fundamental units are called the derived units. Every derived unit comes from some physical laws defining that unit. For example, velocity is distance covered in unit time internal ($v = LT^{-1}$). If meter is the unit of length, then velocity is metres/sec. The derived unit is usually expressed by its dimension, which may be defined as the complete algebraic formula for the derived unit. If the symbols *L, M, T* represent length, mass and time respectively, then the dimensional symbol for unit of velocity is LT^{-1}. The dimensional symbols for accelaration is LT^{-2} and that of force is LMT^{-2}. The French Academy of Sciences, while working on a proposal for a single system of weights and measures, decided, as a first principle to adopt the Universal system of weights and measures that are not man made reference standards, but instead based on permanent measures provided by nature. **The unit of length, the meter was defined as the ten millionth part of the distance from the pole to the equater along the meridian passing through Paris.** The unit of mass, the gram , was defined on the **mass of a cubic centimetre of distilled water at 4°C under normal atmospheric presure (760mm of Hg.).** The unit of time (the second) was defined as 1/86400 of the mean solar day. The search for a truly universal time led astronomers to define a time unit known as ephimeris time (ET). The International Committee of weights and measure **has now defined the second in term of the frequency of cesium transition, giving a value of 9, 192, 631, 770 Hz to the hyperfine transitions of the cesium atom when unperturbed by external fields.**

SI is the abbreviation for "Systeme International d' Units" in French and is the modern form of the metric system introduced at the Eleventh International Conference of Weights and measures 1960. In 1956, India changed over to the metric system of weights and Measures.

In the SI system of units, seven base units are employed, namely the meter, kilogram, second and ampere of the MKSA system as also the units of Kelvin, Candela and mol.

Metre : It is the unit of length and is at present defined as being equal to the distance that light travels in vacuum in a time interval of (1/299 792458) seconds.

Kilogram : The unit of mass, Kilogram is defined as the mass of a cubic decimeter of water at its temperature of maximum density at 4°C. **The International Prototype Kilogram is the mass of the Platinium-Irdium cylinder kept at the office of the International Bureau of Weights and Measures.**

Second : It is defined above.

Ampere : **It is the fundamental unit of electric current and is defined as the constant current which when flowing through two straight parallel conductors of infinite length and negligible circular cross section that are placed at a distance of 1 metre in vaccum, produces a force equal to 2×10^{-7} Newton per metre length between the conductors.**

Kelvin : The unit 'Kelvin' is the unit of temperature. The thermodynamic Kelvin scale is defined as the fundamental scale of temperature. **The magnitude of Kelvin is defined assigning a value of 273.16 Kelvin as the temperature of the triple point of water. The triple point of water is the temperature of thermodynamic equilibrium between ice, water (in liquid form) and water vapour.** (The celsius temperature scale is defined by the equation $t = T - T_0$, where T is the thermodynamic temperature in Kelvins and $T = 273 \cdot 15$ K).

Candela : The standard of luminuous intensity is a black body or Planck radiator at the temperature of solidification of Platinum (2042 Kelvin). **The Candela is defined as the 1/60 (one sixtieth) of the luminuous intensity per cm^2 of the black body radiator at 2042 Kelvin.**

Mole : The mole is the amount of material that possesses as many elementary entities as there are atoms in 0.012 kg of carbon 12.

The basic SI quantities along with its units and symbols is given in Table.

Quantity	*Unit*	*Symbol*
Length	metre	m
Mass	Kilogram	Kg
Time	Second	s
Electric current	Ampere	A
Thermodynamic Temperature	Kelvin	K
Luminuous Intensity	Candela	Cd
Material Content	Mole	mole

In the various system of units, the following decimal multiples and submultiples are employed (vide table below)

***Table* 1·7**

Name	*Symbol*	*Equivalent*
atto	*a*	10^{-18}
femto	*f*	10^{-15}
pico	*p*	10^{-12}
nano	*n*	10^{-9}
micro	μ	10^{-6}
milli	*m*	10^{-3}
centi	*c*	10^{-2}
deci	*d*	10^{-1}
deca	*da*	10
hecto	*h*	10^{2}
kilo	*k*	10^{3}
Mega	*M*	10^{6}
Giga	*G*	10^{9}
Tera	*T*	10^{12}
Peta	*P*	10^{15}
Exa	*E*	10^{18}

Normally, the prefixes for factors higher than unity have Greek origin while the prefixes for factors lower than unity have latin origin. Thus 1 $\mu F = 10^{-6}$ F while 1 $M\Omega = 10^6\ \Omega$.

The particulars of certain electric and magnetic quantities, units and their relationships is given in Table 1·8 below.

Table 1·8 : Certain Electrical and Magnetic Units

Quantity and Symbol	*Equation*	*Name and Symbol*
Electromative force E	$P = E \,.\, I$	Volt, V
Electric Resistance, R	$R = V/I$	Ohm, Ω
Electric current, I	$F = 10^{-7}\, I^2 \dfrac{dN}{dz}$	Amperes, A
Potential, V	$P = V \,.\, I$	Volt, V
Electric charge, Q	$Q = I \,.\, t$	Coulomb, C
Capacitance, C	$C = Q/V$	Farad, F
Electric field, E	$E = V/l$	V/m
Electric flux density, D	$D = Q/l^2$	C/m^2
Permittivity, ε	$\varepsilon = D/E$	F/m
Magnetic Flux, φ	$E = \dfrac{d\phi}{dt}$	Weber, Wb
Magnetic flux density, B	$B = \phi/l^2$	Tesla, T
Magnetic Field Strength, H	$\oint H\, dt = NI$	A/m
Inductance L, M	$M = \phi/I$	Henry, H
Permeability, μ	$\mu = B/H$	H/m

Some supplementary and derived units are also depicted in Table below.

Table 1·9 Supplementary and Derived Units

Quantity and Symbol	*Unit*	*Dimensions*
Solid Angle, Ω	Steradian	$[L^2]^\circ$
angle, θ	radian	$[L]^\circ$
Area, A	square meter	$[L^2]$
Volume, V	cubic meter	L^3
Density, ρ	Kg/m^3	$L^{-3}M$
Velocity, v	m/s	LT^{-1}
Accelaration, a	m/s^2	LT^{-2}
Angular velocity, ω	rad/s	$L^\circ T$
Force, F	Newton	LMT^{-2}
Pressure, P	Newton/m^2	$L^{-1}MT^{-2}$
Work, Energy, W	Joule (N.m)	L^2MT^{-2}

Power, P	Watt	L^2MT^{-3}
Frequency, f	Hertz (c/s)	T^{-1}
Electric Resistance, R	Ohm	$L^2MT^{-3}I^2$
Inductance, L	Henry	$L^2MT^{-2}I^2$
Capacitance, C	Farad	$L^{-2}M^{-1}T^4I^2$
Illumination	Lux (lm/m^2)	

Solved Problems

PROBLEM 1 *What is the amount of current that flows in the filament of a* 100 W *bulb connected to a* 220 V *power line ?*

SOLUTION

$$I = \frac{\text{Wattage}}{\text{Voltage}} = \frac{100\text{ W}}{220\text{ V}} = 0{\cdot}454\text{ Amp.}$$

PROBLEM 2 *Find the amount of power in a circuit where a source of 150 volts produces 6 Amp in a* 36 Ω *resistor.*

SOLUTION

$$P = I^2R$$
$$= (6)^2 \times 36$$
$$= 1296\text{ W}$$

PROBLEM 3 *How much current will be needed for a* 120 Ω *resistor that dissipates 1 Kilowatts ?*

SOLUTION

$$I = \sqrt{\frac{P}{R}} = \sqrt{\frac{1000\ W}{120\ \Omega}} = 2{\cdot}89\text{ Amperes}$$

PROBLEM 4 *A 10 ohm resistor is rated at 4 watts. What is the greatest current that it can safely carry ?*

SOLUTION

Using $P = I^2R$

or $4 = I^2 \times 10$

or $I = \sqrt{0{\cdot}4} = 0{\cdot}632$ Amps

PROBLEM 5 *The speed of a motor of resistance* 2 Ω *is controlled by a circuit connected to a 12 volt battery as shown in Fig. 1.34. The potentiometer is* 10 Ω, *100 watts. Is it a satisfactory arrangement ?*

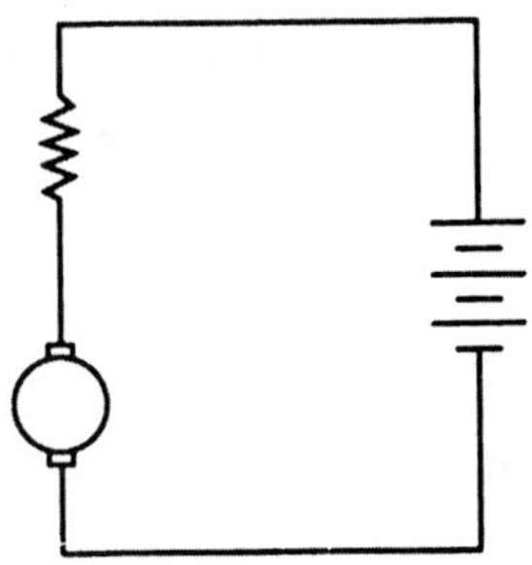

Fig. 1.34

SOLUTION

$$P = I^2R$$

$$100 = I^2 \times 12$$

$$I = \sqrt{\frac{100}{12}} = 2{\cdot}89 \text{ Amps.}$$

When the potentiometer is set to 10 Ω, the total resistance of the circuit is 12 Ω.

$$I = \frac{10\ V}{12\ \Omega} = 0{\cdot}8\text{A}$$

This amount of current is less then 2·89 Amps and the potentiometer is safe.

If the potentiometer value is decreased to say 2 Ω

$$I = \frac{10}{4} = 2{\cdot}5 \text{ Amp.}$$

The potentiometer track is safe.

However when the potentiometer resistance is 1 Ω, $I = \frac{10}{3} = 3{\cdot}3$ Amp. This is greater than 2·89 Amperes. The potentiometer wire will get heated and temperature will rise. For still lower values of potentiometer setting. I will be still higher and the potentiometer will be damaged. It may even be burnt out.

PROBLEM 6 *Find the four band colour codes for each of the following resistors*

(a) 4·7 Ω, 5% (b) 390 Ω, 5%

(c) 22 KΩ, 10% (d) 820 K, 10%

(e) 1 MΩ, 10%

SOLUTION

(a) Yellow, Violet, Gold, Gold

(b) Orange, White, Brown, Gold

(c) Red, Red, Orange, Silver

(d) Gray, Red, Yellow, Silver

(e) Brown, Black, Green, Silver

Exercise Problems

1 What is the current that flows through a filament of 500 W bulk connected to a 230 V power line ?
[Ans : 2·17 A]

2 Find the power in a circuit when the source of 220 V produces 1 ampere in 100 Ω resistor.
[Ans : 100 W]

3 Find the power if the voltage of 220 V produces 4 amperes in 20 Ω resistor.
[Ans : 320 W]

4 How much current is needed for a 400 W, 230 V toaster ?
[Ans : 1·74 Amp.]

5 What is the resistance of 1600 W, 230 V heater ?
[Ans : 3·31 ohm]

6 Find the current in a 16 Ω resistor that dissipates 1000 watts.
[Ans : 7·9 Amps]

7 Using the four band colour code, indicate colours of the bands for each of the following resistors.

(a) 2·7 Ω, 5% (b) 15 Ω, 10%

(c) 470 Ω, 5% (d) 8·2 KΩ, 5%

(e) 220 KΩ, 10% (f) 330 KΩ, 10%

(g) 1·5 MΩ, 10%

8 Using the five band colour code, indicate the colours of the bands for each of the following resistors.

(a) 275 Ω, ± 0·25% (b) 61·3 Ω, ± 0·1%

(c) 824 Ω, ± 0·5% (d) 37·1 KΩ, ± 0·25%

(e) 21·5 KΩ, ± 0·1% (f) 2·78 KΩ, ± 0·25%

9 Write the resistance values of chip resistors if the markings on the chip resistor are

(a) 273 (b) 371

(c) 1052 (d) 395

(e) 324 (f) 782

(g) 623

10 Find the current source equivalent circuit of Fig. 1·35

[Ans : i_S = 10 mA, R_2 = 600 Ω]

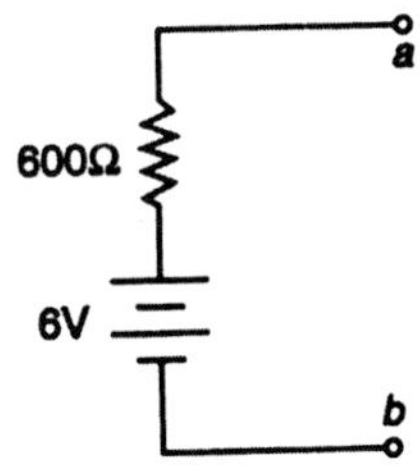

Fig. 1.35. Reference exercise problem 10

11 Find the voltage source equivalent circuit of Fig. 1·36.

[Ans : V = 25 V, R_1 = 100 Ω]

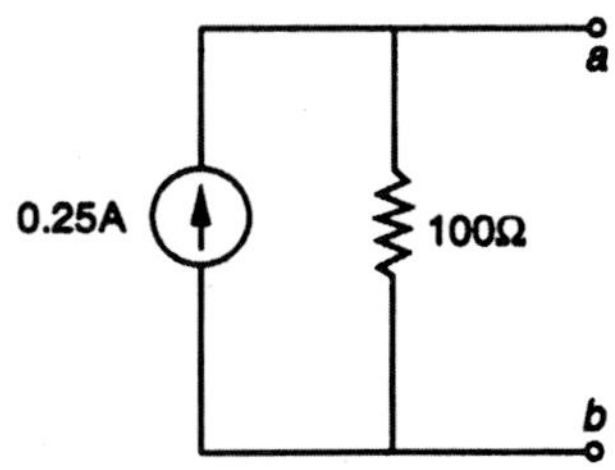

Fig. 1.36. Circuit in exercise problem 11

12 Find the value of the output voltage in the a.c. equivalent circuit of the transistor amplifier in Fig. 1·37. [Ans : 0·142 V]

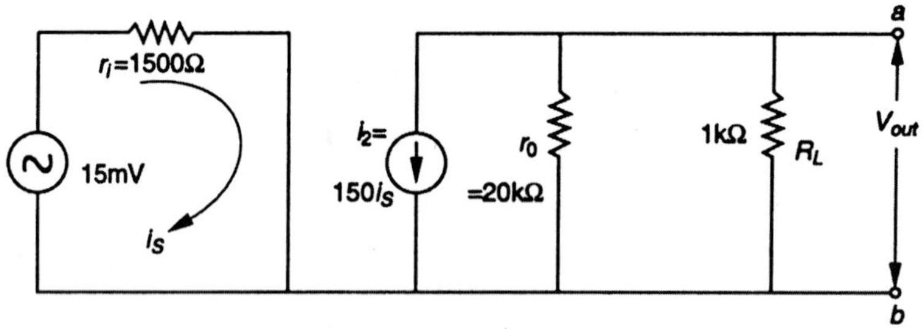

Fig. 1.37

Review Questions

1. What do your understand by electronics ?
2. Mention the role of electronics in everyday life.
3. Mention some important applications of electronics.
4. Discuss the role of electronics in modern day communication and entertainment.
5. Discuss the recent trends in the development of electronics
6. What are the units of resistance, current and electromotive force ?
7. Discuss the typical wattage of carbon composition and wire around resistors. How are wire wound resistors fabricated.
8. What is meant by capacitance ? When does a capacitor breakdown ?
9. Discuss some important applications where capacitors are employed.
10. Discuss the constructional features of a mica capacitor and a ceramic capacitor.
11. Discuss why the capacitance varies when the movable plates move.
12. Discuss the constructional features of an electrolytic capacitor. Can electrolytic capacitors be employed in any type of circuit ? Discuss the range of values obtainable in electrolytic capacitors.
13. Discuss what would happen, if in an electrolytic capacitor the positive marked polarity is connected to the negative voltage terminal and the negative marked polarity to the + of battery ?
14. Discuss the components which are employed as volume control and tuning control in a radio receiver.
15. What is a padding capacitor or padder ? Discuss its constractional features.
16. What is meant by inductance ? Discuss the amount of inductance involved when we say 1 Henry.

17. Discuss the specific functions which are carried out by a transformer in an electronic circuit.

18. What is meant by an autotransformer ? Isolation transformer ? Mention some of their applications.

19. Mention the various types of active components involved in an electronic circuit.

20. Discuss the five basic SI units. Define the SI units of length, mass and time.

21. What is the unit of magnetic flux in electrical field strength ?

22. Name some electrical sources of power.

23. Discuss the structual features of a dry cell.

24. Discuss the constructional features of a nickel cadmium cell.

25. What is an ideal voltage source ? Discuss how a practical voltage source can be represented by an ideal voltage source.

26. What is an ideal current source ? How can a practical current source be represented by an ideal current source.

27. Discuss how a practical voltage source can be represented by its equivalent current source or vice versa.

Objective Type Questions

1. The subject of electronics deals with
(a) *X*-rays and gamma rays (b) motors and generators
(c) transformers (d) electronic devices and circuits

2. An example of a semiconductor active device is
(a) transformer (b) battery
(c) transistor (d) alternator

3. Which of the following is a passive device ?
(a) transistor (b) field effect transistors (*FETs*)
(c) vacuum triode (d) resistors

4. The term IC generally refers to
(a) integrated circuits (b) integral control
(c) industrial combustion (d) internal capacitors

5. One example of a semiconductor device is
(a) transformer (b) resistor
(c) integrated circuit (d) capacitor

6. An electronic circuit requires a value of capacitor of 220 μF. Such a value of capacitor can be realised by using
(a) ceramic capacitor (b) mica capacitor
(c) electrolytic capacitor (d) padding capacitor

7. If in an electronic circuit, the value of the capacitor needed is 220 pF variable, the value can be easily obtained by
(a) poly styrene capacitor (b) mica capacitor
(c) electrolytic capacitor (d) air gang capacitor

8. Radars are employed for
(a) detection and ranging of aircraft/missiles (b) geostationary satellites
(c) welding of detached retinal nerves (d) perform mathematical computations very fast

9. A resistor has a colour band sequence as follows : Brown, Green, Green, Silver. Its value is
(a) 1·5 MΩ, 10% (b) 150 KΩ, 10%
(c) 39 KΩ, 10% (d) 1·5 KΩ, 5%

10. The colour bond sequence of a resistor is as follows-Orange, White, Red, Gold. Its value is
(a) 39 KΩ, 5% (b) 390 KΩ, 5%
(c) 3·9 KΩ, 5% (d) 390 Ω, 5%

11. For a five band resistor, if the colour band sequence is Red, Violet, Green, black and blue, its value is

(a) 27·5 Ω ± 0·5% (b) 27·5 Ω, ± 0·5%

(c) 275 Ω ± 0·25% (d) 275 Ω ± 0·25%

12. For a chip resistor the marking is 394, its value is

(a) 390 Ω (b) $3·9 \times 10^4$ Ω

(c) 39×10^4 Ω (d) 394 Ω

13. For a zero ohm resistor, the resistance value and wattage are typically as follows

(a) 1 W, Zero Ohm (b) $\frac{1}{2}$ W, 0·002 Ω

(c) $\frac{1}{8}$ W, 0·004 Ω (d) $\frac{1}{2}$ W, 0·001 Ω

14. If all six dials in a decade resistance box are 6 the total resistance is

(a) 666·666 Ω (b) 6666·66 Ω

(c) 66666·6 Ω (d) 666666 Ω

15. The resistance of a 400 *W*, 230 *V* toaster is

(a) 1·74 Ω (b) 13·23 Ω

(c) $\frac{1}{1·74}$ Ω (d) $\frac{13·23}{1·74}$ Ω

16. The current needed in a 25 ohm resistor that dissipates 400 W is

(a) 2 A (b) 4 A

(c) 16 A (d) 80 A

17. The temperature of the triple point of water is

(a) 273 K (b) 273·04 K

(c) 273·15 K (d) 0°C

18. The force in units of Newton/metre length produced when a constant current of 1 ampere flows through two straight parallel conductions of infinite length and negligible circular cross section is

(a) 2×10^{-7} (b) 1×10^{-7}

(c) 2×10^{-8} (d) 1×10^{-9}

19. The unit of magnetic field strength is

(a) Henry (b) Tesla

(c) Ampere/metre (d) Weber

Chapter 2

FUNDAMENTAL CONCEPTS : ENERGY BANDS IN SOLIDS

2·1 INTRODUCTION

As we are entering into the new millennium we are entering into an era where information technology will take a lead role. Information technology is an interactive technology made possible by the successful integration of electronics and communication on one hand with computers on the other hand. The role of electronics in the present day society can not be under estimated. Human beings in the present day society all over the global village feel the impact of electronics in the daily life. It may be mentioned here that the Internet and particularly the World Wide Web is now used by over 200 million people worldwide with every type of computer system.

Electronics is that branch of science and technology which deals with the theory and application of a class of devices which depend on the motion of electrons through vacuum, gas or semi conductor. These devices are termed as electron devices. Typical examples of such devices are diodes, transistors, triodes (which are obsolete now), FETs, ICs, LSIs, VLSIs etc.

Electronics can be broadly classified into two fields.

(1) **Physical electronics** which addresses the behaviour of electrons while moving through vaccum, gas or semiconductor.

(2) **Electronics Engineering** that deals with the design, fabrication and application of electron devices.

2·2 ELECTRIC CHARGES

Electric charges can be either positive or negative. It is specified as being a multiple of the electronic charge. The electron is the fundamental charged particle. The mass of the electron while at rest is

$$m_0 = 9{\cdot}11 \times 10^{-28} \text{ gm} = 9{\cdot}11 \times 10^{-31} \text{ kgs}$$

This is called the rest mass. The charge of the electron is

$$e = 1{\cdot}608 \times 10^{-19} \text{ Coulomb}$$

The mass of the electron is a function of the velocity according to the theory of relativity. The mass m of an electron moving with velocity v can be expressed as

$$m = \frac{m_0}{\sqrt{1 - \frac{v^2}{c^2}}} \qquad \ldots(2\cdot1)$$

where c is the velocity of light in free space. c has a value

$$c = 2{\cdot}998 \cdot \times 10^8\ m/\text{sec} \approx 300{,}000\ \text{km/sec}$$

If the numerical value of $v << c$, $m \approx m_0$.

The value of m for different values of velocities is depicted in Table below.

v/c	m/m_0	v/c	m/m_0
0	1	0·97	4·113
0·25	1·016	0·98	5·025
0·5	1·155	0·99	7·089
0·75	1·512	0·995	10·012
0·8	1·666		
0·9	2·294		
0·95	3·203		

An electron which starts from rest and is accelerated through a potential difference of 100 V has a velocity v which $5{\cdot}9 \times 10^6$ m/sec. Take another example. An electron which starts from rest and is accelerated by a potential difference of 1600 volts, acquires a velocity of $2{\cdot}4 \times 10^7$ m/sec. However an electron accelerated through a potential difference of 6·4 KV acquires a velocity of $9{\cdot}48 \times 10^7$ m/sec which is 0·32 c. At this high speed the mass of the electron increases by percent. Normally the change of the mass of the electron may be neglected for potentials below 5KV.

As regards the sizes, the atomic radius is of the order of 10^{-10} m and the radius of the electron is 10^{-15} m. These dimensions being extremely small, they are normally neglected and assumed to have point masses.

The electron volt is the unit of energy. The electron volt (eV) is defined as the amount of kinetic energy acquired by an electron which is initially at rest, while accelarated through a potential difference of 1 volt. Since the charge of electron $e = 1{\cdot}6 \times 10^{-19}$ coulomb and 1 coulomb × 1 Volt = 1 joule, we may write

$$1\ eV = 1{\cdot}6 \times 10^{-19}\ \text{Coulomb} \times 1\ \text{Volt}$$

$$= 1{\cdot}6 \times 10^{-19}\ \text{Joule}$$

2·3 BOHR'S ATOMIC MODEL

In 1913, the Bohr atomic model was proposed. Bohr made the following postulates

1. The electrons rotate in circular orbits around the positively-charged nucleus, which is assumed to be at the centre of the orbits.

2. The centrifugal force acting on the rotating electron is balanced by the electrostatic pull between the positively charged nucleus and the electrons.

3. The electron can not revolve round the nucleus in any arbitrary orbit but in certain discrete orbits.

4. Only certain discrete orbits (stationary orbits) are permitted for which the orbital angular momentum (or moment of the momentum) is equal to an integral multiple of $\hbar$, the modified Planck's constant. That means,

$$\text{the orbital angular momentum} = m\,v\,r = n\,\hbar = n\frac{h}{2\pi} \qquad ...(2\cdot2)$$

where $\hbar = \frac{h}{2\pi}$ and h is the Planck's constant.

(n is called the principal quantum number)

5. The energy of the electron in the orbit is constant. It does not radiate any energy.

6. The electron radiates energy when it moves from a higher energy orbit to a lower energy orbit. Alternatively, the electron may move from a lower energy orbit to a higher energy orbit if energy is supplied from outside. If E_1 and E_2 are the energy values of the electron before and after the transition, then

$$E_2 - E_1 = hf$$

or

$$\Delta E = hf$$

where f is the frequency of the emitted radiation.

Since the stability of the atoms demands

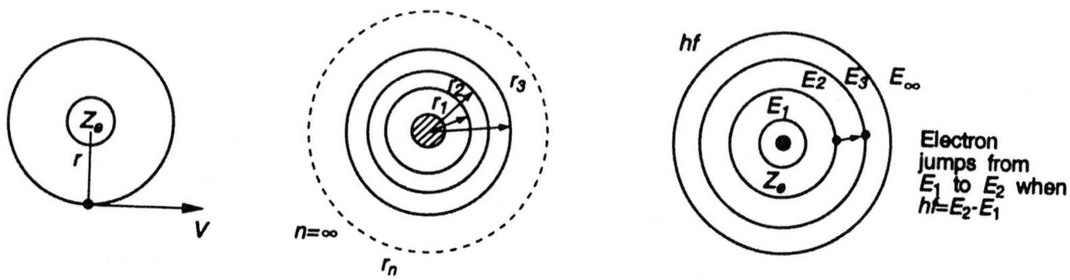

Fig. 2.1. Bohr model of the hydrogen atom

Centrifugal force on the rotating electron

= **Electrostatic force exerted by the nucleus on the electron**, we may write

$$\frac{mv^2}{r} = \frac{Q_1 Q_2}{4\pi\varepsilon_0 r^2}$$

or

$$\frac{mv^2}{r} = \frac{Ze\,.\,e}{4\pi\varepsilon_0 r^2} \qquad ...(2\cdot3)$$

where r is the radius of the orbit, v, the velocity and m the mass of the electron. $Q_1 = Ze$ is the charge of the nucleus and $Q_2 = e$ is the electron charge.

Also by the Bohr's postulate

$$m\,v\,r = \frac{n\,h}{2\pi}$$

and equation (2·3), we get

$$r = \frac{n^2 h^2 \varepsilon_0}{\pi m Z e^2} \qquad \text{...(2·4)}$$

The value of r_n the radius of the electron orbit corresponds to a value n or,

$$r_n = \frac{n^2 h^2 \varepsilon_0}{\pi m Z e^2} \qquad \text{...(2·5)}$$

Now

$$\left.\begin{aligned} r_1 &= \frac{h^2 \varepsilon_0}{\pi m Z e^2} \\ r_2 &= (2)^2 r_1 = 4r_1 \\ r_3 &= (3)^2 r_1 = 9r_1 \\ r_4 &= 16 r_1 \\ r_5 &= 25 r_1 \\ &\ldots\ldots\ldots\ldots \end{aligned}\right\} \qquad \text{...(2·6)}$$

It will be noticed that the values of $r_1, r_2, r_3, \ldots$ are discrete *i.e.* all values of r between r_1 and r_2 is forbidden just like other values of r between r_2 and r_3 is forbidden.

The velocity v_n of the electron is

$$v_n = \frac{Z \cdot e^2}{n \cdot 2 \varepsilon_0 h} = 18\,\pi \times 10^9 \frac{e^2 Z}{nh}$$

$$= 56{\cdot}55 \times 10^9 \frac{e^2 Z}{nh} \qquad \text{...(2·7)}$$

The velocity

$$\left.\begin{aligned} v_1 &= 56{\cdot}55 \times 10^9 \frac{Z e^2}{h} \\ v_2 &= \frac{v_1}{2} \\ v_3 &= \frac{v_1}{3} \\ &\vdots \\ v_n &= \frac{v_1}{n} \end{aligned}\right\} \qquad \text{...(2·8)}$$

The rotational orbital frequency f_n of the electron is

$$f_n = \frac{v_n}{2\,\pi r_n} = \frac{m e^4 Z^2}{4 \varepsilon_0^2 n^3 h^3} \qquad \text{...(2·9)}$$

The kinetic energy of the electron from eauation (2·4)

$$K.E = \frac{1}{2} m v^2 = \frac{e^2 Z}{8\,\pi \varepsilon_0 r} \qquad \text{...(2·10)}$$

Since the potential V at a distance r is

$$V = \frac{Z e}{4\,\pi \varepsilon_0 r} \qquad \text{...(2·11)}$$

the potential energy is

$$P.E = -\frac{Ze^2}{4\pi\varepsilon_0 r} \qquad ...(2\cdot12)$$

Therefore, orbital energy = $K.E + P.E$

$$E_n = -\frac{Ze^2}{8\pi\varepsilon_0 r} \qquad ...(2\cdot13)$$

$$= -\frac{mZ^2e^4}{8\varepsilon_0^2 n^2 h^2} \qquad ...(2\cdot14)$$

The negative sign in equation (2·14) implies that an amount of energy equal in value to that predicted by equation (2·14) is to be supplied from outside to remove the electron from the hydrogen atom $Z = 1$. Thus, one gets from equation (1·14)

$$E_n = \frac{-me^4}{8\varepsilon_0^2 n^2 h^2} = \frac{-21\cdot7\times10^{-19}}{n^2}\text{ Joules}$$

$$= -\frac{13\cdot6}{n^2}eV \qquad ...(2\cdot15)$$

For hydrogen atom, there is only electron and this normally resides in the state corresponding to $n = 1$. This is also called the **ground state**. When energy is supplied from outside, the electron may be raised to higher energy orbits like $n = 2$, $n = 3$ ...... and the electron is said to be in the excited state. However, the electron may stay in the excited state for a short period of time called the lifetime after which it returns back to the ground state. The value of the lifetime is very small typically 2×10^{-8} secs. **If the electron returns from the excited state to the ground state,the difference in energy is released.** If the frequency corresponds to electromagnetic rays, we call radiation is emitted. The process is called radiation. If the electron jumps from a higher energy orbit say $n = 4$ to an orbit $n = 3$, $n = 2$, and then $n = 1$ it may emit different radiation frequencies. The electron ultimately returns to the ground state because here the potential energy is a minimum. In nature every thing tends to remain in a state where the potential energy is a minimum. This is similar in analogy to that of ball thrown upwards. After some time, it returns back because here the potential energy is minimum (corresponding to the ground state).

From equation (2·15) we see that **if an electron is to be removed from a hydrogen atom, an amount of energy equal to the difference between $n = 1$ and $n = \infty$, $E_n - E_1 = 13\cdot6\ eV$ is to be supplied from outside. Once the electron is removed from the atom, we call the atom as "ionized"**

The Bohr model of an atom is depicted in Fig 2·1

The configuration of hydrogen atom is depicted in Fig 2·2.

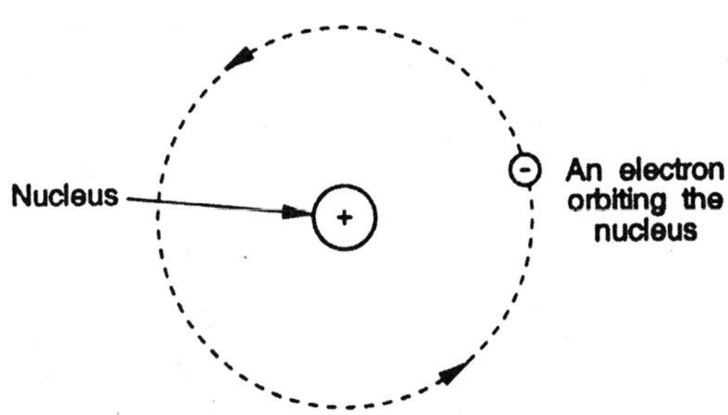

Fig. 2.2. Configuration of the hydrogen atom

In the periodic table of elements the elements are arranged as per the **atomic number** which is the number of electrons in the electrically balanced (neutral) atom. The elements can also be arranged according to their **atomic weight,** which is approximately the number of protons and neutrons in the nucleus. The element hydrogen has an atomic number of one and an atomic weight of one whereas helium (*He*) has an atomic number of 2 and atomic weight of 4. **Silicon has an atomic number of 14 and an atomic weight of 28.** In the normal (or neutral) state, all atoms of an element possesses the same number of electrons and protons. Thus, the positive charges cancel the negative charges and the atom as a whole is electrically neutral (net charge of zero).

2·4 ELECTRON SHELLS AND ORBITS

According to the Bohr's postulates, each of the discrete orbits correspond to certain energy level. In the atom, the orbits are grouped into energy bonds called **shells.** The number of shells in a given atom is fixed. Each of the shells possesses a fixed maximum number of electrons at permissible energy levels (or orbits). The variations in the energy levels (or orbits). The variations in the energy levels within a shell is much smaller compared to the differences in energy between the shells. The shells are designated as *K*, *L*, *M*, *N* and the like. The *K* shell is closest to the nucleus. The *K* and *L* shells are depicted in Fig. 2·3.

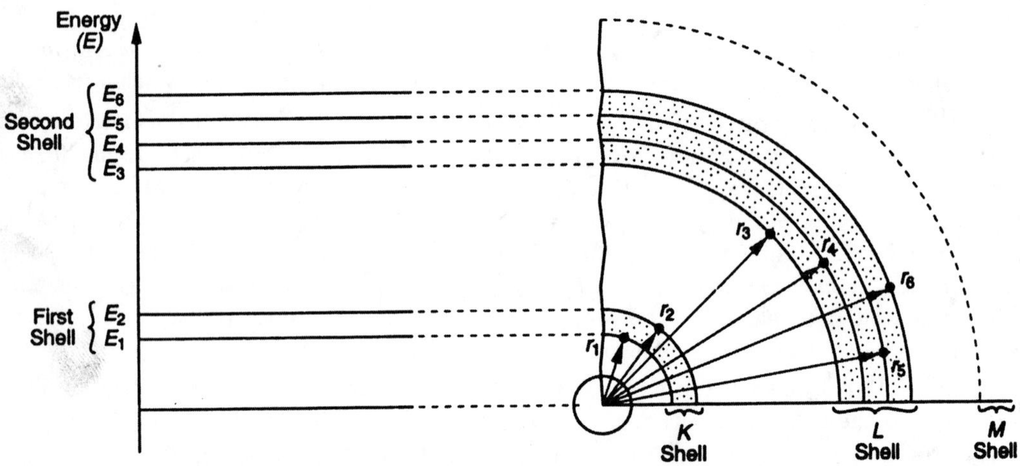

Fig. 2.3. Illustration of the *K* and *L* shells and the corresponding energy levels

The energy state of an electron in an atom is described by four discrete quantum numbers. They are

n, the principal quantum number

l, the azimuthal quantum number

m, the magnetic quantum number

and *s*, the spin quantum number

The principal quantum number n_1 specifies the angular moment of the electron while rotating in a fixed orbit round the nucleus. *n* is an integer having values $n = 1, 2, 3 \ldots$.

The azimuthal quantum number *l* defines the orbital motion of the electron. *l* can possess integer values only. Thus $l = 0, 1, 2 \ldots (n-1)$ the maximum value of *l* being $(n-1)$.

The orbital magnetic quantum number m describes the quantized directions of the plane of the orbit of the electron to the direction of the externally applied magnetic field. It can possess values from $-l$, $-(l-1), -(l-2), \ldots\ldots -2, -1, 0, +1, +2, \ldots\ldots (l-2), (l-1), l$.

The spin quantum numbers defines the direction of the spin of the electron. It can have two possible values $+\frac{1}{2}$ or $-\frac{1}{2}$.

According to Pauli's exclusion principle, no two electrons can possess the same set of values for the four quantum numbers.

The number of possible states in any shell is twice the square of the principal quantum number *i.e.* $2n^2$. Table 2·1 helps to depict the different sub shells and the maximum number of electrons they may accomodate.

TABLE 2.1

Shell	K	L		M			N			
Principal quantum number, n	1	2		3			4			
l	0	0	1	0	1	2	0	1	2	3
sub shell	s	s	p	s	p	d	s	p	d	f
Maximum number of electrons that can be accomodated $= l\,(2l + 1)$	2	2	6	2	6	10	2	6	10	14
Total number of electrons $2n^2$	2	8		18			32			

2·5 ATOMIC ELECTRON CONFIGURATION

This describes the distribution of electrons in various sub shells around the nucleus. This can be found very easily if we remember that each subshell can accomodate a maximum of $2(2l + 1)$ electrons and that the maximum number of electrons in a shell is $2n^2$.

For sodium (Na) atoms, the value of Z, the atomic number is 11. An atom of sodium possesses 11 electrons. The electronic configuration is $1s^2\, 2s^2\, 2p^6\, 3s^1$, the superscripts implying the number of electrons in the sub shells. Thus out of the 11 electrons in a sodium atom 2 electrons reside in sub shell $1s$, 2 in subshell $2s$, 6 in subshell $2p$ and 1 in subshell $3s$. There is this one electron in the outer most subshell and the atom is monovalent.

For copper atoms, the value of the atomic number is 29. The electronic configuration is $1s^2\, 2s^2\, 2p^6\, 3s^2\, 3p^6\, 3d^{10}\, 4s^1$. The atom is also monovalent.

2·6 ENERGY BANDS IN CRYSTALS

A crystal is a solid formed by regular (geometric) arrangement of atoms in space. If the atomic positions in the crystal are represented by points, called lattice points, a crystal lattice is obtained. In the crystal lattice, the distance between the atoms is fixed and is called the lattice constant of the crystal.

From equation (2·11) and (2·12), one can draw the variation of the potential and potential energy of the electron with distance from the nucleus (centre) as depicted in Fig. 2·4 (a) and (b) respectively

Let us now consider that two similiar atoms placed very close together. The net potential energy of an electron is obtained as the algebraic addition of the potential energies of the individual atoms.

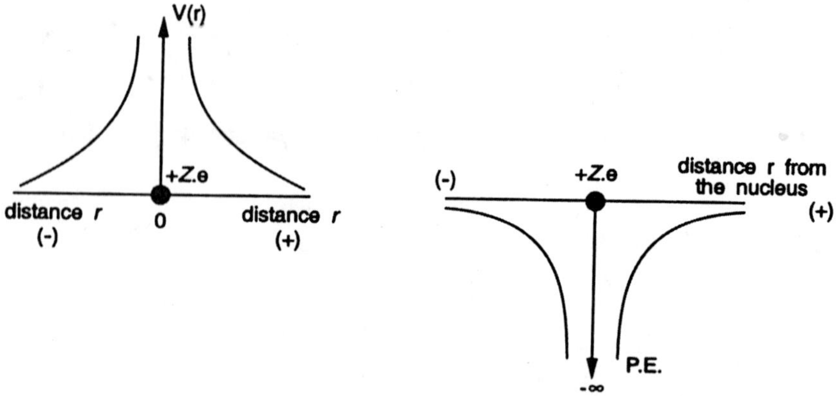

Fig. 2.4. (a) Potential *V*(*r*) variation as a function of *r*, (b) Potential energy variation with distance *r*

In the space between the two nuclei, the net potential energy is lower compared to the potential energy of an isolated nucleus.

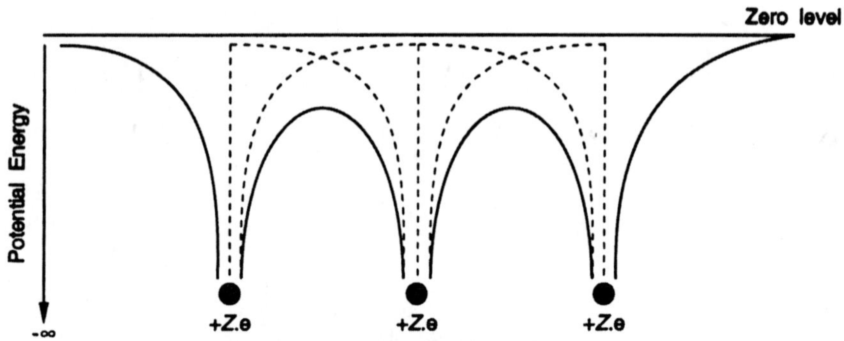

Fig. 2.5. Potential energy plotted as a function of distance between a number of similar nuclei put close together (solid lines). The dotted lines correspond to energy energy variation of isolated nuclei.

The potential energy between the atomic nuclei comprises a series of humps. Near the boundary of the crystal, the potential energy increases and approaches zero at infinity as depicted in Fig. 2·5.

Equation (2·14) gives the total energy of the electron in a solid. When a large number of atoms are closely packed as in a solid, each of the atoms will exert a force to the neighbouring atoms. The crystal may likely to be a single electronic system that obeys Pauli's exclusion principle. Each discrete energy level of the isolated atom splits into as many levels as there are atoms in the crystal, such that Pauli's exclusion principle is satisfied. The **seperation** between the split off energy levels is **extremely small** since the number of electrons N is very large. The densely packed discrete and closely spaced energy levels are termed as **energy band.** In Fig. 2·6, the energy bands are depicted as shaded continuous regions.

In the diagram depicted in Fig. 2·6, it will be noticed that while the lower energy levels form narrow bands the higher energy levels form wider bands. The interatomic spacing although fixed for one crystal

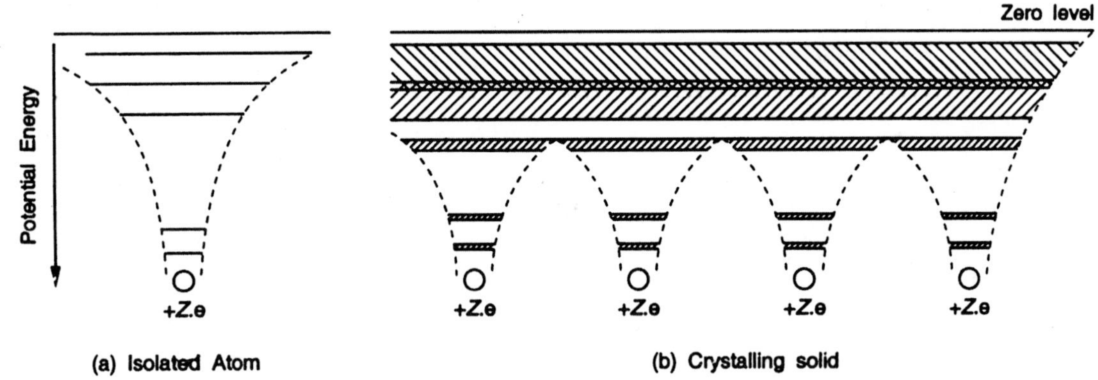

Fig. 2.6. Figure illustrating splitting of energy levels and formation of Energy Bands in a Crystalline Solid

varies from one crystalline solid to another. The width of an energy band is higher if the atoms are closely spaced (dense packing). As the number of atoms in the solid is extremely large ($\approx 10^{29}/m^3$), the spacing between the discrete levels in the band is small and the band can be treated as continuous.

As the electrons always tend to occupy the lowest energy levels, the lower energy levels are normally completely filled. The higher energy bands are either fully empty or partly filled. In a partly filled band, electrons can move freely with minimum expenditure of energy. The gap which seperates two bands is called the forbidden energy gap or forbidden band. No electron can occupy any state corresponding to this band.

It is normally found that most of the important properties of a crystal arises from the region in which the lower band is completely filled and the next higher band is empty or is partially filled. The higher value of the potential energy near the boundary prevents the electrons from escaping out of the crystal. If sufficient amount of energy is imparted to the electron by external mechanism the electron will be capable to cross the surface potential barrier and come out of the crystal surface.

Metals, Insulators and Semi conductors

The energy level diagram of various types of materials is depicted in Fig 2·7.

In the case of a metal, the higher energy band is partially filled or completely filled band and the next empty band overlaps in energy. Thus any electron in the partially filled band or overlapped band can very easily moved with the application of an electric field. The movement of the electrons gives rise to an electric current. Thus a metal is a good conductor of electricity. The higher filled band is called the valence band and the next higher partially filled (or overlapping) band, the conduction band. **Metals like copper, gold, silver possess conductivities in the range of 4×10^7 to $6{\cdot}3 \times 10^7$ mhos/metre. In the case of the conductors, the resistivity increases with increase in temperature.**

In the case of a semiconductor, the forbidden energy gap between the upper most filled band, called the valence band and the lower most empty band, called the conduction band is small ($\approx 0{\cdot}9$–$1{\cdot}5$ e*V*). Germanium and silicon the two most important semiconductor materials have forbidden gaps of the order 0·77 eV and 1·2 *eV* respectively at 0 K. As the forbidden gap is small, some of the valence electrons gains sufficient energy to overcome the forbidden gap and go to the conduction band. **The electrons move freely under the action of the applied electric field. The absence of the electron in the valence band is termed as**

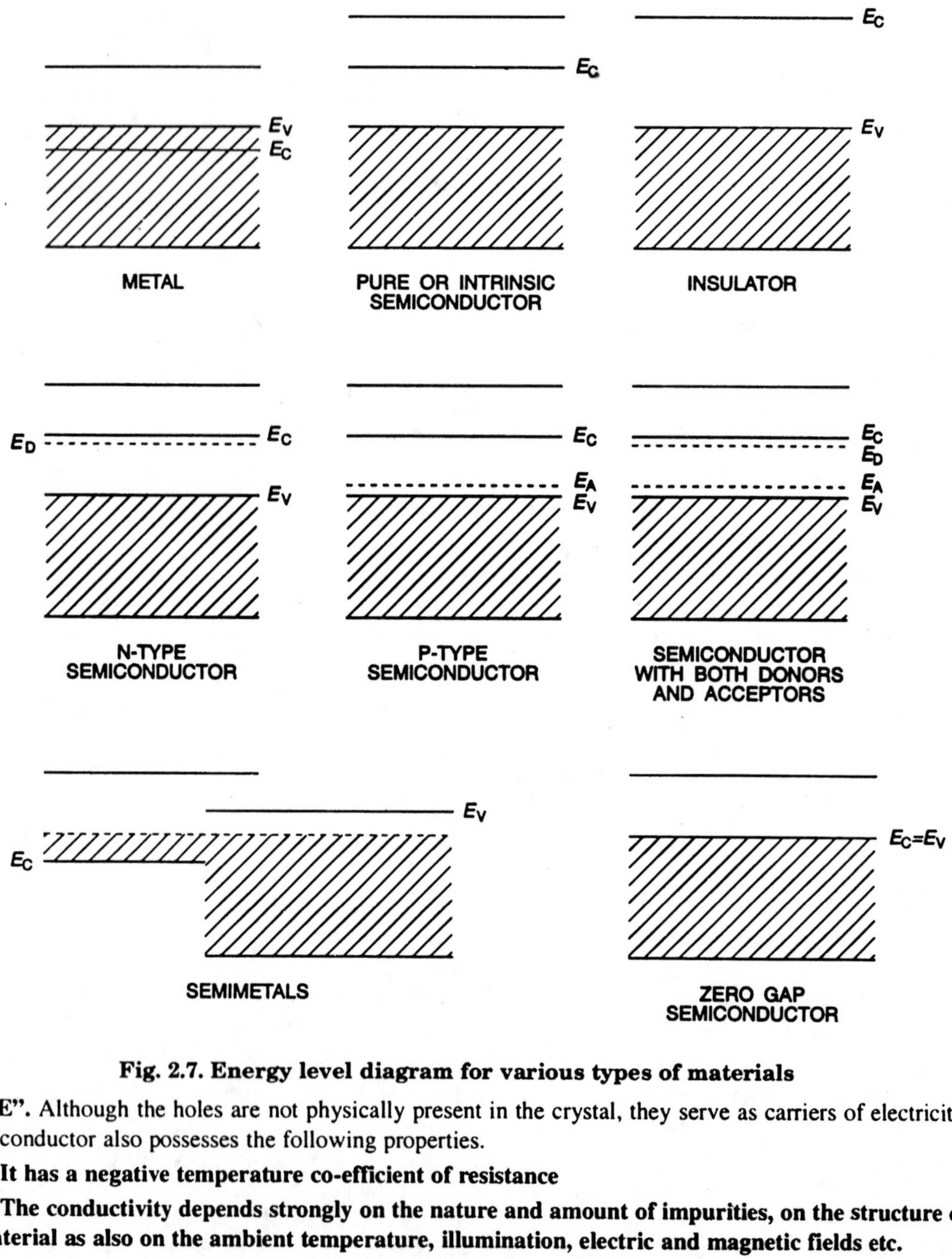

Fig. 2.7. Energy level diagram for various types of materials

"HOLE". Although the holes are not physically present in the crystal, they serve as carriers of electricity. A semiconductor also possesses the following properties.

1. **It has a negative temperature co-efficient of resistance**

2. **The conductivity depends strongly on the nature and amount of impurities, on the structure of the material as also on the ambient temperature, illumination, electric and magnetic fields etc.**

Semiconductors when heavily doped behave like metals. Besides, heavily doped semiconductors exhibit the property of super conductivity at very low temperatures [transition temperature (0·5 K)].

Semiconductors can be either (*i*) **pure or intrinsic** (*ii*) ***n* type or *p*-type extrinsic.** The energy band diagam of *n*- and *p*-type semiconductors are depicted in Fig 2·7 for hidden.

In the case of insulators, the energy band is very wide (≈ 5 eV). In such insulators, only a few electrons may gain enough thermal energy to move from the valence band to the conduction band. Since the number of electrons in the conduction band is very small, the insulator is a very bad conductor of electricity. The energy band structure of insulators is depicted in Fig 2·7.

PROBLEM 2.1 *Find the value of the kinetic, potential and total energy of an electron orbiting in the Bohr orbit in the hydrogen atom.*

SOLUTION

$$m_0 = 9{\cdot}1 \times 10^{-31} \text{ kg}$$

$$e = 1{\cdot}602 \times 10^{-19} \text{ Coulomb}$$

$$\varepsilon_0 = 8{\cdot}854 \times 10^{-12}$$

$$h = 6{\cdot}625 \times 10^{-34} \text{ Joules sec}$$

$$n = 2$$

$$\text{Thus Kinetic energy} = \frac{m_0 e^4}{8 \varepsilon_0^2 \times n^2 \times h^2}$$

$$= \frac{9{\cdot}1 \times 10^{-31} \times (1{\cdot}6 \times 10^{-19})^4}{8 \times (8{\cdot}854 \times 10^{-12})^2 \times 2^2 \times (6{\cdot}625 \times 10^{-34})^2}$$

$$= 5{\cdot}425 \times 10^{-19} \text{ Joules} = 3{\cdot}4 \text{ eV}$$

$$\text{Potential energy} = \frac{-m_0 e^4}{4 \varepsilon_0^2 n^2 h^2} = -10{\cdot}85 \times 10^{-19} \text{ Joules} = -6{\cdot}8 \text{ eV}$$

$$\text{Total energy} = KE + PE = 3{\cdot}4 + (-6{\cdot}8) = -3{\cdot}4 \text{ eV}$$

PROBLEM 2.2 *Find the total number of electron that can reside in the K, L, M shells in an atom.*

SOLUTION According to Pauli's exclusion principles, no two electrons can possess the same set of values for the four quantum numbers.

The total number of electrons that can reside in the various shells is as follows.

		Total no. of electrons
K shell		
	$n = 1, l = 0$, Subshell 1s, $m = 0, s = \pm\frac{1}{2}$	2 (Two)
L shell	$n = 2, l = 0$, subshell 2s, $m = 0, s = \pm\frac{1}{2}$	2
	$n = 2, l = 1$, subshell 2p, $m = -1, 0, +1, s = \pm\frac{1}{2}$	6
		8 electrons
M shell	$n = 3, l = 0, m = 0, s = \pm\frac{1}{2}$ (subshell 3*s*)	2
	$n = 3, l = +1, m = -1, 0, +1, s = \pm\frac{1}{2}$ (subshell 3*p*)	6
	$n = 3, l = 2, m = -2, -1, 0, +1, +2, s = \pm\frac{1}{2}$ (subshell 3*d*)	10
		18 electrons

PROBLEM 2.3 *Calculate the value of the kinetic energy (in units of Joules and eV) gained by an electron at rest when it is accelerated by apotential difference of (i) 100 V (ii) 500 V. Calculate also its velocity.*

SOLUTION (i) The kinetic energy gained by the electron for $V = 100$ V is

$$KE = e \times 100 \text{ Joules}$$
$$= 1{\cdot}602 \times 10^{-19} \times 100 \text{ Joules}$$
$$= 160{\cdot}2 \times 10^{-19} \text{ Joules}$$
$$= 100 \text{ eV}$$

The corresponding velocity is $v = \sqrt{\dfrac{2 \times (K.E)}{m_0}} = \sqrt{\dfrac{2 \times 160{\cdot}2 \times 10^{-19}}{9{\cdot}107 \times 10^{-31}}}$

$$= 5{\cdot}93 \times 10^6 \; m/sec.$$

(ii) When the accelerating potential is 500 V

$$KE = e \times 500 \text{ Joules}$$
$$= 801 \times 10^{-19} \text{ Joules}$$
$$= 500 \text{ eV}$$

The corresponding velocity is

$$v = \sqrt{\frac{2 \times KE}{m_0}} = \sqrt{\frac{2 \times 801 \times 10^{-19}}{9{\cdot}107 \times 10^{-31}}} = 13{\cdot}26 \times 10^6 \text{ m/sec}$$

PROBLEM 2.4 *A positively charged particle has a mass equal to 3674 times the mass of an electron and a charge which is twice that of the electron. Find the K E. and velocity of the particle when it is accelerated from rest through a potential difference of 5000 Volts.*

SOLUTION In this example $m = 3674\, m_0 = 3674 \times 9{\cdot}107 \times 10^{-31}$ kg

charge $q = 2 \times e = 2 \times 1{\cdot}602 \times 10^{-19}$ Coulomb and $V = 5000$ Volts

Thus $$\frac{1}{2} mv^2 = qV$$

or $$KE = qV = 2 \times 1{\cdot}602 \times 10^{-19} \times 5000$$
$$= 1{\cdot}601 \times 10^{-15} \text{ J} = 10{,}000 \text{ eV}$$

velocity $v = \sqrt{\dfrac{2\,q\,V}{m}} = \sqrt{\dfrac{2 \times 2 \times 1{\cdot}602 \times 10^{-19} \times 5000}{3674 \times 9{\cdot}107 \times 10^{-31}}}$

$$= 9{\cdot}78 \times 10^5 \text{ metres/sec}$$
$$= 978 \text{ km/sec}$$

PROBLEM 2.5 *Calculate the mass of an electron when* $v = 0{\cdot}2c$

SOLUTION

$$m = \frac{mo}{\sqrt{1 - \frac{v^2}{c^2}}}$$

$$= \frac{9{\cdot}107 \times 10^{-31}}{\sqrt{1-(0{\cdot}2)^2}} = \frac{9{\cdot}107 \times 10^{-31}}{0{\cdot}979}$$

$$= 1{\cdot}0202 \times 9{\cdot}107 \times 10^{-31} \text{ kg}$$

$$= 9{\cdot}29 \times 10^{-31} \text{ kg}$$

PROBLEM 2.6 *Use Bohr's theory for determining the first four visible lines for the n = 2 orbit (Balmer series) for the hydrogen atom.*

SOLUTION

$$hf_{n2} = E_n - E_2 = -13{\cdot}6\left(\frac{1}{n^2} - \frac{1}{2^2}\right)$$

$n = 3$, $hf_{32} = 1{\cdot}89$ eV, $\lambda = 0{\cdot}656$ μm = 6560 Å (RED)

$n = 4$, $hf_{42} = 2{\cdot}55$ eV, $\lambda = 0{\cdot}486$ μm = 4860 Å (BLUE)

$n = 5$, $hf_{52} = 2{\cdot}85$ eV, $\lambda = 0{\cdot}434$ μm = 4342 Å (BLUISH PURPLE)

$n = 6$, $hf_{62} = 3{\cdot}02$ eV, $\lambda = 0{\cdot}410$ μm = 4103 Å (DEEP PURPLE)

$n = \infty$, $hf_{\infty 2} = 3{\cdot}40$ eV, $\lambda = 3636$ Å (DARK PURPLE (invisible))

2·7 FERMI DIRAC DISTRIBUTION FUNCTION

In the case of a metal, the free electrons are moving around in a region which is more-or-less equipotential. There is a change in potential only near an ion. But these charged regions (ions) are a very small fraction of the total volume available for electron movement and hence the electrons may be assumed to move in a field free region or equipotential space. The free electrons find their source of energy as being the thermal energy. **The average behaviour of the electrons in the ensemble or electron gas (typical densities being 10^{23}/c.c.), are determined by the principles of statistical mechanics.** One quantitity which would be very useful in this connection is the occupancy probability of a given state by the electrons. **The Maxwell Boltzmann (MB) distribution function can be applied if the number of available quantum states is much more compared to the number of particles. The MB distribution function in thus not applicable for free electrons in metals.** The Fermi Dirac (FD) distribution function is useful in this case. For the Fermi Dirac distribution function, the probability $f(E)$ of occupancy of a state with energy E is given by

$$f(E) = \frac{1}{1 + \exp[(E-F)/K_B T]} \quad \text{...(2·16)}$$

where F is called the Fermi energy, K_B is the Boltzmann constant

($K_B = 1{\cdot}38 \times 10^{-23}$ J/K) and T is the absolute temperature.

At absolute zero of temperature $T = 0$ K, and

$$f(E) = 1 \text{ for } E < F \quad \text{...(2·17)}$$

and $$f(E) = 0 \text{ for } E > F \quad \text{...(2·18)}$$

This implies that energy levels below F are occupied by the electrons and all states above F are completely empty. The energy state F (Fermi Energy) gives the maximum energy that can be occupied by the electrons at 0 K.

Fig. 2·8 depicts the variation of $f(E)$ with E/F for various temperatures. It will be noticed that at temperatures $T > 0$ K, $f(E) > 0$ for $E > F$, implying that some of the electrons have sufficient energy so as to move to states above the Fermi level.

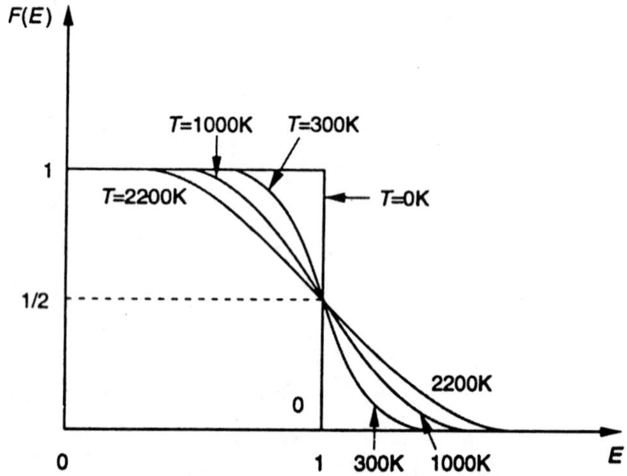

Fig. 2.8. Occupancy probability for Fermi Dirac distribution function plotted as a function of energy *E* for various temperatures.

If $E = F, f(E) = \frac{1}{2}$ for $T > 0$. The Fermi level can thus be defined as the energy state corresponding to which the occupancy probability is $\frac{1}{2}$ at a finite non zero temperature. In a pure semiconductor, the Fermi level stays near, to centre of the forbidden energy gap, whereas for a conductor (metal) the Fermi level lies in the conduction band.

Fermi level in intrinsic semiconductor

In an intrinsic or pure semiconductor, the Fermi level F lies in the middle of the energy band gap in between the conduction and valence band edges.

This can be seen from Fig. 2·9 (*a*)

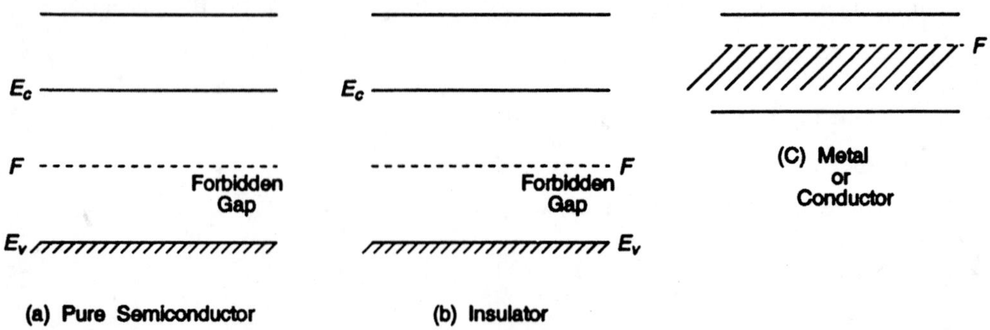

Fig. 2·9 Position of the Fermi levels in semiconductors, metals and Insulators

2·8 SILICON AND GERMANIUM ATOMS

The two important semiconductor materials which are used to manufacture diodes, transistors and other semiconductor devices are germanium and silicon. Silicon possesses 14 protons in its nucleus while germanium posseses 32 protons. However both have four valence electrons. This can be seen from Fig. 2.10.

Covalent Bonds

When a large number of silicon atoms combine, they arrange themselves in a regular geometric pattern called a crystal. In the crystal structure, the atoms are held together by covalent bonds. The covalent bonds are formed due to the interaction of the valence electrons of the atoms (Fig. 2.11a). A solid block of silicon is a crystalline material (also called single crystal).

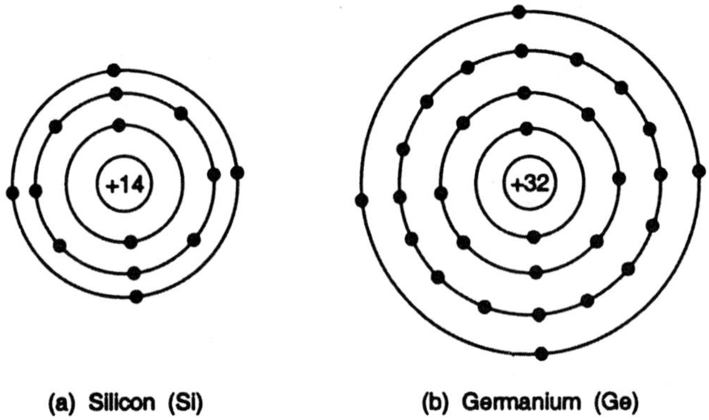

Fig. 2·10 Structure of Silicon and Germanium atoms

In the covalent bond each silicon atom shares itself in a group with four nearby atoms. The silicon atom shares its four valence electrons with each of its four nearest neighbours. In effect, it creates eight valence electrons for each atom and form a state of chemical stability. **This uniform sharing of valence electrons forms what is known as covalent bonds which holds the atoms together. Each of the shared electrons are equally attracted by the two nearest atoms which share it.** The covalent bonding in an intrinsic silicon crystal is depicted in Fig. 2·11. A similar covalent bonding is found in the case of germanium.

In an atom, the electrons can exist within allowed energy levels. The electrons are not permitted to stay in the forbidden energy gaps between the shells. The energy band diagram of silicon at 0 K for an unexcited crystal shows that all the electrons are in the valence band. The silicon crystal thus behaves like an insulator at absolute zero of temperature.

At room temperature ($T = 300$ K) some of the valence electrons will have sufficient thermal energy so as to move to conduction band. An electron in the conduction band is called a free (conduction) electron since it can give rise to conductivity as there are a large number of vacant states available.

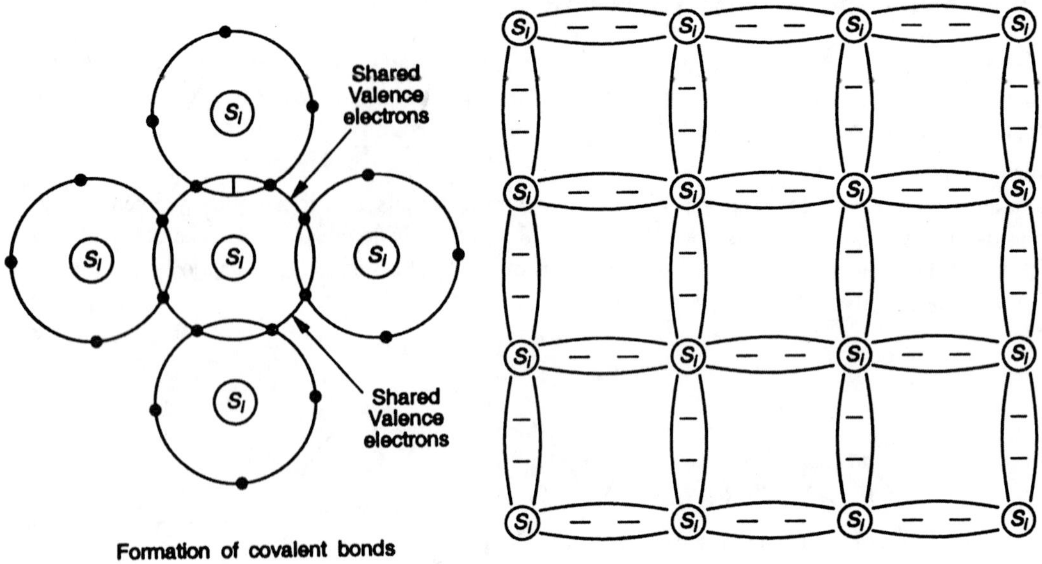

Fig. 2.11 Covalent Bonds in an intrinsic silicon crystal

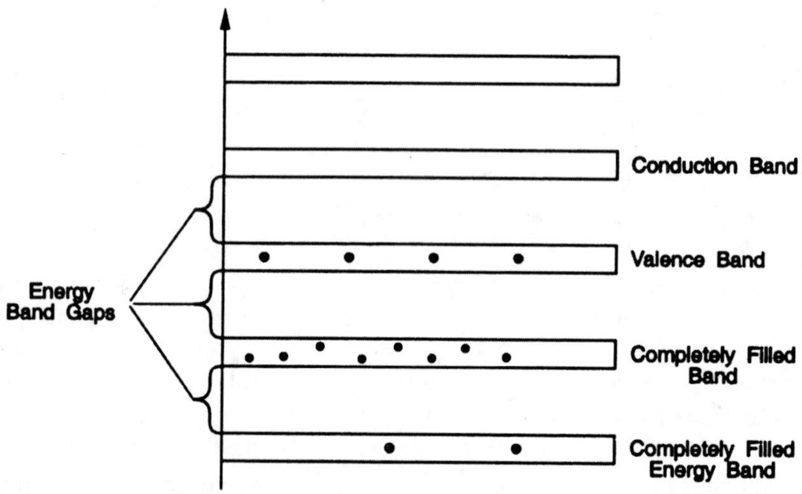

Fig. 2.12 Energy band diagram of silicon at 0 K for an unexcited crystal

It should be remembered in this connection that as soon as an electron moves to the conduction band, it leaves behind a vacant (empty) state in the valence band. The absence of the electron in the valence band is termed as a **hole. This vacant position can now be occupied by a nearby electron in the valence band and this hole shifts to a new position.** The hole is said to move.

The hole is ascribed a positive charge since a hole is created by leaving of a negatively charged electron. **For each and every electron raised to the conduction band by thermal or light energy, an attendant hole is left behind in the valence band, thus creating what is known as electron-hole pair.**

In a metal, the number of free electrons is of the order of the number of atoms in the crystal (of the order of the Avogrado's number $\approx 6 \times 10^{23}$ per cubic cm). In the semiconductor at a temperature of 300 Kelvin, the number of thermally generated electron-hole pairs is of the order of 10^{10}/cubic cm. or approximately 6×10^{13} per crystal atoms. This is a very small ratio. However, this small number of thermal generated electron hole pairs is enough to allow a semiconductor to conduct at a finite temperature. Also, since the number of thermally generated carriers increase with increase in temperature, the conductivity increases with temperature. **For a conductor, the conductivity decreases with increase in temperature since the more vigorous crystal lattice vibrations impede the flow of electrons.**

2·9 Impurities

In the case of the metal (or conductor), the electron concentration is fixed and there are no holes. In contrast, the electron and hole concentrations in a semiconductor can be charged by adding small amounts of impurities (dopants) to the crystal lattice. A semiconductor with added impurities is called extrinsic or impurity semiconductor.

2·9·1 *P*-type extrinsic semiconductor

By adding acceptors, the number of holes created by thermal generation can be supplemented. Acceptor atoms like boroun (B), gallium (Ga) and Indium (In) have three valence electrons in the outermost shell compared to that of the lost silicon atom. When an accept or atom is introduced into the crystal, it gives an unoccupied electron state in the band gap just above the valence band. With a small amount of thermal energy, the electron from the valence band can fill the vacant state.

When this happens, the electron becomes trapped in the covalent bond of the acceptor atom. This creates a fixed immobile, negative ion at the location of the acceptor atom. In effect the electron from the valence band leaves behind a hole that is free to move in the crystal. The hole is created in the valence band without any creation of an attendant free electron in the conduction band. The doping of acceptor impurities in the silicon lattice is depicted in Fig. 2·13.

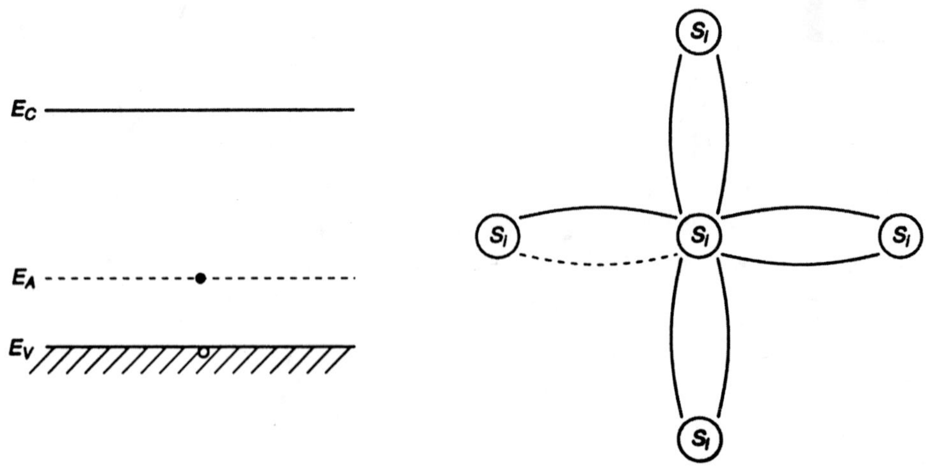

Fig. 2.13 Doping of acceptor impurities

2·9·2 Donor atoms—*N* type Extrinsic Semiconductor

By adding donor atoms like antimony (Sb), phosphorous (P) or arsenic (As) to silicon or germanium lattice, an n-type semiconductor can be obtained. The donor atoms have five valence electrons in the outermost shell of the atom.

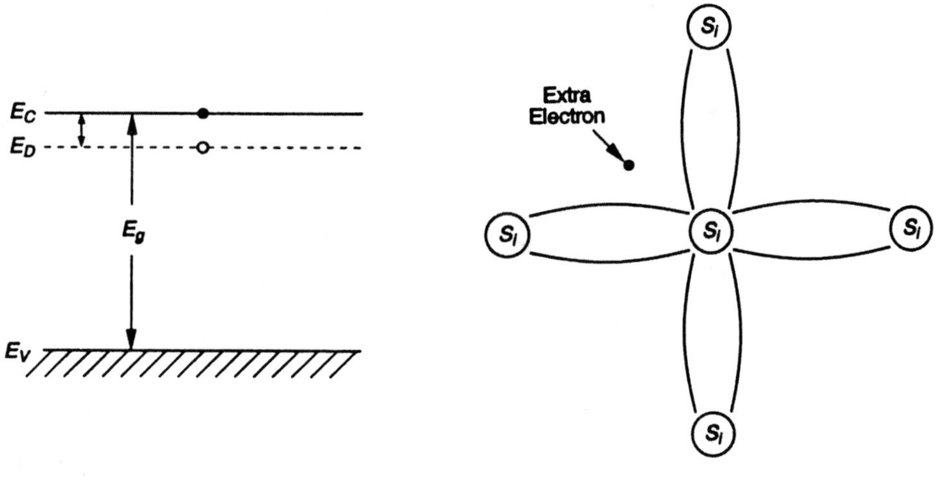

Fig. 2·14 Doping of donor impurities

When a donor atom is introduced into the lattice, the energy level of the extra electron lies in the band gap of the host crystal, just below the conduction band. From there, the extra electron is thermally excited into the conduction band. It provides a new liberated electron (free electron) in the near empty conduction band. The excited electron leaves behind a fixed positive ion core at the site of the donor atom but no associated hole in the valence band. Fig 2·14 depicts the doping of donor impurities to a silicon lattice.

2·9·3 Carrier densities within a semiconductor

In any semiconductor at a finite temperature electrons and holes are constantly generated by thermal agitation. Further, some of the electrons and holes recombine, thus annihilating themselves. Thus at any given temperature, the generation and recombination processes balance precisely so that the net equilibrium hole and electron densities satisfy the relation

$$n_0 \cdot p_0 = n_i^2 \qquad \ldots(2\cdot19)$$

(This equation is also termed as the mass action law)

where n_0 and p_0 are the electron and hole densities and n_i is the intrinsic carrier concentration.

The value of n_i depends on temperature and is given by

$$n_i^2 = AT^3 \exp(-Eg_0/K_BT) \qquad \ldots(2\cdot20)$$

The variation of n_i with temperature is given by the plot shown in Fig. 2·15.

For an intrinsic semiconductor

$$n_0 = p_0 = n_i \qquad \ldots(2\cdot21)$$

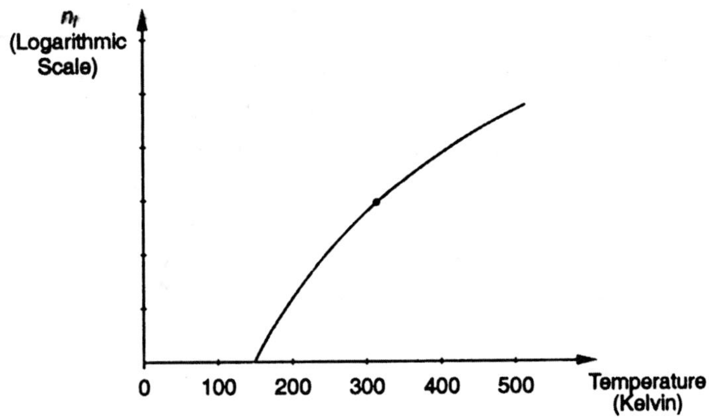

Fig. 2.15. Variation of intrinsic carrier concentration with temperature

For an extrinsic semiconductor, if N_A and N_D are the acceptor and donor ion concentrations the charge neutrality condition is written as

$$\underset{\text{Donor ion Concentration}}{p_0 + N_D} = \underset{\text{Acceptor ion Concentration}}{n_0 + N_A} \quad \text{...(2·22)}$$

where N_A and N_D are the acceptor and donor densities. In equation (2·22) it has been assumed that all the acceptors and donor atoms have been ionized. The charge neutrality condition can be combined with mass action law to determine the hole and electron concentrations in a doped semiconductor.

For example in a p type semiconductor, $N_D = 0$

and $$n_0 = \frac{n_i^2}{p_0} \quad \text{...(2·23)}$$

For $N_D = 0$, then we get from equation (2·22)

$$\frac{n_i^2}{p_0} + N_A = p_0 \quad \text{...(2·24)}$$

or, $$p_0^2 - p_0 N_A - n_i^2 = 0 \quad \text{...(2·25)}$$

or, $$p_0 = \frac{N_A}{2} \pm \frac{(N_A^2 + 4n_i^2)^{1/2}}{2} \quad \text{...(2·26)}$$

2·10 DEFINITION OF SEMICONDUCTORS

From the above discussion we can define semiconductors as follows

Semiconductors are a group of materials which depict the following properties :

1. **It possesses a conductivity lying in the range 10^{-4} to 10^5 siemens/meter.**
2. **Pure semiconductors depict a negative temperature co-efficient of resistance.**

3. When irradiated by an incident light source, a semiconductor typically exhibits a photovoltage or a light dependent resistance.

4. Semiconductors depict high thermoelectric powers which may be either positive or negative, as in the case of metals.

5. A junction between an *n* and a *p*-type semicondunctor exhibits rectification property

6. Semiconductors depict a band gap which lie in the range of 0·2 to 2·5 eV. Due to small band gap, electrons are excited from the valence band to the conduction band at ordinary temperatures.

The two most important semiconductor materials are germanium (Ge) and silicon (Si). Other elementary semiconductors are boron (B), Carbon (C), phosphorous (P), Sulphur (S) selenium (Se) and grey tin (S_n). They all belong to group IV to VI of the periodic table. Some important compound semiconductors are :

1. Binary : Ga As (gallium arsenide)
2. Binary : In P (indium phosphide)
3. Binary : Ga Sb (gallium antimonide)
4. Binary : In As (indium arsenide)
5. Ternary : Ga In As (gallium indium arsenide)
6. Quaternary : Ga In As P (gallium indium arsenide phosphide)
7. Binary : CdS (cadmium sulphide)
8. Ternary : Hg Cd Te (mercury cadmium telluride)

These semiconductors mentioned above have wide spread use in various electronic, optoelectronic and other applications.

A distinction between metals and semiconductors is given in Table 2·2 below.

Table 2·2 : Distinction between Metals and Semiconductors

	Metal		Semiconductor
1.	Crystalline	1.	Crystalline
2.	Electrons are the charge carriers	2.	Both electrons and holes are the charge carriers
3.	exhibits low resistivity (high conductivity)	3.	exhibits moderate conductivity
4.	Resistivity increases with temperature as lattice vibrations make possible more opportunities for electron collision	4.	Resistivity decreases with temperature as more and more electron hole pairs are generated thermally
5.	High carrier density (of the order of Avogrado's number)	5.	Moderate carrier density
6.	Current flows by drift (due to electric field)	6.	Current flows due to both diffusion (concentration gradient) and drift (electric field)

Some important properties of silicon is listed in Table 2·3 below.

Table 2·3 : Properties of Silicon

		Silicon
1.	Crystal structure	Diamond
2.	Melting point	1415 °C
3.	Density	2·33 gm/c.c
4.	Atomic weight	28·1 gm/mole
5.	Atomic density	5×10^{22}/c.c
6.	Energy gap at 300 K	1·11 eV
7.	Energy gap at 0 K	1·15 eV
8.	Intrinsic carrier concentration at 300 K	$1{\cdot}5 \times 10^{10}$ /c.c
9.	Relative dielectric permittivity	11·7

Review Questions

1. What do you understand by the term "electron" ? What is its charge and mass ?
2. Discuss the variation of mass of an electron with velocity.
3. What do you understand by the term "ionization" ?
4. Calculate the value of the ionization energy of an hydrogen atom from first principles.
5. What do you understand by principal quantum number, azimuthal quantum number, magnetic quantum number and spin quantum number ?
6. Draw the energy band diagram of a semiconductor, a metal and an insulator.
7. What is meant by occupation probability of an electron in a semiconductor ? Draw the variation of occupation probability with energy for various temperatures in a semiconductor.
8. What do you understand by the term "Fermi level" ? What is its position in an intrinsic semiconductor ?
9. What is meant by covalent bonding ?
10. Draw the energy band diagram of silicon at 0 K (absolute zero of temperature) for an unexcited crystal.
11. Mention what do you mean by acceptors and donors.
12. What do you understand by a "hole" ?
13. Distinguish between metals and semiconductors.
14. Mention at least three important features of a semiconductor which are not possessed by metals..
15. Mention the names of three compound semiconductors widely used in industry.

Objective Type Questions :

Tick mark (√) the most appropriate answer

1. The charge of an electron is :
   (a) $1{\cdot}6 \times 10^{-19}$ (b) $9{\cdot}1 \times 10^{-28}$
   (c) $9{\cdot}1 \times 10^{-31}$ (d) $1{\cdot}38 \times 10^{-27}$ Coulomb

2. The mass of the electron in kgs is :
(a) 9.11×10^{-28} (b) 9.11×10^{-31}
(c) 1.6×10^{-19} (d) 1.38×10^{-27}

3. One 'eV' unit of energy equals :
(a) 1.6×10^{-19} Joule (b) 9.11×10^{-31} Joules
(c) 1.38×10^{-27} Joules (d) 9.11×10^{-28} Joules

4. The maximum number of electrons which the M shell of an atom can contain is :
(a) 5 (b) 8
(c) 28 (d) 32

5. The electron distribution of a silicon atom is :
(a) 2, 2, 10 (b) 2, 10, 2
(c) 2, 4. 8 (d) 2, 8, 4

6. The maximum number of electrons which the valence shell of an atom can possess is :
(a) 2 (b) 6
(c) 8 (d) 18

7. The total number of electrons in the Nth shell of an atom is :
(a) 2 (b) 8
(c) 18 (d) 32

8. According to Fermi Dirac distribution function, the probability of occupancy of a state with energy E is given by :
(a) $f(E) = \exp(-F/K_BT)$ (b) $f(E) = \dfrac{1}{1 + \exp\left(\dfrac{E-F}{K_BT}\right)}$
(c) $f(E) = \dfrac{1}{1 + \exp\left(\dfrac{F-E}{K_BT}\right)}$ (d) $f(E) = \dfrac{\exp\left(\dfrac{E-F}{K_BT}\right)}{1 + \exp\left(\dfrac{E-F}{K_BT}\right)}$

9. The Fermi level lies near the centre of the band gap in an :
(a) n-type (b) p-type
(c) intrinsic semiconductor

10. The ionization energy level of an acceptor is :
(a) at the centre of the band gap
(b) near the valence band edge
(c) near the conduction band edge

11. The following material is not an example of a semiconductor :
(a) silicon (b) germanium
(c) gallium arsenide (d) copper

12. The following materials act as acceptor (p-type) in silicon :
(a) arsenic (b) phosphorous
(c) antimony (d) boron

13. The following material acts as a donor (n-type) in silicon :
(a) gallium (b) indium
(c) antimony (d) aluminium

14. The number of valence electrons in the outermost shell of silicon is :
(a) 2 (b) 3
(c) 4 (d) 5

15. In an extrinsic p-type semiconductor, the number of electrons is :
(a) equal to (b) less than
(c) greater than the number of holes

16. In an intrinsic semiconductor, the number of electron is :
(a) equal to (b) less than

(c) greater than the number of holes

17. Which of the following statements is correct ?
(a) Conductivity decreases in a metal as the number of electron-hole pairs generated increases
(b) The conductivity increases in a semiconductor as the lattice vibrations create more opportunities for electronic collisions
(c) The conductivity decreases in a semiconductor as the lattice vibrations create more opportunities for electronic collisions
(d) drift and diffusion currents are both important in metals.

18. Which of the following is incorrect ?
(a) conductivity decreases in a conductor as lattice vibrations create more opportunities for electronic collisions.
(b) conductivity decreases in a conductor as more electron-hole pairs are thermally generated
(c) the carrier density is of the order of the Avogrado's number in metals
(d) metals are crystalline

19. Which of the following is incorrect ?
(a) the acceptor energy level can not be located in the energy band gap of silicon as it is forbidden
(b) the donor energy level can not be located in the energy band gap of silicon as it is forbidden
(c) the intrinsic carrier concentrotion increases with temperature for semiconductors
(d) the intrinsic carrier concentration decreases with temperature for semiconductors.

Exercises Problems

1. Find the values of the potential energy, kinetic energy and total energy of an electron orbiting the Bohr orbit of the helium atom.

2. Find the total number of electrons that can reside on the Nth shell of the hydrogen atom.

3. An electron is accelarated from rest. If the velocity the electron attains is :
(a) 10^6 m/sec
(b) 2×10^6 m/sec, find the potential difference through which it was accelarated. Determine also the kinetic energy of the electron.

4. A positively charged particle has a mass equal to 1837 times the mass of an electron and a charge which is the same as that of the electron. Find the kinetic energy and velocity of the particle when it is accelarated from rest through a potential difference of 2500 volts.

5. Find out the mass of an electron when
$v = 0{\cdot}25\, c$
[Ans. $m = 1{\cdot}016\, m_0$]

6. In an n-type semiconductor, the value of the acceptor concentration $N_D = 10^{11}$/c.c. If the intrinsic carrier concentration is $1{\cdot}5 \times 10^{10}$, find the value of the electron concentration and hole concentration.
[**Ans.** $n_0 = 1 \times 10^{11}$/cc, $p_0 = 2{\cdot}2 \times 10^9$/c.c.]

Chapter 3

SEMICONDUCTOR DIODES AND MISCELLANEOUS DEVICES

3.1 INTRODUCTION

A semiconductor *p–n* junction diode is formed when a wafer of the semiconductor material (say silicon) is doped so that one region is *n*-type and the other region is *p*-type. The *n* region possesses an abundant number of electrons (majority carriers) and a small number of thermally generated holes and the *p*-region possesses a large number of holes (majority carriers) and a small number of thermally generated electrons (minority carriers). This is depicted in Fig 3·1 above. The behaviour of the semiconductor is fundamental to the understanding of the behaviour of diodes, transistors and various other solid state devices.

3.2 THE DEPLETION LAYER FORMATION

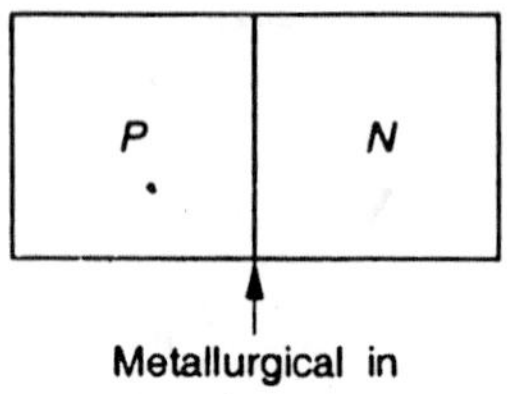

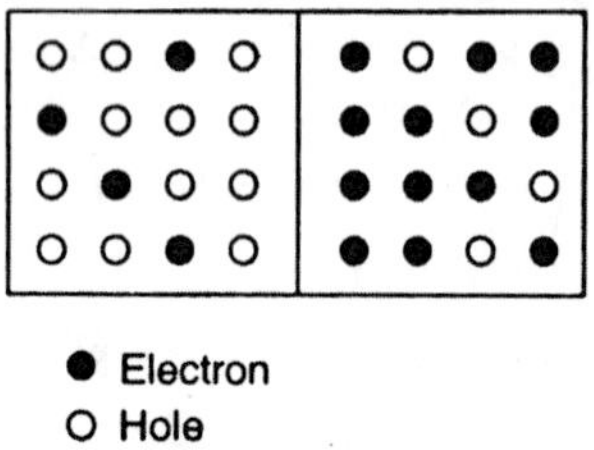

Fig. 3.1. *p-n* junction

When the external applied voltage is zero, the conduction (free) elecrons in the *n*-type region drift randomly in all directions. When the junction is formed, some of the electrons in the *n* region near the junction diffuse to the *p*-type region and recombine with holes near the junction. When an electron recombines with a hole, a penta valent atom on the *n* side acquires a net positive charge (and becomes a positive ions while a trivalent atom on the *n*-side becomes negatively charged (and becomes a negative ion)

Due to the recombination process a large number of positively charged ions collect on the *n*-side and a large number of negatively charged ions (trivalent) collect on the *p*-side of the *p–n* junction. As this building up of charged ions continue, the free electrons in the *n*-side must now overcome the attraction of both positive ions and repulsion of the negative ions in order to migrate to the *p*-region. Thus as the collection or build up of positive ions on the *n*-side and negative ions on the *p*-side continue, the area becomes essentially depleted of free electrons on the *n*-side and the holes on the *p*-side and this forms what is known as the **depletion layer.** Under equilibrium conditions, no further electrons can cross the junction. This is depicted in Fig 3·2 below.

The build up of positive ions and negative ions on either side of the junction creates a barrier potential (V_B) across the depletion layer. The barrier potential has a value of 0·7 V for silicon and 0·3 V for germanium.

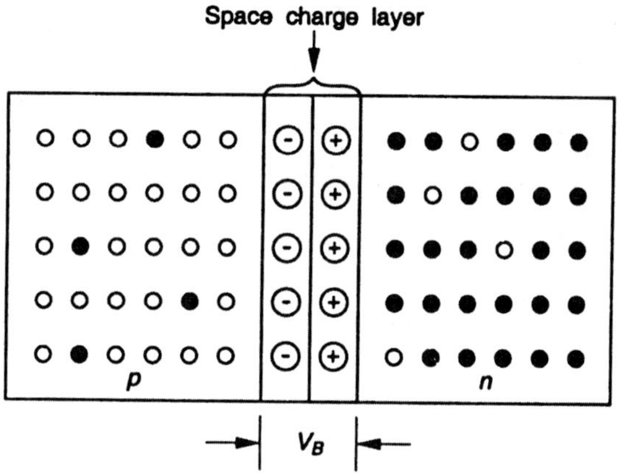

Fig. 3·2. P-N junction equilibrium. The small number of electrons in the *p* region (•) and few holes in the *n*-region (o) are thermally generated electron hole pairs.

With increase in junction temperature, the barrier potential is lowered. The barrier potential is a significant parameter because it determines the amount of external voltage that must be applied across the the junction to produce the necessary electrical current.

3.3 ENERGY BAND DIAGRAM

The energy band diagram of the *p–n* junction is depicted in Fig 3·3 above. Let us turn our attention to the *p–n* jnction at the instant of its formation. The energy bands of the trivalent impurity atoms in the *p*-type material are placed at a bit higher energy compared to the energy bands of the pentavalent impurity atoms. This is so due to the fact that the core attraction of the valence electrons in the pentavalent atom are more than the core attraction of the valence electrons (+3) in the trivalent atom. The trivalent valence electrons are thus in a somewhat higher orbit (and higher energy level).

It will be noticed that there is some overlap in the conduction bands in the *p* and *n* regions and also some overlap in the valence bands in the *p* and *n* regions. This means that some of the electrons near the top of the conduction band will diffuse to the *p*-region and then recombine with holes in the valence band. With the diffusion process, the depletion layer is built up. Further the energy bands in the *n*-region shift down, more the electrons with higher energy are lost to diffusion. Ultimately, when the top of the conduction band in the *n* region is at the same level as the bottom of the *p*-region conduction band, the diffusion process stops and the equilibrium condition is reached. This is depicted pictorially in Fig. 3·4.

It will be noticed that there is an energy gradient across the depletion layer instead of an abrupt change in energy level.

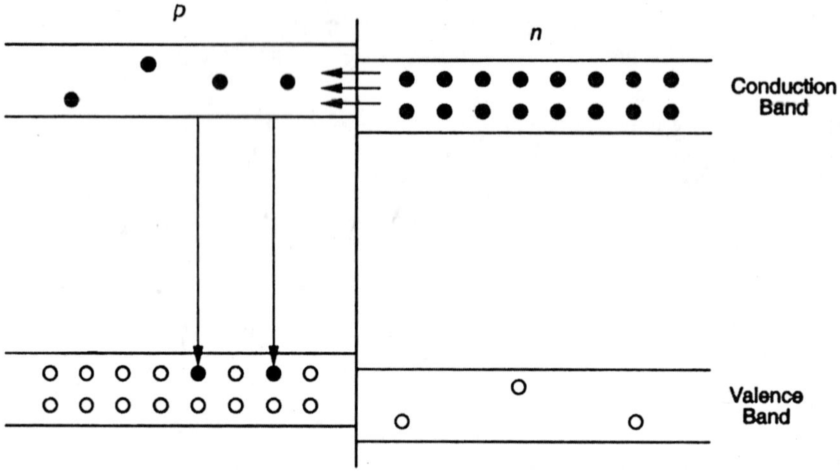

Fig 3.3. Energy band gap diagram of a *p–n* junction as diffusion takes place across the junction

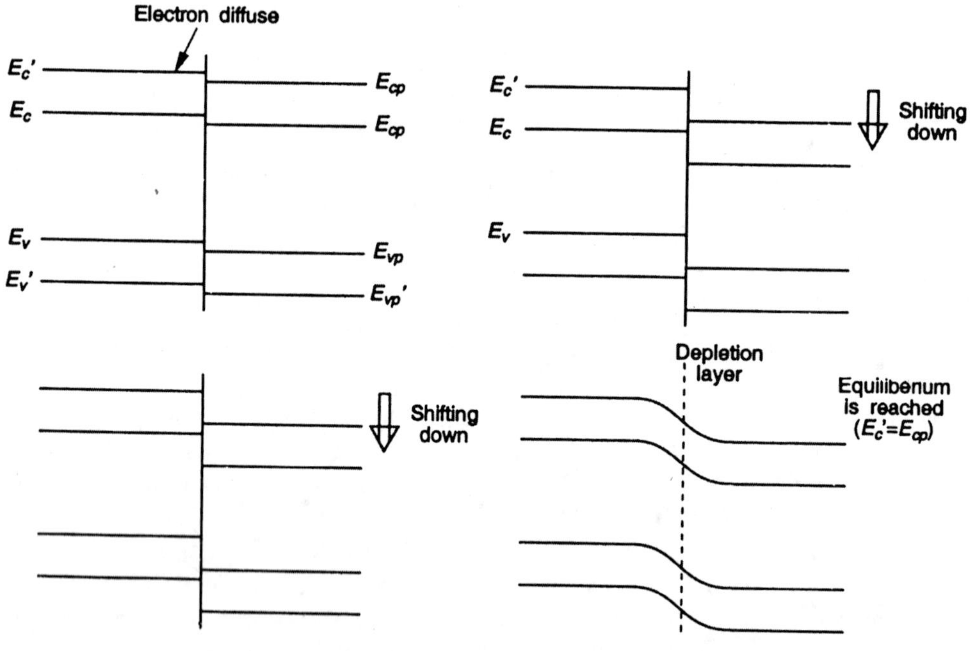

Fig 3.4. Figure depicting the energy band diagram as the electron diffusion process continues

Forward and Reverse bias

By the term bias, we imply a fixed d.c. voltage that is applied to set the reference operating point of a semiconductor device. By the term forward bias, we imply that the positive (+) terminal of the battery be connected to the *p*-type side and the negative (–) terminal of the battery be connected to the *n*-type side of the diode as depicted in Fig. 3.5.

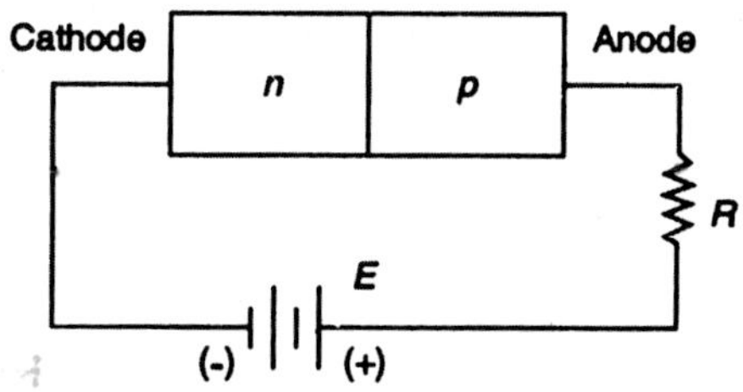

Fig 3.5. Connection for forward bias. The resistor *R* restricts the amount of current flow through the device and protects it from overcurrent

When the diode is forward biased, free electrons in *n*-region are pushed towards the *p–n* junction while holes in the *p*-region are pushed towards the *p–n* junction, because like charges repel. When the free electrons from the *n*-type region overcome the barrier potential, the external voltage source *E* gives the free electrons enough energy to penetrate the depletion layer and cross the junction. In the *p*-region the electrons come across the holes and recombine with them. As the electrons cross the junction, more electrons flow from the negative terminal of the battery. In the *n* type semiconductor, the current flows due to movement of free electrons towards the junction.

Once the free or conduction electrons are in the *p*-region and are in a position combine with holes, they become valence electrons. They, then, travel as valence electrons from hole to hole towards the positive terminal of the battery. The travel of the valence electrons, in fact, gives rise to hole movement in a direction opposite to the electron movement. Thus, the movement of current in the *p*-region is essentially the movement of holes towards the junction. This is shown pictorially in Fig 3·6.

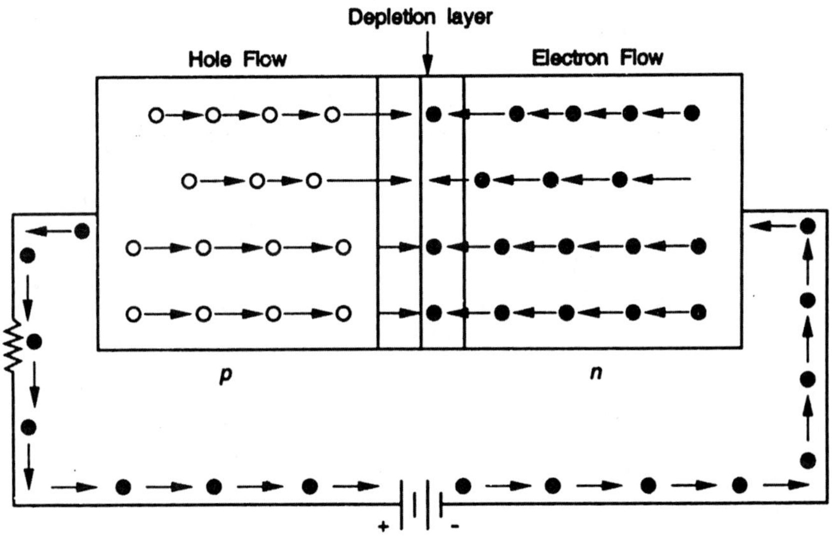

Fig 3.6. Electron flow in a *p-n* junction diode
(small dots are representing electrons and open circles are holes)

The potential barrier can be thought of as a small battery that oppose bias as is illustrated in Fig 3·7.

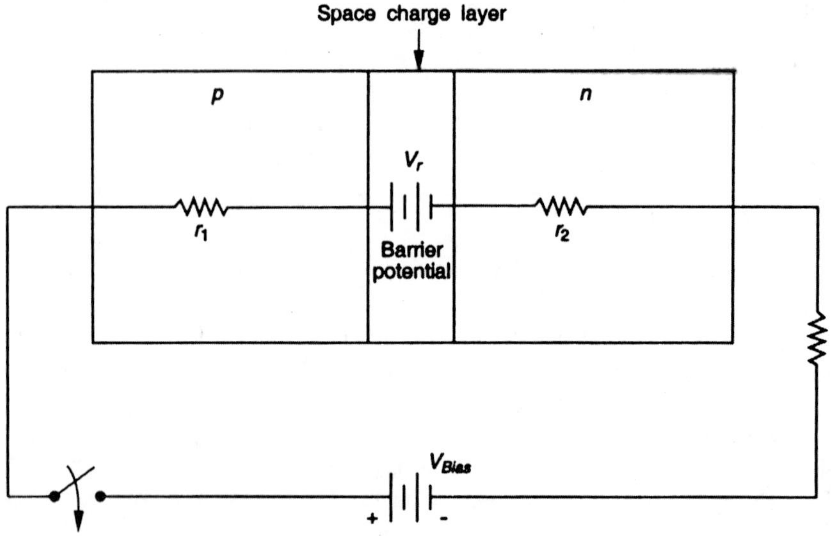

Fig 3.7. P-N junction diode barrier potential and its bulk resistance equivalent (r_1 and r_2 are the bulk resistances, V_{vr} is the barrier potential, V_r = 0·7 V for Si. The diode conducts if $V_{\text{bias}} > V_r$ while diode does not conduct if $V_{\text{bias}} < V_r$)

It should be noted in this connection that the barrier potential is not a voltage source and cannot be measured by a volt meter. The potential barrier has the effect of a battery in that the external voltage source must overcome the barrier voltage before the diode conducts. At 25°C, the barrier voltage has a value of 0·7 V for silicon and 0·3 V for germanium. When the diode conducts in forward bias, the voltage across the diode remains at approximately the barrier voltage. Further, the drop in voltage changes very minutely for forward current changes. The bulk resistances of the *p* and *n* region are very small (a few). The energy band diagram of a forward biased *p–n* junction is depicted in Fig 3·8(a) below.

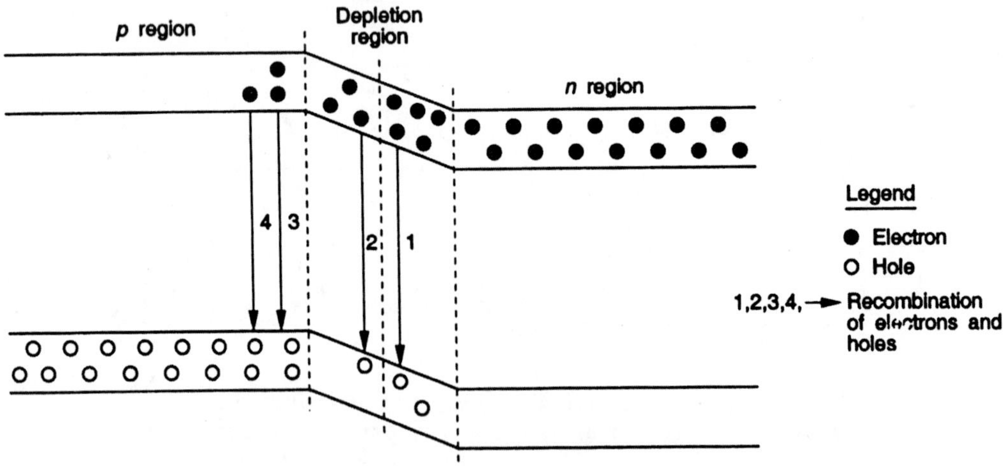

Fig 3.8. (a) Energy band diagram of a forward biased *p–n* junction

Reverse Bias

Under reverse bias, the positive of the battery is connected to the n-terminal of the diode and the negative of the external battery is connected to the p-type side of the p–n junction. The purpose of **reverse bias is to block or prevent current flow through the p–n junction.** The positive terminal of the battery attracts electrons away from the junction while the negative terminal of the battery attracts holes in the p-region away from the junction. The depletion layer width thus increases. There will be more negative ions in the p-region and more positive ions in the n-region.

The depletion region increases and increases until the potential difference across it is equal to the applied voltage. The electrons and holes then stop from moving away any further and the majority current ceases. When the diode is reverse biased the wide depletion region effectively behaves as an insulator. The initial current very quickly becomes zero when reverse bias is applied. For silicon, the reverse current is in the μA or nA range while for germanium the reverse current is higher.

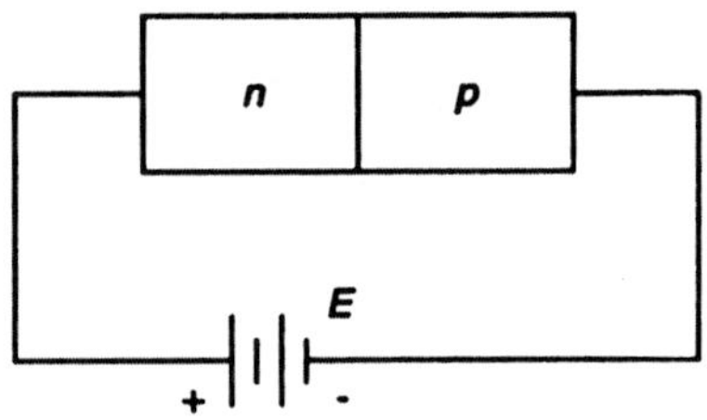

Fig 3.8(b). A reverse biased p–n junction

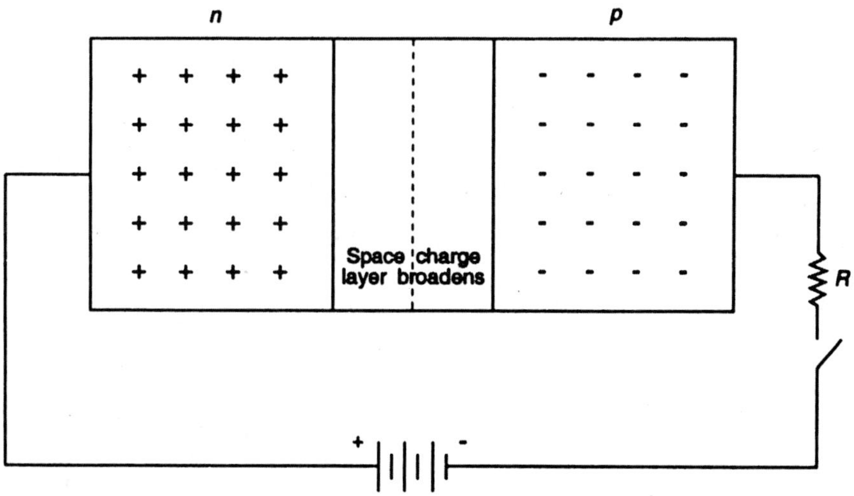

Fig 3.8(c). Widening of the depletion region under reverse bias. The wide depletion region acts as an insulator. In the figure '+' indicates ionized donors and '–' indicates ionized acceptors.

Reverse Breakdown

If the externally applied voltage is very high, a process known as avalanche effect may take place and the diode may ultimately breakdown in a process called avalanche breakdown.

In the avalanche effect, when the magnitude of the reverse voltage is very high, a minority conduction electron acquires enough energy from the external source to accelerate it towards the positive end of the diode. During the movement, the electron collides with an atom and knocks out one of the valence electrons of the atom by imparting sufficient energy. These two electrons (the initial electron and the electron newly created) now move at high velocity and once again collides with two more atoms and knocks out two of their valence electrons.

Thus four electrons are obtained. These four electrons travel with high energy and collide with atoms knocking out four electrons from the valence shell of the atoms. Thus a total of eight atoms are obtained. This rapid multiplication of the conduction band electrons increases the reverse current appreciably. This is known as avalanche effect. Normally, diodes are not operated in reverse breakdown region because they will be damaged by the excessive power. However, Zener diodes are operated in the reverse breakdown region.

3.4 V-I CHARACTERISTICS CURVE

The diode current-voltage characteristics curve is depicted in Fig. 3·9.

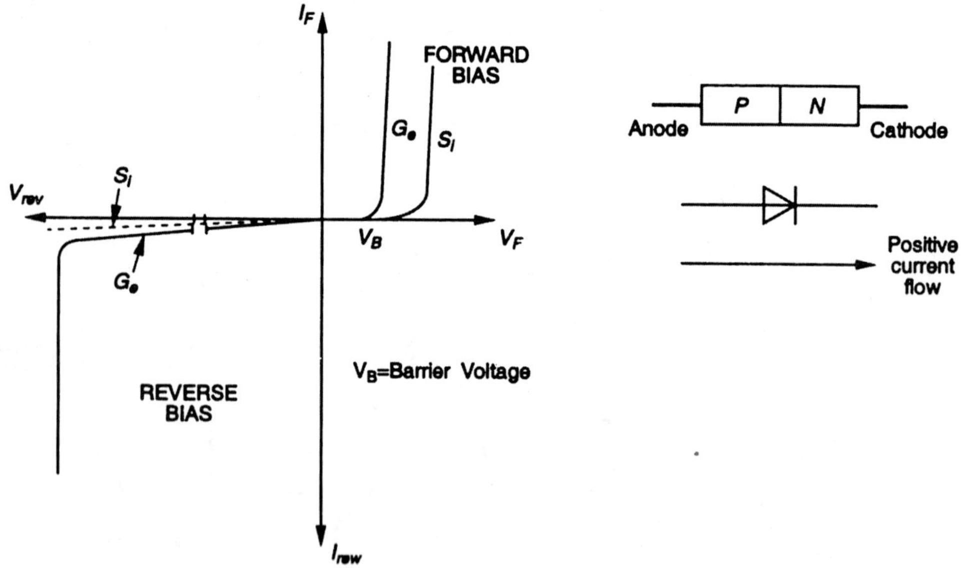

Fig 3.9. Characteristics curve of a *p–n* junction diode

A diode conducts current under forward bias when the bias voltage exceeds the barrier voltage while it blocks current flow when the reverse voltage is below the breakdown voltage. It will be seen that the forward curent I_F is nearly zero for voltage $V_F < V_B$, (the barrier voltage) while the current rises very fast once the barrier voltage V_B is exceeded. When the diode is conducting ($V < V_B$) the current must be restricted by an external resistor (called current limiting resistor). The diode voltage drop increases very slightly when the forward diode current is high.

In the standard diode symbol, the arrow head shows the direction of positive current flow. The terminals are called the cathode and the anode. When the anode is positive with reference to the cathode, the diode is called forward biased and current flows from anode to cathode. When the anode is negative with respect to the cathode, the diode is reverse biased. Figure 3·10 (a) depict a diode under forward and reverse bias.

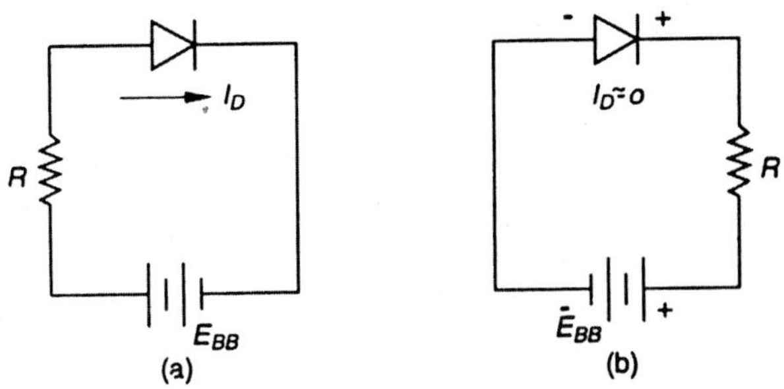

Fig 3.10. A diode circuit showing (a) forward and (b) reverse biasing

The resistance of a diode is low (below 100 Ω) when forward biased and it is high (above 1 M Ω) when reverse biased. If a multimeter reads very low resistance under both forward and reverse bias, it indicates a short circuited (shorted) diode while if a multimeter indicates very high resistance under both the conditions, it means an open circuited diode. Fig 3·11 (a) and (b) shows the characteristics of an ideal diode and the approximate equivalent circuit of a practical diode.

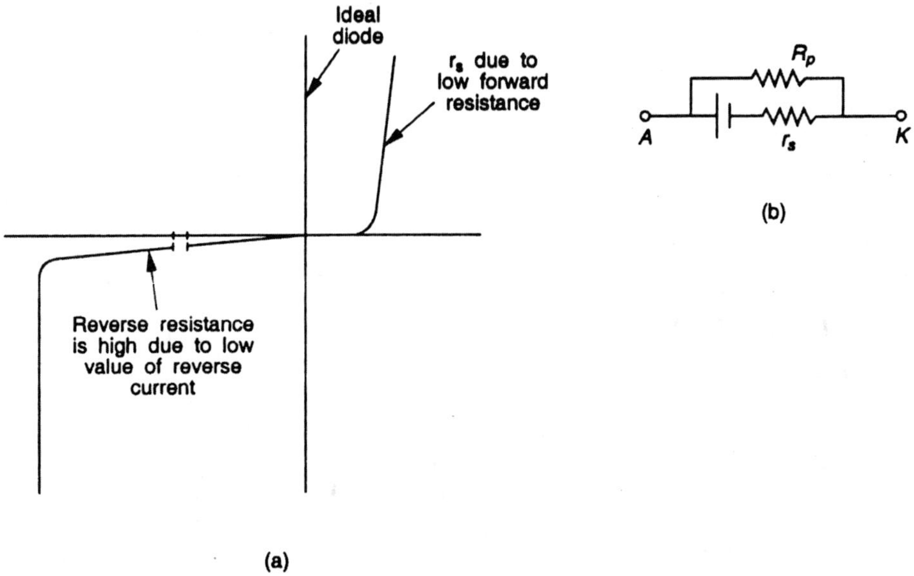

Fig. 3.11. (a) Diode characteristics and (b) approximate equivalent circuit of a diode

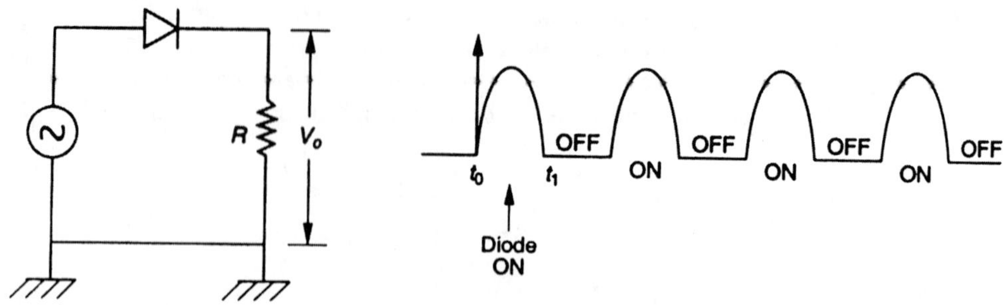

Fig 3.12. A Half wave rectifier

The diode may be thought of as a switch. Under forward bias, the diode acts like a closed switch while under reverse bias it acts like an open switch.

3.5 HALF WAVE RECTIFIER

A rectifier is a device which converts a.c. voltage to a d.c. voltage. The half wave rectifier is the simplest example.

During the positive half cycle the diode is forward biased, it conducts and output voltage V_0 appears across the diode. During the negative half cycle, the diode is reverse biased (it acts like an open switch) and there is no current flow through the load resistance R. The output voltage is thus zero between $t = T/2$ to T, where T is the time period. The average value of the rectified voltage is

$$V_{av} = \frac{V_p}{\pi} \qquad \text{...(3·1)}$$

where V_p is the peak voltage

For a peak value $V_p = 25$ V, V_{av} is

$$V_{av} = 7{\cdot}96 \text{ volts}$$

In an actual diode (Si), the rectified output voltage is less than the input voltage by the diode voltage drop (which is 0·7 V for a silicon diode).

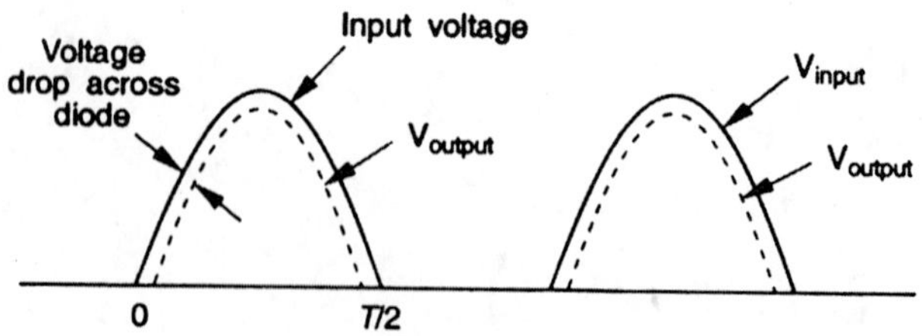

Fig 3.12 A. Effect of diode voltage drop on the output voltage of a diode

Half wave rectifier with transformer coupled input

Figure 3·13 depicts the circuit diagram of a half wave rectifier with transformer coupled input.

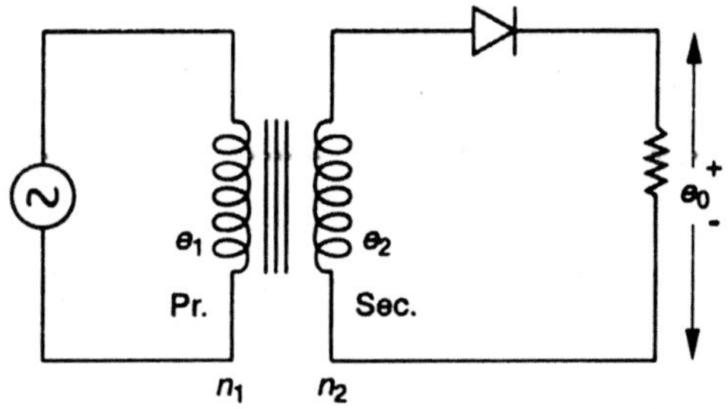

Fig 3.13. Half wave rectifier circuit with transformer coupled input

In this example, the diode gets its input voltage from the secondary of the transformer. The input voltage e_1 at the primary of the transformer is multiplied by a factor $\left(\frac{n_2}{n_1}\right)$ so that the secondary voltage e_2 becomes

$$e_2 = \left(\frac{n_2}{n_1}\right) e_1 \qquad \text{...(3·2)}$$

EXAMPLE 3·1 *Find the peak value of the output voltage of Fig.* 3·14.

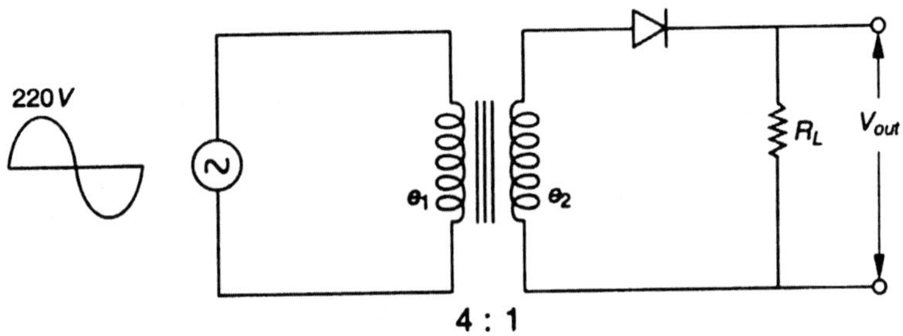

Fig. 3.14

SOLUTION.

The turns ratio $= \frac{n_2}{n_1} = \frac{1}{4} = 0·5$

The secondary peak voltage $= \left(\frac{n_2}{n_1}\right) e_1 = 0·25 \times 220 = 55$ V

The peak value of the rectified output voltage is

$$(V_{out})_{peak} = 55 - 0·7 = 54·3 \text{ V}$$

3.6 FULL WAVE RECTIFIERS

In a full wave rectifier the unidirectional current flows for the entire input cycle. Figure 3·15 depicts the schematic diagram of a full wave rectifier and the input and output voltages for the full wave rectifier.

In the full wave rectifier circuit, two rectifier diodes are arranged such that one of the diodes D_1 conducts for one half of the input cycle and the other diode conducts for the next half cycle. The current through the load resistance thus flows for the entire cycle of the input voltage. In the circuit diagram depicted in Fig. 3·15, the transformer consists of the primary winding *P* and the secondary winding is centre tapped.

When the end *a* of the secondary is positive with respect to point O, the diode D_1 conducts and current i_1 flows through it. During this half cycle diode D_2 is OFF since the end *b* is negative with respect to point

O. During the next half cycle end b is positive, diode D_2 conducts (current = i_2) and diode D_1 is OFF. The end c of the load resistor is positive during both the half cycles and an unidirectional current flows through the load. The various current and voltage wave forms are depicted in Fig 3·15.

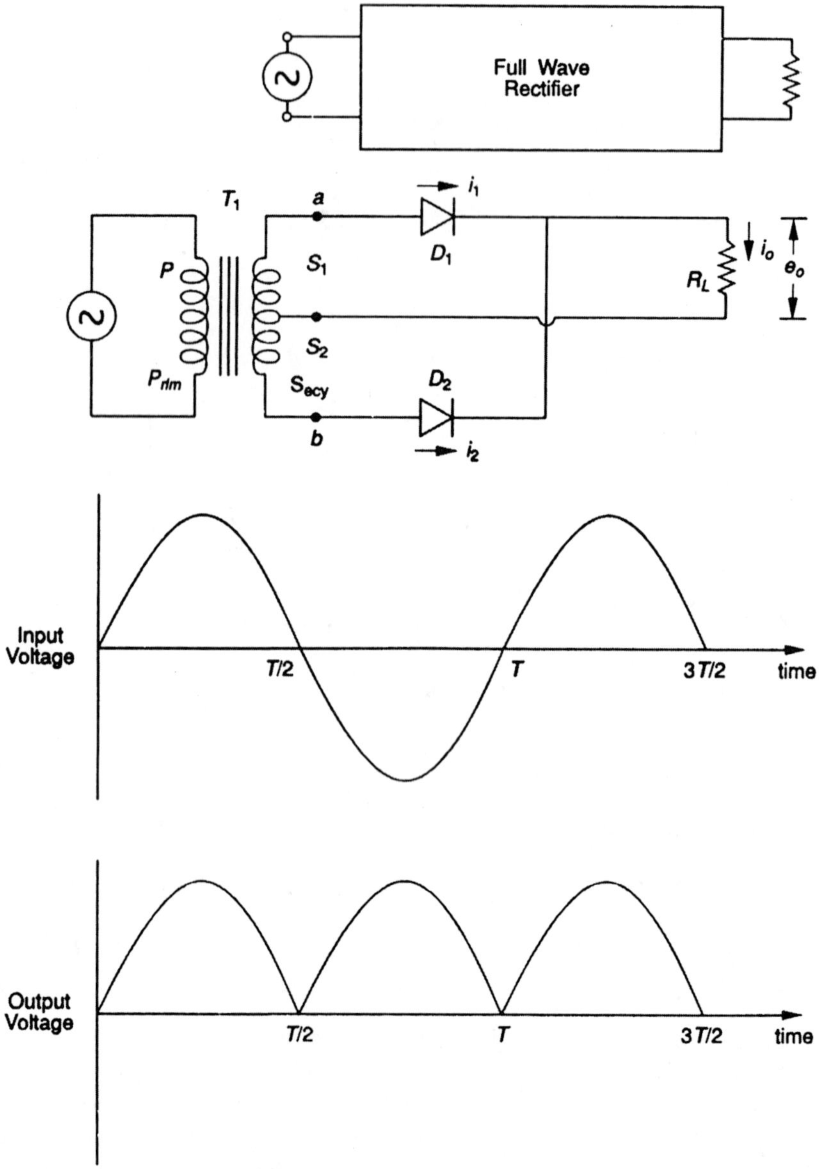

Fig 3.15. Full wave rectifier and its input and output voltage wave forms

3·7 BRIDGE RECTIFIER

Full wave rectification is also accomplished when a bridge circuit as shown in Fig 3·16 is employed. In the bridge circuit there are four diodes D_1, D_2, D_3 and D_4.

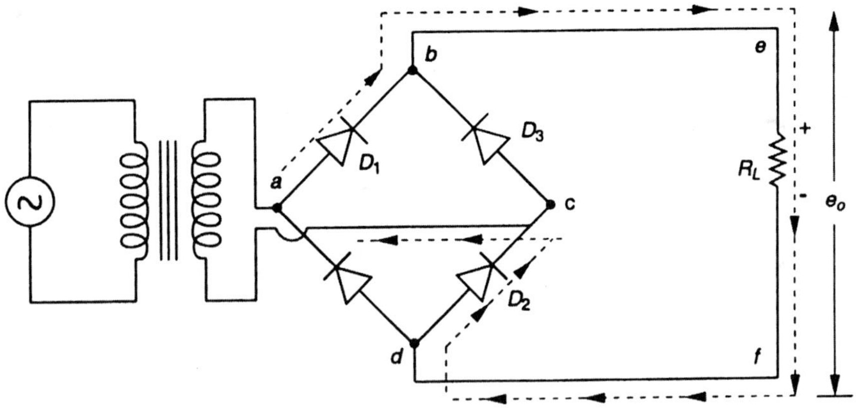

Fig 3.16. Bridge rectifier circuit (The dotted line shows the current flow path during the instant when end *a* is positive w.r.t. end *c*)

During the positive half cycle, the current flows via $a - D_1 - b - R_L - d - D_2 - c$ (diodes D_1 and D_2 conduct). The current enters the load resistor at terminal e which is positive. During the negative half cycle (end c is positive w.r.t end a), the current flows via path $c - D_2 - b - e - R_L - f - d - D_4 - a$. When diodes D_2 and D_4 conduct (while D_1 and D_3 is OFF). For the full cycle, the load current is unidirectional. The output voltage is similar to that in Fig 3·15.

In the bridge circuit the PIV (Peak Inverse Voltage) is the maximum voltage of the transformer secondary.

3.8 REGULATION CHARACTERISTICS

The variation of the average value of output voltage (d.c. voltage) with d.c. current I_{dc} is called regulation. The graphical plot showing the variation of V_{dc} with I_{dc} is called the voltage regulation characteristics of the rectifier. In an ideal rectifier, V_{dc} is independent of I_{dc}. However, in actual practice, the variation of V_{dc} with I_{dc} is as depicted in Fig. 3·17 below. The behaviour of the actual rectifier is measured by means of a parameter called the percentage voltage regulation. The percentage voltage regulation is expressed as

$$\% \ VR = \frac{V_{NL} - V_{FL}}{V_{FL}} \times 100 \ \%$$

where V_{NL} is **no load voltage** while V_{FL} is **full load voltage.**

A rectifier converts a.c. to d.c. If a load resistor R_L is connected across sinusoidal a.c. supply, the load current is also sinusoidal. If the load resistor is connected across a pure d.c. source or battery, the load current will not vary with time. This is depicted in Fig 3·18 (b). However if a load resistor is connected across either a half wave or full wave rectifier, the load current waveform will be as depicted in Figs. 3·18 (c) and (d) respectively.

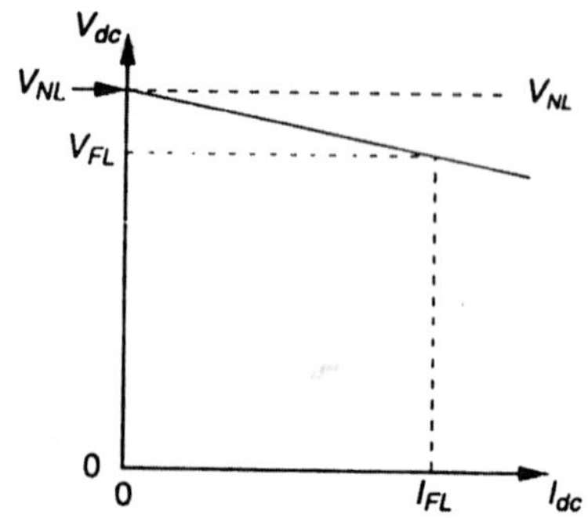

Fig 3.17. Regulation Characteristic

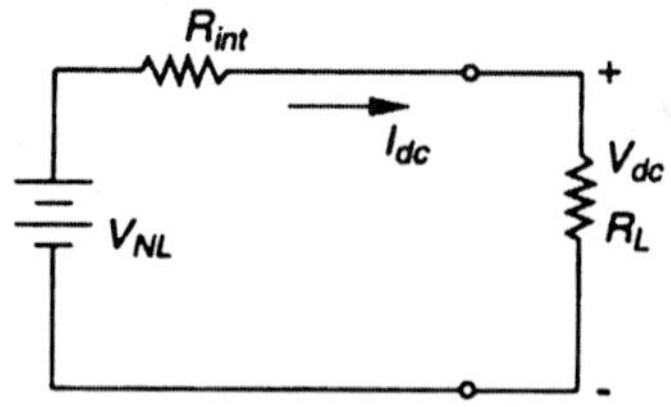

Fig. 3.17A. Thevenin Equivalent circuit

If the load is connected to the output of a half wave rectifier, the current will be unidirectional and will consist of half wave sinusoidal caps each extending for π radians (neglecting diode drops) and no current for the angle π to 2π radians. The average value of the load current will be small as depicted by the dotted line in Fig. 3·18 (c). For the full curve rectifier, the output voltage and the output current wave shape will be (vide Fig 3·18 (d)) the two half sinusoidal caps each extending for π radians per cycle. The average value of the load current will be as depicted in Fig. 3·18 (d). A unidirectional fluctuating waveform may be thought of as comprising a number of components. It has an average or d.c. value which is superimposed over a number of a.c. sinusoidal components (fundamental and harmonics). The undesired alternating components are called ripples.

For a half wave rectifier, the ripple frequency is the same as the power line frequency. For a full wave rectifier, however, the ripple frequency is twice the mains supply frequency. For the full wave rectifier case, the current (or voltage) pulses varies after each $\omega t = \pi$ radians of the mains supply frequency.

For a 50 Hz mains supply, the ripple frequency is 50 Hz for half wave rectifier and the ripple frequency is 100 Hz for full wave rectifier.

The effectiveness with which the rectifier converts a.c. power to d.c. power is measured by the following parameters.

The ripple factor (RF) is defined as

$$RF = \frac{\text{r.m.s. value of the components of the wave}}{\text{average or d.c. value of the wave}} \qquad \ldots(3\cdot3)$$

The rectification efficiency gives as to what fraction of the total input a.c. power is converted to d.c. output power. The rectification efficiency is

$$\eta = \frac{\text{d.c. power delivered to the load}}{\text{a.c. input power from transformer secondary}}$$

$$= \frac{P_{dc}}{P_{ac}} \qquad \ldots(3\cdot4)$$

Performance of half wave rectifier

The half wave rectified current wave can be described mathematically as follows :

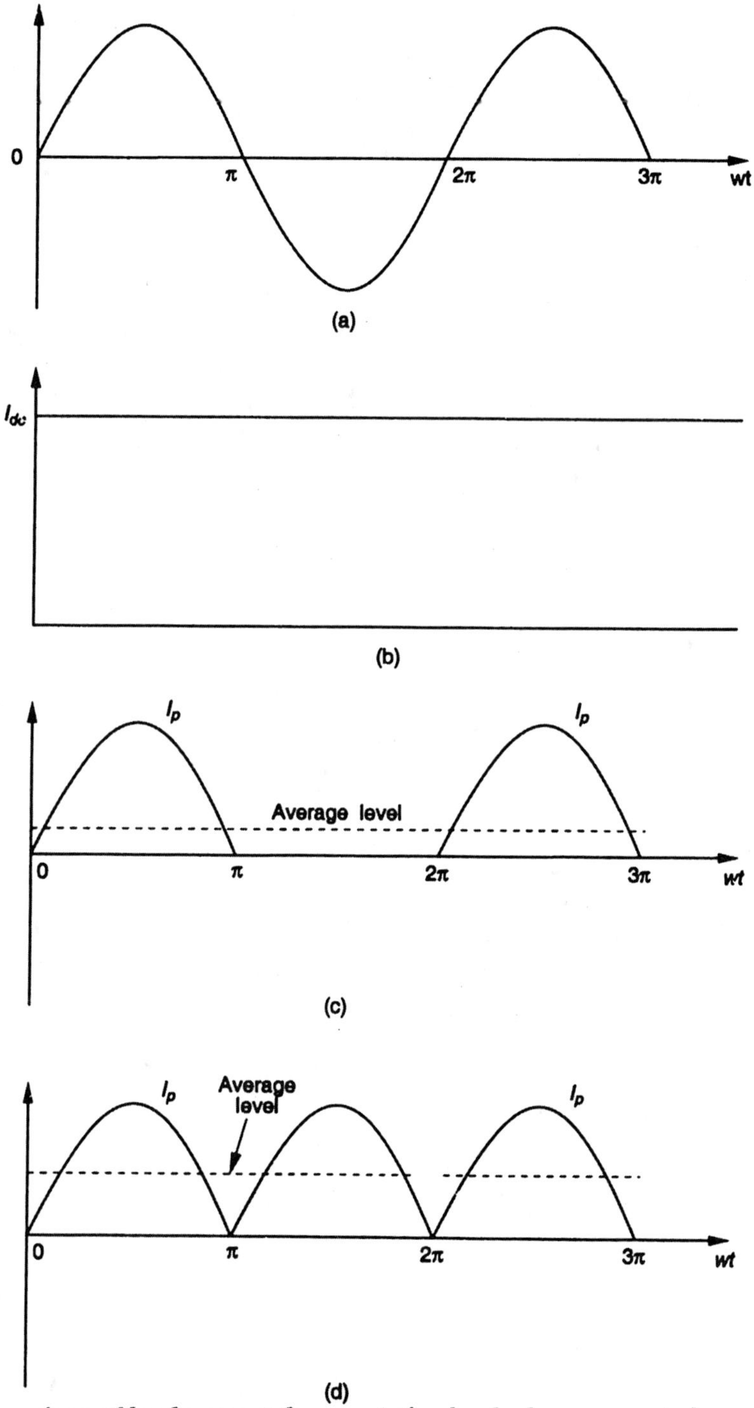

Fig 3.18. Comparison of load current for a resistive load when connected across (a) a.c. supply (b) d.c. supply (c) half wave rectifier (d) full wave rectifier

$$i_L = I_p \sin \omega t \qquad \text{for } 0 \le \omega t \le \pi \qquad \ldots(3{\cdot}5)$$

and

$$i_L = 0 \qquad \text{for } \pi \le \omega t \le 2\pi \qquad \ldots(3{\cdot}6)$$

The average value of the current flowing through the load is

$$I_{av} = \frac{1}{2\pi}\int_0^{2\pi} i_L \, d(\omega t) = \frac{1}{2\pi}\int_0^{\pi} I_p \sin \omega t \, d(\omega t) \qquad \ldots(3{\cdot}7)$$

$$= (I_p/\pi)$$

The r.m.s. or effective value of current flowing through the load is given by

$$I_{rms} = \sqrt{\frac{1}{2\pi}\int_0^{2\pi} i_L^2 \, d(\omega t)} \qquad \ldots(3{\cdot}8)$$

$$= \sqrt{\frac{1}{2\pi}\left[\int_0^{\pi} I_p^2 \sin^2 \omega t \, d(\omega t) + \int_0^{2\pi} 0 \, d(\omega t)\right]}$$

$$= \sqrt{\frac{I_p^2}{2\pi}\int_0^{\pi} \frac{(1 - \cos 2\,\omega t)}{2} \, d(\omega t)}$$

$$= \sqrt{\frac{I_p^2}{2 \times 2\pi}\left\{\omega t - \frac{\sin^2 \omega t}{2}\right\}\Bigg|_0^{\pi}}$$

$$= \sqrt{\frac{I_p^2}{4\pi} \times (\pi)} = \frac{I_p}{2} \qquad \ldots(3{\cdot}9)$$

The r.m.s. value of the a.c. components of output current

$$I_{ac} = \sqrt{\frac{1}{2\pi}\int_0^{2\pi} (i_L - I_{av})^2 \, d(\omega t)} \qquad \ldots(3{\cdot}10)$$

$$= \sqrt{\frac{1}{2\pi}\int_0^{2\pi} (i_L^2 + I_{av}^2 - 2\, i_L I_{av}) \, d(\omega t)}$$

$$= \sqrt{I_{rms}^2 + I_{av}^2 - 2I_{av}^2}$$

$$= \sqrt{I_{rms}^2 - I_{av}^2} \qquad \ldots(3{\cdot}11)$$

The ripple factor is given as

$$RF = \frac{I_{ac}}{I_{av}} = \frac{\sqrt{I_{rms}^2 - I_{av}^2}}{I_{av}} \qquad \ldots(3{\cdot}12)$$

$$= \sqrt{\left(\frac{I_{rms}}{I_{av}}\right)^2 - 1} \qquad \ldots(3{\cdot}13)$$

$$\frac{I_{rms}}{I_{av}} = \frac{I_p/2}{I_p/\pi} = 1{\cdot}57 \qquad \ldots(3{\cdot}14)$$

Thus, the ripple factor is given by

$$RF = \sqrt{(1{\cdot}57)^2 - 1} = 1{\cdot}21 \qquad \ldots(3{\cdot}15)$$

Since the ripple factor is greater than unity, it implies that the ripple current (or voltage) is greater than the d.c. or average value of current (or voltage). Thus the half wave rectifier exhibits high ripples.

The d.c. power delivered to the load is

$$P_{dc} = I_{dc}^2 R_L = \left(\frac{I_p}{\pi}\right)^2 R_L \quad \text{...(3·16)}$$

The total input a.c. power is

$$P_{ac} = I_{rms}^2 (R_L + r)$$

$$= \frac{I_p^2}{4}(R_L + r) \quad \text{...(3·17)}$$

The rectification efficiency is

$$\eta = \frac{P_{dc}}{P_{ac}} = \frac{(I_p/\pi)^2 R_L}{(I_p/2)^2 (R_L + r)} \times 100\%$$

$$= \frac{40{\cdot}5}{1 + \dfrac{r}{R_L}} \% \quad \text{...(3·18)}$$

If r, the diode (forward bias) resistance is neglected in comparison to R_L, the efficiency becomes 40·5 %. This implies, that under the condition of no diode loss, 40·5 % of the a.c. input power is converted to d.c. power. The rest of the power is a.c. power in the load. The form factor (FF) is given as the ratio of the r.m.s. component I_{rms} of the output current and the d.c. value I_{dc}. We can, thus, write

$$FF = \frac{I_{rms}}{I_{dc}} = 1{\cdot}57 \quad \text{...(3·19)}$$

The transformer utilisation factor (TUF) or the transformer secondary utilisation factor (SUF) is defined as the ratio of the d.c. output power and the transformer secondary power.

$$TUF = \frac{P_{dc}}{V_s \cdot I_s} \quad \text{...(3·20)}$$

where V_s and I_s are the transformer secondary voltage and current ratings.

The angle between the fundamental components of the input current and voltage is called the displacement angle. If α represents the displacement angle, then $\cos\alpha$ is called the displacement factor. *i.e.*

$$DF = \cos\alpha \quad \text{...(3·21)}$$

The harmonic factor (HF) of the input current is defined as

$$HF = \left[\frac{I_s^2 - I_{sf}^2}{I_{sf}^2}\right]^{1/2}$$

$$= \left[\left(\frac{I_s}{I_{sf}}\right)^2 - 1\right]^{1/2} \quad \text{...(3·22)}$$

where I_{sf} is the fundamental component of the input current I_s.

Performance parameters for a full wave Rectifier

The full wave rectified current wave with resistive load is depicted in Fig. 3·19. Mathematically, the current wave shape can be described by

$$i_L = I_p \sin \omega t \quad \text{...(3·23)}$$

where $\omega = (2\pi f)$ is the angular frequency of input a.c. supply.

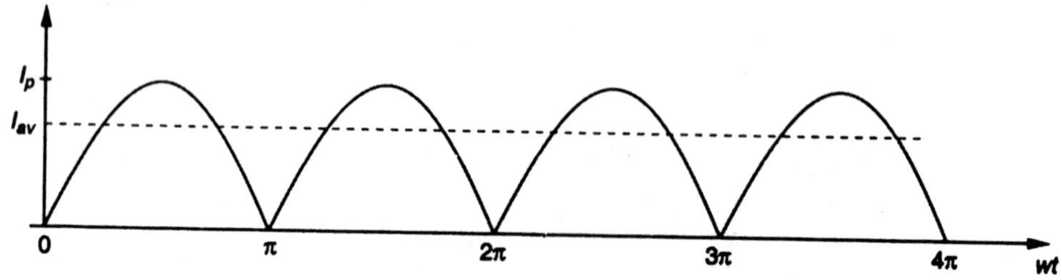

Fig. 3.19. Full wave rectified current wave

The d.c. or average value of the current is

$$I_{dc} = I_{av} = \frac{1}{2\pi}\int_0^{2\pi} i_L \, d(\omega t) \quad \text{...(3·24)}$$

$$= \frac{1}{\pi}\int_0^{\pi} i_L \, d(\omega t) = \frac{1}{\pi}\int_0^{\pi} I_p \sin \omega t \, d(\omega t) \quad \text{...(3·25)}$$

$$= \frac{2 I_p}{\pi}$$

The effective or r.m.s value of the current is given by

$$I_{rms} = \sqrt{\frac{1}{2\pi}\int_0^{2\pi} i_L^2 \, d(\omega t)} \quad \text{...(3·26)}$$

$$= \sqrt{\frac{1}{\pi}\int_0^{\pi} i_L^2 \, d(\omega t)}$$

$$= \sqrt{\frac{1}{\pi}\int_0^{\pi} I_p^2 \sin^2 \omega t \, d(\omega t)}$$

$$= \sqrt{\frac{I_p^2}{\pi}\int_0^{\pi}\left(\frac{1 - \cos 2\,\omega t}{2}\right) d(\omega t)}$$

$$= \sqrt{\frac{I_p^2}{\pi}\left\{\frac{\omega t}{2} - \frac{\sin 2\omega t}{4}\right\}\Bigg|_0^{\pi}}$$

$$= \sqrt{\frac{I_p^2}{\pi} \cdot \frac{\pi}{2}} = \frac{I_p}{\sqrt{2}} \quad \text{...(3·27)}$$

The ripple factor for the full wave rectifier is given as

$$RF = \sqrt{\left(\frac{I_{rms}}{I_{av}}\right)^2 - 1} \qquad \ldots(3{\cdot}28)$$

$$= \sqrt{\frac{I_{rms}^2}{I_{av}^2} - 1}$$

$$= \sqrt{\frac{I_p^2}{2} \times \frac{\pi^2}{4I_p^2} - 1}$$

$$= 0{\cdot}4834 \qquad \ldots(3{\cdot}29)$$

The rectification efficiency for the full wave rectifier is calculated from

$$\eta = \frac{P_{dc}}{P_{ac}} \qquad \ldots(3{\cdot}30)$$

where $P_{dc} = I_{av}^2 \cdot R_L$

$$= \left(\frac{2I_p}{\pi}\right)^2 R_L \qquad \ldots(3{\cdot}31)$$

and $P_{ac} = I_{rms}^2 (R_L + r)$

$$= \left(\frac{I_p}{\sqrt{2}}\right)^2 (R_L + r) \qquad \ldots(3{\cdot}32)$$

The rectification efficiency is

$$\eta = \frac{P_{dc}}{P_{ac}} = \left(\frac{2I_p}{\pi}\right)^2 \cdot R_L \Big/ \left(\frac{I_p}{\pi}\right)^2 (R_L + r) \qquad \ldots(3{\cdot}33)$$

$$= \frac{0{\cdot}812}{1 + \frac{r}{R_L}}$$

The percentage efficiency is, thus,

$$\eta = \frac{81{\cdot}2}{1 + \frac{r}{R_L}} \% \qquad \ldots(3{\cdot}34)$$

The form factor (FF) is

$$FF = I_{rms}/I_{dc}$$

$$= \frac{0{\cdot}707}{0{\cdot}6366} = 1{\cdot}11 \qquad \ldots(3{\cdot}35)$$

The ripple frequency is twice the supply frequency.

The Table 3·1 below depicts the comparison between different types of rectifiers.

***Table* 3·1**

	Half wave Rectifier	*Full wave centre tap transformer type*	*Full Wave Rectifier Bridge type*
Number of diodes	1	2	4
Secondary Voltage Peak	V_p	V_p	V_p
Peak Inverse Voltage	V_p	$2V_p$	V_p
Peak load current	$V_p/(R_L + r)$	$V_p/(R_L + r)$	$V_p/(R_L + 2r)$
Average current	I_p/π	$2I_p/\pi$	$2I_p/\pi$
RMS current	$I_p/2$	$I_p/\sqrt{2}$	$I_p/\sqrt{2}$
Ripple factor	1·21	0·4834	0·4834
Maxm. Rectification efficiency	40·6 %	81·2 %	81·2 %
Ripple frequency	f	$2f$	$2f$

3.9 FILTERS

The output of a rectifier consists of a unidirectional voltage which is pulsating in nature. It has both a d.c. component and a.c. components. The a.c. variations can be filtered or smoothed out from the rectified voltage. In the following sections, we shall discuss various types of filter circuits.

3.9.1 Series Inductor filter

An inductance has the basic property of opposing any change in current flowing through it. This feature of the inductance is employed in a series inductor filter. A back e.m.f. is induced in the inductor, whenever the

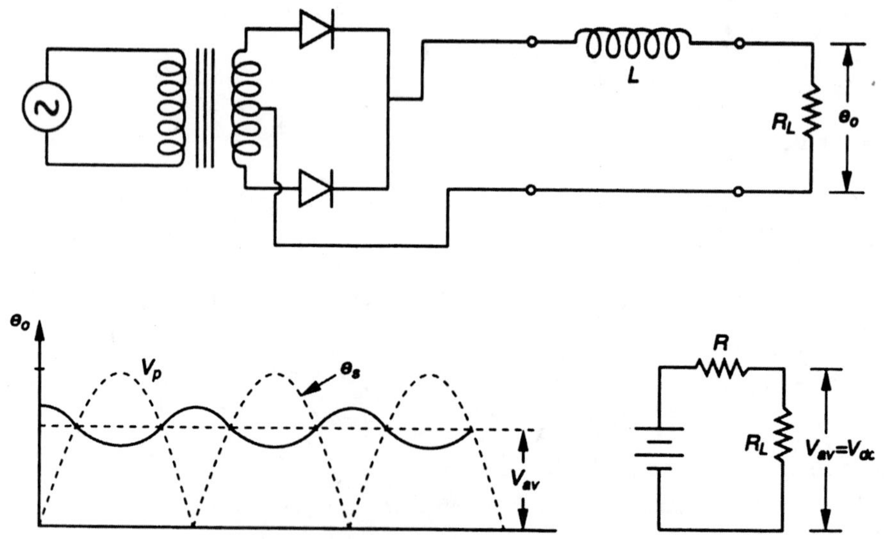

Fig 3.20(a). Series Inductor Filter in a full wave rectifier

current through the inductor tends to vary, the induced back e.m.f. prevents the current from altering its value. An abrupt or sudden variation of current in the circuit is filtered or smoothed out by the presence of the inductor.

Considering the conduction of diode D_1, we may use the Kirchhoffs voltage laws to the circuit

$$e = (r_s + r + r_l)\, i + L\frac{di}{dt} \qquad \ldots(3\cdot36)$$

where r_s = resistance of one half of the secondary coil

r = dynamic resistance of the diode

r_l = series d.c. resistance of the choke

The resistances r_1, r_s and r_l are very small and L is high. Thus

$$e = L\frac{di}{dt} \qquad \ldots(3\cdot37)$$

or

$$i = \frac{1}{L}\int_0^t e\, dt \qquad \ldots(3\cdot38)$$

If the applied voltage is

$$e = E_p \sin \omega t \qquad \ldots(3\cdot39)$$

Peak value of current $I_p = \frac{E_p}{R_L}$ ...(3·40)

The output of a full wave rectifier can be expressed by Fourier series as follows

$$e = \frac{2E_p}{\pi} - \frac{4E_p}{3\pi}\cos 2\omega t - \frac{4E_p}{15\pi}\cos 4\omega t \ldots\ldots \qquad \ldots(3\cdot41)$$

Thus, the average or d.c. value of current can be expressed as

$$I_{dc} = \frac{2E_p/\pi}{R_L} \qquad \ldots(3\cdot42)$$

For the full wave rectifier circuit, the output ripple frequency is twice the supply frequency and at this frequency, the impedance Z is

$$Z = \sqrt{R_L^2 + (2\omega L)^2} \qquad \ldots(3\cdot43)$$

Thus

$$|Z| = \sqrt{R_L^2 + 4\omega^2 L^2}$$

The peak current

$$I_p = \frac{E_p}{|Z|} = \frac{E_p}{\sqrt{R_L{}^2 + 4\omega^2 L^2}} \qquad \ldots(3\cdot44)$$

Thus the output or load current can be written as

$$i = \frac{2E_p}{\pi R_L} - \frac{4E_p}{3\pi\sqrt{R_L^2 + 4\omega^2 L^2}}\cos(2\omega t - \phi) \ldots\ldots \qquad \ldots(3\cdot45)$$

where ϕ indicates the phase angle.

The ripple factor is

$$RF = \frac{\dfrac{4E_p}{3\pi\sqrt{2}}\left[\dfrac{1}{\sqrt{R_L^2 + 4\omega^2 L^2}}\right]}{\dfrac{2E_p}{\pi R_L}} \quad \ldots(3\cdot46)$$

$$= \frac{\sqrt{2}}{3}\left[\frac{1}{\sqrt{1 + \dfrac{4\omega^2 L^2}{R_L^2}}}\right] \quad \ldots(3\cdot47)$$

Since the quantity ωL is quite large, the ripple factor is

$$RF = \frac{1}{3\sqrt{2}} \cdot \frac{R_L}{\omega L} \quad \ldots(3\cdot48)$$

$$= \frac{R_L}{1333\ L} \text{ at } \omega = 2\pi f \text{ where } f = 50 \text{ Hz} \quad \ldots(3\cdot49)$$

The ripple factor is thus dependent on the value of the load resistance. This L or inductor filter is useful for small resistances (heavy load).

The output voltage for the series inductor type filter is

$$E_{av} = E_{dc} = \frac{2E_p}{\pi} \quad \ldots(3\cdot50)$$

It assumes a perfect inductor with no resistance and the other resistances are quite small. In practice, the output d.c. voltage, if the resistance in the circuit is r

$$E_{av} = \frac{2E_p}{\pi} \cdot \frac{R_L}{r + R_L} \quad \ldots(3\cdot51)$$

It will be seen, that with a proper choke, the load current will never drop to zero.

3.9.2 Shunt capacitor or C filter

A full wave rectifier with shunt capacitor filter is depicted in Fig. 3·20(b). The capacitor C is connected between the positive voltage point d and the centre tap b. The capacitance C is such that it has a low reactance.

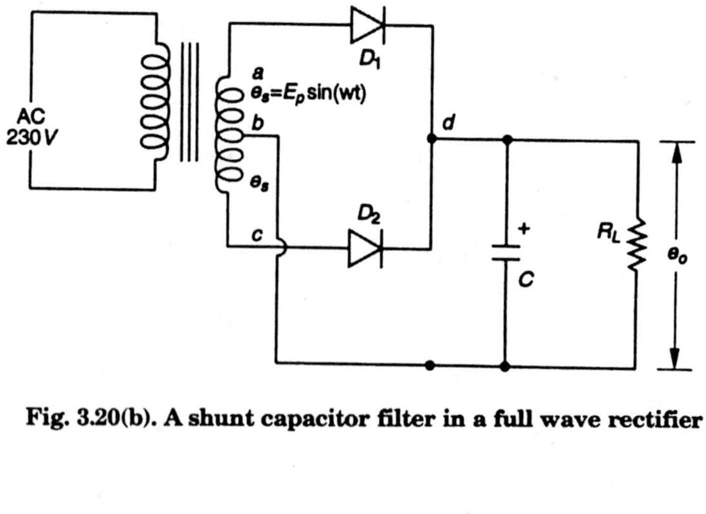

Fig. 3.20(b). A shunt capacitor filter in a full wave rectifier

Thus the alternating components find a low reactance path through C and a small portion of the alternating components flows through R, producing a small ripple voltage. Let us consider an instant of time when terminal a is positive with respect to b. The diode D_1 conducts and the voltage V_d across D_1 is very small. The voltage at point d is $(e_s - V_d) \approx e_s$ for $e_s >> V_d$. The capacitor, thus, charges to the full value with end d positive. The capacitor stores the electrical energy and the voltage V_c across C attains the peak value of input voltage E_p.

As the voltage at point d falls, the capacitor delivers the electrostatic energy by discharging through the load. The voltage e_0 at the load would thus be maintained to a value higher than zero. The load voltage would have fallen to zero in each half cycle. This will be apperent from the output voltage waveform as depicted in Fig. 3·20(c).

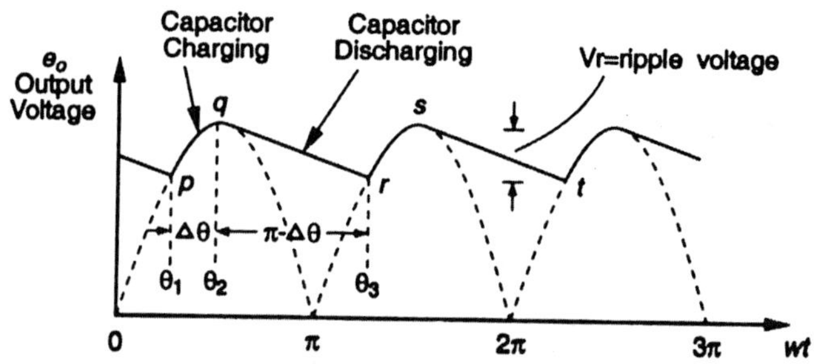

Fig 3.20(c). Output voltage waveform in a shunt capacitor type filter.

It will be seen that the capacitor charges to the peak value of the secondary voltage e_s within a few cycles. In Fig. 3·20(c) it will be noticed that the capacitor charges between angle p and the charging continues till q, the peak voltage point. At q the sinusoidal voltage drops but the capacitor starts discharging through the load R_L with a time constant $R_L \cdot C$. The discharging continues till r when the input supply voltage at d is higher and the capacitor begins to charge up once again and continues till s. In this way the output voltage across the load will be as depicted by the solid line due to capacitor charging and discharging. The capacitor charges for a period $\Delta\theta$ and discharges for $(\pi - \Delta\theta)$. The ripple voltage is (V_r).

The d.c. or average value of load current is the same as the average value of the capacitor discharge current over an interval of time $\left(\frac{\pi - \Delta\theta}{\omega}\right) = t_2$, where t_1 is the charging time and t_2 is the discharging time for the capacitor.

$$t_1 = \frac{\Delta\theta}{\omega} \qquad \text{...(3·52)}$$

$$t_2 = \frac{\pi - \Delta\theta}{\omega} \qquad \text{...(3·53)}$$

The amount of charge lost by the capacitor is

$$Q_{\text{discharge}} = I_{dc} \times t_2 \qquad \text{...(3·54)}$$

This loss in charge is made up by the charging during interval t_1. Thus

$$Q_{ch} = V_r \times C \qquad ...(3\cdot55)$$

Since $Q_{dch} = Q_{ch}$, we may write

$$I_{dc} \,.\, t_2 = V_r \,.\, C \qquad ...(3\cdot56)$$

or $$V_r = \frac{I_{dc}}{C} t_2 \qquad ...(3\cdot57)$$

Since $$R_L >> \frac{1}{\omega C}$$

and $$t_2 >> t_1$$

$$V_r = \frac{I_{dc}}{C} \cdot \frac{1}{2f} \qquad ...(3\cdot58)$$

The r.m.s. value of the a.c. voltage wave form is

$$(V_r)_{rms} = \frac{V_r}{2\sqrt{3}} \qquad ...(3\cdot59)$$

Also $$I_{dc} = \frac{E_{dc}}{R_L} \qquad ...(3\cdot60)$$

Thus $$(V_r)_{rms} = \frac{I_{dc}}{2fC} \times \frac{1}{2\sqrt{3}} \qquad ...(3\cdot61)$$

$$= \frac{E_{dc}}{2fR_L \,.\, C} \times \frac{1}{2\sqrt{3}} \qquad ...(3\cdot62)$$

The ripple factor is thus

$$RF = \frac{(V_r)_{rms}}{V_{dc}} = \frac{1}{4\sqrt{3} f C R_L} \qquad ...(3\cdot63)$$

$$= \frac{2886}{R_L C} \qquad ...(3\cdot64)$$

where R_L is in ohms, C is in μF and $f = 50$ Hz

The average or d.c. voltage is

$$E_{av} = E_{dc} = E_p - \frac{V_r}{2} \qquad ...(3\cdot65)$$

where $$V_r = \frac{I_{dc}}{2fC} \qquad ...(3\cdot66)$$

and $$I_{dc} = \frac{E_{dc}}{R_L} \qquad ...(3\cdot67)$$

Thus $$E_{av} = E_p - \frac{I_{av}}{4fC} \qquad ...(3\cdot68)$$

or $$E_{av} = E_p \left[\frac{4f R_L \,.\, C}{1 + 4f R_L C} \right] \qquad ...(3\cdot69)$$

3.9.3 The *LC* filter (or Choke Input Filter)

The ripple factor *RF* increases with increase in resistive load for an inductor input filter while the ripple factor decreases with increase in resistive load for a capacitor filter. Thus a combination of these two filters may make the ripple factor independent of load (R_L). Such a filter has been depicted in Fig. 3·21.

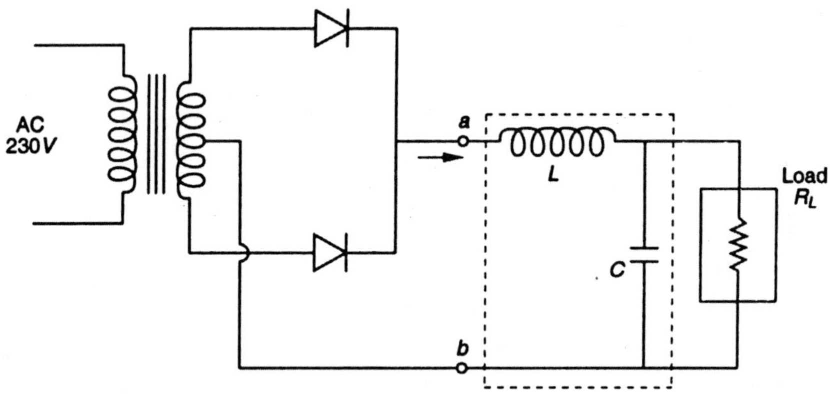

Fig 3.21. *LC* Filter connected to a full wave rectifier

In Fig. 3·21, if the value of *L* is very small, the voltage drop across it will be small. The capacitor will then charge up to almost the peak value of the voltage E_p. Thus, it would cut off the diodes allowing only short pulses of current. As the value of the inductor *L* is increased, the current pulses are smoothed out. They will then flow for longer time intervals but with smaller amplitudes. A critical value of inductance *LC* may be calculated for which either of the diodes D_1 and D_2 will be always conducting and the current will not fall to zero.

Mathematical Analysis

For a full wave rectifier

$$E_{av} = \frac{2E_p}{\pi} \qquad ...(3·70)$$

However, if *L* represents the series resistance

$$E_{av} = \frac{2E_p}{\pi} - I_{dc} \cdot R \qquad ...(3·71)$$

The ripple current that flows through *L* is bypassed mostly through *C* because of its low reactance. The voltage across R_L is, thus, much free from the ripple that passes through *L*.

For an *LC* filter

$$R_L << \frac{1}{\omega C}$$

$$\omega L << \frac{1}{\omega C}$$

The r.m.s. value of the alternating component is

$$(I_r)_{rms} = \frac{4R_p}{3\pi\sqrt{2}\,.\,\omega L} \qquad ...(3\cdot72)$$

$$= \frac{\sqrt{2}}{3\omega L} E_{av} \qquad ...(3\cdot73)$$

The ripple voltage developed across C is

$$(E_r)_{rms} = (I_r)_{rms} \cdot \frac{1}{\omega C} \qquad ...(3\cdot74)$$

Thus
$$(E_r)_{rms} = \frac{\sqrt{2}}{3}\left[\frac{1}{4\omega^2 L\,.\,C}\right] E_{av} \qquad ...(3\cdot75)$$

The ripple factor is

$$RF = \frac{(E_r)_{rms}}{E_{av}}$$

$$= \frac{\sqrt{2}}{3} \cdot \frac{1}{4\omega^2 L\,.\,C} \qquad ...(3\cdot76)$$

considering the ripple frequency as $2f$, where f is the supply frequency.

Thus
$$RF = \frac{\sqrt{2}}{3} \cdot \frac{1}{4\,(2\pi \times 50)^2\, L\,.\,C} = \frac{1\cdot19}{L\,.\,C} \qquad ...(3.77)$$

where L is expressed in Henry and C in μF.

Critical Inductance, L_C

For $L = L_C$. either of the diodes D_1 or D_2 is in conduction and the current does not decrease to zero.

The current entering at point a in the inductor has two components, a d.c. component $i_{dc} = i_{av} = \frac{E_{dc}}{R_L}$ and a sinusoidal component whose peak value is $i_p = \frac{4E_p}{3\pi\,\omega L}$.

In order to maintain the above condition (of non zero current), the d.c. component i_{dc} must exceed i_p

or
$$i_{dc} \geq i_p \qquad ...(3\cdot78)$$

Stated mathematically, this means

$$\frac{E_{dc}}{R_L} \geq \frac{4E_p}{3\pi\omega L_C} \qquad ...(3\cdot79)$$

where L_C is the critical values of the inductance. This yields

$$L_C \geq \frac{R_L}{3\omega} \qquad ...(3\cdot80)$$

or
$$L_C \geq \frac{R_L}{942\cdot5} \qquad ...(3\cdot81)$$

or
$$L_C \geq 1\cdot06\, R_L \times 10^{-3} \qquad ...(3\cdot82)$$

Equation (3·81) suggests that for light load (large R_L) the value of L_C should be high and the ripple is not high. Further

$$I_{av} = I_{dc} = \frac{E_{dc}}{R_L} = \frac{2E_p/\pi}{3\omega L_C} \qquad ...(3\cdot83)$$

$$= \frac{2E_p}{\pi} \times \frac{1}{3{\cdot}2\,\pi f L_C} = \frac{E_p}{3\pi^2 f L_C} \quad \ldots(3{\cdot}84)$$

$$I_{av} = {\cdot}0338\, E_p / f L_C \quad \ldots(3{\cdot}85)$$

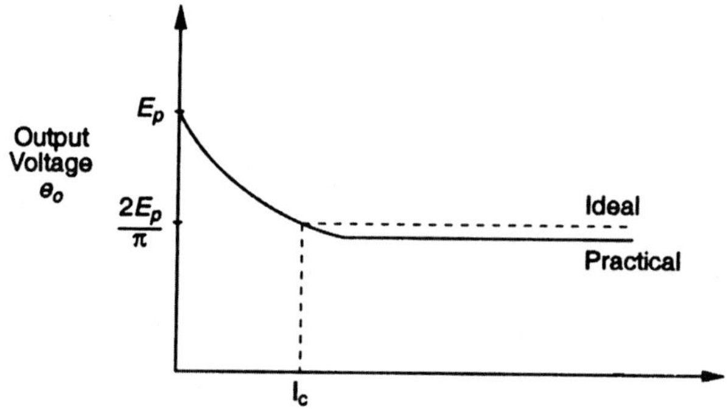

Fig 3.22. Variation of d.c. output voltage with d.c. load current

This current value is known as the critical current. However, when L_C is small or reduced, the critical value of L_C is also small.

The variation of the output or load voltage with load current is depicted in Fig. 3·22. At zero current (NO LOAD), the output voltage is E_p while at some minimum current I_C, the current drops to $\frac{2E_p}{\pi}$ and then it remains constant for an ideal situation if the diode resistance, choke resistance and the transformer winding resistances are zero. However, in actual practice, the presence of the resistances of the diode, inductor L and the transformer secondary gives the practical curve as depicted above.

If $L > L_C$, then a bleeder resisor R_{bleeder} must be permanently connected across the capacitor C such that a minimum current I_C always flows through it and the output voltage e_0 will not rise to E_p even if the load resistance is removed.

Thus

$$R_{\text{Bleeder}} \leq 900\, L_{\max} \quad \ldots(3{\cdot}86)$$

The bleeder resistor provides a path for the capacitor to discharge when the power supply is switched off.

This type of filter is employed for current values in excess of 100 mA for greater flexibility in selecting filter components.

3.9.4 *C–L–C* (or π-filter)

The *C–L–C* filter comprises a capacitor filter followed by an *L–C* filter.

When an input capacitor filter instead of an *L*-section filter is employed in a full wave diode rectifier, the output voltage is high. At small load (R_L), the average voltage on C_1 falls below E_p.

During the conduction period of diode D_1, the capacitor C_1 charges to the peak value E_p of the voltage $e_s = E_p \sin \omega t$. This charging process will continue as long as D_1 conducts. When diode D_1 turns OFF, the

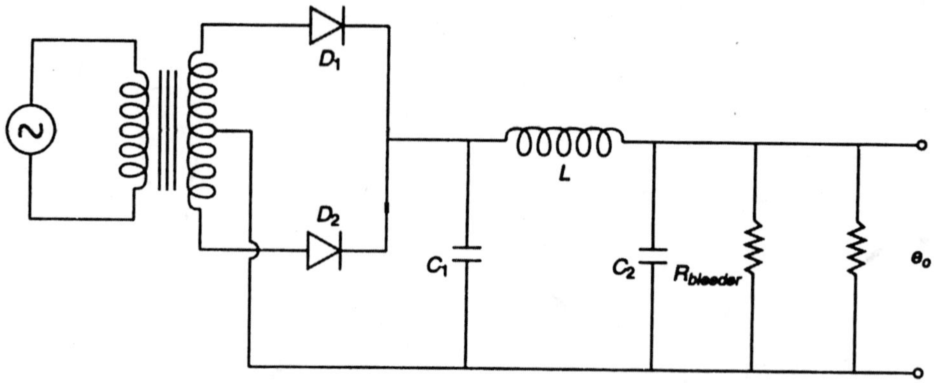

Fig 3.23. *C–L–C* or π type filter connected with a full wave rectifier

capacitor discharges through the inductor L and load R_L until D_2 conducts (as in the case of a capacitor type filter). Let us assume that the discharge current of C_1 to be constant at I_{dc} and also that the conduction period is smaller compared to the discharge period. The ripple voltage can be expressed as

$$V_r = \frac{I_{av}}{2f C_1} \quad \ldots(3{\cdot}87)$$

The output d.c. voltage can be written as

$$V_{dc} = V_{av} = E_p - \frac{I_{av}}{2fC_1} - I_{av} R_1$$

$$= E_p - I_{av}\left\{R_1 + \frac{1}{2fC_1}\right\} \quad \ldots(3{\cdot}88)$$

The triangular ripple waveform can be expressed in a Fourier series given by

$$V_R = -\frac{V_r}{\pi}\sin 2\omega t + \frac{V_r}{2\pi}\sin 4\omega t - \frac{V_r}{3\pi}\sin 6\omega t \ldots\ldots \quad \ldots(3{\cdot}89)$$

This wave is applied to L–C_2 section.

If $$R >> \frac{1}{4\pi f C_2}$$

where R is effective load including bleeder resistance,

the ripple factor can be shown to be

$$RF = \frac{1}{\sqrt{2}\,(\omega R C_1)}\left[\frac{1}{4\omega^2 L C_2 - 1}\right]$$

For $f = 50$ Hz, RF $$= \frac{5700}{C_1 C_2 L R} \quad \ldots(3{\cdot}90)$$

When C_1, C_2 are in μF, L in Henry and R in ohms. It will be seen from equation (3·90) that the ripple factor RF in inversely proportional to the effective load resistor R. The ripple factor decreases with increasing R. The input capacitor gives a means of a higher voltage for a given transformer rating than is possible with an L_C filter. It also has a lower ripple.

Comparison of various types of Filters

	Inductor Filter (L filter)	Shunt capacitor filter	Choke Input or L–C filter	C–L–C filter (π) filter
Ripple factor	Increases with R_L	Decreases with R_L	Independent of R_L	Decreases with effective load R

EXAMPLE 3.2 *In a half wave rectifier circuit, the diode resistance is $r = 1{\cdot}0\ \Omega$ and a load resistance $R = 100\ \Omega$. The input supply voltage is 30 volts peak value. Find (a) the peak, r.m.s and average values of current (b) the d.c. output power (c) the a.c. input power (d) efficiency of the rectifier.*

SOLUTION

(a) the peak current $= I_p = \dfrac{E_p}{r + R_L} = \dfrac{30}{100 + 1}$

$= 297$ mA

r.m.s. current $= I_{rms} = \dfrac{I_p}{2} = \dfrac{297}{2} = 148{\cdot}5$ mA.

average (or d.c.) current $= I_{av} = \dfrac{I_p}{\pi} = \dfrac{297}{\pi} = 94{\cdot}5$ mA

(b) d.c. output power $= P_{dc} = I_{dc}^2 R_L$

$= (94{\cdot}5 \times 10^{-3})^2 \times 100 = 0{\cdot}893$ watts

(c) the a.c. input power $= (I_{rms})^2 \times (r + R_L)$

$= (148{\cdot}5 \times 10^{-3})^2 \times 101$

$= 2{\cdot}22\ W$

(d) Rectification efficiency $= \eta = \dfrac{P_{dc}}{P_{ac}} = \dfrac{0{\cdot}893}{2{\cdot}22}$

$= 40\%$

EXAMPLE 3.3 *In a centre tap full wave rectifier, the load resistance $R_L = 470\ \Omega$. Each of the diodes has a dynamic resistance of 2 ohms. The voltage across one half of the transformer secondary winding is 50 sin 314t. Calculate (a) the peak current value (b) the average or d.c. value of current (c) the r.m.s. value of current (d) the ripple factor and (e) the rectification efficiency.*

SOLUTION

$e_s = 50 \sin 314t$

(a) the peak value of voltage $= 50\sqrt{2} = 70{\cdot}7$ Volts

Peak value of current $= I_p = \dfrac{E_p}{r + R_L} = \dfrac{70{\cdot}7}{2 + 470} = 0{\cdot}1498\ A$

$= 149{\cdot}8$ mA

(b) average or d.c. value of current $= I_{dc} = \dfrac{2I_p}{\pi} = \dfrac{2 \times 149{\cdot}8}{\pi} = 95{\cdot}36$ mA

(c) r.m.s current $= I_{rms} = \dfrac{I_p}{\sqrt{2}} = \dfrac{149{\cdot}8}{\sqrt{2}} = 105{\cdot}92$ mA

(d) ripple factor $= RF = \sqrt{\left(\frac{I_{rms}}{I_{dc}}\right)^2 - 1} = \sqrt{\left(\frac{105 \cdot 92}{95 \cdot 36}\right)^2 - 1} = \sqrt{0 \cdot 2338}$

$= 0 \cdot 4834$

(e) rectification efficiency $\eta = \frac{P_{dc}}{P_{ac}}$

$$P_{dc} = (I_{dc})^2 \,.\, R_L = (95 \cdot 36 \times 10^{-3})^2 \times 470 = 4 \cdot 273 \text{ W}$$

$$P_{ac} = (I_{rms})^2 \, (r + R_L) = (105 \cdot 92 \times 10^{-3})^2 \times (470 + 2) = 5 \cdot 293 \text{ W}$$

$$\eta = \frac{P_{dc}}{P_{ac}} \times 100\% = \frac{4 \cdot 273}{5 \cdot 293} \times 100\% = 72 \cdot 07\%$$

EXAMPLE 3.4 *In a half wave rectifier circuit with shunt capacitor filter, the load resistance is 2k Ω and the input supply voltage is 50 sin 314t. Find the size of the filter capacitor such that the peak to peak ripple voltage is 6 percent of the d.c. output voltage.*

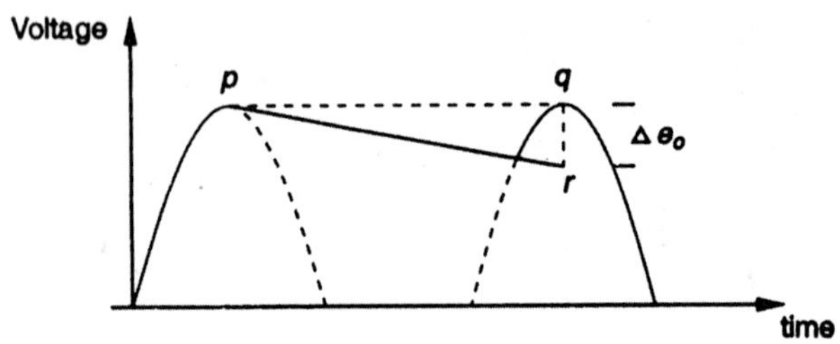

Fig. E3·4. Reference Example 3·4

SOLUTION

$$\frac{V_r}{V_{dc}} = 0 \cdot 06$$

From the right angled triangle $p\,q\,r$ we get,

$$\frac{\Delta e_0}{T} = \frac{E_p}{C\,R_L}$$

or $$\Delta e_0 = \frac{E_p \,.\, T}{R_L\,C}$$

Thus $$\frac{\Delta e_0}{E_p} = \frac{T}{R_L\,C} = \frac{1}{50 \times 2000 \times C}$$

or $$0 \cdot 06 = \frac{1}{50 \times 2000 \times C}$$

or $$C = \frac{1}{50 \times 2000 \times 0 \cdot 06} = 166 \cdot 6 \ \mu\text{F}$$

EXAMPLE 3.5 *Find the average or d.c. value of the output voltage when the filter capacitor is connected in the full wave rectifier.*

SOLUTION

$$(e_0)_{av} = (e_0)_{dc} = E_p\left(1 - \frac{\Delta e_0}{2}\right)$$

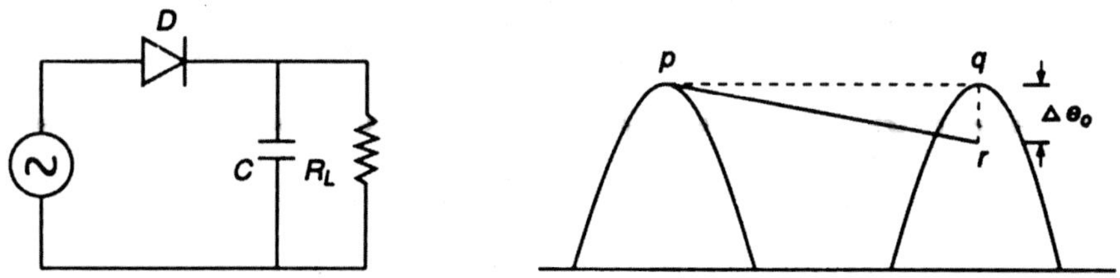

Fig E3.5. (a) and (b) Reference Example 3·5

$$= 50\left(1 - \frac{0{\cdot}06}{2}\right)$$

$$= 50 \times 0{\cdot}97$$

$$= 48{\cdot}5 \text{Volts.}$$

EXAMPLE 3.6 *Calculate the value of the ripple factor for the full wave rectifier circuit of Fig. E 3·6.*

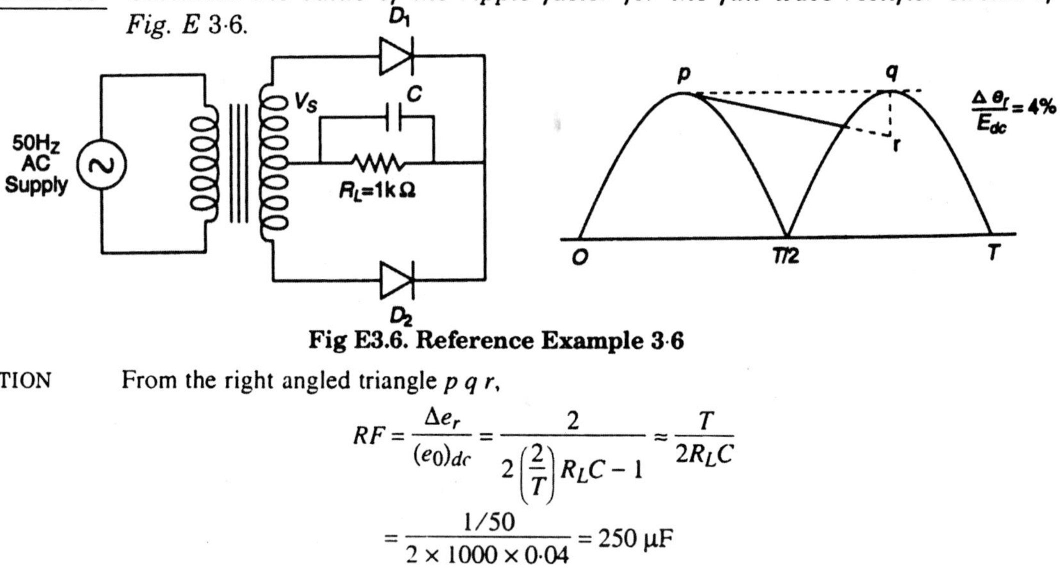

Fig E3.6. Reference Example 3·6

SOLUTION From the right angled triangle $p\ q\ r$,

$$RF = \frac{\Delta e_r}{(e_0)_{dc}} = \frac{2}{2\left(\frac{2}{T}\right)R_L C - 1} \approx \frac{T}{2R_L C}$$

$$= \frac{1/50}{2 \times 1000 \times 0{\cdot}04} = 250\ \mu\text{F}$$

EXAMPLE 3.7 *Find the value of the shunt capacitor (needed in a full wave rectifier circuit, such that the ripple voltage is 5 percent of the average value if R_L = 200 Ω and the source frquency is 50 Hz.*

SOLUTION

$$C = \frac{1}{2f \, . \, R_L \times (0{\cdot}05)} = \frac{1}{2 \times 50 \times 200 \times {\cdot}05}$$

$$= 1000\ \mu\text{F}$$

EXAMPLE 3.8 *In a full wave rectifier circuit, the input supply voltage across each of the transformer secondary halves is $e_s = 50\sqrt{2}\sin 314t$, R_L = 220 Ω, V_{dc} = 30 V. The diodes have a forward conduction resistance of 4 ohms and a reverse resistance of infinity. Find the turns ratio of the transformer.*

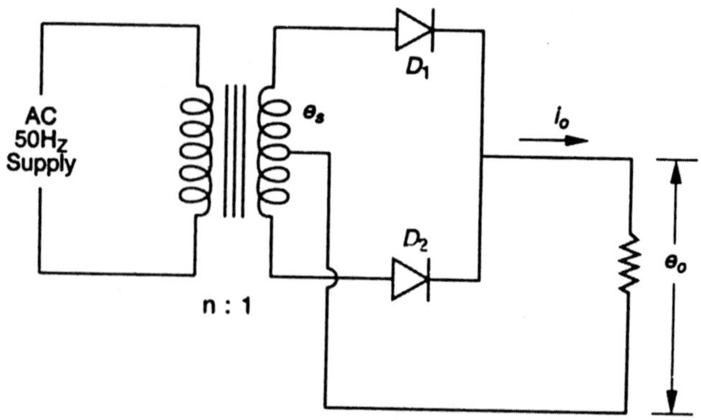

Fig E3.8. Reference Example 3.8

SOLUTION

The average output voltage $= \dfrac{1}{2\pi}\displaystyle\int_0^{2\pi} \dfrac{V_m}{n} \sin \omega t \, d(\omega t)$

$$= \frac{2V_m}{n\pi}$$

where n = Turns ratio of the transformer

Thus $$(e_0)_{dc} = \frac{2V_m}{n\pi} = \frac{2 \times 50\sqrt{2}}{n\pi}$$

and $$(i_0)_{dc} = \frac{30}{2/\pi} \sin(314t) = \frac{47{\cdot}12}{220} \sin(314t)$$

$$= 214 \sin(314t) \times 10^{-3} \text{Amp.}$$

Voltage across conducting diode

$$v_{D_1} = R_d \, i_L(t) = 4 \times 214 \times 10^{-3} \sin(314t)$$

$$= {\cdot}856 \sin(314t)$$

$$\frac{e_s}{n} = v_{D_1} + e_{out} = (0{\cdot}856 + 47{\cdot}08) \sin(314t)$$

$$\frac{e_s}{e_s/n} = \frac{50\sqrt{2}}{47{\cdot}936} = 1{\cdot}475$$

EXAMPLE 3.9 *A bridge rectifier has a load resistance of* 1500 Ω. *If the input voltage is* $110\sqrt{2}$ sin (314*t*) *and the forward conduction resistance of* 10 Ω, *find (a) the output voltage (b) the ripple voltage at the output (c) the voltage regulation.*

SOLUTION

$$(I_0)_{av} = \frac{2I_p}{\pi}$$

where I_p = Peak current $= \dfrac{E_p}{R_L + 2R_f} = \dfrac{110\sqrt{2}}{1500 + 2 \times 10}$

$$= \frac{110\sqrt{2}}{1520} = \frac{155{\cdot}56}{1520} = 102{\cdot}34 \text{ mA}$$

$$(I_0)_{av} = \frac{2I_P}{\pi} = \frac{2 \times 102{\cdot}34}{\pi} = 63{\cdot}66 \text{ mA}$$

DC load voltage $(E_0)_{dc} = (I_0)_{av} \times R_L = 63{\cdot}66 \text{ mA} \times 1500$

$= 95{\cdot}49$ Volts

(b) Output ripple voltage is

$$V_r = (I_{rms}^2 - I_{av}^2)^{1/2} R_L$$

$$= \left\{ \left(\frac{102{\cdot}34}{\sqrt{2}} \times 10^{-3} \right)^2 - (63{\cdot}66 \times 10^{-3})^2 \right\}^{1/2} \times 1500$$

$= \{ 5{\cdot}236 \times 10^{-3} - 4{\cdot}053 \times 10^{-3} \}^{1/2} \times 1500$

$= \{1{\cdot}173 \times 10^{-3}\}^{1/2} \times 1500$

$= 51{\cdot}3$ Volts

(c) The percentage regulation is

$$\frac{2R_f}{R_L} \times 100 = \frac{2 \times 10}{1500} \times 100 = 1{\cdot}33\%$$

3.10 BREAKDOWN DIODES

The breakdown diodes are *p–n* junction devices which operate in the breakdown region of the reverse voltage characteristics. The breakdown mechanisms that may take place are (*i*) Avalanche breakdown, and (ii) Zener breakdown. The power dissipation capabilities of breakdown diodes are such that they can operate in the breakdown region within the limits prescribed.

Avalanche Breakdown

As the reverse voltage across the *p–n* junction increases, the electric field across the junction increases, (Electric field = Voltage/Distance). When the reverse bias magnitude reaches a certain level, the electric field imparts sufficiently high energy to a thermally generated carrier crossing the junction. **This highly energetic carrier while colliding a crystal ion on its way disrupts a covalent bond and generates another electron-hole pair. These carriers, also, can acquire sufficient energy from the electric field and make collisions with other crystal ions to generate further electron-hole pairs.** In this way the number of carriers are multiplied. Within a very small time interval, a large number of carriers (electrons and holes) are generated. The mechanism of carrier multiplication by this process is known as Avalanche multiplication. The avalanche multiplication process generates a large reverse current. This is depicted in Fig. 3·24 (the dashed portion). The diode is known to work in Avalanche breakdown region.

Zener Breakdown

The breakdown mechanism because of internal field emission is called Zener breakdown after *C*. Zener who first proposed the idea in 1934.

In the Zener breakdown process, doping concentrations on either side of the *p–n* junction being very high, the corresponding electric field is also very high. The high electric field forces a bound electron to come out of its covalent bond. And, a free electron results. **This process generates a large number of carriers and a correspondingly high reverse current. The process is called as Zener breakdown.** It

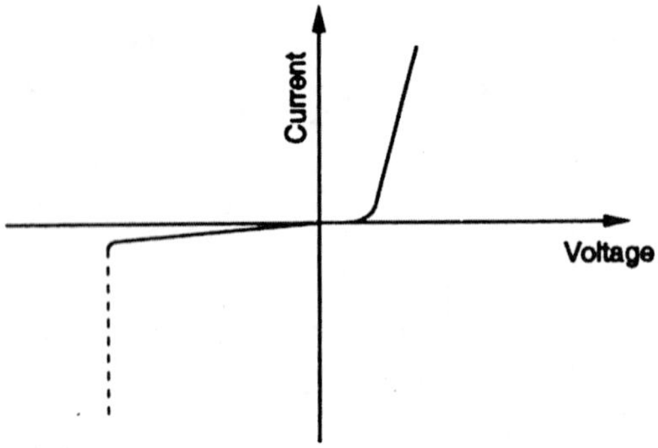

Fig 3.24. I–V characteristics of a *p–n* junction diode

should be noted in this connection, that whereas **in avalanche breakdown, the source of carriers is the collision process of carriers with crystal ions, the Zener breakdown occurs due to rupture of covalent bonds.**

For lightly doped diodes, the breakdown voltage is higher and avalanche multiplication is the dominant process. The Zener breakdown occurs for heavily doped diodes below 6 V. Although the physical mechanisms of operation of the two diodes are different, both of them are commonly called as Zener diodes.

It should also be noted here that the temperature co-efficient of **Zener breakdown voltage is negative while that for avalanche breakdown voltage the temperature co-efficient is positive.** This may be explained as follows. In the Zener breakdown process, the doping in the *p* and *n*-regions is heavy, the depletion region is thin (small). As the temperature increases, the energy of the valence electron increases such that a smaller applied voltage can pull the electrons out of the covalent bonds. Thus the Zener breakdown voltage decreases and the temperature co-efficient of Zener breakdown voltage is negative. For avalanche breakdown process, the doping is not heavy (light) and the depletion region width is large. As temperature increases, the crystal ions vibrate with larger amplitude, thereby increasing the possibility of collision of the carriers and the crystal ions. The carriers now have less opportunity to gather enough energy between

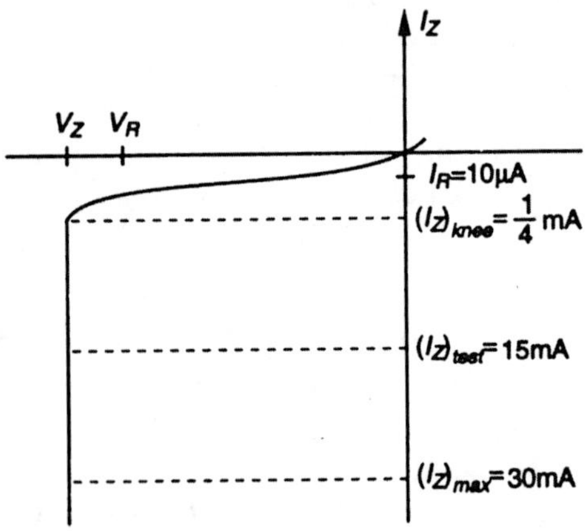

Fig 3.25. Zener diode chracteristics

collisions to trigger the avalanche process. **Thus, the avalanche breakdown voltage (typically having a value greater than 6 V) increases with increase in temperature.** The temperature co-efficient is thus **positive.** Typically, the temperature co-efficient of the breakdown voltage is ± 0·1% per degree Kelvin rise in temperature.

Figure 3·25 depicts the Zener diode test characteristics of a typical Zener diode. Zener diodes are available having Zener voltage from 1·8 V to 200 V with power ratings as small as $\frac{1}{4}$ W to a value as high as 50W. Silicon is the material typically employed for Zener diode fabrication. The electrical equivalent circuit of a Zener diode is depicted in Fig 3·26.

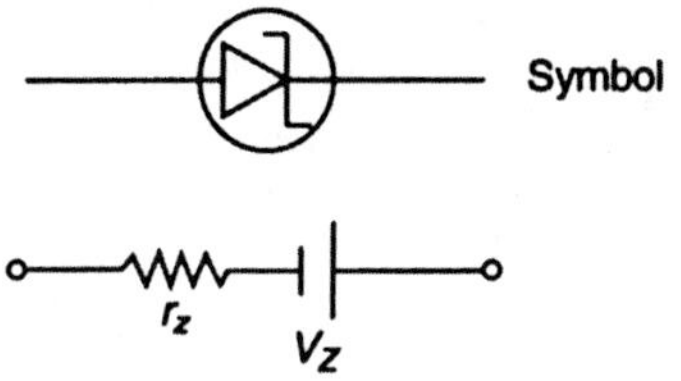

Fig 3.26. Equivalent circuit of a Zener diode.

In the Figure of 3·26, Z_{ZK} refers to the maximum dynamic impedance at $I = I_{ZT}$. Typically its value is below 10 Ω. Normally $Z_{ZT} = r_Z$ Z_{ZK} is the maximum knee impedance (≈ 600 Ω). The Zener voltage nominal is about 10V. The test current I_{ZT} is the magnitude of current corresponding to $\frac{1}{4}$ W of power level and Z_{ZT} is the dynamic impedance at this current level. The temperature co-efficient reflects the percentage change of V_Z with temperature. It is given by

$$T_C = \frac{\Delta V_Z}{V_Z (T_2 - T_1)} \times 100\% \ /^{\circ}C \qquad ...(3\cdot91)$$

where ΔV_Z is the change in nominal Zener voltage as temperature changes from T_1 to T_2.

EXAMPLE 3.10 *In a typical* $\frac{1}{2}$ *W Zener diode,* $V_Z = 12$ V *at 25°C. Find the value of the nominal* V_Z *at* $T = 90°C$ *if* T_C*, the temperature co-efficient is* ± 0·075%/°C.

SOLUTION

$$\Delta V_Z = \frac{V_Z \cdot T_C (T_2 - T_1)}{100}$$

In this example, $V_Z = 12$ V, $T_C = 0{\cdot}075\%/°C$,

$$T_2 - T_1 = 90 - 25 = 65°C.$$

Thus

$$\Delta V_Z = \frac{12 \times \cdot 075 \times (90 - 25)}{100} = \frac{0{\cdot}9 \times 65}{100}$$

$$= 0{\cdot}585 \text{ Volts.}$$

Since the temperature co-efficient is positive, the value of V_Z at $T = 90°C$ is

$$V_Z = \Delta V_Z + V_Z = 12 + 0{\cdot}585$$

$$= 12{\cdot}59 \text{Volts.}$$

3.10.1 Biasing of Zener diodes

The correct operation of a Zener involves placing it in a circuit such that

(1) it is reverse biased

(2) the biasing voltage is higher than V_Z

(3) the current does not exceed I_Z max

The Zener diodes are employed for applications like

(1) voltage regulator

(2) voltage limiters or peak clippers

(3) wave form shaping circuits

(4) fixed reference voltage source or for calibration of voltmeters

(5) protection of meters against damages which may result from application of high voltages.

EXAMPLE 3.11 *Assuming the Zener diode in Fig. 3·27 to be ideal, find whether it is correctly biased.*

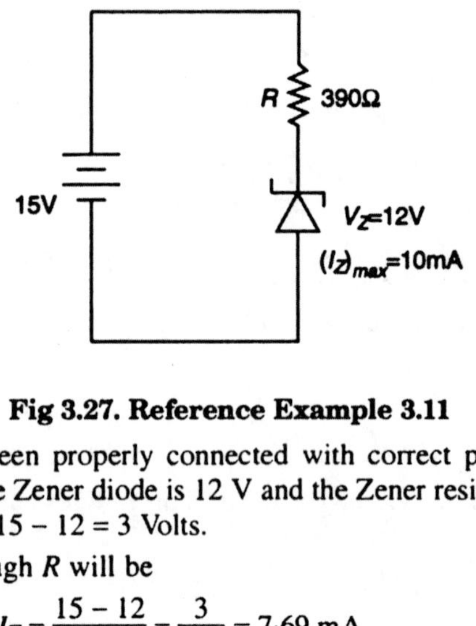

Fig 3.27. Reference Example 3.11

SOLUTION The diode has been properly connected with correct polarity to the battery. Since the voltage across the Zener diode is 12 V and the Zener resistance is zero (ideal), the voltage drop across R is 15 – 12 = 3 Volts.

The current through R will be

$$I_Z = \frac{15 - 12}{390} = \frac{3}{390} = 7{\cdot}69 \text{ mA}$$

The current is less than (I_Z) max specified for the Zener. Thus it has been properly biased.

EXAMPLE 3.12 *Repeat problem 3·11 above, assuming that the Zener diode has a resistance of r_Z = 10 Ω. Also calculate the power dissipated by the diode.*

SOLUTION In the circuit of Fig. 3·27

$$I_Z = \frac{15 - 12}{390 + 10} = \frac{3V}{400} = 7{\cdot}5 \text{ mA}$$

The dissipated power = $V_Z \cdot I_Z$ = 12 V + 7·5 mA

= 90 mW.

EXAMPLE 3.13 *Calculate, whether the Zener diode in the circuit of Fig. 3·28 is correctly biased. Find I_Z and the power dissipated by the diode.*

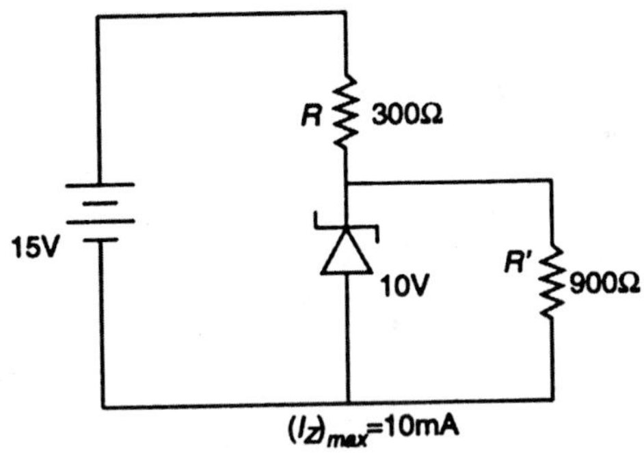

Fig 3.28.

SOLUTION From Fig. 3·28, it is apparent that the Zener diode is connected with proper polarity to the battery.

Now $V_Z = 10$Volts.

Drop across 300 Ω = 5V

Current through 300 Ω resistor $= \frac{5}{300} = 16{\cdot}67$ mA

Current through 900 Ω resistor $= \frac{10V}{900} = 11{\cdot}11$ mA

Current through Zener diode $I_2 = 16{\cdot}67 - 11{\cdot}11 = 5{\cdot}56$ mA

Since $I_Z < I_{Z\,max}$, the Zener diode is properly connected.

Now power dissipated $= V_Z \,.\, I_Z = 5{\cdot}56 \times 10 = 55{\cdot}6$ mW.

EXAMPLE 3.14 *Determine the value of output voltage V_{out} in the circuit of Fig. 3·29. Assume that the Zener diode operates at a temperature of 40°C. The specification for the Zener are V_Z = 10V at 20°C, T_C = temperature co-efficient of Zener voltage = ·075%/°C.*

SOLUTION V_Z' at 40°C = V_Z at 20°C + ΔV_Z

Now $$\Delta V_Z = \frac{V_Z \,.\, T_C \,.\, (T_2 - T_1)}{100}$$

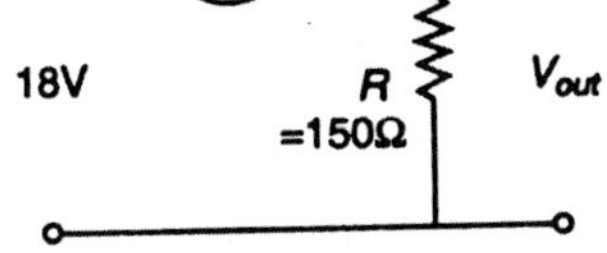

Fig 3.29. Reference Example 3.14

$$\Delta V_Z = \frac{10 \times (\cdot 075)(40-20)}{100} = \cdot 15 \text{ Volts.}$$

$\therefore$ V_Z at $40°C = 10{\cdot}15$ Volts.

Since $V_{in} = 18$ Volts, the drop across resistance R will be $V_R = 18 - 10{\cdot}15 = 7{\cdot}85$ Volts.

3.11 VOLTAGE REGULATION CIRCUITS

All of us are familiar with voltage regulation circuits. In a voltage regulation circuit, the basic aim is to keep the output voltage constant even when the input voltage or the load current varies.

Let us consider the circuit in Fig. 3·30 and see how the output voltage can be kept fixed even when either V_{in} or I_L changes.

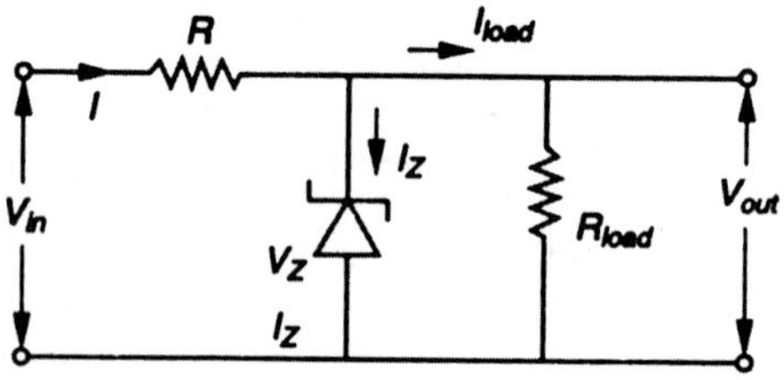

Fig 3.30. A voltage regulator circuit using a Zener diode.

In the circuit diagram, the Zener diode anode is connected to the negative terminal of the battery while the cathode is connected to the positive terminal of the battery. Obviously $V_{in} > V_Z$. In the circuit

$$V_{in} = IR + V_Z \qquad \ldots(3{\cdot}92)$$

and $$I = I_Z + I_{load} \qquad \ldots(3{\cdot}93)$$

where I is the current through R,

I_Z is the current through Zener ($I_Z < I_{Z\ max}$)

I_{load} is the load current.

Let us now examine how this circuit acts as a voltage regulator.

(a) When V_{in} is changed :

When the input V_{in} increases (say), the drop across R which is $V_R = IR$ increases such that V_Z is constant. The increase in V_R increases I, the current through R. Since I_{load} is constant, the increase in current flows through the Zener diode. **Thus the Zener diode keeps the output voltage constant by allowing a higher current to flow through it. Similarly the Zener diode takes a lower current when the input voltage V_{in} drops.**

(b) When the load current is changed :

When the load current, let us say increases while V_{in} is constant, the following things takes place. The voltage drop across R is constant because V_{in} and V_Z are both constants in equation (3·92). This means that current I is also fixed. Thus $I_Z = I - I_{load}$ decreases when I_{load} increases. The Zener diode keeps the output voltage fixed by accepting a lower current to flow through it. Similarly, the Zener diode keeps the output voltage fixed by allowing a higher current to flow through it, when the load current decreases. ($I_Z = I - I_{load}$ increases).

We may, also, write down that

$$V_{in} = V_{out} + (I_Z + I_{load})\, R \qquad \text{...(3·94)}$$

or,

$$R = \frac{V_{in} - V_{out}}{(I_Z + I_{load})} \qquad \text{...(3·95)}$$

In this equation when $I_Z = (I_Z)_{max}$, the maximum value of Zener diode current, $I_{load} = 0$. A practical example of a circuit having two reference voltage levels is depicted in the circuit of Fig. 3·31.

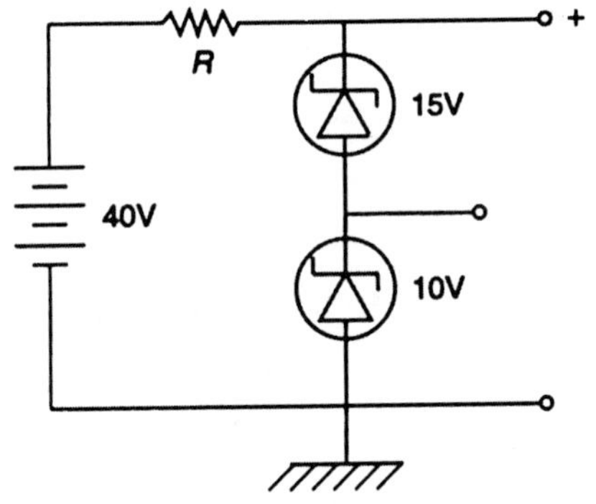

Fig 3.31. A voltage regulator circuit having two reference voltage levels.

EXAMPLE 3.15 *Find the battery current I, the Zener current I_Z and load current I_{load} in the circuit of Fig. 3·32. What will be the effect on the above values if V_{in} increases to 75 Volts. Assume the Zener resistance to be negligible.*

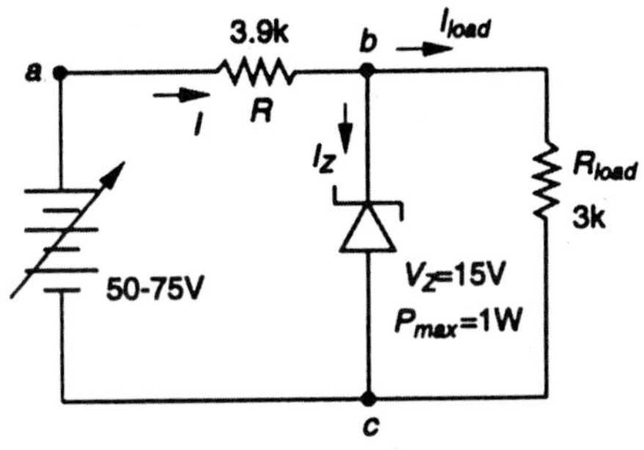

Fig 3.32. Reference Example 3.15

SOLUTION (a) When $V_{in} = 50$ V

$$V_{bc} = 15\text{V},$$

So $$V_{ab} = 50 - 15 = 35 \text{ V}$$

Thus $$I = \frac{35V}{3{\cdot}9\ K} = 8{\cdot}97 \text{ mA}$$

$$I_{load} = \frac{V_Z}{R_{load}} = \frac{15\text{V}}{3\text{K}} = 5 \text{ mA}$$

So $\quad I_Z = I - I_{load} = 8{\cdot}97 - 5 = 3{\cdot}97$ mA.

Now $\quad P_{max} = V_Z \cdot (I_Z)_{max}$

or $\quad (I_Z)_{max} = \frac{1W}{15V} = 66{\cdot}67$ mA

The diode current of 3·97 mA is very much within $(I_Z)_{max}$, the limit.

(b) When $V_{in} = 75$V,

drop across $R = 75 - 15 = 60$V.

$I = 60/3{\cdot}9$ K $= 15{\cdot}38$ mA.

$I_{load} = 5$ mA

So $\quad I_Z = 15{\cdot}38 - 5 = 10{\cdot}38$ mA.

EXAMPLE 3.16 *In Fig 3·33 find the Zener current if the load resistance R_L is (i) 30 kΩ (ii) 10 kΩ (iii) 3 kΩ.*

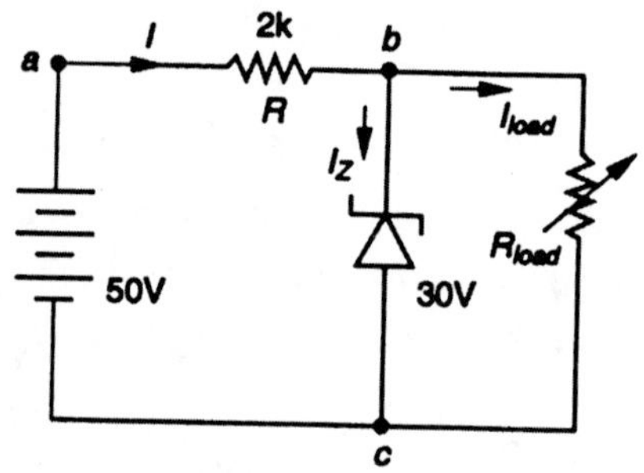

Fig 3.33. Reference Example 3.16

SOLUTION (i) When $R_L = 30$ kΩ,

$V_{bc} = 30$V,

$V_R = V_{ab} = 50 - 30 = 20$V.

$$I = \frac{20\text{V}}{2\text{ K}} = 10 \text{ mA},$$

$$I_{load} = \frac{V_{bc}}{R_{load}} = \frac{30}{30\text{ k}} = 1 \text{ mA}$$

$I_Z = I - I_{load} = 10 - 1 = 9$ mA.

(ii) When $R_L = 10$ kΩ

$$I = 10 \text{ mA}, I_{load} = \frac{30}{10\text{ k}} = 3 \text{ mA}$$

$I_Z = I - I_{load} = 10 - 3 = 7$ mA

(iii) When $R_L = 3$ kΩ

$$I_{load} = \frac{30V}{3K} = 10 \text{ mA}$$

Thus $I_Z = I - I_{load} = 10 - 10 = 0$ mA.

Since $I_Z = 0$, the diode will come out of the break down region and the diode will no longer act as a voltage regulator.

EXAMPLE 3.17 *A 12V reference supply is to be fabricated using a Zener diode and resistor connected in series. The d.c. supply voltage is 24 Volts. Find a suitable diode and calculate the required resistor value. Find the new current level when the supply voltage drops to 20 Volts.*

SOLUTION From Fig 3·30 above

$$R = \frac{V_{in} - V_Z}{I_Z} = \frac{24 - 12}{20 \text{ mA}} = \frac{12}{20 \text{ mA}} = 600 \ \Omega$$

(Use 680 Ω standard resistor)

When $V_{in} = 20$V,

$$I_Z = \frac{20 - 12}{680} = \frac{8}{680} = 11{\cdot}8 \text{ mA}$$

EXAMPLE 3.18 *Design a Zener diode voltage regulator with an output of approximately 9V to operate from a 20V supply. Calculate the maximum load current for the circuit (Texas Instruments 1N 757 Zener diode has a reference voltage of 9·1 Volts and P = 400 mW).*

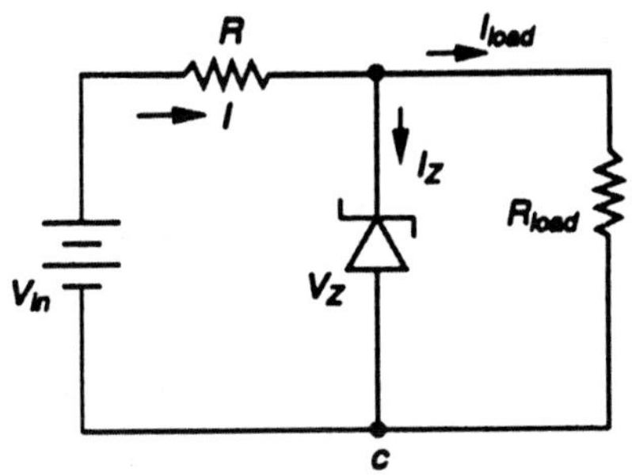

Fig 3.34. Reference Example 3.18

SOLUTION

$$(I_Z)_{max} = \frac{400 \text{ mW}}{9{\cdot}1 \text{ V}} = 43{\cdot}96 \text{ mA}$$

∴ $(I_{load})_{max} + I_{Z\ min} = 43{\cdot}96$ mA

Now
$$R = \frac{V_{in} - V_Z}{(I_{load})_{max} + (I_Z)_{min}}$$

$$= \frac{20 - 9{\cdot}1}{43{\cdot}96 \text{ mA}} = \frac{10{\cdot}9}{43{\cdot}96 \text{ mA}} = 247{\cdot}95 \ \Omega$$

(Use 270 ohm standard resistor).

Thus maximum load current is

$$(I_{load})_{max} = 43{\cdot}96 \text{ mA} - 5 \text{ mA}$$
$$= 38{\cdot}96 \text{ mA}.$$

EXAMPLE 3.19 *Find the range of values of the input voltage such that the Zener diode will remain ON in Fig 3·35.*

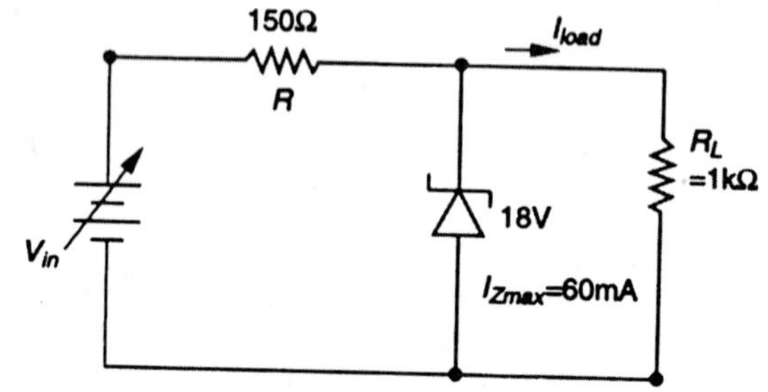

Fig 3.35. Reference Example 3.19

SOLUTION

$$V_{load} = V_Z = \frac{R_L \cdot V_{in}}{R_L + R}$$

$$(V_{in})_{min} = \left(\frac{R_L + R}{R_L}\right) V_Z = \frac{(1000 + 150)}{1000} \times 18\text{V} = 20{\cdot}7 \text{ Volts.}$$

$$I_{load} = \frac{V_{load}}{R_{load}} = \frac{18 \text{ V}}{1 \text{ K}\Omega} = 18 \text{ mA}$$

Now $(I_Z)_{max} + I_L = (I)_{max} = 60 \text{ mA} + 18 \text{ mA} = 78 \text{ mA}$

$$(V_{in})_{max} = V_Z + (I)_{max} \times R = 18 + 78 \text{ mA} + 150\ \Omega$$
$$= 18 + 117 \text{ V} = 29{\cdot}7 \text{ Volts.}$$

So the input volage range is from 20·7 to 29·7 Volts.

When two Zener diodes are connected back to back, they can be used as an a.c. regulator as depicted in Fig. 3·36.

When input voltage V_{in} is + 9 V say, the Zener diode Z is forward biased and behaves as a short circuit while Zener diode Z_2 behaves as an open circuit, so the output voltage is the same as the input voltage. When V_{in} is –9 V, the Zener diode Z_1 is reverse biased (Open) while Zener Z_2 is forward biased. The output voltage is –9 V. The input and output voltages will be the same until V_{in} reads 15 V. The Zener diode Z_2 will then "TURN ON" while Z_1 will be like a short circuit (with a very small resistance). Thus the output voltage will be the same as the Zener diode voltage 15 V. Similarly, when the input is –15 V, the Zener diode Z_1 will "TURN ON" (as a Zener diode) and Z_2 will be like a short circuit. And the output voltage will be –15 V (the Zener Z_1 voltage). For a full cycle of sinusoidal a.c. voltage the output will be a clipped waveform as depictied in Fig. 3·36 (a).

A simple square wave generator may be obtained if $\frac{V_{in}}{V_Z}$ is very high. For example, if V_{in} is 60 V and the Zener voltages are 9 V, then the obtained output wave shape is almost square wave (Fig. 3·37).

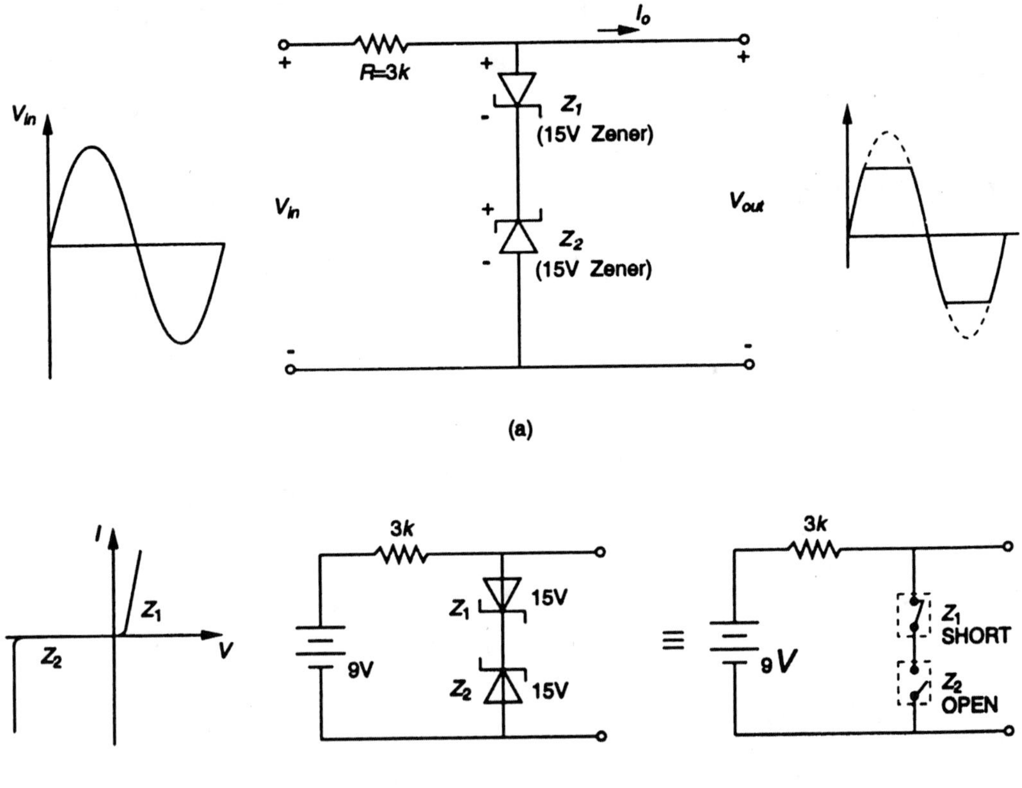

(b)

Fig 3.36. (a) 30 V peak to peak ac regulator (b) Shows circuit behaviour when V_{in} = 9 V

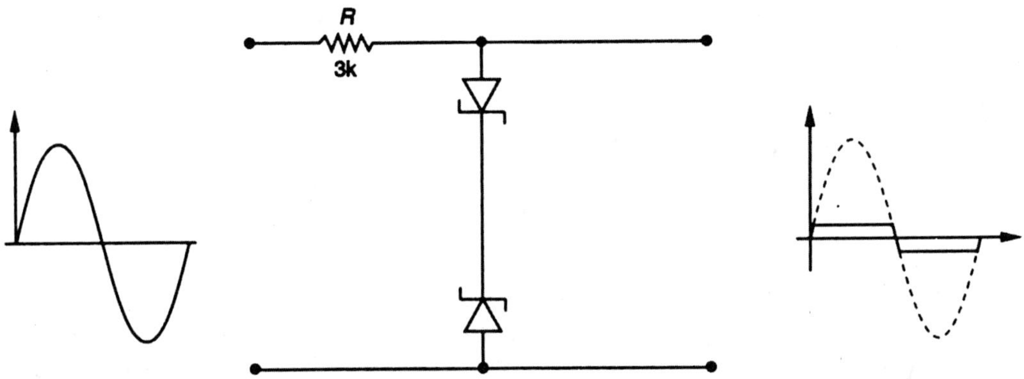

(a)

Fig 3.37. Square wave generator circuit

A Zener diode can also be used in meter protection circuits as depicted in Fig. 3·38 (a) and (b).

(a) One polarity (b) bidirectional polarity.

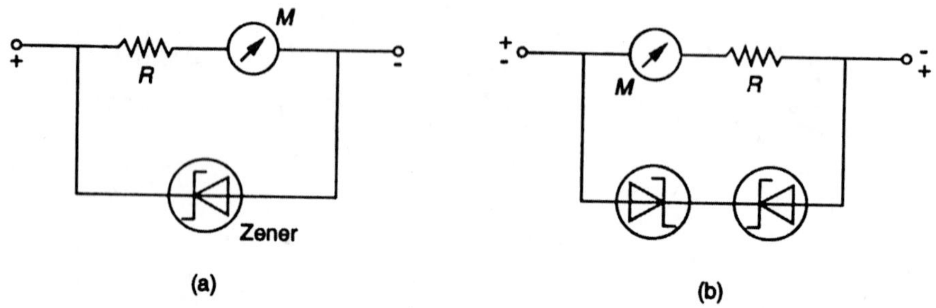

Fig 3.38. Overload protection circuits for a meter.

In Fig. 3·38 (a) the overload protection circuit of a *V–O–M* (volt-ohm-meter) is shown for the indicated polarity. When the applied voltage is higher by say 2·0 V the Zener diode will take the increased current and will save the meter. Two Zener diodes may be connected in series and connected across the meter to protect against overload for bidirectional polarity.

3.12 VARACTOR DIODES

The **varactor diode** is also popularly known as **varicap** (variable capacitance), **voltacap or voltage variable capacitor (VVC)** diode. In effect, the varactor diode is a reverse biased *p–n* junction diode whose transition capacitance (C_T) is a function of the reverse voltage applied across the junction. Figure 3·35 depicts the principle of operation of a varactor diode. With changing value of the reverse voltage applied across the junction, the width of the depletion region (of uncovered charges) varies. When the reverse bias voltage is high, the depletion layer width increases. The *p–n* junction capacitance then decreases with increasing reverse bias voltages.

d = Depletion layer width for low reverse bias

d' = Depletion layer width for large reverse bias.

The transition capacitance of uncovered charges is given by

$$C_T = \varepsilon \frac{A}{d} \qquad \ldots(3{\cdot}96)$$

where ε is the semiconductor material permittivity, A is the area of cross section and d is the depletion layer width. The normal range of variation of the reverse voltage of a *p–n* junction diode is typically limited to 20 Volts. In terms of the applied reverse bias, the transition capacitance can be written approximately as

$$C_T = \frac{K}{(V_T + V_R)^n} \qquad \ldots(3{\cdot}97)$$

where V_T is the knee voltage.

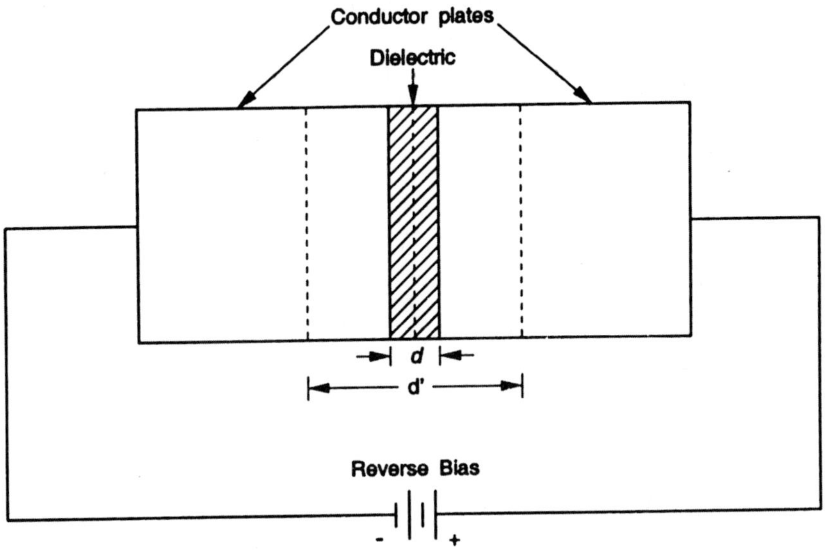

Fig 3.39. Principle of operation of a voltage variable capacitor (VVC).

Both abrupt junction and hyper abrupt junction devices are employed for *VVC* diodes [Fig. 3·40 depicts the doping profiles for *VVC*] Fig. 3.40A shows capacitance variation with reverse bias.

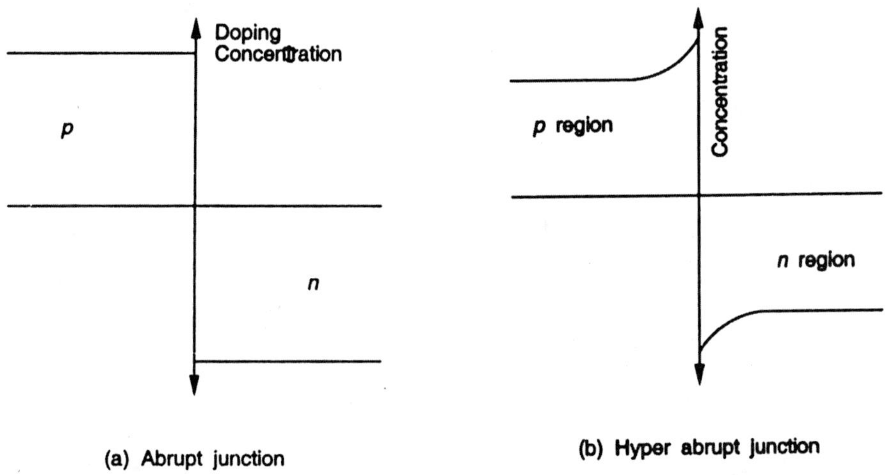

Fig 3.40. Doping profiles of *VVC* diodes.
In the hyper abrupt junction, the doping concentration is highest at the junction.

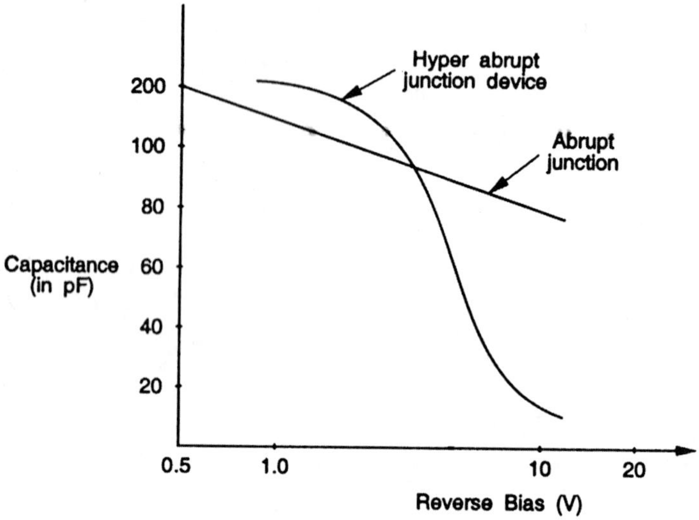

Fig 3.40 A. Capacitance Variation with reverse bias

The symbols and complete equivalent circuit of a *VVC* or varactor diode is depicted in Fig. 3·41 (a) and (b) respectively.

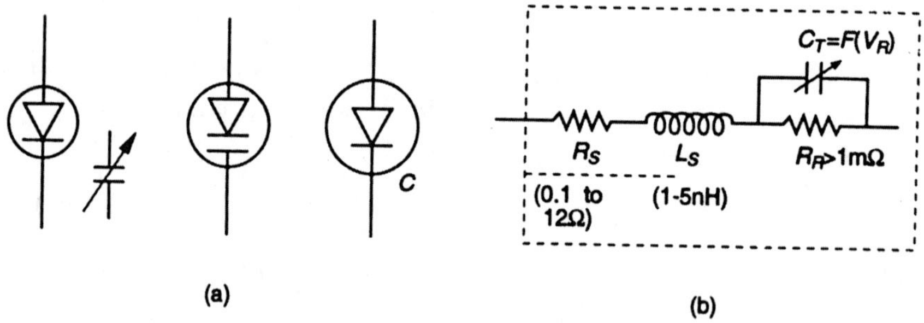

Fig 3.41. (a) Symbols
(b) electrical equivalent circuit of the reverse bias region of a varactor diode

The Q factors of the *VVC* diode can be as high as 600 at a frequency of 60 MHz. However, the value of Q is dependent on bias voltage and frequency. *VVCs*' are available with nominal device capacitances ranging from 6 pF to 600 pF. The capacitance tuning ratio (*TR*) of the *VVC* is defined as

$$TR = \frac{C_T\,(\text{small reverse voltage})}{C_T\,(\text{large reverse voltage})} \qquad \ldots(3{\cdot}98)$$

The value of *TR* may range from 2 to 15 depending on the doping profile of the device.

In hyper abrupt junctions, the doping density is highest near the p–n junction. The high concentration near the junction makes the depletion layer width very thin and the junction capacitor has a very high value. It also makes possible the depletion region width to be more sensitive to bias voltage variations and thus *TR* becomes very high (typical value = 15).

A varactor diode can be made part of a resonant circuit and the voltage applied across the *VVC* may be varied so as to obtain a change of the resonant frequency of the circuit.

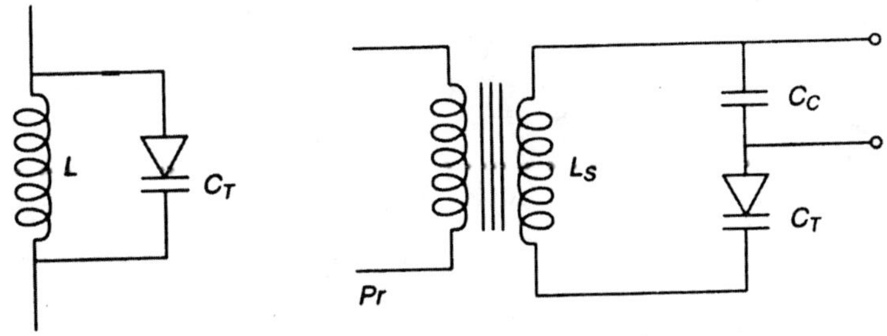

(a) $f_P = \frac{1}{2\pi\sqrt{LC_T}}$ (b) $f_P = \frac{1}{2\pi\sqrt{L_S(C_C + C_T)}}$

Fig 3.42. Typical resonant circuits employing varactor diodes.

The varactor diodes are employed for such applications as

(1) FM modulators (2) Automatic frequency control devices

(3) Adjustible band pass filters (4) Parametric amplifiers.

3.13 TUNNEL DIODES

The tunnel diode was invented by Leo Esaki in 1958. It is also known as an Esaki Diode after its inventor. A tunnel diode is a two terminal negative resistance device, which can be utilized as an amplifier, an oscillator or a switch. It is a high frequency device because of its very fast response to inputs. The tunnel diodes require very small bias voltages and low load resistances than other electronic devices.

Tunnel diodes are fabricated using semiconductors like germanium (Ge), gallium arsenide (Ga As) and gallium antimonide (Ga Sb). When the doping density of impurity atoms is extremely high (about 10^{24} or 10^{25} m^{-3}) in the n and p-regions of an abrupt p–n junction, the depletion region width becomes exprermely thin (of the order of 10 mm).

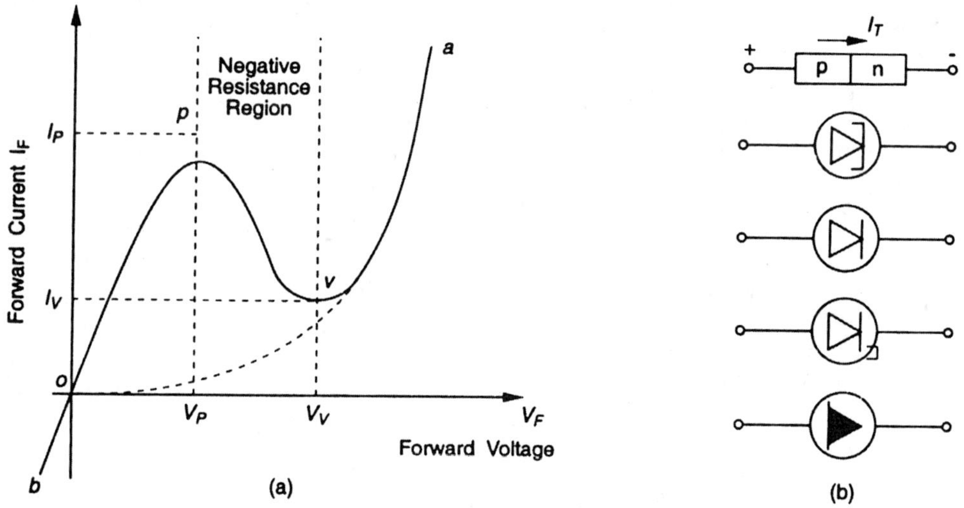

Fig 3.43. (a) Current voltage characteristics of a tunnel diode. (b) Symbols of a Tunnel diode

This high doping density also reduces the reverse breakdown voltage to a very small value (nearly zero), and the diode characteristics appears to have been broken down for any reverse voltage. The current voltage characteristics (Fig. 3·43 *a*) depicts a **negative resistance region** when the diode is **forward biased.** The **negative resistance region can be explained by Quantum mechanical tunnelling and the diode is called a tunnel diode.**

The Quantum mechanical tunnelling can be explained with the help of Fig. 3·44 which shows the energy band diagram of the *p–n* junction for various bias voltages. At the very high doping levels prevalent in the

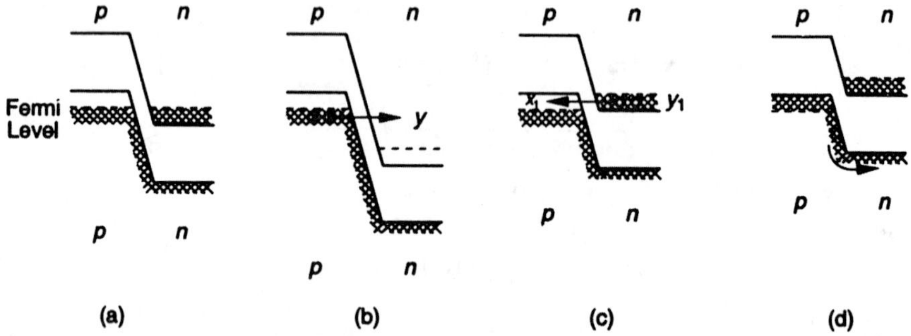

Fig 3.44. Energy band diagram of a tunnel diode for (a) zero bias (b) reverse bias (c) forward bias and (d) high forward bias

semiconductor, the Fermi levels lie within the energy bands at absolute zero of temperature. When no voltage is applied (zero bias condition corresponding to Fig 3·44 *a*), the Fermi level is constant throughout at the *p* and *n*-regions across the junction at constant temperature. **Since the Fermi level is the highest occupied energy level, no current flows through the junction.**

When the *p–n* junction is reverse biased, the Fermi level on the *p* side moves up compared to that on the *n*-side. The electrons on the valence band or the *p* side (position *x*) are now seperated from available empty states on the *n*-side conduction band (shown as position *y*) by an extremely thin depletion region (≈ 10 nm). These electrons can now tunnel quantum mechanically from the *p*-side of the valence band (position *x*) to the *n*-side of the conduction band (position *y*). And this constitutes the tunnel current. The larger the magnitude of the reverse bias, the larger the number of electrons that cross the narrow depletion region. This explains the portion *ob* in the reverse direction of the *I–V* characteristics.

Let us now see what happens when the junction becomes forward biased. Under forward biasing condition (Fig. 3·44 c), the Fermi level on the *n*-side goes up compared to the *p*-side. The electrons in the conduction band of *n*-side (position y_1) now sees empty allowed energy states on the *p*-side (position x_1) on the other side of the thin depletion layer. The electrons can now tunnel from the *n*-side (position y_1) to the *p*-side (position x_1). The magnitude of this tunnel current is much larger compared to the normal current across the junction. When the forward bias magnitude is increased, the tunnel current is also increased because more available empty states are exposed to the electrons. This behaviour is depicted by the section *op* in the *I–V* characteristics(Fig 3·43 a).

The tunnelling current is maximum when the conduction band edge in the *p*-side is in line with the Fermi level on the *p*-side. If the forward voltage is increased further, the tunnel current decreases. This is so because a part of the electrons in the conduction band on the *n*-side now faces the forbidden region. The tunnelling current now drops as is evident from portion *pv* in Fig. 3·43 (a). The tunnel current stops completely when the conduction band edge on the *n*-side is exacly in line with the valence band edge on the *p*-side. The electrons on the *n*-side of the *p–n* junction now face the forbidden band region across the junction. This corresponds to point *v* on the *I*-V characteristics.

With increase in forward bias, the normal p–n junction behaviour takes place. The normal p–n junction current flows and increases exponentially with the applied voltage as depicted by portion v–a of the characteristic curves (Fig. 3·43 a).

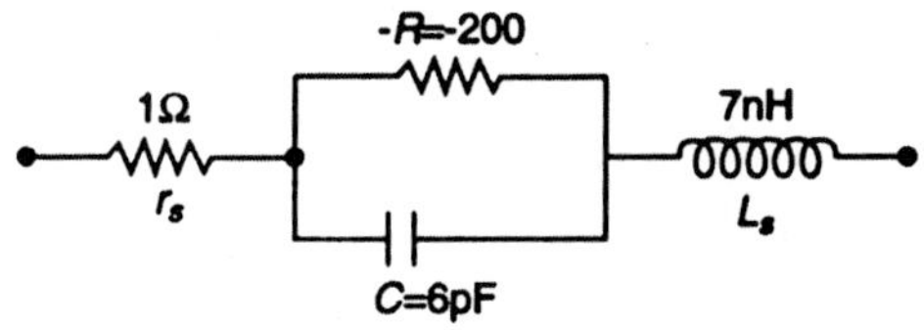

Fig 3.45. Electrical equivalent circuit of a tunnel diode.

From the V–I characteristics (Fig. 3·43 a) the differential resistance may be obtained as the reciprocal of the slope of the V–I characteristics. The negative slope of the V–I characteristics (portion pv), gives a negative differential resistance for the tunnel diode. This negative differential resistance may be employed for fabrication of amplifiers, oscillators or switches. The tunnelling process is very fast and tunnel diodes may be used at very high frequencies (10 GHz).

The typical equivalent circuit of a tunnel diode is depicted in Fig. 3·20. In the figure,

r_s = resistance of the leads, ohmic contact at the lead-semiconductor junction and semiconductor bulk material

L_s = inductance due to the terminal leads

C = Junction diffusion capacitance and

$-R$ = negative differential resistance

In a typical tunnel diode, $\frac{I_p}{I_v} = \frac{1 \cdot 0 \text{ mA}}{0 \cdot 1 \text{ mA}} = 10$

3.13.1 Tunnel diode Parameters

$$\text{The negative resistance} = -\frac{dV}{dI}$$

Its value is dependent on the semiconductor material. It has a typical value between –5 Ω to –200 Ω.

The typical values of I_p/I_v and $-R_N$ for tunnel diodes of various semiconductor materials is listed in Table 3·2

Table 3.2 Values of $\frac{I_p}{I_v}$ and $-R_N$ for tunnel diodes

	$\frac{I_p}{I_v}$	$-R_N$
Silicon	3	$-200/I_p$
Germanium	6	$-120/I_p$
Ga As	20	$-200/I_p$
Ga Sb	12	$-60/I_p$

Tunnel Diode Biasing

The tunnel diode is normally biased so that the Q point is in the negative resistance region (p v) of its I–V characteristics for a negative resistance oscillator.

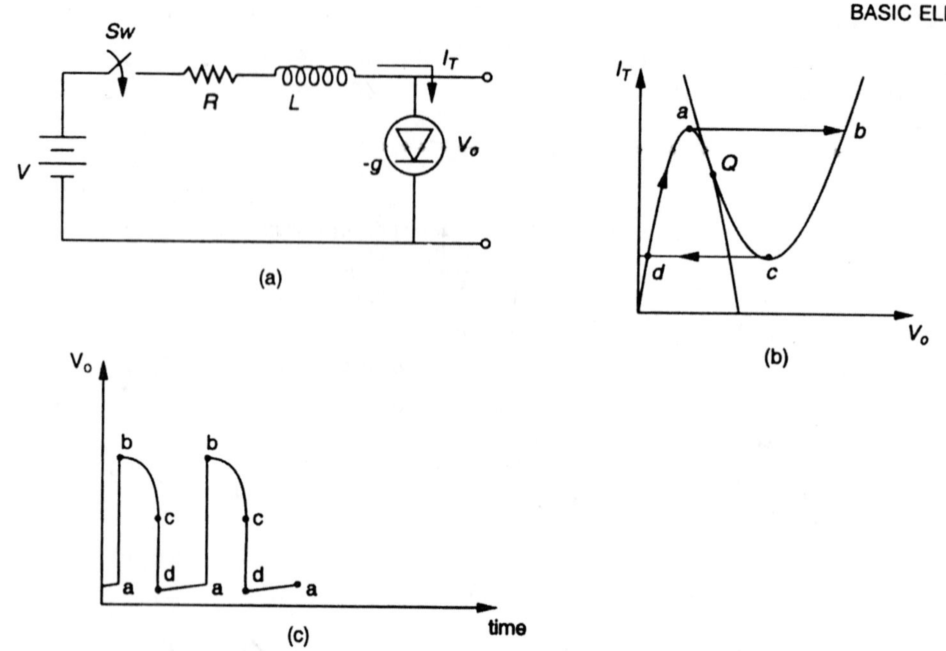

Fig 3.46. Tunnel diode negative resistance oscillator (a) circuit operation (b) operation of the circuit (c) Output wave form

In can be explained very easily that the Q point will be unstable and the point of operation will shift from $a \to b \to c \to d \to a$ ... etc. so as to satisfy Kirchhoff's voltage law and other basic laws. The above waveform is used in computer logic circuitry. A practical sinusoidal oscillator can be fabricated using a circuit as depicted in Fig. 3·47.

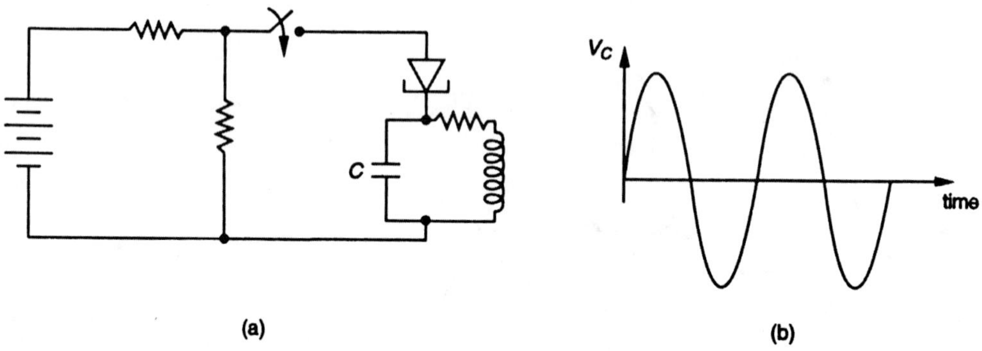

Fig 3.47. (a) A sinusoidal oscillator (b) the generated wave form

In the circuit, the tunnel diode negative resistance compensates the resistive characteristics of the diode so as to provide undamped response characteristics.

Tunnel diode applications include

(1) oscillators (2) switching networks (3) pulse generators and (4) amplifiers.

3.14 PIN DIODES

A PIN diode consists of an intrinsic (*I*) layer sandwitched between a *p*-and an *n*-layer as depicted in Fig. 3·48.

The intrinsic (or undoped) layer has a very high resistance. The PIN diode possesses the following advantages.

1. The capacitance C_{pn} has a lower value because of the intrinsic layer seperating the *p* and *n*-regions. The diode has a fast response. Typically the PIN diodes are used at high frequencies (above 250 MHz).

When a PIN diode is forward biased, it provides a resistance of magnitude $r_{ac} \approx 50/I_{mA}$ when I_{mA} is the current expressed in mA. When reverse biased, the PIN diode offers an infinite resistance like an 'open' circuit.

PIN diodes can be employed for high frequency switching purposes. Its use as a high frequency switch is depicted in Fig. 3·49.

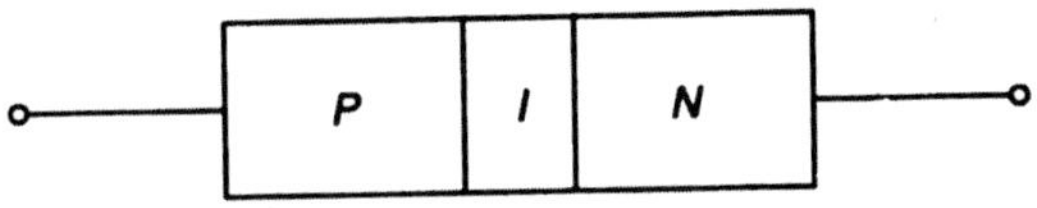

Fig. 3.48 PIN diode

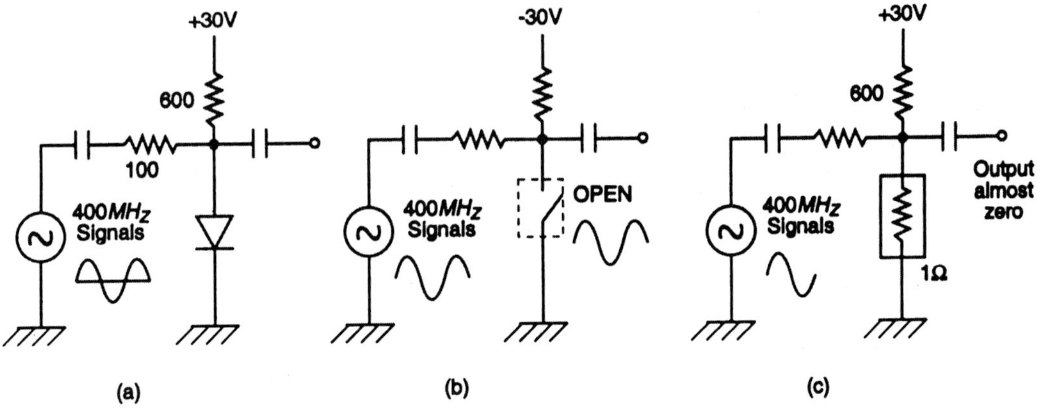

Fig 3.49. Illustrating the use of a PIN diode as a switch in a high frequency (400 MHz) ckt. (a) Circuit (b) Diode reverse biased (c) diode forward biased

When the diode is connected to the negative voltage –30 V, the diode resistance is very high (infinite) and the entire a.c. signals (400 MHz) appear at the output (Fig 3·49 b). When the diode is forward biased by connecting to the + 30 V supply, the biasing current is $I = \frac{30}{600} = 50$ mA and the PIN diode resistance is of the order of $\frac{50}{50 \text{ mA}} = 1\ \Omega$.

Thus the diode appears as a short and the entire 400 MHz signal appears across the 100 Ω resistor. The switching of the diode biasing supply from –30 V to + 30V can be accomplished with a transistor. Thus a very high frequency signal (in the MHz range) can be switched ON and OFF with the speed of a transistor switching circuit.

The PIN diode can also be employed as an *AM* modulator as will be clear from the circuit of Fig. 3·50.

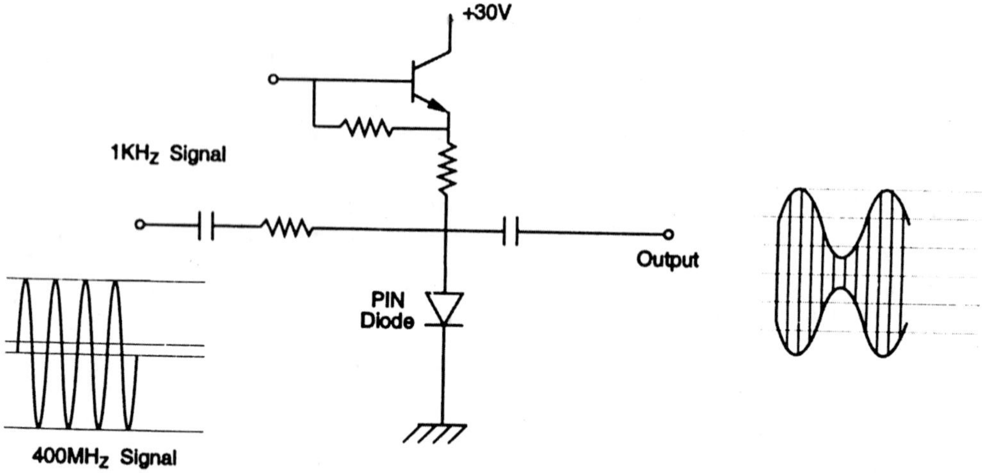

Fig 3.50. An *AM* modulator using PIN diode.

A 400 MHz signal is being modulated by a 1 kHz signal. The 1 kHz signal is applied to the base of a transistor where the d.c. output current is changed at the same rate.

3.15 SCHOTTKY DIODE

The Schottky barrier diode or hot carrier diode possesses rectification properties. However it is constituted as a **metal-semiconductor junction.** It can be used in clipping and clamping circuits, mixing, computer gating and detection circuits. It is widely used in communication receivers and radar units.

In the Schottky diode, the metal (such as gold, silver, platinum or tungsten) form a junction with an *n*-type layer. Since electrons are the carriers in the Schottky diode (no holes in the metallic layer), there is no presence of depletion layer or stored charges.

When no external voltage is applied, the electrons on the *n*-side which possess lower energy levels than the electrons in the metal cannot overcome the barrier (called Schottky barrier) and thus can not cross the junction.

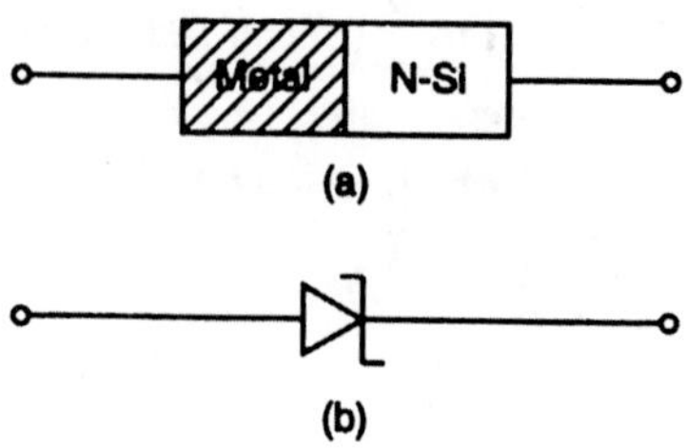

Fig. 3.51. (a) A Schottky diode (b) Symbol

However, when the diode is forward biased with the positive of the battery connected to the metal and the negative of the battery connected to *n–Si*, the free electrons on the *n–Si* side overcome the barrier and come to the metal. Since the electrons enter the metal with very large energy, they are also called as **hot carriers.** The diode is also termed as the **hot carrier diode.**

One of the important features of the Schottky diode is that it can switch very fast since there is no depletion layer or stored charges. Schottky diodes can be employed to rectify signals which can operate at frequencies upto about 300 MHz. The present maximum current rating of the device is 300 Amps. **Schottky diodes posseses a very low *ON* state voltage and very smal recovery time, typically nanoseconds.** The leakage current increases with the voltage rating and their ratings are limited to 100 V, 300 A. A typical 40 V/60 A Schottky diode has a switching time of 0·23 μ secs and an *ON* state resistance of 0·01 ohm. **The Schottky diode has also a very low noise figure which is important in communication receivers and radar systems.**

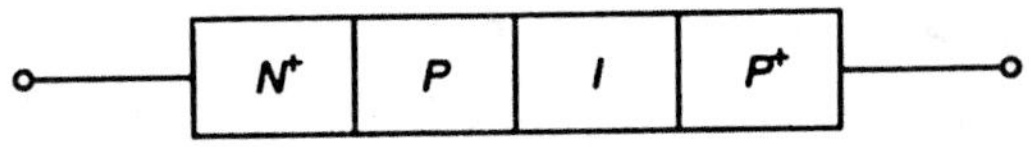

Fig 3.52. Structure of a IMPATT diode (Read diode)

IMPATT Diode : The term IMPATT stands for **Impact Avalanche and Transit time** diode. The device possesses a negative conductance such that oscillations can be obtained at microwave frequencies. A simplified version of the IMPATT diode (called the Read Diode) consists of $N^+ PIP^+$ device. The N^+ and P^+ regions are heavily doped and the I region is intrinsic (undoped). The diode can be biased such that avalanche multiplication begins at the N^+P junction and the avalanche generated holes drift towards the right into the I region. The device can be suitably operated and placed in a resonant cavity timed to a proper frequency. In this way microwave oscillations are generated. Impatt diodes are used over a frequency range of 10 to 100 GHz. **Impatt diodes are fabricated using silicon and GaAs**. They are the most powerful solid state sources for microwave power.

3.16 GUNN DIODE

The Gunn diode also possesses a negative conductance. A typical structure of a GUNN diode is shown in Fig. 3·53.

In a GaAs Gunn diode, current pulses are created by a mechanism which transfers electrons from one region of the conduction band into another. In GaAs, the forbidden gap is 1·43 eV *i.e.* the conduction electrons

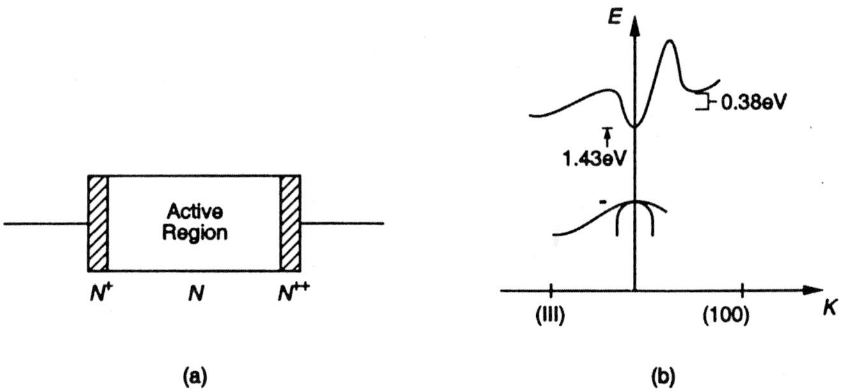

Fig 3.53. (a) Structure of a GUNN Diode (b) Energy Band diagram for GaAs

reside in a band which is 1·43 eV above the valence band. Besides there is a satellite conduction band valley which lies 0·3 eV above the minimum conduction band. Although the higher band is not usually of interest, it will be seen that electrons can be raised to the satellite valley by applying a large electric field. **The electrons in the higher band has a much smaller mobility than the lower conduction band.** Thus electrons when transferred to the low-mobility satellite band, they actually slow down. At a critical threshold field E_{Th}, **the electrons start to be transferred to the upper band where the velocity is much smaller. This transfer leads to a negative differential conductance which may be employed for microwave generation. The device thus produces a voltage controlled differential negative resistance (DNR).** The threshold field E_{th} is about 3·2 kV/cm in GaAs and 10·5 kV/cm in In P.

As Oscillators, Gunn diodes operate from 5 GHz at 120 mW output to 40 GHz at 1 mW output. Efficiencies of 3 to 5 percent are possible at present. Gunn diodes generate microwave oscillations with less noise than Impatt diodes. Gunn diodes can be used as local oscillators at frequencies between 1 to 100 GHz. Besides they are used in radars and microwave test experimental setup.

3.17 LEDs OR LIGHT EMITTING DIODES

The radiation emitted by a light emitting diode takes place due to **recombination of charge carriers in the junction region of a forward biased *p–n* junction.**

The *LED* lamp comprises an encapsulated *LED* chip in a plastic package with a suitable lens. ***LED* lamps are small, light weight and mechanically rugged. Their operating temperatures are low and they can be easily modulated. They can also be interfaced directly with *TTL* digital logic famillities due to compatible operating voltage and current levels. They have a long lifetime (25000 hours).**

The *LEDs* emit in a narrow wavelength band. The emitted wavelength depends on the semiconductor material used.

The direct band gap material GaAs is used for fabrication of *LEDs* in the infrared emitting region (*IRED*). The material Ga $As_y P_{1-y}$, a ternary semiconductor, is utilised for fabrication of visible *LEDs*. In the mixed semiconductor materials, the energy gap of GaAsP can be increased from E_g = 1·44 eV for GaAs (y = 1) to E_g = 2·2 eV for Ga P (y = 1) by changing the phosphorous concentration. As a result, the wavelength of peak emission is also shifted from near infrared to the green area of the visible spectrum. The semiconducting compound Ga $As_{0.6} P_{0.4}$ which is a direct band gap material is used for fabrication of red *LEDs*. Some common *LED* materials are listed in Table 3·3 below.

Table 3·3 Common LED materials

Material	Wavelength of Peak emission
Ga As	9000 Å (IRED)
$Al_x Ga_{1-x} As$	8200 to 8500 Å (IRED)
$In_x Ga_{1-x} As_y P_{1-y}$	1·2 μm (IRED)
Ga P (Zn, oxygen doped)	7000 Å RED
Ga P (Nitrogen doped)	5650 Å Green/Yellow
Ga $As_y P_{1-y}$	5900–6650 Å (Yellow) (Red)

The application of *LEDs* are innumerable. They are used as (1) indicator lamps (2) Seven segment or alphanumeric displays (3) Telephones (4) Data Systems (5) Opto couplers (6) Infrared range

metering (7) fiber optic communication (8) Infrared remote control (9) Consumer application (10) Optoelectronic integrated circuits (OEIC) (11) pocket calculators (12) picture phone (13) solid state visual displays.

Theory of Operation :

The light emitting diodes work on the principle of luminescence under current injection. When the semiconductor p–n junction is forward biased, the negative charge carriers are injected into the active region where they recombine. If the recombination is radiative, a photon is emitted. The recombination may also be nonradiative, without an accompanying proton. **The radiative fraction of the recombination leads to light emission (See Fig. 3·54).**

A rough idea as to the emission wave length may be obtained from energy

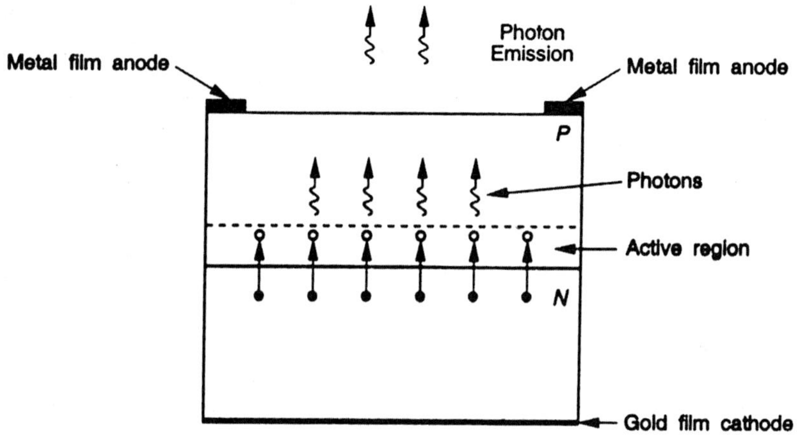

Fig 3.54. Illustrating light emission from an LED

$$E = hf = \frac{hc}{\lambda} \qquad \text{...(3·99)}$$

Where h is the Planck's constant, f is the frequency, c, the speed of light in vaccum (3×10^8 m/sec) and λ the wavelength. The value of h is $6{\cdot}625 \times 10^{-34}$ Joule-sec.

Thus $$\lambda = \frac{hc}{E} = \frac{6{\cdot}625 \times 10^{-34} \times 3 \times 10^8}{E}$$

$$= \frac{19{\cdot}88 \times 10^{-26}}{E} \qquad \text{...(3·100)}$$

where E is expressed in Joules.

If E is in electron volts, then λ can be calculated as

$$\lambda = \frac{19{\cdot}88 \times 10^{-26}}{E \times 1{\cdot}6 \times 10^{-19}} = 12{\cdot}42 \times 10^{-7}/E_{metres} \qquad \text{...(3·101)}$$

For Ga As, $E_g = 1{\cdot}43$ eV and thus

$$\lambda = \frac{1{\cdot}242}{1{\cdot}43} = 8685 \text{ Å}$$

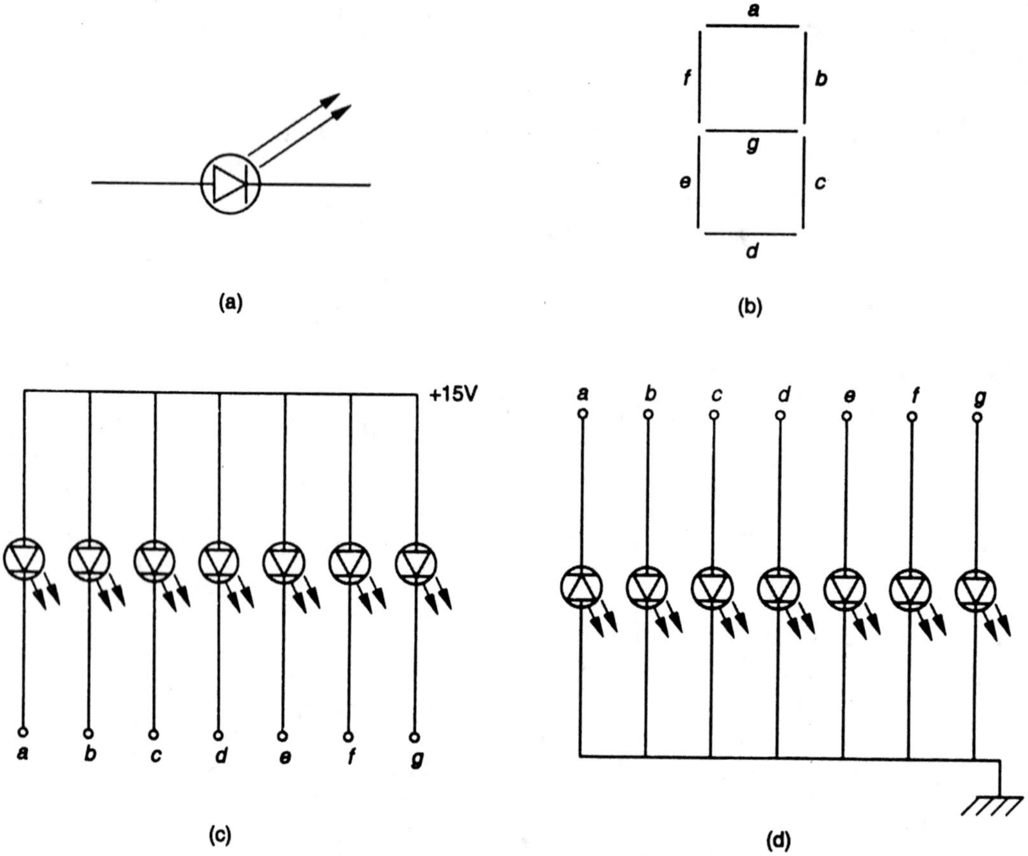

Fig 3.55. (a) Circuit symbol of *LED* (b) Seven segment arrangement (c) Common anode connection in seven segment display (d) Common cathode connection in seven segment display

Fig. 3·55 (a) shows the symbol of an *LED*. The typical arrangement of a seven segment *LED* is depicted in Fig. 3·55 (b). A desired numeral from 0 then 9 can be displayed by passing current through appropriate segments. Since the *LED* is very small, to enlarge the lighted surface, solid plastic light pipes are often used. The forward voltage of an *LED* is typically 1·2 V and a current of 20 mA is typical.

In the common anode connection, all the anodes are connected together to + 5 Volts and the ground terminal is used to light the individual segments.

3.18 LIQUID CRYSTAL DISPLAYS

In an ordinary liquid, the molecules are randomly oriented. In a liquid crystal the molecules are oriented in a definite crystalline pattern. The liquid crystal material is an organic compound like cholesteryl nonanoate or *p*-azoxyanisole. These compounds show the optical properties of a solid while maintaining the fluidity of a fluid.

If an electric field is applied to the liquid crystal, the molecules intend to align themselves perpendicular to the field. The charge carriers that flow through the liquid disturbs the arrangement and cause a turbulence

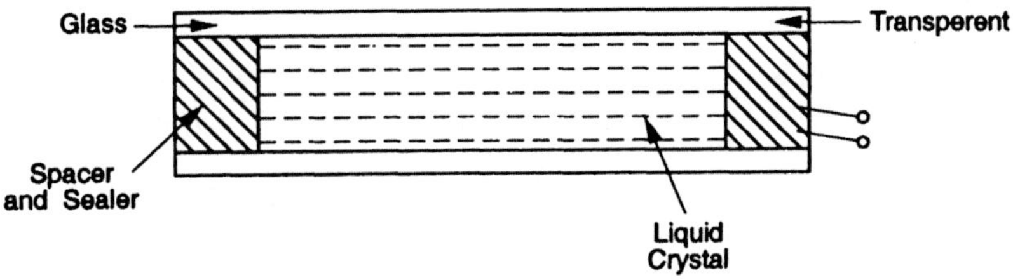

Fig 3.56. Structure of a liquid crystal cell

in the liquid. When the field is not applied, the liquid crystal is transparent. When excited, the molecular turbulence causes the light to be scattered in all directions such that the activated areas appear to be light. This scattering principle is known as dyanamic scattering.

The liquid crystal cell comprises a thin layer of the organic compound (liquid crystal material) placed between glass sheets having a transparent metal film electrode deposited on the inside faces. When the glass sheets on either sides are transparent, the *LCD* is called transmittive type. Alternatively, when one of the glass sheets have a transparent coating and the other a reflective coating, the *LCD* is called reflective type.

Typically, in a dynamic scattering *LCD*, a 30 V peak-to-peak square wave with frequency of 60 Hz is used as the voltage source. The field effect type cell employs 8 Volts peak to peak. A square wave is applied to the back plane which is common to all the *LCD* cells. The same square wave either in phase or anti-phase is applied to all other terminals. If the square wave applied to the terminal and the back plane are in phase, no voltage is developed across that cell. On the other hand, the cell terminals which have been supplied with the square wave in antiphase to that of the back terminal, have an a.c. voltage applied across them. These cells are energized and appear bright.

***LED* displays are small in size.** However ***LCD* displays can be fabricated almost to any convenient size. The power consumption in *LCD* is very small, typically 20 μW for a segment. Thus for a seven segment *LCD* display it is only 140 μW if all the segments are excited at the same time. The *LCD* does not emit any light as *LEDs* but scatter or reflect the ambient light.** The comparison between *LEDs* and *LCDs* is given in Table 3·4.

Table 3·4 : Comparison between LEDs and LCDs

LEDs	LCDs
1. *LEDs* emit light	1. *LCDs* only reflect or scatter the ambient light
2. *LEDs* depend on Forward biased *p–n* junction diodes	2. *LCDs* operate when an electrical field is applied to *LCD* cells of certain organic compounds (liquid cells)
3. Typical semiconductor materials for *LEDs* are Ga As, Ga P, Ga As P	3. Typical liquid crystals (organic compound) are cholesteryl nonanoate and *p*-azoxyanisole
4. The supply voltage for *LEDs* 1·5 V d.c.	4. *LCDs* are supplied with 30 V peak-to-peak square wave with frequency of 60 Hz
5. Power consumption is high (typically 400 mW for a seven segment *LED* display	5. Power consmption is very small (typically 140 μW per seven segment display when all the segments are excited
6. *LEDs* are small	6. *LCDs* can be fabricated to almost any convenient size.

3.19 JUNCTION PHOTODIODES

A very common photodetector is the junction photodiode. The basic principle of operation of a junction photodiode can be explained with the help of Fig. 3·57. The semiconductor *p–n* jnction photodiode works under **reverse biasing** conditions.

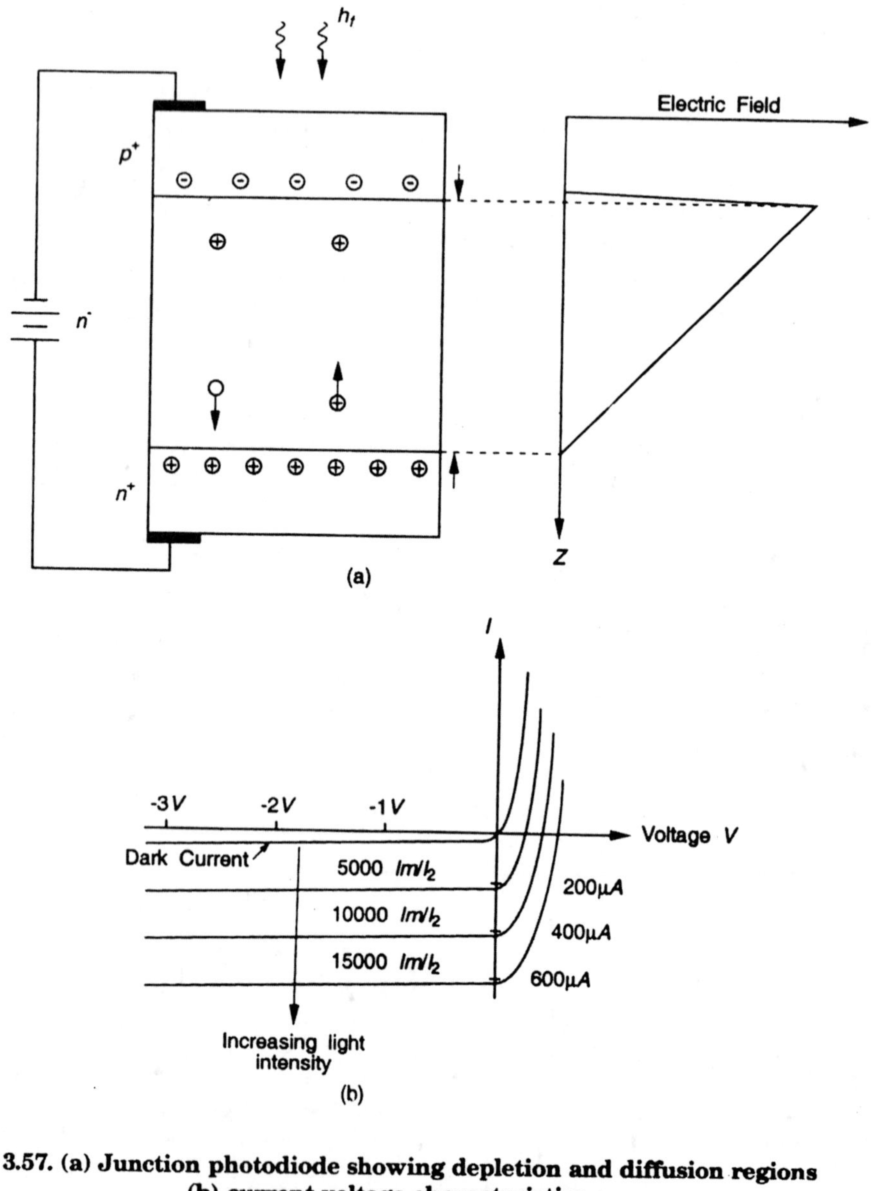

Fig 3.57. (a) Junction photodiode showing depletion and diffusion regions (b) current-voltage characteristics

The depletion region is this very wide and a very small current, the reverse saturation current flows through the device. When light is incident on the upper surface, photons are absorbed in the depletion region and also in the bulk regions, particularly at the top. Electron-hole pairs will also be generated due to excitation by light. The light generated carriers (electrons and holes) will move in opposite directions due to the reverse bias and a current will flow. The magnitude of the current depends on the **quantum efficiency**. If 100 photons produce 90 electron-hole pairs, the quantum efficiency is 90%. The electrons and holes which are generated within are diffusion length of the depletion layer, on both sides of it also move to the depletion region and are then seperated by the reverse bias, producing the photocurrent. A wide depletion layer is essential so that more photons can be absorbed. The current voltage characteristics of the photodiode is depicted in Fig 3·57 (b). It will be found that with increasing light intensity, the current increases under reverse biasing conditions. The reverse resistance can be altered by a factor of 20–25. It will be noted that the photodiode can turn ON and OFF in a few nanoseconds. It has a fast response. Its applications include detection, demodulation, switching, character recognition, optical communication etc.

3·20 PHOTOCONDUCTORS (OR PHOTOCONDUCTIVE CELLS)

When light is incident on the surface of a material, electrons may be broken away from their parent atoms. These electrons can take part in conduction and the resistance is reduced. The circuit symbols and structure of a suitable photoconductive cell is depicted in Fig 3·58. Typical light sensitive materials are **cadmium sulphide** (*CdS*) **and cadmium selenide** (*CdSe*). These material are arranged in the form of a long strip on a disc shaped base with protective sides. The two ends of the strip are connected to terminals at the base. A glass cover is normally given for added protection.

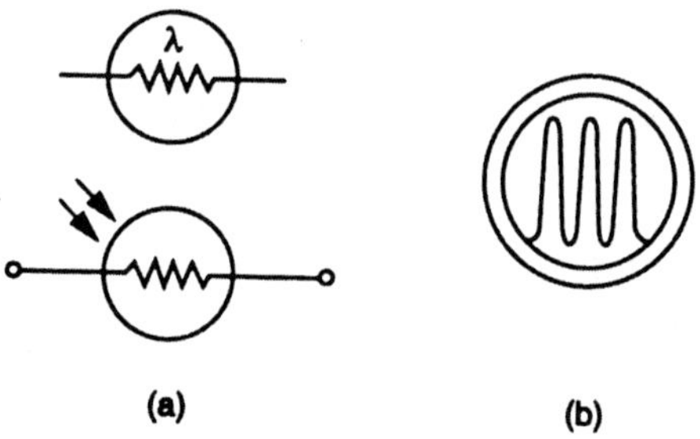

Fig 3.58. (a) Symbol of a photoconductor (b) Structure of a photoconductor

Fig. 3·59 shows the variation of resistance with light intensity. The resistance is highest when there is no illumination. With increasing illumination the resistance decreases as shown in Fig 3·59. The material cadmium selenide has response time of the order of 100 m sec while cadmium sulphide has a response time of 100 millisecs.

The application of a photoconductive cell in a relay control circuit is depicted in Fig. 3·60. When no light falls on the cell (dark) the cell resistance is high and the current flowing is too low to energize the relay. When the cell is illuminated, the cell resistance falls, the current increases and the relay is energized.

EXAMPLE 3.11 *In Fig. 3·34, the relay is energized when a current of* 10 *m A flows through it. The cell is illuminated with light of intensity* 400 lm/m^2. *Find the required series resistance R and the magnitude of dark current*

[Given cell resistance = 100 kΩ *when dark and cell resistance* = 1200 Ω *when illumination is* 500 lm/m^2*]*

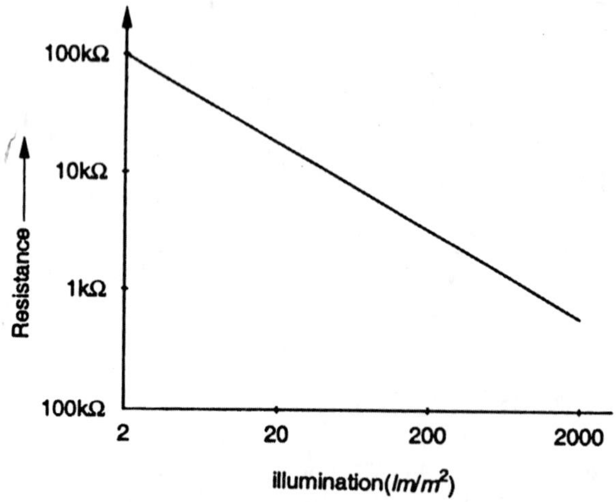

Fig 3.59. Resistance variation with illumination for a photoconductive cell.

SOLUTION

$$\text{Current } I = \frac{30\ V}{R + R_C}$$

where R_C = Photoconductor cell resistance

$$\therefore \quad R = \frac{30}{I} - R_C.$$

$$= \frac{30}{10\text{ mA}} - 1200 = 3\text{ k}\Omega - 1200\ \Omega$$

$$= 1800\ \Omega$$

Now cell dark resistance = 100 kΩ

$$\text{Dark current} = \frac{30\text{ V}}{1800 + 100\text{ k}} = 2{\cdot}95 \times 10^{-4}\text{Amp}$$

$$\approx 0{\cdot}295\text{ mA}$$

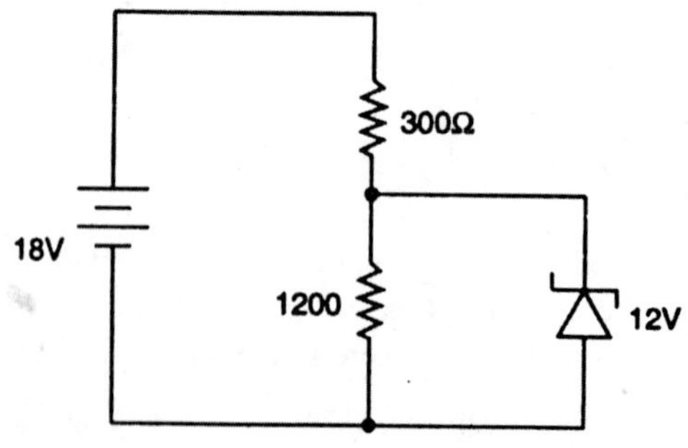

Fig 3.60. Relay control circuit using photoconductive cell.

Review Questions

1 What is meant by Avalanche breakdown ? Distinguish between the Avalanche and Zener breakdown mechanisms.

2 Draw the characteristics of a Zener diode.

3 Mention some applications of Zener diodes.

4 Draw a voltage regulator circuit using a Zener diode and explain its principles of operation.

5 Discuss how a square wave generator can be built using two Zener diodes in series.

6 What do you understand by the term "varactor diode" ?

7 Draw the current voltage characteristics of a tunnel diode.

8 Mention some application of a varactor diode.

9 Draw the electrical equivalent circuit of a tunnel diode.

10 What is a *p–i–n* diode ?

11 What is a Schottky diode ? Mention its important characteristics.

12 What is an IMPATT diode ?

13 What is a GUNN diode ? Mention its important characteristics.

14 What is an *LED* ? Mention three semiconductor materials which can be used for fabricating *LEDs*.

15 Mention some appications of *LEDs*.

16 What is an *LCD* ? Discuss how it operates.

17 What is a photodiode ? Mention some of its uses.

18 What is a photoconductor ?

Objective Type Questions

Select the correct answer or most suitable alternative(s)

1 Zener diodes are normally fabricated using
(a) GaAs (b) Ge
(c) Si (d) CdS

2 A Zener diode, while in use is always
(a) forward biased (b) series connected
(c) reverse biased (d) connected in parallel

3 The temperature co-efficient of the avalanche breakdown voltage is
(a) positive (b) negative
(c) zero (d) can be either positive or negative

4 In a varactor diode, with increasing reverse bias voltage, the junction capacitance
(a) increases (b) decreases
(c) remains constant (d) is zero

5 The primary reason why electrons can tunnel through a *p–n* junction is that
(a) the barrier potential height is small (b) they possess large energy
(c) impurity level is low (d) the depletion layer is extremely thin

6 Which of the following devices do not possess negative resistance properties ?
(a) Tunnel diode (b) Zener diode
(c) Impatt diode (d) Gunn diode

7 Which of the following devices can not be fabricated using silicon ?
(a) Zener diode (b) Varactor diode
(c) Gunn diode (d) light emitting diode

8 A Schotty diode depends for its operation on
(a) majority carriers only (b) minority carriers only
(c) both majority and minority carriers (d) voltage variable capacitance

9 A varactor diode
(a) has a voltage variable capacitance (b) has always a uniform doping profile
(c) can be utilized for microwave oscillation generation
(d) has extremely thin depletion layer

10 The devices which may be employed for an oscillator in the frequency range of 10 to 100 GHz is
(a) Schottky diode (b) Impatt diode
(c) Zener diode (d) photodiode

11 A diode which uses a metal semiconductor junction is called
(a) tunnel diode (b) Gunn diode
(c) varactor diode (d) Schottky diode

12 Zener diodes are typically used as
(a) amplifiers (b) oscillators
(c) rectifiers (d) voltage regulators

13 A liquid crystal display (*LCD*) can be fabricated employing
(a) Silicon (b) Ga As
(c) Chloresterol nonanoate (d) In Ga As P

14 Light emission takes place in the following devices
(a) *LCDs* (b) Read diode
(c) Photodiodes (d) *LEDs*

15 In which of the following devices, the resistance changes when an incident light falls on it ?
(a) Photodiodes (b) Photoconductors
(c) *LCDs* (d) PIN diode

16 The material CdS can be used for fabrication of
(a) *LEDs* (b) *LCDs*
(c) Photoconductors (d) Gunn diodes

Exercise Problems

1 In a typical $\frac{1}{4}$ W Zener diode, $V_Z = 9{\cdot}1$ V at 20°C. Find the value of the nominal V_Z at $T = 100$°C if T_C, the temperature co-efficient of resistance is ± 0·075%/°C.

2 Determine whether the Zener diode in the circuit of Fig. *E* 3·2 is correctly biased. Find I_Z and the power dissipated by the diode.

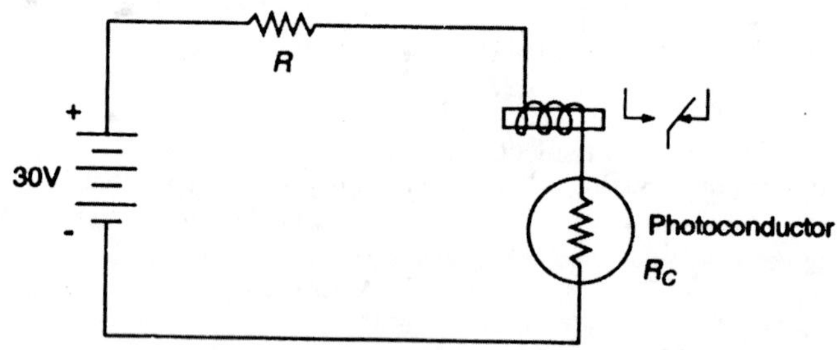

Fig EP.3.2. Reference Exercise Problem 2

3 Repeat problem 2 above if the Zener diode is operated at a temperature of 100°C while the nominal Zener voltage is 12 V at 20°C. Assume the temperature co-efficient of breakdown voltage as ± 0·075%/°C.

4 Find the value of the output voltage V_{out} in the circuit of Fig. E 3·4. Assume that the Zener diode operates at a temperature if 60°C. The specification for the Zener diode is V_Z = 12 V at 20°C and T_C = 0·075%/°C.

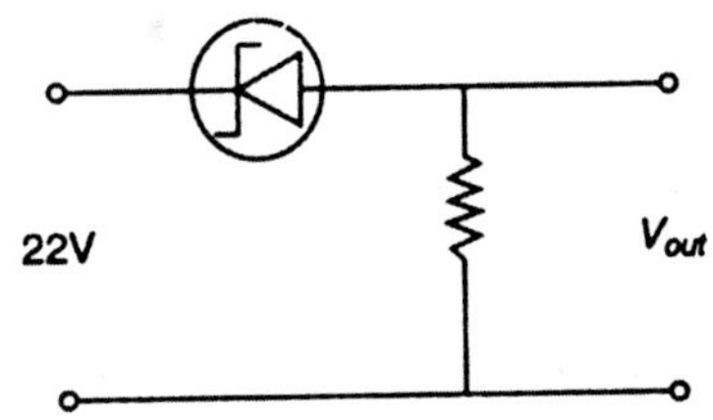

Fig EP.3.4. Reference Exercise Problem 4

5 Find the battery current I, the Zener current I_Z and the load current I_{load} in the circuit of Fig. E 3·5. What will be the effect on then above values if V_{in} rises to 75 volts. Consider the Zener resistance to be negligible.

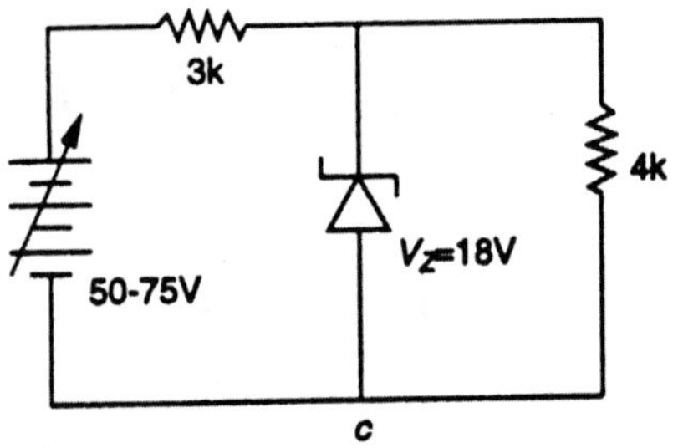

Fig EP.3.5. Reference Exercise Problem 5

6 In Fig E 3·6, find the Zener current if the load resistance R_L is
(i) 25 KΩ (ii) 10 KΩ
(iii) 2·5 KΩ

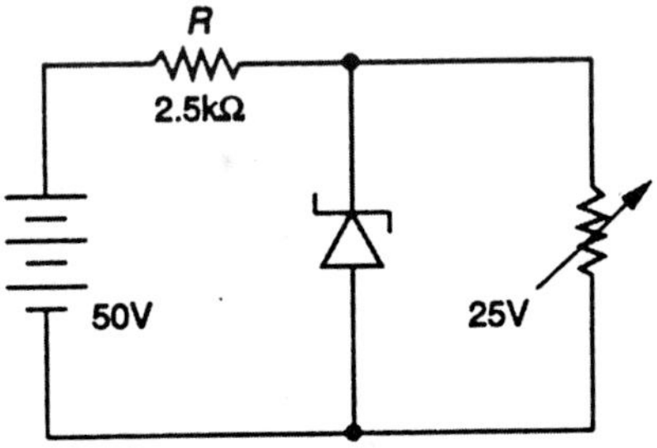

Fig EP.3.6. Reference Exercise Problem 6

Chapter 4

BIPOLAR JUNCTION TRANSISTORS

4.1 INTRODUCTION

Three consecutive *p* and *n* layers of doped semiconductors (*pnp* or *npn*) formed by diffusion or epitaxial method and seperated by two *p-n* junctions constitutes a bipolar junction transistor. By the term **"transistor", we imply a semiconductor amplifier.** Using transistors, we can build amplifiers that give voltage (or current) gain, a high input impedance or a high (or low) output impedance. Transistors are active devices with highly non linear characteristics. The use of the term bipolar arises due to the fact that both majority and minority carriers (electrons and holes) take part in the conduction process.

Two different types of bipolar transistors are *p-n-p* and *n-p-n*. Fig. 4.1 depicts a basic bipolar transistor construction. In the *p-n-p* type, two *p*-type semiconductor regions are seperated by the *n*-region, while in the *n-p-n* transistor, two *n*-type regions are seperated by a *p*-type region. The three layers of the device are called the emitter (*E*), the base (*B*) and the collector (*C*) regions. **The emitter region is heavily doped, the base region is moderately doped while the collector region has light doping.** In the conventional manufacturing of a *BJT*, the base region is kept physically quite thin.

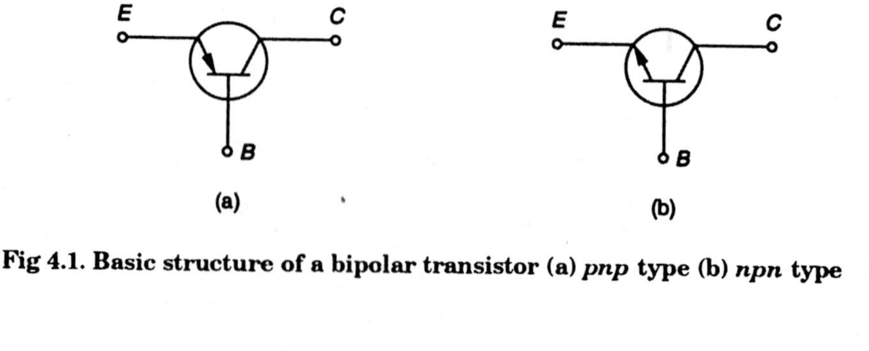

Fig 4.1. Basic structure of a bipolar transistor (a) *pnp* type (b) *npn* type

4.2 TRANSISTOR OPERATION

The transistor has two *p-n* junctions constructed in a special way in series back-to-back. The transistor junctions are the emitter-base junction (*EBJ*) and the collector-base junction (*CBJ*). Three different modes of operation of a transistor depending on the bias conditions (foward or reverse) are possible. They are

Table 4.1. Different modes of operation of the BJT

Mode	*EBJ*	*CBJ*
Active	Forward Bias	Reverse Bias
Cut off	Reverse Bias	Reverse Bias
Saturation	Forward Bias	Forward Bias
Reverse	Reverse Bias	Forward Bias

For operation of the transistor as an amplifier, the active mode is employed while for switching applications both the cut off and saturation modes are utilized.

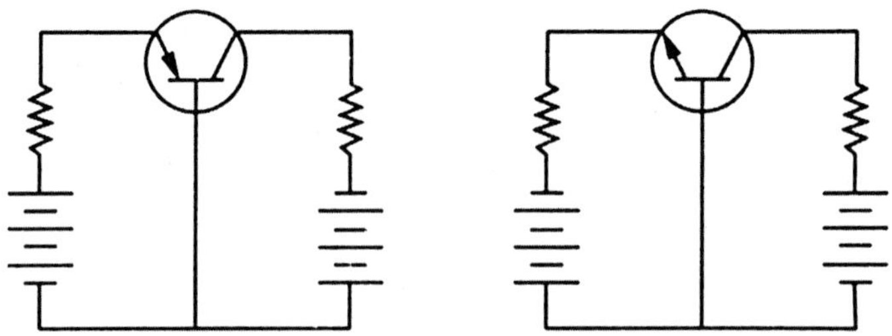

(i) pnp (ii) npn

Fig 4.2. Biasing of bipolar transistor

Let us see in Fig. 4.3 as to what takes place in an actual transistor when the device acts in the active mode. The emitter base junction is forward biased while the collector-base junction is reverse biased. The forward bias between the base and the emitter decreases the emitter-base depletion layer and the reverse bias from base to collector increases the base-collector depletion layer, as in shown in Fig. 4.3 (a). The *n*-type emitter region is full of conduction band (or free) electrons which diffuse across the forward biased base-emitter junction into the *p*-type base region. The action is similiar to a forward biased diode. In an actual transistor, the base region is **very narrow** and is lightly doped so that it has a very limited number of holes. Therefore, only a small percentage of all the electrons flowing across the base-emitter junction can recombine with the few holes available. Thus a small current flows out of the base lead as valence electrons, forming the small base current.

Many of the electrons flowing from the emitter into the thin base region do not recombine and diffuse into the base-collector (B–C) depletion layer. After the electrons enter the base-collector depletion region, they are attracted by the field of the collector base voltage. This is depicted in Fig. 4.3 (c). The electrons now move through the collector, out through the collector terminal to the positive of the V_{CC} battery. This is the collector current as shown. The collector current is dependent on the magnitude of the base current but is independent of the V_{CC} voltage.

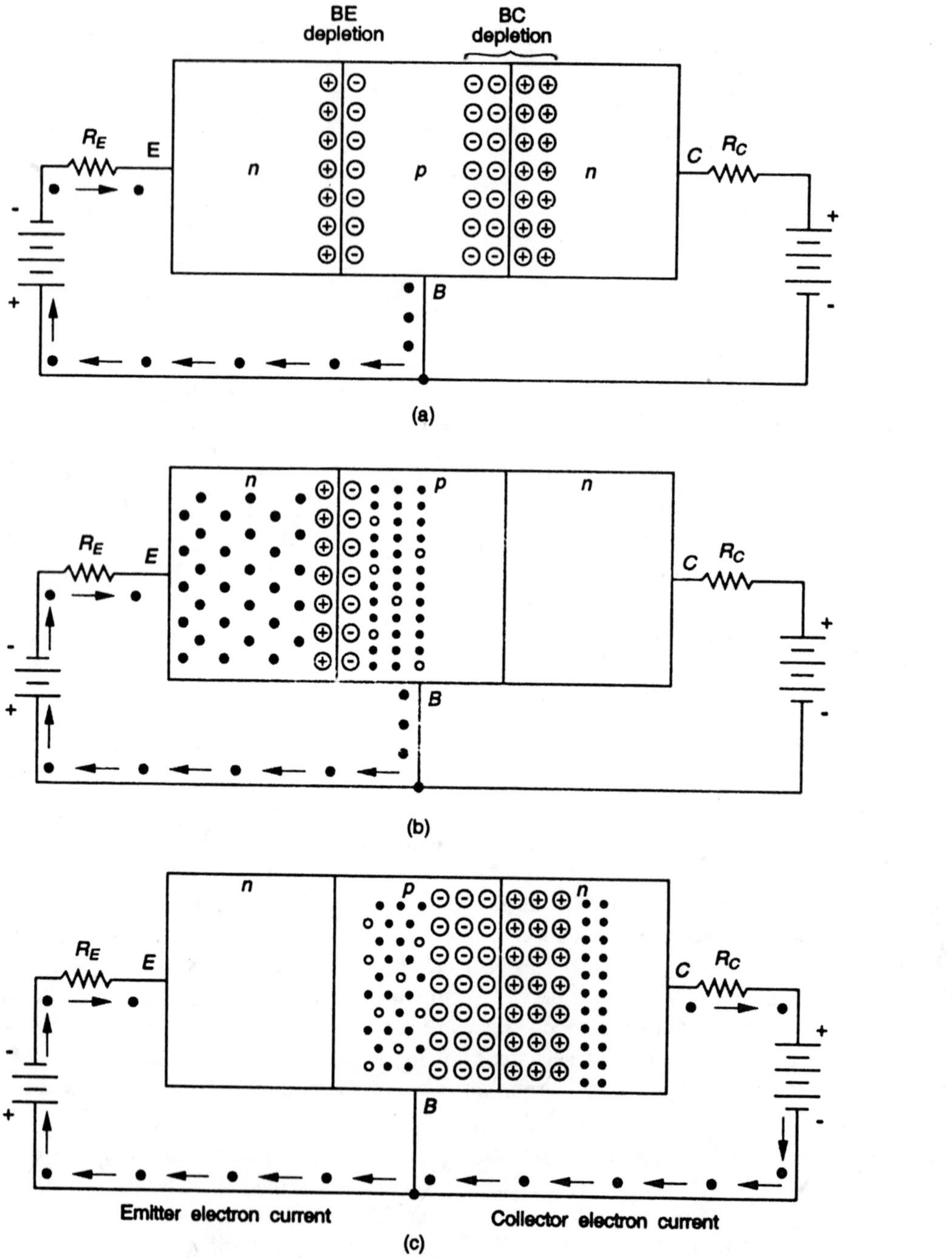

Fig 4.3. Illustration of the action in a *BJT* (*npn*). In an ideal transistor, the base region is very narrow. But in the Fig. it is shown as wide for the sake of clarity

In an actual transistor, the base is very narrow (0.0025 cm). The base is lightly doped whereas the collector is very lightly doped. The emitter is however, heavily doped. This is shown schematically in Fig. 4.4.

A transistor is a semiconductor device used for amplification purposes. The amplifying action of the transistor occurs due to its ability to transfer the signal current from a low resistance circuit to a high resistance circuit. The two terms **transfer** and **resistor** can be combined together to result in the name "**transistor**".

transfer + resistor → transistor

It will also be seen later on that in the equivalent circuit of a transistor, either voltage generator or current generator is needed to describe its action **(semiconductor amplification). The transistor can also be decribed as a semiconductor amplifier.**

The base-collector junction is reverse biased .Its resistance is high.

The base-emitter junction is forward biased .Its resistance is small.

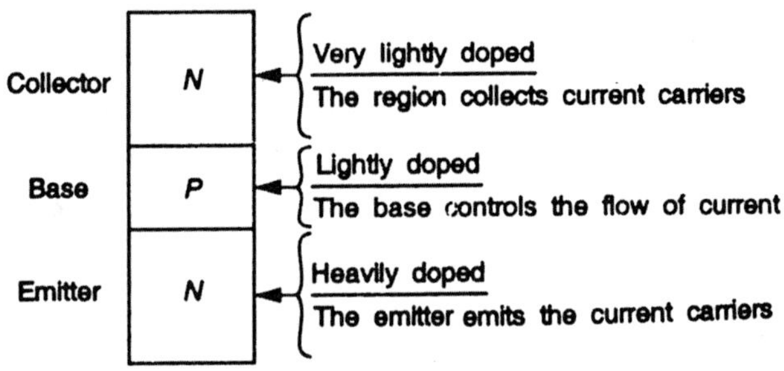

Fig 4.4. Schematic diagram for a transistor.

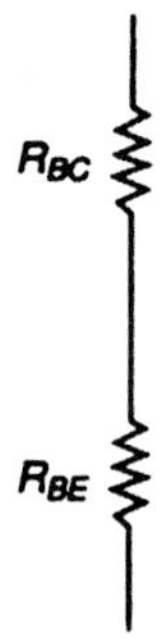

Fig 4.5. Junction resistances in a transistor

4.2.1 Characteristics Curves

The collector characteristic curves of an npn transistor is shown in Fig. 4·7 (a). This can be plotted by using the circuit of Fig. 4·6. If V_{BB} and V_{CC} are both adjustible, I_B and V_{CE} will both be changed. The value of I_B depends on V_{BB}. If V_{BB} produces a specific (non-zero) value of I_B and $V_{CC} = 0$ then $I_C = 0$ and $V_{CE} = 0$. If V_{CC} is now gradually increases. V_{CE} will increase and also I_C will increase.

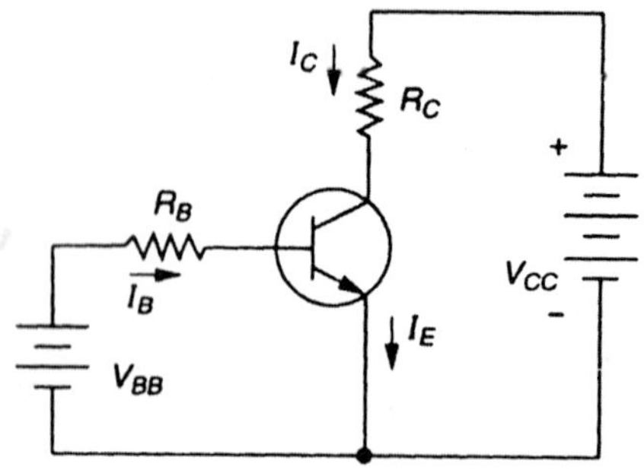

Fig. 4.6

When the value of V_{CE} becomes $0{\cdot}7V$ or higher, the base-to-collector junction becomes reverse biased and the value of the collector current becomes the full value determined by the relation $I_C = \beta_{dc} \cdot I_B$. The value of I_C at this point reaches a constant level. The value of I_C remains constant even it V_{CE} is now increased. In practice, however, I_C increases using slightly as V_{CE} increases due to the widening of the base collector depletion layer. This leads to a condition in which in the number of holes for recombination in the base region is low. For other I_B values, similiar I_C-versus V_{CE} curves can be generated. The various $I_C - V_{CE}$ curves for different I_B values form a family of collector curves for a transistor for the npn transistor [Fig. 4.7 a]. For a pnp transistor a similar set of curves can be obtained [vide Fig. 4.7 b]

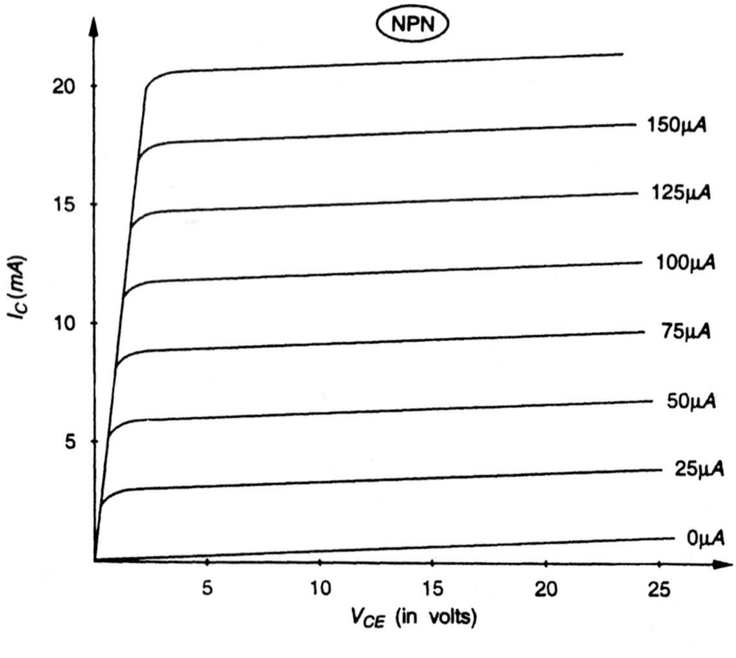

Fig. 4.7 (a)

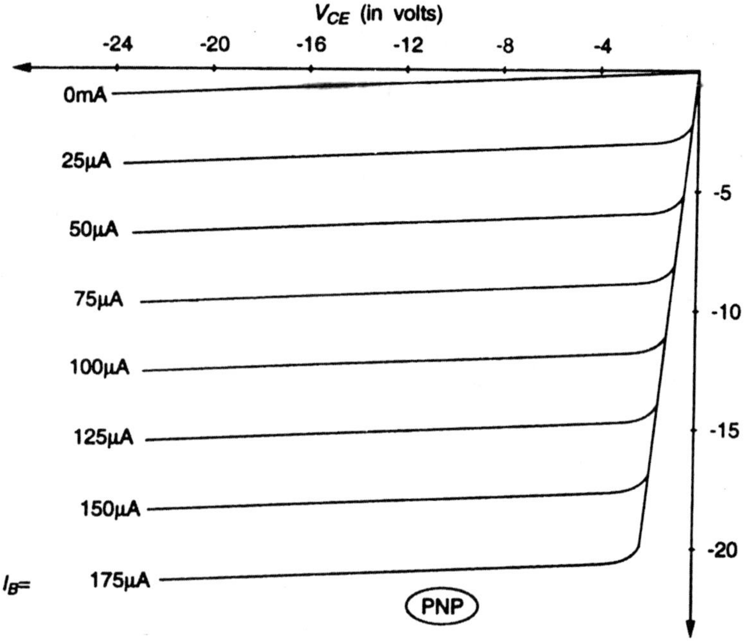

Fig. 4.7 (b) Collector Family of curves for *npn* and *pnp* transistors.

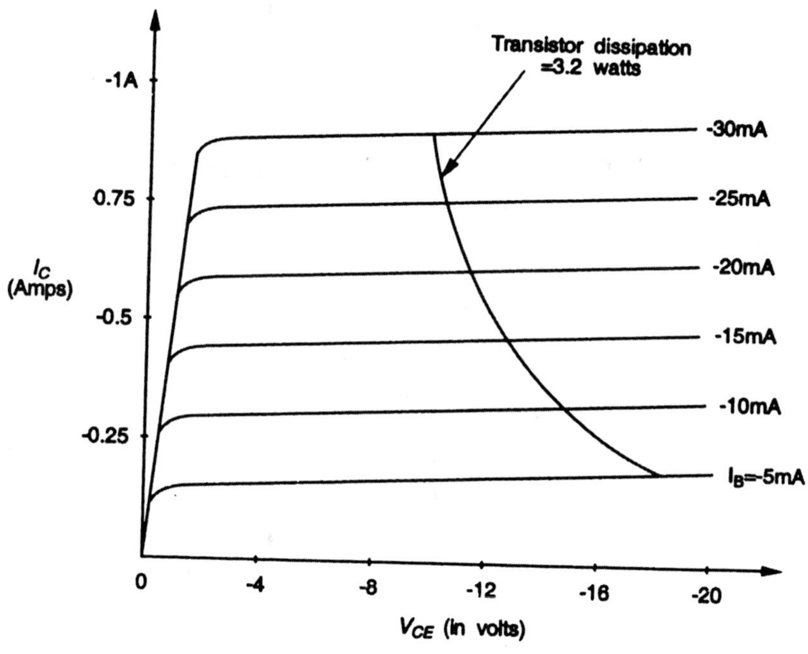

Fig. 4.8 Constant Power curve superimposed on the collector characteristics

For a transistor, the product of V_{CE} and I_C must not exceed the maximum power dissipation. V_{CE} and I_C cannot be the maximum at the same time. If I_C is maximum, V_{CE} will be smaller in value. If V_{CE} is maximum, I_C will have a lower value. For I_C maximum value

$$V_{CE} = \frac{P\ diss\ (max)}{I_C}$$

For V_{CE} maximum value

$$I_C = \frac{P\ diss\ (max)}{V_{CE}}$$

The (I_C, V_{CE}) points for $P\ diss\ (max) = I_C\ V_{CE}$ are shown by the curve *bc* [vide Fig. 4·8]. The maximum value of I_C in the limiting rating between points *a* and *b*, V_C in the maximum rating between points *c* and *d*. *P* diss (max) is maximum rating between points *b* and *c*.

The collector current *vs* collector voltage curves are shown in Fig. 4.9.

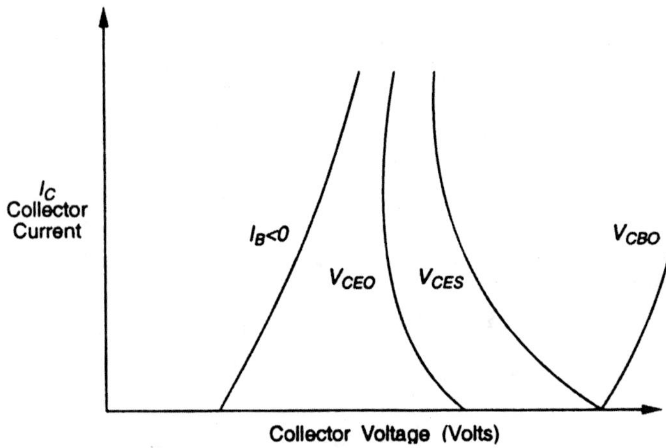

Fig 4.9. Collector current vs. collector voltage curves

The value of V_{CE} is smaller compared to V_{CEO} when the emitter base junction is forward biased and the base current I_B has the same sign as the collector current I_C. In the special case, when the base-collector junction is reverse biased, the emitter is shorted to the base. The characteristics is then called V_{CES}. In this case, the emitter base junction offers a high resistance at low currents and most of the current flows through the base lead making $V_{CB} = V_{CBO}$. As the current through the device is increased, the voltage drop across forward biases the base-emitter junction so that I_B is negligibly small and the reverse voltage approaches V_{CEO}.

The current I_E in the emitter is related to the collector current I_C by the equation

$$I_E = I_C + I_B \qquad \ldots(4.1)$$

where I_B is the base current.

The d.c. current gain β_{dc} is given by

$$\beta_{dc} = \frac{I_C}{I_B} \qquad \ldots(4.2)$$

The value of d.c. beta ranges between 20 and 200. β_{dc} is specified as h_{FE} in the transistor data sheets. The current gain α_{dc} is defined as

$$\alpha_{dc} = \frac{I_C}{I_E} \qquad \text{...(4.3)}$$

The value of α_{dc} of a transistor lies between 0.95 and 0.99.

Relationship between α_{dc} and β_{dc}

Since $I_E = I_C + I_B$, we may write

$$\frac{I_E}{I_C} = 1 + \frac{I_B}{I_C} \qquad \text{...(4.4)}$$

or,

$$\frac{1}{\alpha_{dc}} = 1 + \frac{1}{\beta_{dc}} \qquad \text{...(4.5)}$$

or,

$$\alpha_{dc} = \frac{\beta_{dc}}{1 + \beta_{dc}} \qquad \text{...(4.6)}$$

From equation (4.5), we may also find

$$\beta_{dc} = \frac{\alpha_{dc}}{1 - \alpha_{dc}} \qquad \text{...(4.7)}$$

EXAMPLE 4.1 *In an npn BJT, the base current $I_B = 40\ \mu A$ and collector current $I_C = 4.25\ mA$. Find the value of β_{dc} and α_{dc}.*

SOLUTION

$$\beta_{dc} = \frac{I_C}{I_B} = \frac{4{\cdot}25\text{ mA}}{40\ \mu\text{A}} = 106{\cdot}25$$

and

$$\alpha_{dc} = \frac{\beta_{dc}}{1 + \beta_{dc}} = \frac{106}{1 + 106} \simeq 0{\cdot}9906$$

EXAMPLE 4.2 *In a BJT, $\beta_{dc} = 175$ and $I_B = 40\ \mu A$. Find the values of I_C and α_{dc}.*

SOLUTION

$$I_C = I_B \cdot \beta_{dc} = 40\ \mu\text{A} \times 175 = 7\text{ mA}$$

and

$$\alpha_{dc} = \frac{\beta_{dc}}{1 + \beta_{dc}} = \frac{175}{1 + 175} = 0{\cdot}9943.$$

EXAMPLE 4.3 *In an npn bipolar junction transistor, $I_C = 7{\cdot}5\ mA$ and $\alpha_{dc} = 0{\cdot}9914$. Find the values of I_B and β_{dc}.*

SOLUTION

$$I_E = \frac{I_C}{\alpha_{dc}} = \frac{7{\cdot}5\text{ mA}}{0{\cdot}9914} = 7{\cdot}565\text{ mA}$$

Thus

$$I_B = I_E - I_C = (7{\cdot}565 - 7{\cdot}5)\text{ mA}$$
$$= 65\ \mu A$$

and

$$\beta_{dc} = \frac{\alpha_{dc}}{1 - \alpha_{dc}} = \frac{0{\cdot}9914}{1 - 0{\cdot}9914} = \frac{0{\cdot}9914}{0{\cdot}0086}$$
$$= 115{\cdot}38$$

DC analysis

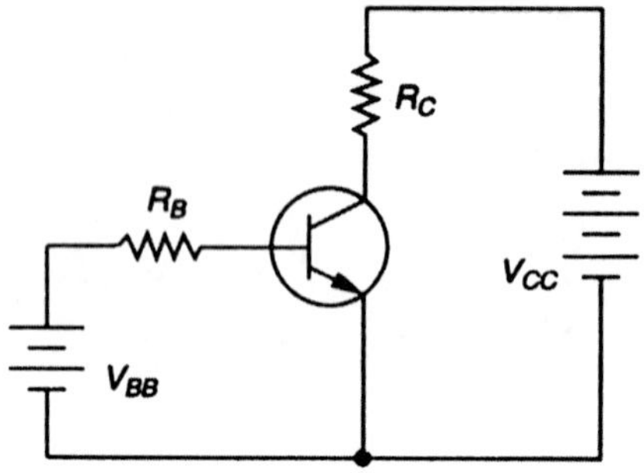

Fig 4.10. DC analysis for an *npn* transistor

Let us consider the circuit of Fig. 4.10.

In the circuit,

$$V_{BE} = 0{\cdot}7 \text{ V} \quad \text{...(4.8)}$$

The voltage drop across R_B is

$$V_{RB} = I_B \,.\, R_B \quad \text{...(4.9)}$$

where $$I_B = \frac{V_{BB} - V_{BE}}{R_B} \quad \text{...(4.10)}$$

Also $$I_E = \frac{I_C}{\alpha_{dc}} \quad \text{...(4.11)}$$

and $$I_C = I_B \,.\, \beta_{dc} \quad \text{...(4.12)}$$

The voltage drop across the resistor R_C is

$$V_{RC} = I_C \,.\, R_C \quad \text{...(4.13)}$$

Further $$V_{CE} = V_{CC} - I_C R_C \quad \text{...(4.14)}$$

The voltage drop between collector and base

$$V_{CB} = V_{CE} - V_{BE} \quad \text{...(4.15)}$$

4.3 CUTOFF AND SATURATION

If the base current is zero, the transistor remains in cut off. The condition $I_B = 0$ is equivalent to having an open base lead. The collector current with open base lead is designated as I_{CEO} where the subscript *CEO* stands for collector to emitter with base open. The amount of current I_{CEO} is very very small. It is also called as the collector leakage current. I_{CEO} is produced partly by thermally generated carriers and partly by surface leakage current.

As the value of I_B is increased, the value of I_C increases and V_{CE} decreases since the voltage drop across R_C rises. When V_{CE} reaches the saturation value, V_{CE} (sat), the base to collector junction becomes forward biased and I_C can not increase any more even with increase in I_B. The relationship $I_C = \beta_{dc} \cdot I_B$ do not hold

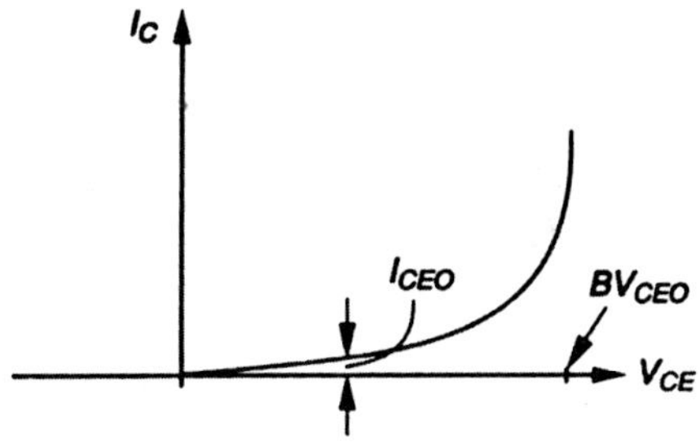

Fig 4.11. Collector current vs. collector-to-emitter voltage characteristic showing I_{CEO}

under saturation conditions. V_{Ce} (sat) for a transistor takes place below the knee of the collector curves. The value of V_{CE} (sat) is only a few tenths of a volt for silicon transistors.

EXAMPLE 4.4 *Find whether or not the transistor in the circuit of Fig. 4.12 is in saturation. Assume V_{CE} (sat) = 0·25 V.*

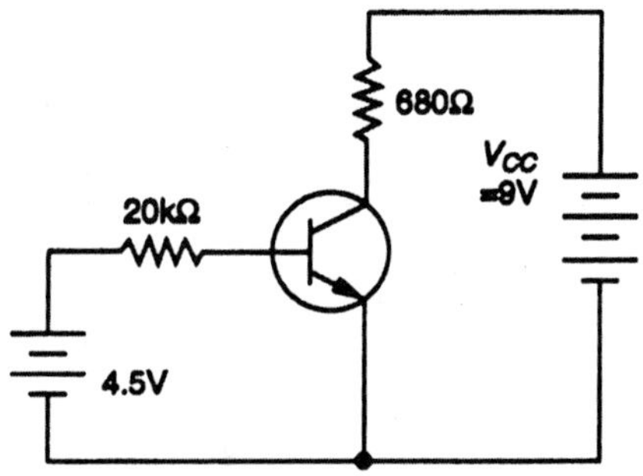

Fig 4.12. Reference Example 4.5

SOLUTION

$$I_C\,(\text{sat}) = \frac{V_{CC} - V_{CE}\,(\text{sat})}{680} = \frac{9 - 0{\cdot}25\ \text{V}}{680}$$

$$= 12{\cdot}86\ \text{mA}.$$

Now $$I_B = \frac{V_{BB} - V_{BE}}{R_B} = \frac{4{\cdot}5\ \text{V} - 0{\cdot}7\ \text{V}}{20\ \text{k}\Omega} = 0{\cdot}19\ \text{mA}.$$

$$I_B = \beta_{dc}\,.\,I_B = 100 \times 0{\cdot}1\ \text{mA} = 19\ \text{mA}.$$

Since I_C (calculated) = 19 mA is greater than I_C (sat), the transistor is in saturation. The collector current of 19 mA is never reached. If you increase I_B further, the collector current is at the saturation value.

EXAMPLE 4.5 *Find whether the transistor is saturated if $\beta_{dc} = 50$, $V_{BB} = 3$ V, $R_B = 15$ kΩ, $R_C = 1$ kΩ, $V_{CC} = 12$ V and V_{CE} (sat) = 0·25 V.*

SOLUTION

$$I_C\,(\text{sat}) = \frac{V_{CC} - V_{CE}\,(\text{sat})}{R_C} = \frac{12 - 0{\cdot}25}{1\ k\Omega}$$

$$= 11{\cdot}75\ \text{mA}.$$

Let us now find whether I_B is large enough to produce I_C (sat).

$$I_B = \frac{V_{BB} - V_{BE}}{R_B} = \frac{3\text{ V} - 0{\cdot}7\text{ V}}{15\text{ k}} = \frac{2{\cdot}3\text{ V}}{15\text{ k}} = 0{\cdot}1533\text{ mA.}$$

$$I_C = \beta_{dc} \cdot I_B = 50 \times 0{\cdot}1533 = 7{\cdot}66\text{ mA.}$$

The transistor is not in saturation.

4.4 VARIATION OF β_{dc}

The parameter β_{dc} of a transistor is not a constant. It depends on both the collector current and temperature. If temperature is held constant, increase of I_C causes β_{dc} to increase to a peak value. A further increase in I_C decreases β_{dc}. If I_c is kept constant and temperature is varied, β_{dc} changes directly with temperature. If temperature increases, β_{dc} goes up and vice versa. This is illustrated by the curves in Fig. 4.13.

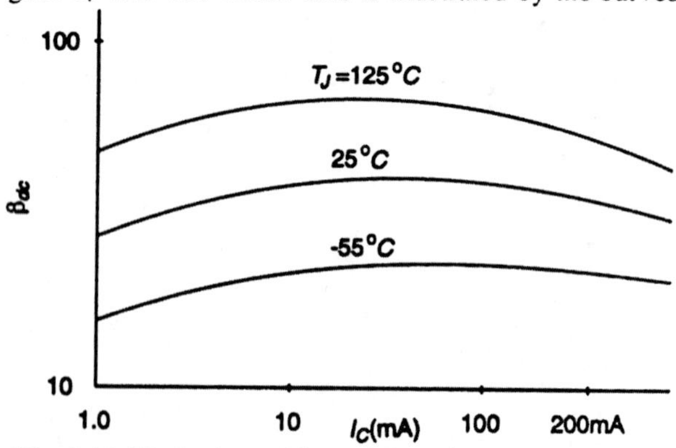

Fig 4.13. Variation of β_{dc} with collector current

For the transistor 2N 3904, the maximum ratings are

$$V_{CEO} = 40\text{ V}$$
$$V_{CBO} = 60\text{ V}$$
$$V_{EBO} = 6\text{ V}$$
$$I_C = -200\text{ mA dc}$$
$$P_D = 310\text{ mW.}$$

The power dissipation $P_D = V_{CE} \times I_C$.

With $V_{CE} = 20$ V, $I_C = 10$ mA, $P_D = 200$ mW.

This is less than 300 mW. As with breakdown voltages, a good design should include a safety factor for long operating life of the transistor. Safety factors of 2 or more are common. For example for 300 mW maximum power rating, safety factor requires a power dissipation below 150 mW.

4.5 TRANSISTOR AS A VOLTAGE AMPLIFIER

In an amplifier, the magnitude of the electrical signal is amplified. The transistor shows current gain (called β). The principle of transistor amplification can be understood from the Fig. 4.14.

When the input voltage changes by ΔV_{in}, the collector current also changes by ΔI_C. The change in collector current gives rise to a change in output voltage

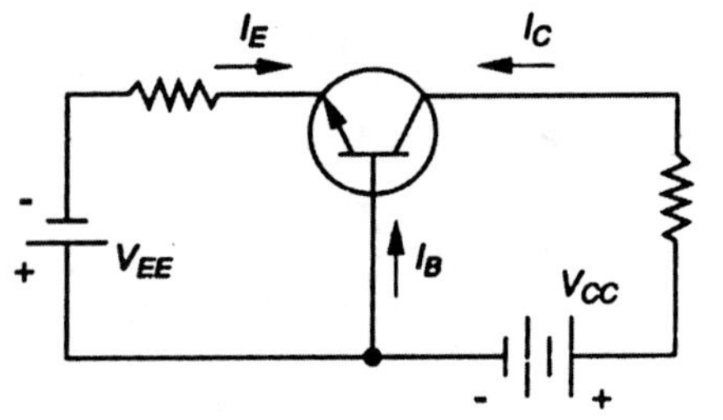

Fig. 4.14 A simple transistor amplifier circuit

$$\Delta V_C = -R_L \cdot \Delta I_C \quad ...(4.16)$$

If ΔI_E is the emitter current change, $\Delta V_{in} = r_e \, \Delta I_E$ ...(4.17)

The voltage gain G_v of the transistor amplifier is

$$G_v = \frac{\Delta V_C}{\Delta V_{in}} = \frac{-R_L}{r_e} \cdot \frac{\Delta I_C}{\Delta I_E} = +\frac{R_L}{r_e} \cdot G_i \quad ...(4.18)$$

where G_i is the current gain. $G_i = -\dfrac{\Delta I_C}{\Delta I_E}$ ...(4.19)

Since when ΔI_E increases, ΔI_C decreases, the quantity $\dfrac{-\Delta I_C}{\Delta I_E}$ is positive.

Thus voltage gain G_v is positive. And the input and output voltages are in phase.

Although $G_i < 1$, R_L can be very high compared with r_e. Thus G_v may be high. If in an example, $G_i = -\dfrac{\Delta I_C}{\Delta I_E} = 0{\cdot}97$, $R_L = 2500\ \Omega$ and $r_e = 25\ \Omega$,

$$G_v = 0{\cdot}97 \times \frac{2500}{25} = 97.$$

The power gain G_p is given by $G_p = \dfrac{\text{Output power}}{\text{Input power}}$ ...(4.20)

where output power $P_0 = (\Delta I_C)^2 \cdot R_L$ ...(4.21)

and input power $P_{in} = (\Delta I_E)^2 \cdot r_e$ ...(4.22)

Thus
$$G_p = \left(\frac{\Delta I_C}{\Delta I_E}\right)^2 \cdot \frac{R_L}{r_e} \quad ...(4.23)$$
$$= G_i \cdot G_v. \quad ...(4.24)$$

The current that flows in an *npn* transistor is depicted below (Fig. 4.15).

EXAMPLE 4.6 *Find the voltage gain and output voltage in Fig. 4.16 if $r_e = 40\ \Omega$.*

SOLUTION The voltage gain is $G_v \simeq \dfrac{R_L}{r_e} = \dfrac{1\ \text{k}\Omega}{40\ \Omega} = 25$

The output voltage is thus $V_{out} = G_v\, V_{in} = 25 \times 50\ \text{mV} = 1{\cdot}25\ \text{V}_{rms}$

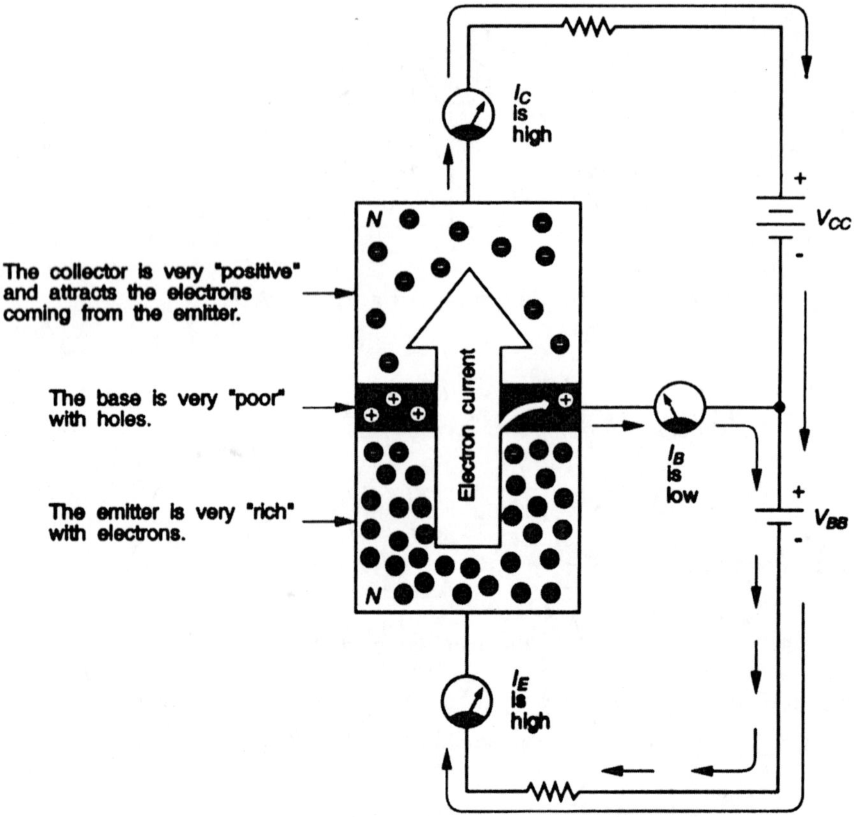

Fig 4.15. Current flow in an npn transistor

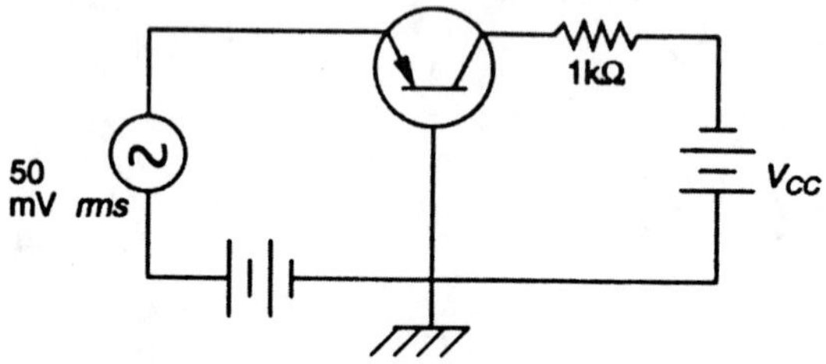

Fig 4.16.

4.6 THE TRANSISTOR AS A SWITCH

A transistor is cut off when the base-emitter junction is not forward biased. It is equivalent to an open switch. A transistor is in saturation when the base emitter junction is forward biased and the base current is sufficient to drive the collector current to saturation. This is equivalent to a closed switch between collector and emitter. These two states are depicted in Fig. 4.17.

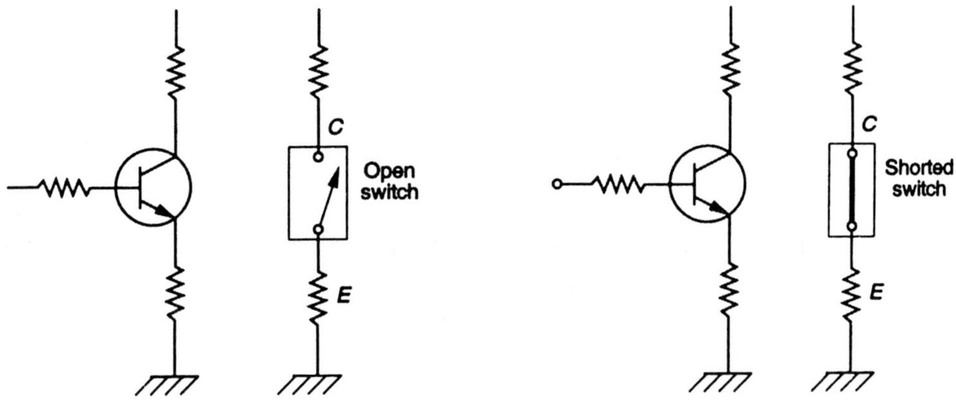

(a) Cut off state (b) Saturation state

Fig 4.17. Switching action of a transistor

In practice, in the state of saturation, the actual voltage drop is a few tenths of a volt.

Cut off conditions

At cut off, neglecting leakage currents, and assuming other currents to be zero,

$$(V_{CE})_{\text{cut off}} = V_{cc} \qquad \ldots(4.25)$$

Saturation conditions

In saturation, the collector current $(I_C)_{sat}$ is given by

$$(I_C)_{sat} = \frac{V_{CC} - (V_{CE})_{sat}}{R_L} \qquad \ldots(4.26)$$

Since $(V_{CE})_{sat} \approx 0$ V

$$(I_C)_{sat} \simeq \frac{V_{CC}}{R_L} \qquad \ldots(4.27)$$

The minimum value of base current needed for saturation is given by

$$(I_B)_{min} = \frac{(I_C)_{sat}}{\beta_{dc}} \qquad \ldots(4.28)$$

Normally I_B is kept much greater than $(I_B)_{min}$ such that the transistor is well in saturation.

4.7 THE r_e TRANSISTOR MODEL

The bipolar junction transistors are called as current controlled devices. In the r_e model, a diode and a current controlled source are employed in the transistor equivalent circuit to explain the behaviour.

Fig. 4.18 depicts the common base bipolar junction transistor amplifier. In the circuit diagram, input impedance is

$$Z_i = r_e \qquad \ldots(4.29)$$

Typically the value of Z_i ranges between 0 and 50 ohms. Similarly the output impedance

$$Z_0 \simeq \infty \text{ ohms} \qquad \ldots(4.30)$$

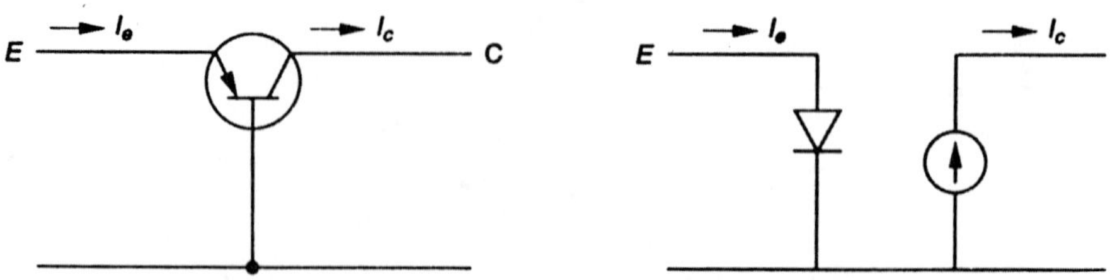

Fig 4.18. Common base bipolar junction transistor amplifier

The output voltage $\quad v_0 = -I_0 R_L = -(-I_C)\, R_L = \alpha\, I_e R_L$ ...(4.31)

The input voltage $\quad v_i = I_e\, Z_i = I_e\, r_e$ ...(4.32)

Thus voltage gain $\quad G_v = \dfrac{v_0}{v_i} = \dfrac{\alpha\, I_e\, R_L}{I_e\, r_e} \simeq \alpha\, \dfrac{R_L}{r_e}$ ...(4.33)

The current gain $\quad G_i = \dfrac{I_0}{I_i} = \dfrac{-I_C}{I_e} = -\dfrac{\alpha I_e}{I_e}$

$= -\alpha \simeq -1$ ...(4.34)

Common-emitter configuration

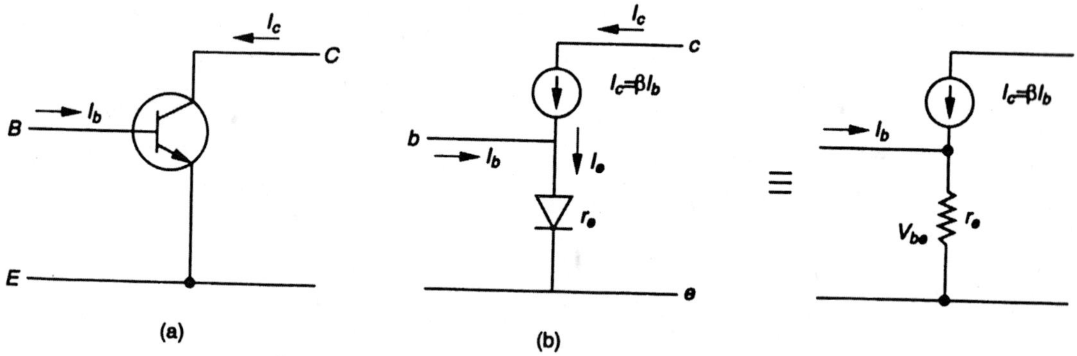

Fig 4.19. (a) Common emitter type *BJT* amplifier (b) Equivalent circuit using r_e transistor model

In the circuit diagram of Figure 4.19, the collector current is given by

$I_c = \beta I_b$ ...(4.35)

Also, emitter current $\quad I_e = (\beta + 1)\, I_b$ ...(4.36)

$\approx \beta I_b$ ...(4.37)

Input impedance $\quad Z_i = \dfrac{V_i}{I_i} = \dfrac{V_{be}}{I_b}$ ...(4.38)

where $\quad V_i = V_{be} = I_e\, r_e \simeq \beta I_b\, r_e$ ...(4.39)

Thus $\quad Z_i = \dfrac{V_{be}}{I_b} = \dfrac{\beta I_b\, r_e}{I_b}$

$\approx \beta r_e$ ...(4.40)

In the common emitter configuration, the values of input impedance $Z_i = \beta r_e$ ranges between a few hundred ohms (minimum value) to a maximum of about 7 to 8 kΩ.

The above model does not include an output impedance (r_0), but if the value of r_0 is available from manufacturer's data sheets, the output circuit can be shown as depicted in Fig. 4.20.

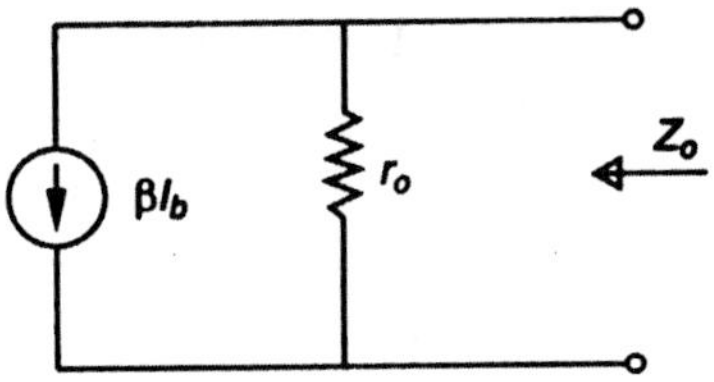

Fig 4.20. Inclusion of r_0 in the transistor equivalent circuit

The output impedance for the circuit will this be

$$Z_0 = r_0 \quad \text{...(4.41)}$$

The voltage gain is $$G_v = \frac{V_0}{V_i} = \frac{-\beta I_b R_L}{I_b \beta r_e} \quad \text{...(4.42)}$$

$$= \frac{-R_L}{r_e} \quad \text{...(4.43)}$$

The current gain is $$G_i = \frac{I_0}{I_i} = \frac{\beta I_b}{I_b} = \beta \Big|_{CE,\, r_0 = \infty} \quad \text{...(4.44)}$$

EXAMPLE 4.7 *For a common base circuit, $I_E = 2$ mA, $\alpha = 0{\cdot}97$ and an ac signal of 1 mV is applied between the base and emitter terminals. (a) Find the input impedance (b) Find the voltage gain if a load of 500 Ω is connected to the output terminals. (c) Find the output impedance and current gain.*

SOLUTION

(a) $$r_e = \frac{26\text{ mV}}{I_E} = \frac{26\text{ mV}}{2\text{ mA}} = 13\ \Omega$$

(b) $$I_i = I_e = \frac{V_i}{Z_i} = \frac{1\text{ mV}}{13\ \Omega} = 76{\cdot}9\ \mu\text{A}$$

Thus $$V_0 = I_C R_L = \alpha I_e R_L = 0{\cdot}97 \times 76{\cdot}9\ \mu\text{A} \times 500\ \Omega$$
$$= 37{\cdot}3\text{ mV}$$

Voltage gain $$= G_v = \frac{V_0}{V_i} = \frac{37{\cdot}3\text{ mV}}{1\text{ mV}} = 37{\cdot}3$$

For $Z_0 = \infty$, G_i = Current gain $= -\alpha = -0{\cdot}97$

EXAMPLE 4.8 *For a common emitter configuration circuit, $\beta = 140$ and $I_E = 2$ mA, $r_0 = \infty$. Find (a) input impedance (b) voltage gain if a load of 2 kΩ is applied (c) current gain for a 1·2 kΩ load.*

SOLUTION

(a) $$r_e = \frac{26\text{ mV}}{I_E} = \frac{26\text{ mV}}{2\text{ mA}} = 13\ \Omega$$

$$Z_{in} = \beta r_e = (140) \times 13 = 1820\ \Omega$$

(b) Voltage gain $$= \frac{-R_L}{r_e} = -\frac{1{\cdot}2 \times 10^3}{13} = -92{\cdot}3$$

(c) Current gain $$= \frac{I_0}{I_i} = \beta = 140$$

4.8 HYBRID PARAMETERS

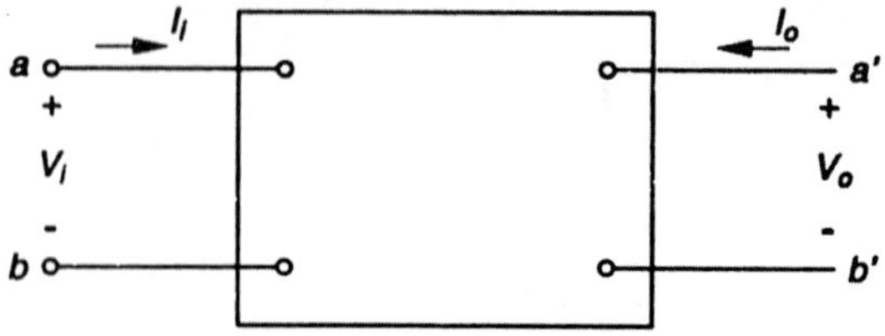

Fig 4.21. A two port system shown as black box

In Fig. 4.21, if the input port is ab and the output port is $a'b'$ and if I_i, V_i are the currents and voltages at the input port and if I_0, V_0 represent the current and voltage at the output port, respectively, then, they may be related by

$$V_i = h_{11} I_i + h_{12} V_0 \quad \ldots(4.45)$$

$$I_0 = h_{21} I_i + h_{22} V_0 \quad \ldots(4.46)$$

In the above equations, the parameters (h_{11}, h_{12}, h_{21} and h_{22}) are called hybrid (or h) parameters. They represent respectively,

$h_{11} = h_i$ = input resistance (in ohms)

$h_{12} = h_r$ = reverse transfer voltage ratio (unitless)

$h_{21} = h_f$ = forward transfer current ratio (unit less)

$h_{22} = h_0$ = output conductance (in Siemens)

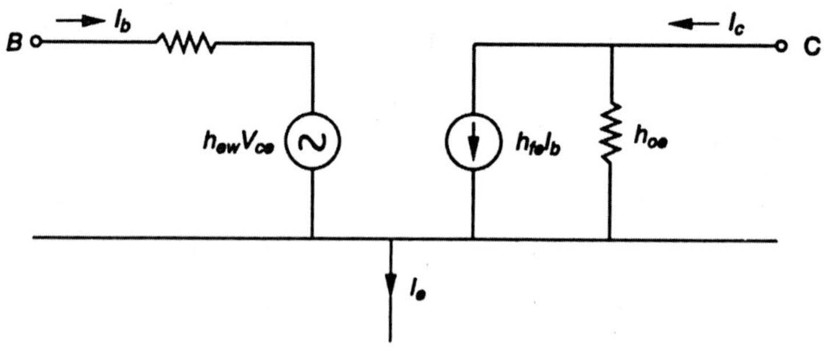

Fig 4.22. Hybrid parameter equivalent circuit

For a common emitter transistor configuration.

The parameters are

$$h_{ie} = \frac{\partial V_i}{\partial i_i} = \frac{\partial V_{be}}{\partial i_b} \simeq \frac{\Delta v_{be}}{\Delta i_b}\Bigg|_{V_{CE} = \text{const.}} \quad \ldots(4.47)$$

$$h_{re} = \frac{\partial v_i}{\partial v_0} = \frac{\partial v_{be}}{\partial v_{ce}} \simeq \frac{\Delta v_{be}}{\Delta v_{ce}}\Bigg|_{I_B = \text{const.}} \quad \ldots(4.48)$$

$$h_{fe} = \frac{\partial i_0}{\partial i_i} = \frac{\partial i_c}{\partial i_b} \simeq \frac{\Delta i_c}{\Delta i_b}\Bigg|_{V_{CE} = \text{const.}} \quad \ldots(4.49)$$

$$h_{ol} = \frac{\partial i_0}{\partial v_0} = \frac{\partial i_c}{\partial v_{ce}} \simeq \frac{\Delta i_c}{\Delta i_{ce}}\Bigg|_{I_B = \text{const.}} \quad \ldots(4.50)$$

The parameters h_{ie} and h_{re} can be determined from the i_B *vs*. v_{BE} characteristics (for different V_{ce} values). Similarly, the hybrid parameters h_{fe} and h_{oe} can be determined from the i_C *vs* . V_{CE} characteristic curves (for different i_B values).

The hybrid and r_e models for a *BJT* in the common emitter and common base configurations can be depicted as shown in Fig. 4.23 below.

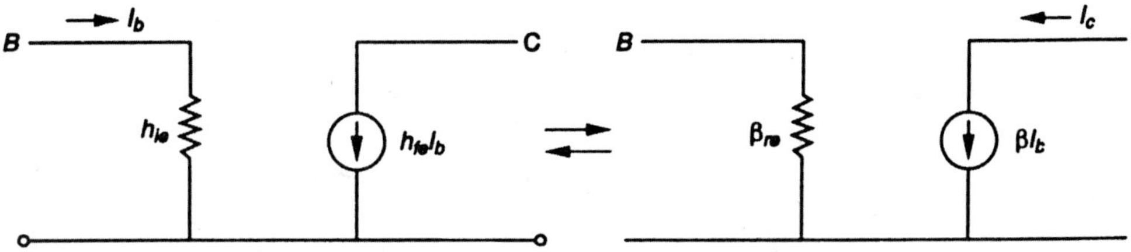

(a) Common emitter circuit equivalents

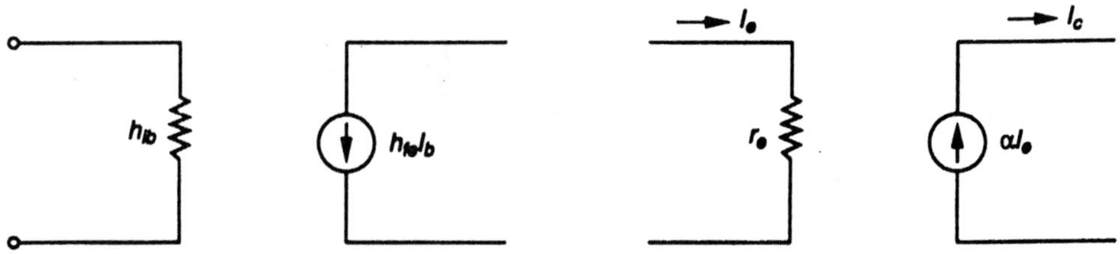

(b) Common base circuit equivalents

Fig 4.23. Bipolar junction transistor circuits (hybrid vs. r_e model)

EXAMPLE 4.9 *For a common emitter BJT circuit, h_{fe} = 120, h_{oe} = 20 μs, h_{oe} = 1 μs, I_e = 2 mA. Determine (a) the common emitter hybrid model (b) the common base r_e model.*

SOLUTION

(a) $$r_e = \frac{26 \text{ mV}}{I_e} = \frac{26 \text{ mV}}{2 \text{ mA}} = 13\ \Omega$$

$$h_i = \beta r_e = 120 \times 13 = 1{\cdot}56 \text{ k}\Omega$$

$$r_0 = \frac{1}{h_{oe}} = \frac{1}{20\ \mu\text{s}} = 50 \text{ k}\Omega$$

So the circuit is shown below

(b) $$r_e = 13\ \Omega$$

$$\alpha \simeq 1$$

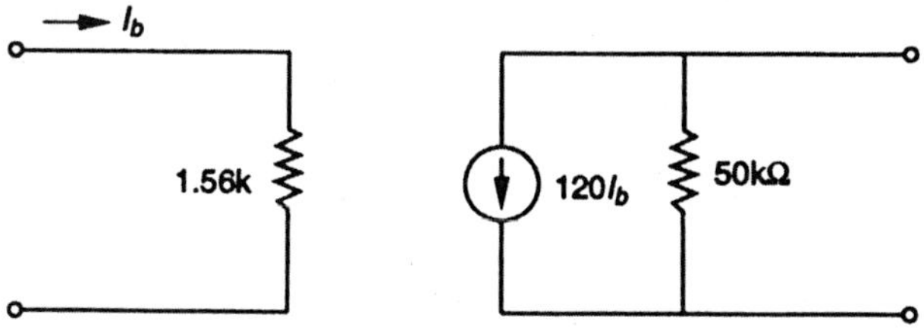

Fig 4.24.

$$r_0 = \frac{1}{h_{ob}} = \frac{1}{1\ \mu s} = 1\ M\Omega$$

Thus the circuit is shown in Fig. 4.25.

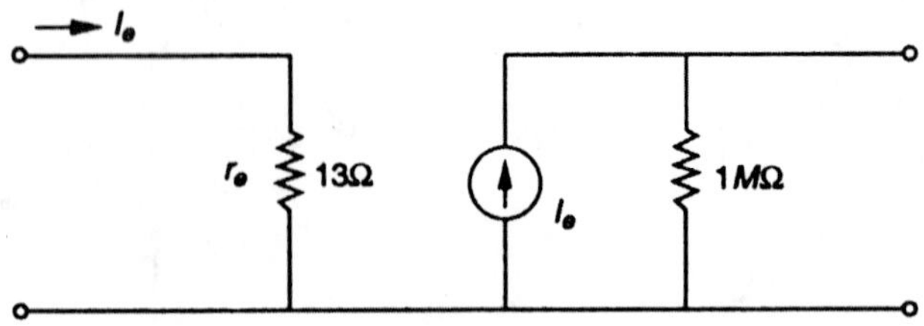

Fig 4.25. Common base r_e model

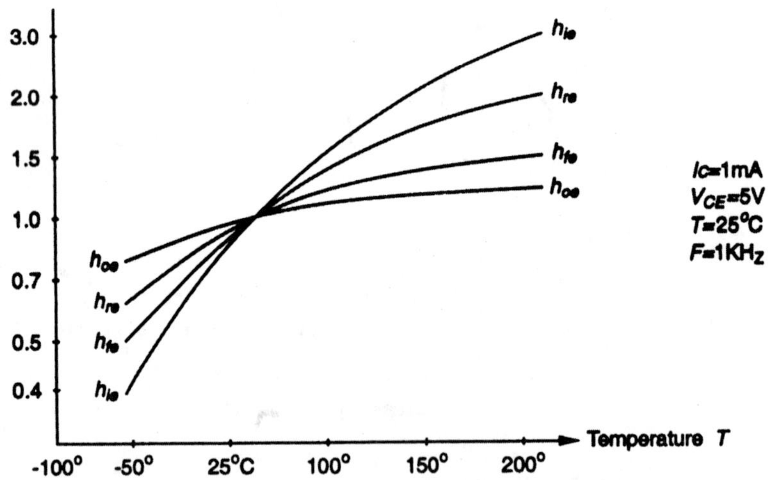

Fig 4.26. Variation of *h* parameters with temperature

The typical parameter value for the common emitter, common collector and common base configurations are depicted in Table below

Parameter	*CE*	*CC*	*CB*
h_i	1000 Ω	1000 Ω	20 Ω
h_r	2×10^{-4}	$\simeq 1$	3×10^{-4}
h_f	50	- 51	- 0·98
h_o	25 μA/V	25 μS	0·5 μS

The hybrid parameters vary with collector current, collector-to-emitter voltage and temperature. The variation of the hybrid parameters with temperature is shown in Fig. 4.26.

Review Questions

1. Draw the basic structure of (a) *pnp* and (b) *n-p-n* type bipolar junction transistor.
2. What are the different modes of operation of the bipolar junction transistor ?
3. Illustrate with suitable diagram, the operation of current flow in an *npn BJT*.
4. Draw the characteristics curves of an *n-p-n BJT*.
5. Draw the constant power curve on the $I_C - V_{CE}$ characteristics of the *BJT*.
6. Define α_{dc} and β_{dc} for a *BJT*.
7. Derive the relationship between α_{dc} and β_{dc}.
8. Discuss the variation of β_{dc} with collector current.
9. Derive equations (4.18) and (4.23).
10. Derive equation (4.34).
11. Draw the equivalent circuit using r_e transistor model of a common emitter *BJT*.
12. Draw the hybrid parameter equivalent circuit for a common emitter transistor.
13. Define the hybrid parameters h_{ie}, h_{re}, h_{fe} and h_{oe} for a common emitter transistor.
14. Discuss the variation of the common emitter transistor *h* parameters with temperature.

Objective Type Questions

Tick mark (√) the correct answer.

1. The three terminals of a bipolar junction transistor are called
   (a) *p, n, p* (b) *n, p, n*
   (c) anode, cathode and gate (d) emitter, base and collector16.
2. For an *n-p-n* transistor, the *n*-regions are
   (a) emitter and base (b) base and collector
   (c) emitter and collector
3. For operation of *p-n-p* amplifier, the base of the amplifier must be
   (a) 0 V (b) positive with respect to collector
   (c) negative with respect to collector (d) greater than the collector current
4. The bias condition of a transistor for being used as an amplifier is
   (a) forward-reverse (b) reverse-reverse
   (c) reverse-forward (d) forward-forward
5. If β_{dc} is 50, the value of α_{dc} is
   (a) 51 (b) 50 (c) $\frac{50}{51}$ (d) $\frac{1}{50}$
6. The relationship between α_{dc} an β_{dc} is given by
   (a) $\beta_{dc} = \frac{\alpha_{dc}}{1 + \alpha_{dc}}$ (b) $\beta_{dc} = \frac{\alpha_{dc}}{1 - \alpha_{dc}}$ (c) $\alpha_{dc} = \frac{1 + \beta_{dc}}{\beta_{dc}}$ (d) $\alpha_{dc} = \frac{\beta_{dc}}{\beta_{dc} - 1}$
7. In cut off, the value V_{CE} is
   (a) 0 V (b) V_{CC} (c) minimum (d) $V_{BE} = 0{\cdot}6$ V
8. The saturation value of V_{CE} is
   (a) equal to V_{CC} (b) 0 V (c) 0·7 V (d) $\frac{V_{CC}}{2}$

9. If the emitter-base junction is open, the value of collector voltage is
(a) 0 V (b) 0·2 V (c) floating (d) V_{CC}

10. If in a transistor $(I_C)_{sat} < I_B \beta_{dc}$ the transistor is in
(a) cut off (b) saturation
(c) amplification state

11. As temperature is increased, the h parameters h_{ie}, h_{re}, h_{fe} and h_{oe} of a transistor
(a) decreases (b) increases
(c) remains constant

Execise Problems

1. In a *npn* transistor, $I_B = 30$ μA and $I_C = 3$ mA. Find the value of β_{dc} and α_{dc}.
$\left[\textbf{Ans. } 100, \frac{100}{101}\right]$

2. In a *BJT*, $\beta_{dc} = 150$ and $I_B = 50$ μA. Find the values of I_C and α_{dc}.
[Ans. 7·5 mA]

3. In an *npn* bipolar junction transistor, $I_C = 6$ mA and $\alpha_{dc} = 0{\cdot}985$. Find the values of I_B and β_{dc}.

4. In the circuit of Fig. 4.9, $\beta_{dc} = 120$, $V_{BB} = 6$ V, $V_{CC} = 12$ V, $R_B = 12$ kΩ and $R_C = 12$ Ω. Find I_B, I_C, I_E, V_{CE} and V_{CB}.

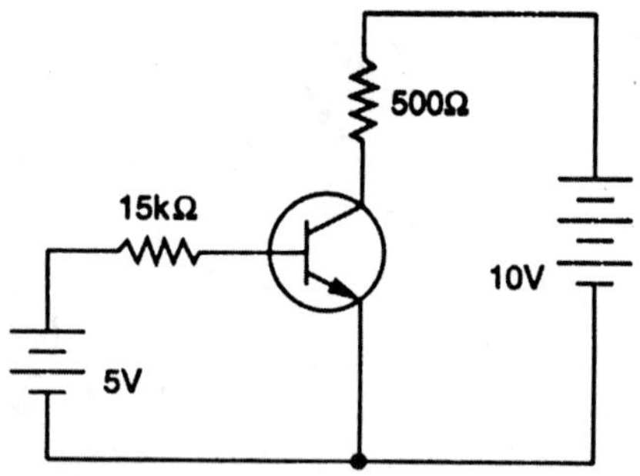

Fig EP.4.5.

5. Find whether or not the transistor in the circuit of Fig. alongside is in saturation. Assume $(V_{CE})_{sat} = 0{\cdot}2$ V.

6. Find if the transistor is saturated if $\beta_{dc} = 60$, $V_{BB} = 3$ V, $R_B = 15$ kΩ, $R_C = 1{\cdot}2$ kΩ, $V_{CC} = 12$ V and $(V_{CE})_{sat} = 0{\cdot}2$ V.

7. For a common base *BJT* amplifier, $I_E = 2{\cdot}5$ mA, $\alpha = 0{\cdot}98$ and an ac. signal of 1 mV is applied between the base and emitter terminals. Determine
(i) the input impedance
(ii) the voltage gain if a load of 400 Ω is connected between the output terminals.
(iii) the output impedance and current gain.

8. For a common emtter *BJT* amplifier, $\beta = 120$ and $I_E = 2{\cdot}5$ mA, $r_0 = \infty$. Determine
(a) input impedance
(b) voltage gain if a load of 2 kΩ is applied
(c) current gain for a 1·2 kΩ load.

Chapter 5

BIPOLAR TRANSISTOR BIASING

5.1 INTRODUCTION

As we already know, a transistor must be d.c. biased in order to operate as an amplifier. The d.c. operating point must be established such that any variation in input signal is amplified and accurately reproduced at the output terminal. **Establishing or setting the bias point** in a transistor means that a certain fixed current and voltage condition is specified. The d.c. operating point is also called the quiscent (or Q) point.

The analysis and design of an electronic amplifier has two components-the d.c. component and the a.c. component. Here, the superposition theorem can be applied and the d.c. behaviour can be separated from

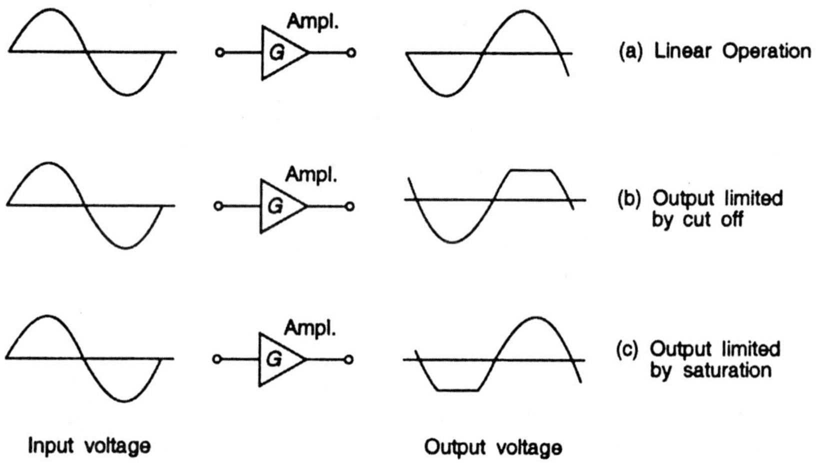

Fig 5.1. Linear and non-linear modes of operation for an amplifier

the a.c. response. It should, however, be kept in mind that the d.c. parameters will affect the a.c. components. Similarly the a.c. components will affect the d.c. parameters. The primary objective of this chapter is to be familiarize with the *BJT* transistor such that the d.c. analysis of a typical *BJT* amplifier (say) can be quickly made.

5.2 CHARACTERISTICS CURVE

Let us confine our attention to Fig 5.2.

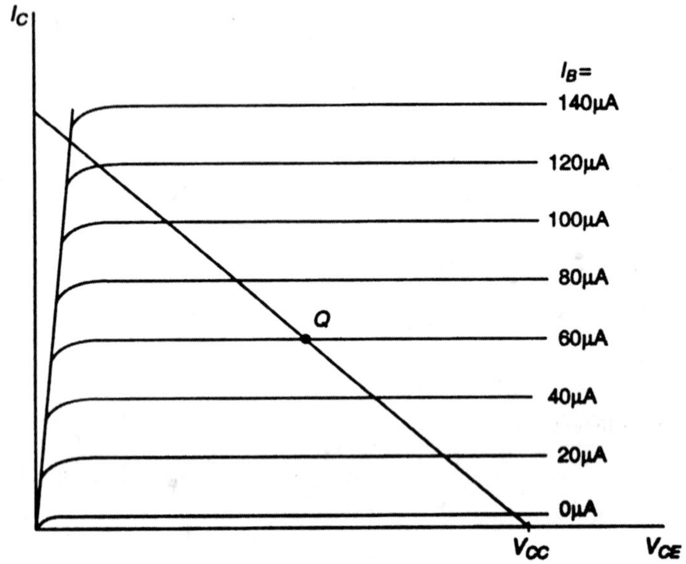

Fig 5.2. Collector current vs. collector-to-emitter voltage characteristics of *BJT* amplifier

The Quiscent or Q-point should be within the active region. The maximum permissible ratings for a BJT are indicated by the characteristics of Fig. 5.2 ($I_C - V_{CE}$ characteristics). The maximum permissible collector to emitter voltage is $(V_{CE})_{max}$. The maximum permissible collector current is $(I_C)_{max}$ and the maximum permissible power is $(P_C)_{max}$. Towards the lower ends are the saturation region where $V_{CE} \leq (V_{CE})_{sat}$ and cut off region where $I_B \leq 0$ μA. The BJT device may be damaged if the operating region is not confined within the permissible limits.

Although various operating points may be selected, the chosen point should not be near the limits or boundaries. It is highly desirable that the preferred point (Q point) is such that the gain of the BJT is more or less constant of the entire input signal swing. It will be seen from Fig. 5.2 that point c is most suitable to be the preferred Q-point. Points a, b or d is near the boundaries and this may give rise to cut off, saturation or other nonlinearities.

Further the Q-point should be such that the changes in temperature result in minimum changes of the operating point. The parameter current gain (β_{dc}) and leakage current (I_{CEO}) are very much temperature sensitive. A quantity S (called the stability factor determines the degree of change of operating point due to changes in temperature.

(1) For linear BJT operation the following two factors should be fulfilled
   (i) Base-Emitter (BE) Junction should be **forward biased**
   (ii) Collector-Base (CB) Junction must be **reverse biased**

(2) For cut off region operation
   Base Emitter junction is reverse biased.

(3) Saturation region operation
(i) Base-Emitter junction forward biased
(ii) Collector-Base Junction also forward biased

Emitter base circuit

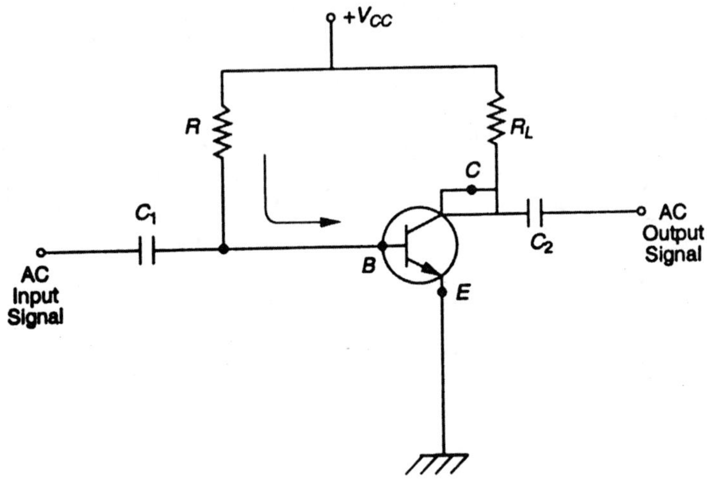

Fig 5.3. A typical biasing circuit

In the circuit diagram, using Kirchhoff's voltage laws, one gets

$$+V_{CC} - I_B R - V_{BE} = 0 \quad \ldots(5.1)$$

or,
$$I_B = \frac{V_{CC} - V_{BE}}{R} \quad \ldots(5.2)$$

Collector-emitter circuit

Here $I_C = \beta I_B$ ...(5.3)

where I_C = collector current and I_B, the base current. Also applying Kirchhoff's voltage laws one gets

$$V_{CE} + I_C R_L - V_{CC} = 0 \quad \ldots(5.4)$$

or,
$$V_{CE} = V_{CC} - I_C R_L \quad \ldots(5.5)$$

where V_{CE} is the collector to emitter voltage drop. Thus

$$V_{CE} = V_C - V_E = V_C \quad \ldots(5.6)$$

and
$$V_{BE} = V_B - V_E = V_B \quad \ldots(5.7)$$

EXAMPLE 5.1 *Calculate for the circuit in Fig. 5.4 the following quantities for the Q point (i) I_B and I_C (ii) V_{CE} (iii) V_C and V_B (iv) V_{CB}*

SOLUTION At the quiscent operating point

(i) $I_B = \frac{V_{CC} - V_{BE}}{R} = \frac{15\text{ V} - 0{\cdot}7\text{ V}}{180\text{ k}} = \frac{14{\cdot}3}{180\text{ k}} = 79{\cdot}44\ \mu\text{A}$

$I_C = \beta\, I_B = 100\ (79{\cdot}44)\ \mu\text{A} = 7{\cdot}94\text{ mA}$

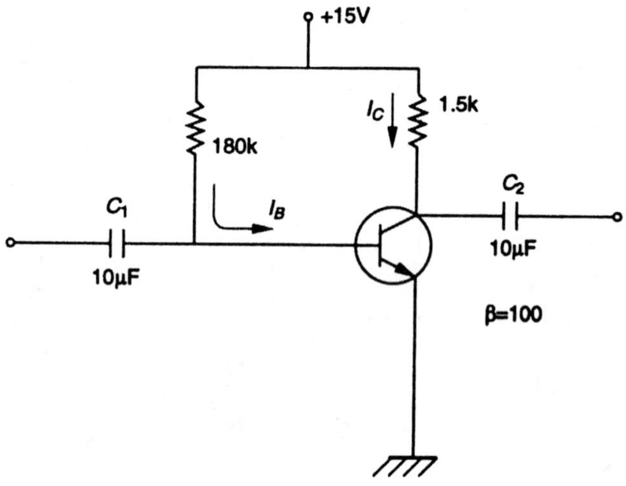

Fig 5.4. Reference Example 5.1

(ii) $V_{CE} = V_{CC} - I_C R_L = 15\text{ V} - (7{\cdot}94\text{ mA})\ (1{\cdot}5\text{ k})$

$= 15 - 11{\cdot}9 = 3{\cdot}1\text{ V}$

(iii) $V_B = V_{BE} = 0{\cdot}7\text{ V}$

$V_C = V_{CE} = 3{\cdot}1\text{ V}$

(iv) $V_{CB} = V_C - V_B = 3{\cdot}1\text{ V} - 0{\cdot}7 = 2{\cdot}4\text{ V}.$

The positive value for V_{CB} implies that $V_{BC} = -\,2{\cdot}4$ volts *i.e.* the collector base junction is reverse biased.

5.3 SATURATION

In the saturation region, the collector to emitter voltage V_{CE} is either equal to or less than $(V_{CE})_{sat}$. When the transistor operates in saturation, the collector base junction is reverse biased. If, in the saturation curve of Fig. 5.5, the Q point is approximated with Q', then I_C may be calculated very easily. Near the saturation point Q', we get

$$V_{CE} = 0\text{ V} \quad \ldots(5.8)$$

$$I_C = (I_C)_{sat} \quad \ldots(5.9)$$

$$R_{CE} = 0\ \Omega \quad \ldots(5.10)$$

From equation (5.4), we may write

$$V_{CC} = (I_C)_{sat} \cdot R_L \quad \ldots(5.11)$$

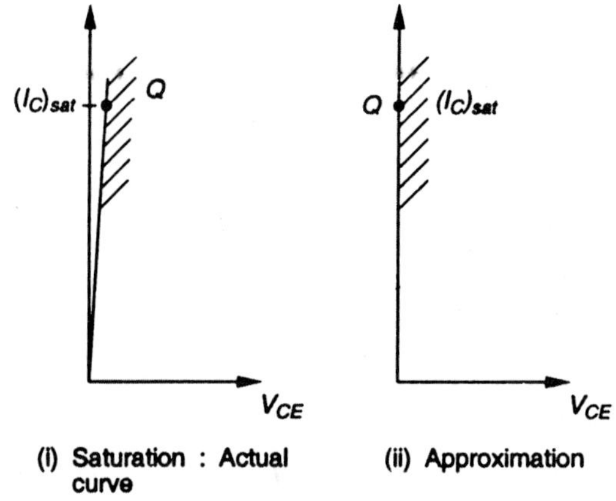

Fig 5.5. Saturation : (i) Actual curve, (ii) approximation

or, $$(I_C)_{sat} = \frac{V_{CC}}{R_L} \qquad ...(5.12)$$

EXAMPLE 5.2 *Find the saturation current $(I_C)_{sat}$ in Fig. 5.4 above.*

SOLUTION $$(I_C)_{sat} = \frac{V_{CC}}{R_L}$$

$$= \frac{15\text{ V}}{1{\cdot}5\text{ k}\Omega} = 10\text{ mA}$$

Load Line

From the circuit of Fig. 5.2, we can write

$$V_{CE} = V_{CC} - I_C R_L \qquad ...(5.4)$$

This equation is very important in that it describes the relationship between V_{CE} and I_C. If we put in equation $I_C = 0$, we get

$$V_{CE} = V_{CC} \qquad ...(5.13)$$

This defines a point A in the characteristic curve (Fig. 5.6). Also if we put $V_{CE} = 0$, we get the equation for I_C.

$$I_C = \frac{V_{CC}}{R_L} \qquad ...(5.14)$$

This latter point gives another point B in the output characteristics curve. When we join A and B we get a straight line that is called the **Load line** since it is defined by the value of the load resistor. By solving for I_B, the actual Q point may be obtained.

If the base resistor R is changed, the value of I_b, the base current, changes and the Q point shifts as depicted in Fig. 5.7. Similarly, if the value of V_{CC} changes, the Q point shifts as depicted in Fig. 5.8. A change in the value of the load resistor also shifts the Q point as depicted in Fig. 5.9.

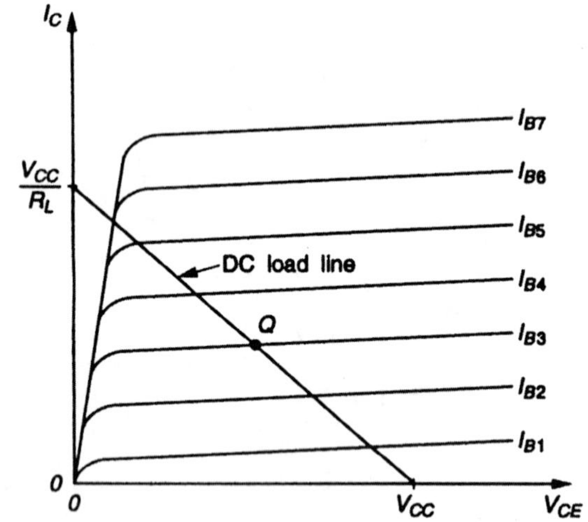

Fig 5.6 DC load line characteristics

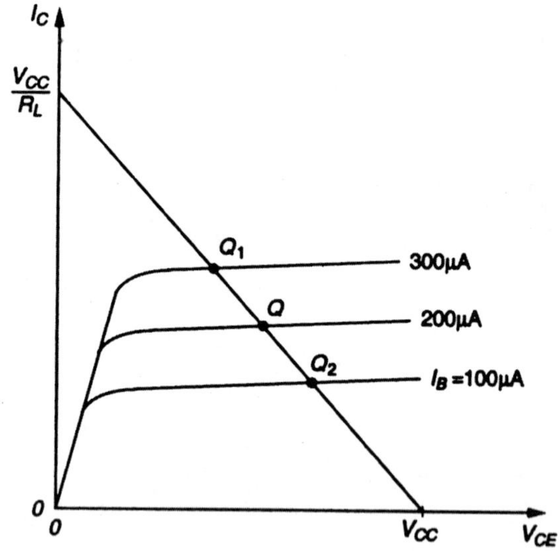

Fig 5.7 Variation of Q point with base current

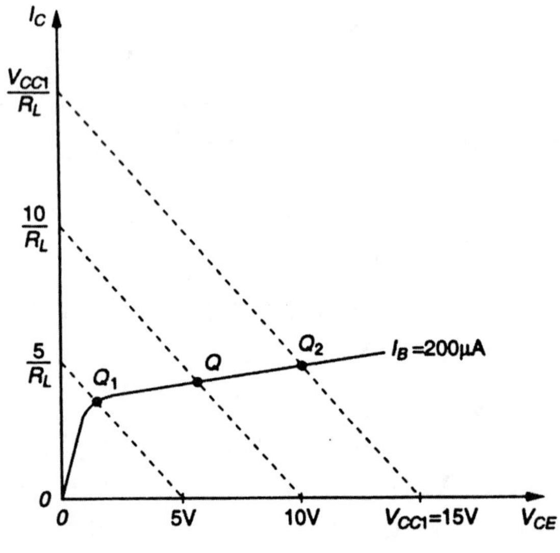

Fig 5.8 Variation of Q point with collector supply voltage

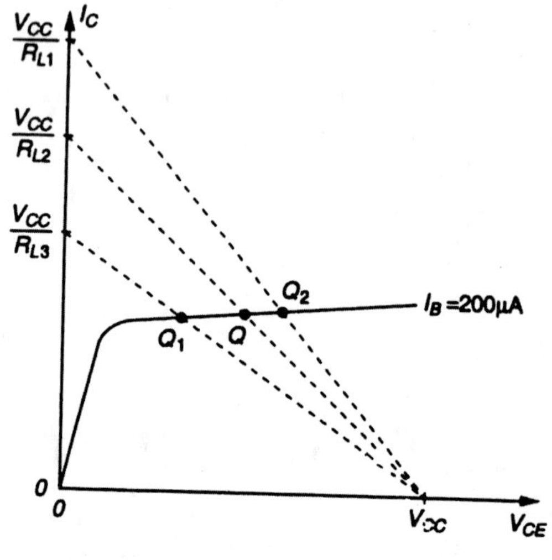

Fig. 5.9 Shift of Q point with load resistance

EXAMPLE 5.3 *Given the load line and the defined Q point for the circuit of Fig. 5.10, find the values of V_{CC}, R and R_L under fixed bias configuration.*

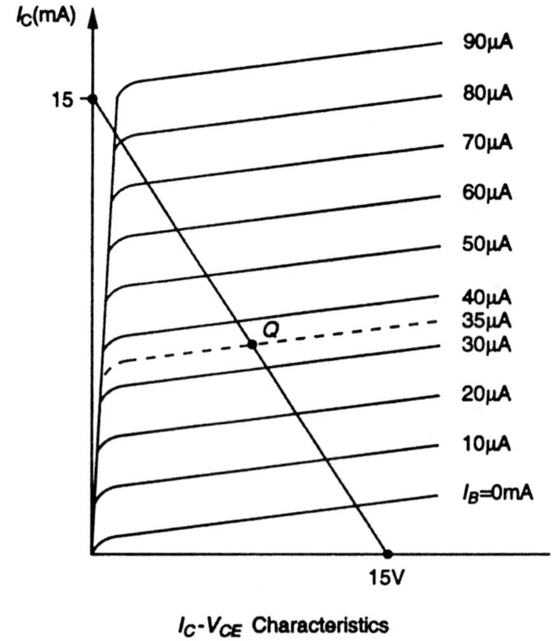

Fig 5.10. Reference Example 5.3

SOLUTION From Fig. 5.10,

$V_{CE} = V_{CC} = 15$ V at $I_C = 0$ mA

$$I_C = \frac{V_{CC}}{R_L} \quad \text{at } V_{CE} = 0 \text{ V},$$

$$R_L = \frac{V_{CC}}{I_C} = \frac{15 \text{ V}}{15 \text{ mA}} = 1 \text{ k}\Omega$$

$$R_B = \frac{V_{CC} - V_{BE}}{I_B} = \frac{(15 - 0{\cdot}7) \text{ V}}{35 \ \mu\text{A}} = 408{\cdot}6 \text{ k}\Omega$$

5.4 EMITTER STABILIZED BIAS CIRCUIT

Applying Kirchhoff's voltage law to the base emitter loop of Fig. 5.11, one gets

$$+ V_{CC} - I_B R - V_{BE} - I_E R_E = 0 \quad \text{...(5.15)}$$

Now $$I_E/I_B = (1 + \beta) \quad \text{...(5.16)}$$

From equation (5.15) and (5.16), we may write,

$$V_{CC} - I_B R - V_{BE} - (1 + \beta) I_B R_E = 0 \quad \text{...(5.17)}$$

or, $$- I_B \{R + (\beta + 1) R_E\} + V_{CC} - V_{BE} = 0 \quad \text{...(5.18)}$$

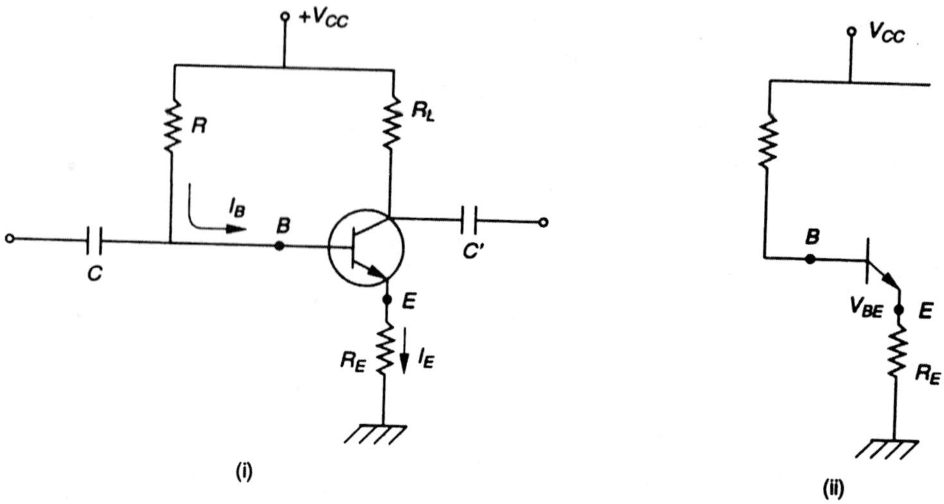

Fig 5.11. (i) Bias circuit with emitter resistor (ii) base emitter portion

or, $$I_B = \frac{(V_{CC} - V_{BE})}{R + (\beta + 1) R_E} = \frac{V_{CC} - 0{\cdot}7}{R + (\beta + 1) R_E} \qquad \text{...(5.19)}$$

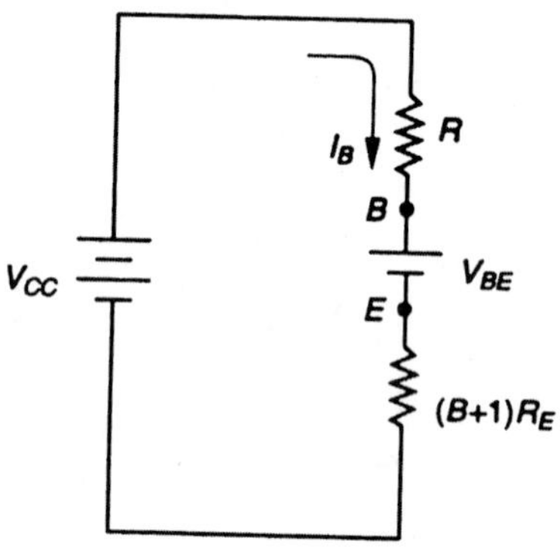

Fig 5.12. Circuit derived from equation (5.19)

An interesting circuit can be drawn from the equation (5.19). Besides the voltage V_{BE} the resistor R_E is **reflected back to** the input base circuit by a factor $(\beta + 1) R_E$. This is the same thing as saying that the emitter resistor R_E which is part of the collector-emitter loop appears as $(\beta + 1) R_E$ in the emitter-base loop.

Collector-emitter loop

The collector emitter circuit can be simplified as drawn in Fig. 5.13.

Writing Kirchhoff's voltage laws, we may express

$$I_E R_E + V_{CE} + I_C R_L - V_{CC} = 0 \qquad \text{...(5.20)}$$

or, $$V_{CE} = V_{CC} - I_E (R_L + R_E) \qquad \text{...(5.21)}$$

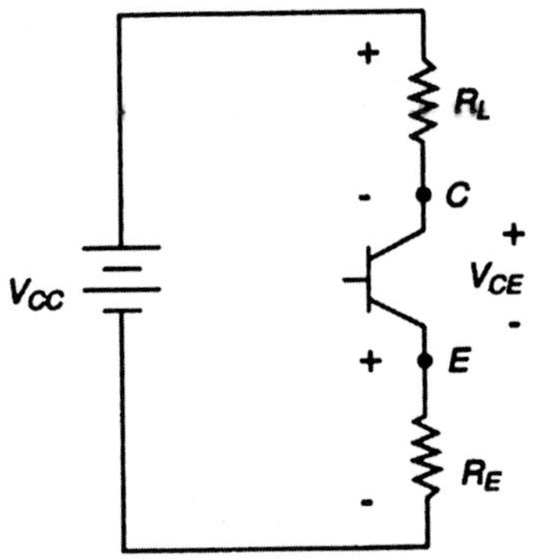

Fig 5.13. Collector emitter circuit

when $\quad I_E \simeq I_C$

Also $\quad V_C = V_{CC} - I_C R_L \qquad \ldots(5.22)$

and $\quad V_B = V_{BE} + V_E \qquad \ldots(5.23)$

EXAMPLE 5.4 *For the emitter-bias stabilized circuit of Fig. 5.14. Find (i) I_B (ii) I_C (iii) V_{CE} (iv) V_C (v) V_E (vi) V_B (vii) V_{CB}*

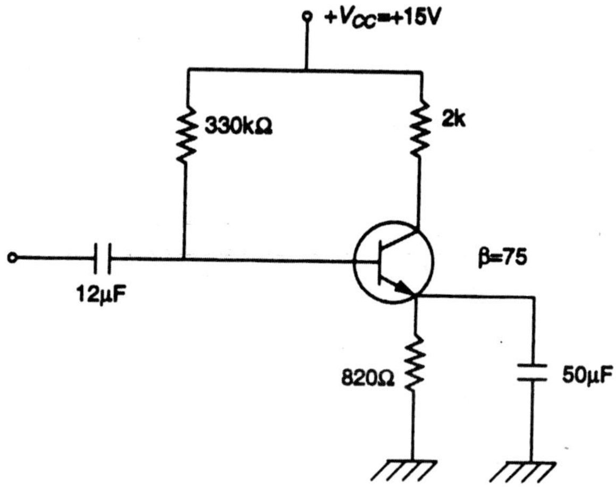

Fig 5.14. Reference Example 5.4

SOLUTION

(i) $$I_B = \frac{V_{CC} - 0{\cdot}7}{330\,k + 76\,(820)} = \frac{14{\cdot}3}{392320} = 36{\cdot}45\ \mu\text{A}$$

(ii) $I_C = \beta\, I_B = (75) \times 36{\cdot}45 = 2{\cdot}73$ mA

(iii) $V_{CE} = V_{CC} - I_C\,(R_L + R_E)$

$= 15 - (2{\cdot}73\text{ mA})\,(2{\cdot}82\text{ k}\Omega)$

$= 7{\cdot}3$ V

(iv) $V_C = V_{CC} - I_C R_L = 15 - (2{\cdot}73\text{ mA})\,(2\text{ k}\Omega)$

(v) $V_E = V_C - V_{CE} = 9.54 - 7.3 = 2.24$ V

(vi) $V_B = V_{BE} + V_E = 0.7$ V $+ 2.24$ V $= 2.94$ V

(vii) $V_{CB} = V_C - V_B = 9.54$ V $- 2.94$ V $= 6.6$ V

V_{BC} is negative indicating a reverse biased collector base junction.

The maximum collector current or the collector saturation level for an emitter bias design can be obtained by placing a "short" between the collector and emitter terminals.

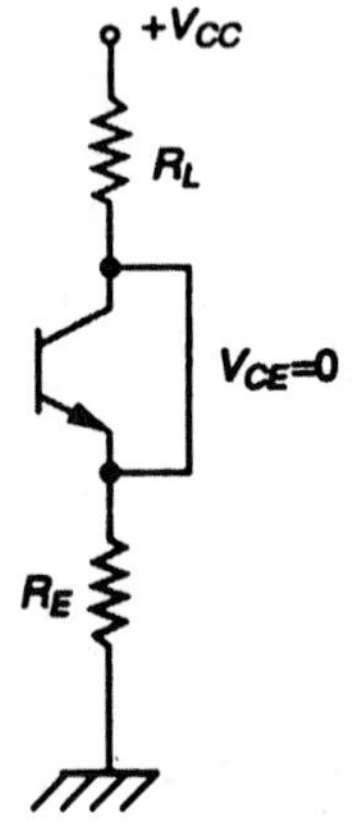

Fig 5.15. Calculation of $(I_C)_{sat}$ for the emitter stabilized bias circuit

Thus $$(I_C)_{sat} = \frac{V_{CC}}{R_L + R_E} \quad ...(5.24)$$

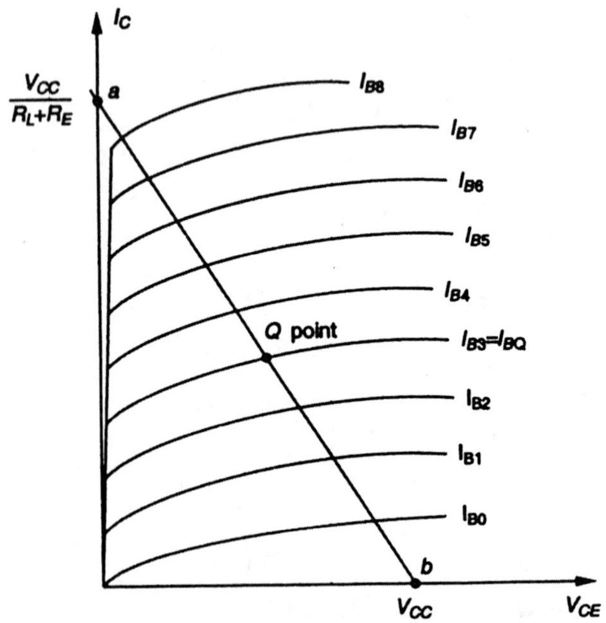

Fig 5.16. The load line

The load line can be drawn on $I_C - V_{CE}$ characteristic from the equation

$$V_{CE} = V_{CC} - I_C (R_L + R_E) \quad \ldots(5.25)$$

The load line passes through points a $\left(0, \frac{V_{CC}}{R_C + R_E}\right)$ and $b\,(V_{CC}, 0)$. The Q point passes through $I_{BB} = I_{BQ}$ curve.

Voltage divider bias circuit

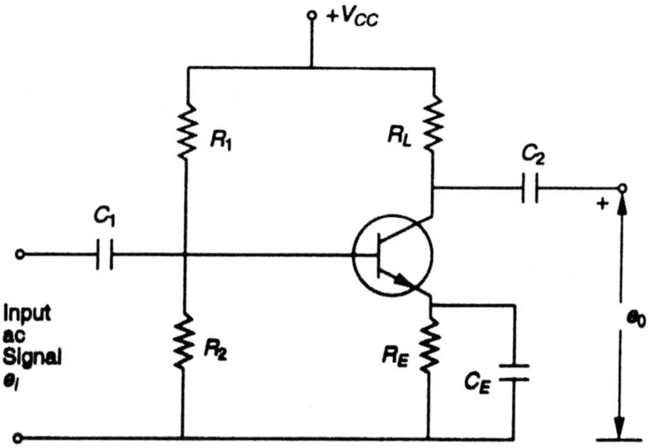

Fig 5.17. Voltage divider bias circuit for the common emitter mode

The self bias or voltage divider bias circuit for the common emitter mode is depicted in Fig. 5.17. The circuit ensures a stable operating point. The improvement in stability can be explained as follows.

If the temperature increases, I_{CO}, the leakage current increases. An increased value of I_{CO} increases the collector current I_C. The voltage drop across R_E increases which means that the forward bias of the base emitter junction is lowered. This decrease in forward bias, decreases the base current. The collector current I_C is thus lowered. The rise in the value of the collector current is thus less than what it would have been if R_E was absent.

The input a.c. signal, when applied to the circuit will appear across R_E. This will give a negative feedback (Chapter 9) and the gain will decrease. The capacitor C_E is introduced so that the a.c. signal is effectively short circuited. The capacitors C_1, C_2 and C_E does not have a say in estabilishing the Q-point.

5.5 CALCULATION OF THE Q-POINT

Let us apply the Thevenin theorem to the input side. (vide Fig. 5.18).

From Fig. 5.18 (b), one gets

$$R_{Th} = \text{Thevenin resistance} = R_1 \parallel R_2 = \frac{R_1 R_2}{R_1 + R_2} \quad \ldots(5.26)$$

Similarly Fig. 5.18 (c) gives,

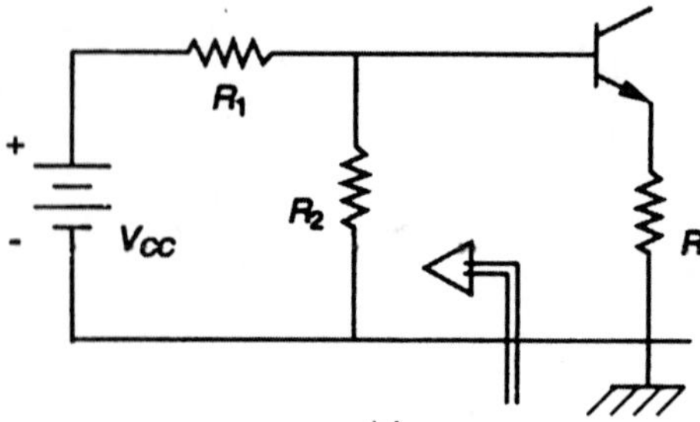

(a)

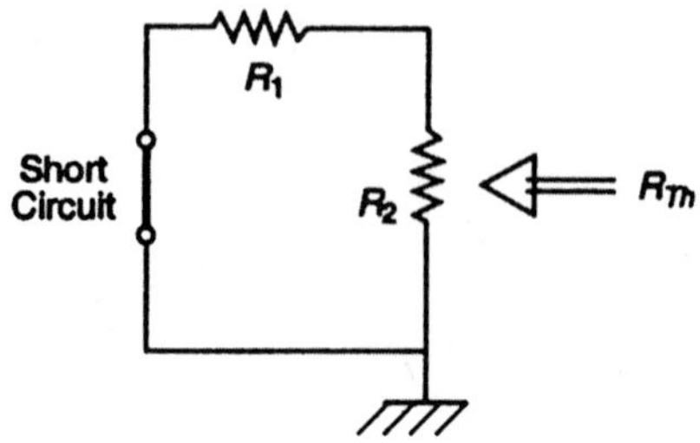

(b)

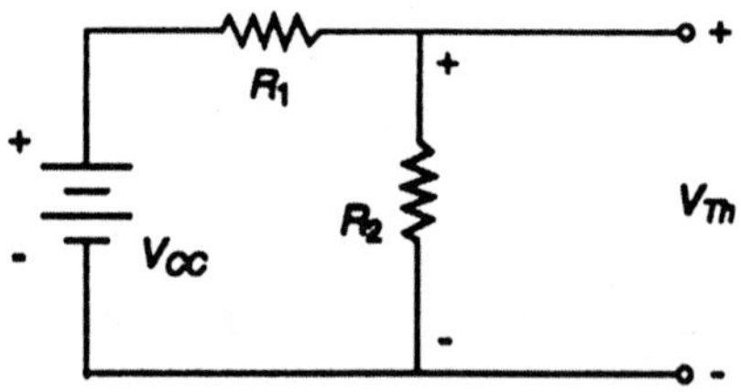

(c)

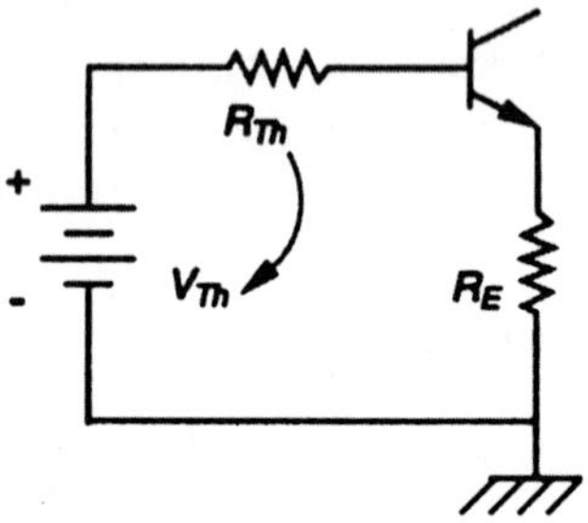

(d)

Fig 5.18. Input portion of the self bias circuit (a) input loop (b) determination of the Thevenin resistance (Battery is replaced by 'SHORT') (c) finding the Thevenin voltage V_{Th} and (d) input circuit Theveninized

$$V_{Th} = \frac{V_{CC} \cdot R_2}{R_1 + R_2} \qquad \ldots(5.27)$$

Putting the Thevenin equivalent circuit to the input side and applying Kirchoff's voltage laws one gets

$$V_{Th} - I_B R_{Th} - V_{BE} - I_E R_E = 0 \quad ...(5.28)$$

With the help of the equation $\quad I_E = (\beta + 1)\, I_B \quad ...(5.29)$

One can express $$I_B = \frac{V_{Th} - V_{BE}}{R_{Th} + (\beta + 1)\, R_E} \quad ...(5.30)$$

We can also write for Fig. 5.17 $\quad V_{CE} = V_{CC} - I_C\,(R_C + R_E) \quad ...(5.31)$

EXAMPLE 5.5 *Find the values of I_C and V_{CE} for the d.c. bias circuit shown in Fig. 5.19.*

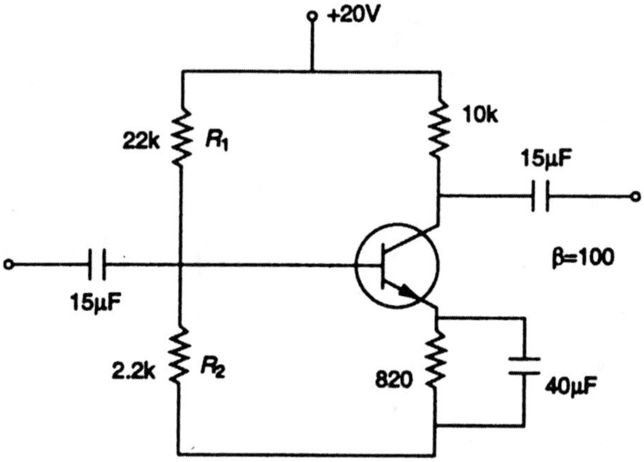

Fig 5.19. Reference Example 5.5

SOLUTION For the circuit shown $R_{Th} = \dfrac{R_1 R_2}{R_1 + R_2} = \dfrac{22\text{ k} \times 2{\cdot}2\text{ k}}{(22\text{ k} + 2{\cdot}2\text{ k})}$

= Thevenin Resistance = 2 k

Also, the Thevenin voltage $V_{Th} = \dfrac{V_{CC} \cdot R_2}{R_1 + R_2}$

$$= \frac{20 \times 2{\cdot}2\text{ k}}{(22\text{ k} + 2{\cdot}2\text{ k})} = 1{\cdot}67\text{ V}$$

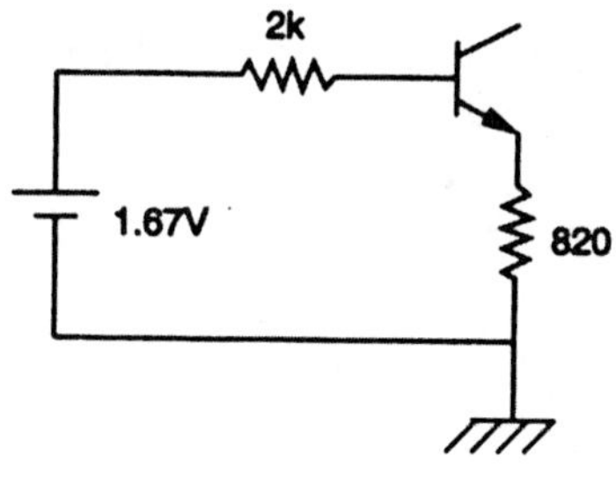

Fig 5.20. Reference Example 5.5

$$\text{Base current} = I_B = \frac{V_{Th} - V_{BE}}{R_{Th} + (\beta + 1)\,R_E}$$

$$= \frac{1{\cdot}67 - 0{\cdot}7}{2\text{ k} + (101) \times 820}$$

$$= 11{\cdot}4\ \mu\text{A}$$

Collector current $I_C = \beta I_B = 100 \times 11{\cdot}4 = 1{\cdot}14$ mA

$$V_{CE} = V_{CC} - I_C\,(R_L + R_E)$$

$$= 20 - 1{\cdot}14\,(10\text{ k} + 820)$$

$$= 20 - 12{\cdot}335$$

$$= 7{\cdot}665\text{ V}$$

5.6 STABILITY FACTORS

From equations (5.28) and (5.31), we may write

$$I_C\left(1 + \frac{\beta R_E}{R_{Th} + R_E}\right) = \beta\frac{V_{Th}}{R_{Th} + R_E} - \beta\frac{V_{BE}}{R_{Th} + R_E} - (1 + \beta)\,I_{CO} \qquad \ldots(5.32)$$

The stability factor

$$S = \frac{\partial I_C}{\partial I_{CO}} = \frac{\beta + 1}{1 + \dfrac{\beta R_E}{R_{Th} + R_E}} = (\beta + 1)\cdot\frac{(1 + R_{Th}/R_E)}{(1 + \beta + R_{Th}/R_E)} \qquad \ldots(5.33)$$

$$S' = \frac{\partial I_C}{\partial V_{BE}} = \frac{-\beta}{R_{Th} + R_E + \beta R_E} \qquad \ldots(5.34)$$

and

$$S'' = \frac{\partial I_C}{\partial \beta} = \frac{1}{\beta(1 + \beta)}\left[I_C\frac{(R_{Th} + R_E)(1 + \beta) - \beta SR_E}{R_{Th} + R_E} - SI_{CO}\right] \qquad \ldots(5.35)$$

Equation (5.33) expresses that S varies from a value of unity to $(1 + \beta)$ as the ratio $\frac{R_{Th}}{R_E}$ is enhanced from very small to very large values. **For good stability against I_{CO}, the value of the resistance R_{Th} should not be much highter than R_E.**

Regarding the stability factor S' (which depicts variation of I_C with V_{BE}), it will be found that as $\beta >> 1$, S' is approximately equal to $-\left(\frac{1}{R_E}\right)$. **Thus a higher value of R_E is required for stabilisation with respect to V_{BE}. The stability factor S'' shows that high β transistor gives better stability than low β transistors.**

It should be noted in this connection that networks which are very stable and relatively insensitive to changes in temperature possess low stability factors. The higher the value of the stability factor, the more susceptible is the network to changes in that parameter.

EXAMPLE 5.6 *Calculate the value of the stability factors S, S′ and S″ for the circuit in Fig. 5.19. [Assume $I_{CO} = 1\ \mu A$]*

SOLUTION
$$S = \frac{\partial I_C}{\partial I_{CO}} = (\beta + 1) = \frac{1 + \dfrac{R_{Th}}{R_E}}{1 + \beta + \dfrac{R_{Th}}{R_E}}$$

$$= (100 + 1) = \frac{1 + \dfrac{2000}{820}}{1 + 100 + \dfrac{2000}{820}}$$

$$= 101 \times \frac{3{\cdot}44}{103{\cdot}44} = 3{\cdot}358$$

$$S' = \frac{\partial I_C}{\partial V_{BE}} = \frac{-\beta}{R_{Th} + R_E + \beta R_E}$$

$$= \frac{-100}{2000 + 820 + 100 \times 820} = -\ 1{\cdot}179 \times 10^{-3}$$

$$S'' = \frac{\partial I_C}{\partial \beta} = \frac{1}{\beta(1+\beta)}\left[I_C \cdot \frac{(R_{Th} + R_E)(1+\beta) - \beta S R_E}{R_E + R_{Th}} - S I_{CO}\right]$$

$$= \frac{1}{100(1+100)}\left[1{\cdot}14 \times 10^{-3}\, \frac{(2000 + 820)(101) - 100 \times 820 \times 3{\cdot}358}{2000 + 820} - 3{\cdot}358 \times 1 \times 10^{-6}\right]$$

$$= \frac{1}{100 \times 101}\left[1{\cdot}14 \times 10^{-3}\, \frac{(2820 \times 101) - 275356}{2820} - 3{\cdot}36 \times 10^{-6}\right]$$

$$= \frac{1}{100 \times 101}\left[1{\cdot}14 \times 10^{-3} \times \left(\frac{-246874}{2820}\right) - 3{\cdot}36 \times 10^{-6}\right]$$

$$= \frac{1}{100 \times 101}[-\ 99{\cdot}8 \times 10^{-3} - 3{\cdot}36 \times 10^{-6}]$$

$$= -\ 9{\cdot}88 \times 10^{-6}$$

Collector-to-base bias

Fig. 5.21 depicts the collector-to base bias circuit.

Applying Kirchhoff's laws, one gets

$$V_{CE} = I_B R + V_{BE} \quad \text{...(5.36)}$$

$$V_{CC} = I_C' R_L + V_{CE} \quad \text{...(5.37)}$$

or, $$V_{CC} = I_C' R_L + I_B R + V_{BE} \quad \text{...(5.38)}$$

or, $$I_B = \frac{V_{CC} - V_{BE} - I_C R_L}{R + R_L} \quad \text{...(5.39)}$$

The collector current is

$$I_C = \beta I_B + (1 + \beta) I_{CO} \quad \text{...(5.40)}$$

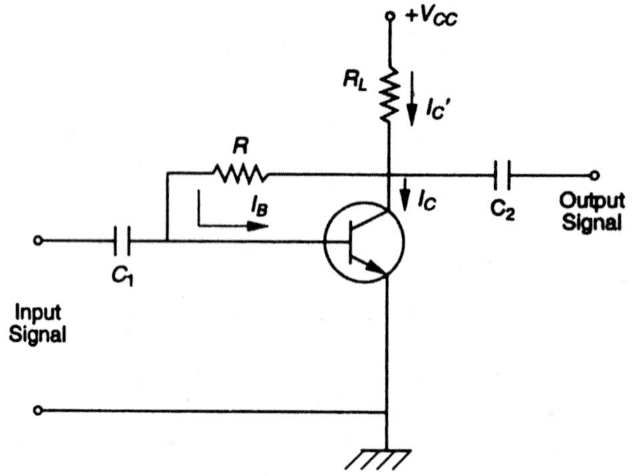

Fig 5.21. Circuit for collector-to-base bias

Also. $I_C' = I_C + I_B$ ...(5.41)

From equations (5.39) and (5.40), one gets

$$I_C\left(1 + \frac{\beta R_L}{R + R_L}\right) = \frac{\beta (V_{CC} - V_{BE})}{R + R_L} + (\beta + 1) I_{CO} \quad ...(5.42)$$

The stability factors are

$$S = \frac{\partial I_C}{\partial I_{CO}} = \frac{\beta + 1}{\frac{\beta R_L}{R + R_L} + 1} \quad ...(5.43)$$

$$S' = \frac{\partial I_C}{\partial V_{BE}} = \frac{-\beta}{R + (\beta + 1) R_L} \quad ...(5.44)$$

and

$$S'' = \frac{\partial I_C}{\partial \beta} = \frac{V_{CC} - V_{BE} - I_C R_L + (R + R_L) I_{CO}}{R_L (1 + \beta) + R} \quad ...(5.45)$$

Since $V_{BE} << V_{CC}$, one gets

$$S'' \approx \frac{V_{CC} - I_C R_L + (R + R_L) I_{CO}}{R + R_L (1 + \beta)} \quad ...(5.46)$$

The stability of the base bias circuit can be understood as follows. If the temperature increases then I_C also increases. This means that the voltage drop across R_L will increase. The base emitter junction will be forward biased lower and the base current I_B will decrease. A decrease in I_B decreases I_C also. The circuit thus has a self balancing tendency when I_C increases.

A slight variation of the circuit is depicted in Fig. 5.22 where a resistor R_E is also placed

Applying Kirchhoff's voltage laws to the circuit in Fig. 5.22, one can write

$$V_{CC} - I_C' R_L - I_B R - V_{BE} - I_E R_E = 0 \quad ...(5.47)$$

Also $I_C' = I_C + I_B$ ...(5.48)

so we may write

$$V_{CC} - \beta I_B R_L - I_B R - V_{BE} - \beta I_B R_E = 0 \quad \text{...(5.49)}$$

or,
$$V_{CC} - V_{BE} - \beta I_B (R_L + R_E) - I_B R = 0 \quad \text{...(5.50)}$$

Solving for the base current, we get

$$I_B = \frac{V_{CC} - V_{BE}}{\beta (R_L + R_E) + R} \quad \text{...(5.51)}$$

Also, applying Kirchhoff's voltage laws to the collector emitter loop, one may write,

$$I_E R_E + V_{CE} + (I_C + I_B) R_L - V_{CC} = 0 \quad \text{...(5.52)}$$

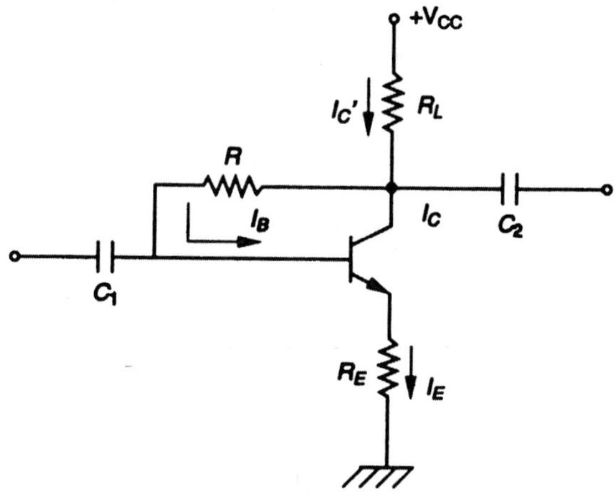

Fig 5.22. Another form of collector-to-base bias circuit

Since $I_B << I_C$,

$$I_C' \simeq I_C \simeq I_E \quad \text{...(5.53)}$$

Thus
$$I_C (R_L + R_E) + V_{CE} - V_{CC} = 0 \quad \text{...(5.54)}$$

and
$$V_{CE} = V_{CC} - I_C (R_E + R_L) \quad \text{...(5.55)}$$

EXAMPLE 5.7 *Find the quiscent values of current I_C and voltage V_{CE} for the circuit in Fig. 5.23.*

SOLUTION For this circuit

$$I_B = \frac{V_{CC} - V_{BE}}{\beta (R_L + R_E) + R} = \frac{11 - 0{\cdot}7}{75\,(5{\cdot}6\text{ k} + 1{\cdot}5\text{ k}) + 220\text{ k}}$$

$$= \frac{10{\cdot}3}{752{\cdot}5\text{ k}} = 13{\cdot}68\ \mu\text{A}$$

$$(I_C)_Q = \beta I_B = 75\,(13{\cdot}68\ \mu\text{A}) = 1{\cdot}03\text{ mA}$$

$$(V_{CE})_Q = V_{CC} - I_C (R_E + R_L)$$

$$= 11\text{ V} - (1{\cdot}03\text{ mA}) \times (5{\cdot}6\text{ k} + 1{\cdot}5\text{ k})$$

$$= 11 - 7{\cdot}31 = 3{\cdot}69\text{ V}.$$

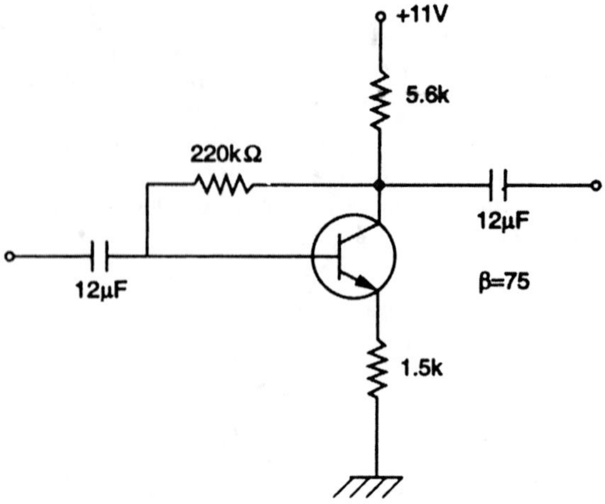

Fig 5.23. Reference Example 5.7

EXAMPLE 5.8 *Repeat example 5.6 if* $\beta = 120$

SOLUTION

The base current is $I_B = \dfrac{V_{CC} - V_{BE}}{\beta(R_L + R_E) + R}$

$$= \frac{(11 - 0{\cdot}7)\text{ V}}{220\text{ k}\Omega + 120\,(5{\cdot}6\text{ k}\Omega + 1{\cdot}5\text{ k}\Omega)}$$

$$= \frac{10{\cdot}3}{1072\text{ k}} = 9{\cdot}6 \times 10^{-3}\text{ A}$$

$$(I_C)_Q = \beta\, I_B = 120\,(9{\cdot}6 \times 10^{-3}) = 1{\cdot}15\text{ mA}$$

$$(V_{CE})_Q = V_{CC} - I_C\,(R_E + R_C)$$

$$= 11\text{ V} - (1{\cdot}15 \times 10^{-3})\,(5{\cdot}6\text{ k} + 1{\cdot}5\text{ k})$$

$$= (11 - 8{\cdot}19)\text{ V} = 2{\cdot}71\text{ V}$$

It will be seen in this example that even if β has increased by a factor of 60%, the value of I_C has increased by 11·6% and $(V_{CE})_Q$ decreased by a factor of 27%.

EXAMPLE 5.9 *Find the d.c. level of* I_B and V_C *for the network in Fig. 5.24.*

SOLUTION

For d.c. operation, the resistor consists of series connection of three resistance values 120 k, 90 k and 60 k. Thus $R = (120 + 90 + 60)\text{ k} = 270\text{ k}$.

$$I_B = \frac{V_{CC} - V_{BE}}{R + \beta\,(R_L + R_E)} = \frac{20\text{ V} - 0{\cdot}7\text{ V}}{270\text{ k} + 100\,(3{\cdot}9\text{ k} + 0{\cdot}41\text{ k})}$$

$$= \frac{19{\cdot}3\text{ V}}{270\text{ k} + 431\text{ k}} = 27{\cdot}5\ \mu\text{A}$$

$$I_C = \beta\, I_B = 100\,(27{\cdot}5\ \mu\text{A})$$

$$= 2{\cdot}75\text{ mA}$$

$$V_C = V_{CC} - I_C'\, R_L$$

$$= 20 - (2{\cdot}75\text{ mA})\,(3{\cdot}9\text{ k})$$

$= 20 - 10{\cdot}73$

$= 9{\cdot}27$ V.

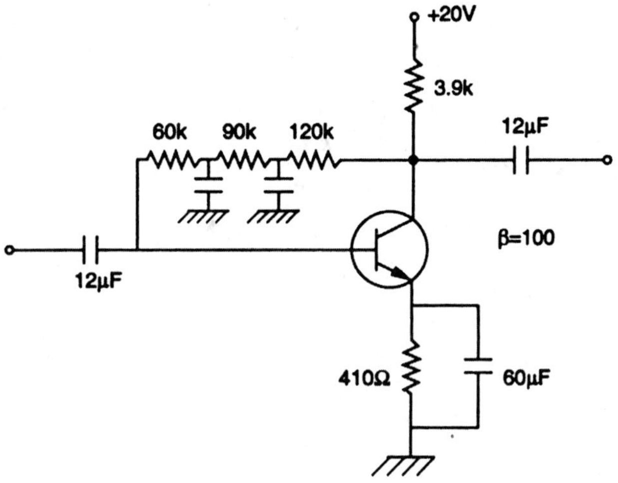

Fig 5.24. Reference Example 5.9

5.7 DESIGNING A BIAS NETWORK

The design process implies as to where a current and/or voltage should be set along with the components required to establish the designated levels. It requires a proper understanding of the basic circuit laws like Kirchhoff's voltage law, Ohm's law etc. If the transistors and supplies are mentioned, the designing involves determining the values of the required resistors. Sometimes, the manufacturer's specification (spec) sheets provide necessary suggestion regarding the operating point (or operating region). Moreover, other factors like current swing, voltage swing etc. may have to be considered.

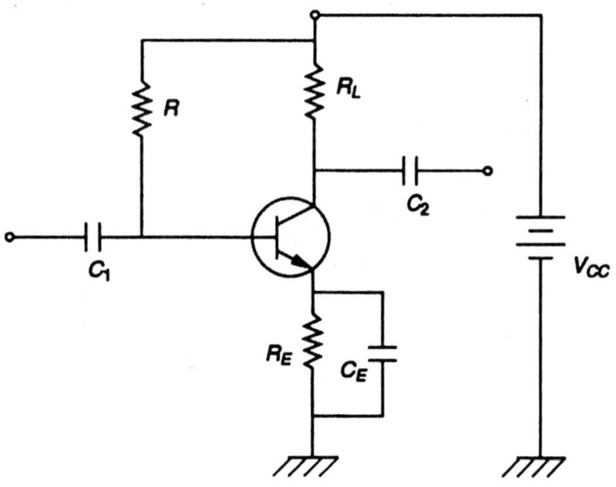

Fig 5.25. Emitter stabilized bias circuit

Let us consider a emitter stabilized bias circuit. In the circuit diagram (Fig. 5.25), the emitter resistor is used for d.c. bias stabilization such that collector current changes (due to temperature variations or other parameter changes) do not cause large changes in the quiscent operating point. The emitter resistor can not be very large because in that case the input voltage swing should be limited. Normally, the voltage drop across R_E is one-fourth to one tenth of the supply voltage.

EXAMPLE 5.10 *Design a bias circuit for the amplifier to establish a current $I_E = 1{\cdot}5mA$ using a power supply $V_{CC} = +15$ V (Assume $\beta_{DC} = 100$)*

SOLUTION As an approximation, we allocate one third of the supply voltage to the voltage drop across R_2. Also, the voltage drop across R_L is $\frac{1}{3}$rd of the supply voltage such that possible signal swing will be $\frac{1}{3}V_{CC}$.

Here $V_B = +5$ V

$$V_E = 5 - 0{\cdot}7 = 4{\cdot}3 \text{ V}$$

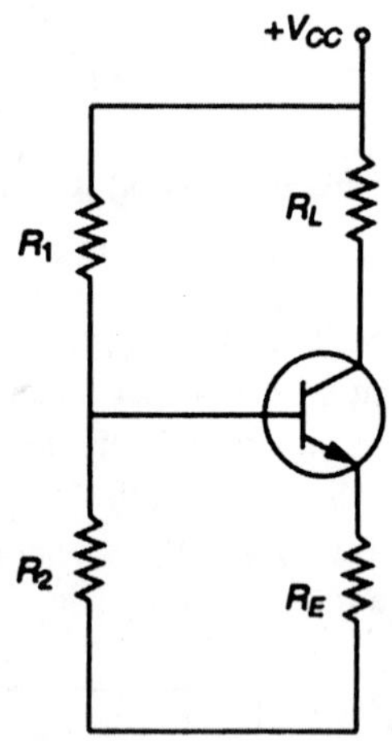

Fig E5.10. Reference Example 5.10

$$R_E = \frac{4{\cdot}3 \text{ V}}{1{\cdot}5 \text{ mA}} = 2{\cdot}86 \text{ k}\Omega$$

The voltage divider circuit is set as $0{\cdot}1\ I_E$. Neglecting base current, we find

$$R_1 + R_2 = \frac{15}{0{\cdot}1\ I_E} = \frac{15}{1{\cdot}5 \times 0{\cdot}1} = 100 \text{ k}\Omega$$

Also $\frac{R_2}{R_1 + R_2} V_{CC} = 5$ V

Thus $R_2 = \frac{5}{V_{CC}} \times (R_1 + R_2)$

$$= \frac{5}{15} \times 100 \text{ k} = 33{\cdot}3 \text{ k}\Omega$$

So $\quad R_1 = 66{\cdot}7\ \text{k}\Omega$

At this point it is desirable to find a more accurate estimate of I_E, taking into account the non-zero base current.

The emitter current I_E is given by

$$\frac{V_{BB} - V_{BE}}{R_E + \left(\dfrac{R}{\beta + 1}\right)} = \frac{4{\cdot}3}{2{\cdot}86\ \text{k} + \dfrac{22{\cdot}22\ \text{k}}{100}} = 1{\cdot}40\ \text{mA}$$

Since $\quad R = R_1 \parallel R_2 = \dfrac{R_1 R_2}{R_1 + R_2} = \dfrac{33{\cdot}3\ \text{k} \parallel 66{\cdot}7\ \text{k}}{33{\cdot}3\ \text{k} + 66{\cdot}7\ \text{k}}$

$$= 22{\cdot}22\ \text{k}$$

A more closer design to the desired 1·5 mA may be obtained by designing with exact equations.

EXAMPLE 5.11 *Find the stability factor for the collector to-base bias circuit of Fig. 5.26.*

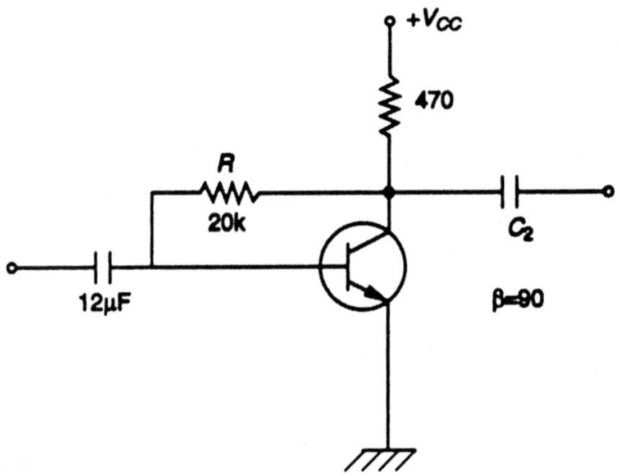

Fig 5.26. Reference Example 5.11

SOLUTION

$$S = \frac{\beta + 1}{1 + \dfrac{\beta R_L}{R_L + R}}$$

$$= \frac{90 + 1}{1 + \dfrac{90 \times 470}{470 + 20{,}000}}$$

$$= 29{\cdot}35$$

EXAMPLE 5.12 *What would be the stability factor if $R_L = 10\ \Omega$ (d.c. resistance of the primary of an output transformer).*

SOLUTION

$$S = \frac{\beta + 1}{1 + \dfrac{\beta R_L}{R_L + R}} = \frac{90 + 1}{1 + \dfrac{90 \times 10}{10 + 20{,}000}} = 87{\cdot}08$$

EXAMPLE 5.13 *The bias network shown in Fig. 5.27 below operates at the Quiscent point Q* ($I_C = 20$ mA, $V_{CE} = 6$ V). *The base bias current is 2 mA. Find the stability factor if* $\beta = 90$.

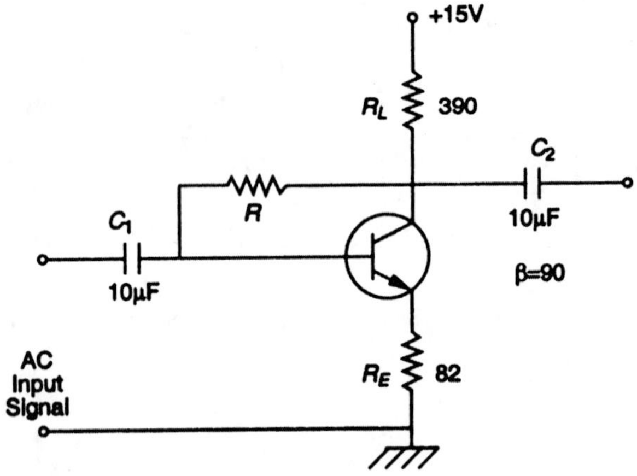

Fig 5.27. Bias network in Example 5.13

SOLUTION The Q point co-ordinates are $I_C = 20$ mA, $V_{CE} = 6$ V,

Therefore,

$$R = \frac{V_{BB} - V_{BE} - I_C R_E}{I_B}$$

$$= \frac{15 - 0{\cdot}7 - 20 \times 10^{-3} \times 82}{2 \text{ mA}}$$

$$= \frac{14{\cdot}7 - 1{\cdot}64}{2} = 6{\cdot}53 \text{ k}\Omega$$

$$S = \frac{\beta + 1}{1 + \dfrac{\beta R_E}{R_E + R}} = \frac{90 + 1}{1 + \dfrac{90 \times 82}{82 + 6{\cdot}53 \times 10^3}}$$

$$= \frac{91}{2{\cdot}116} = 43$$

EXAMPLE 5.14 *Find the Q point values for the circuit in Fig. 5.28.*

SOLUTION The collector current $I_C = \dfrac{V_{CC} - V_{BE}}{R_L + \dfrac{R}{\beta}}$

$$= \frac{12 - 0{\cdot}7}{6{\cdot}8 \text{ k} + \dfrac{82 \text{ k}}{75}} = 1{\cdot}43 \text{ mA}$$

Since the voltage at the emitter is zero,

$V_{CE} = V_C \approx V_{CC} - I_C R_L$

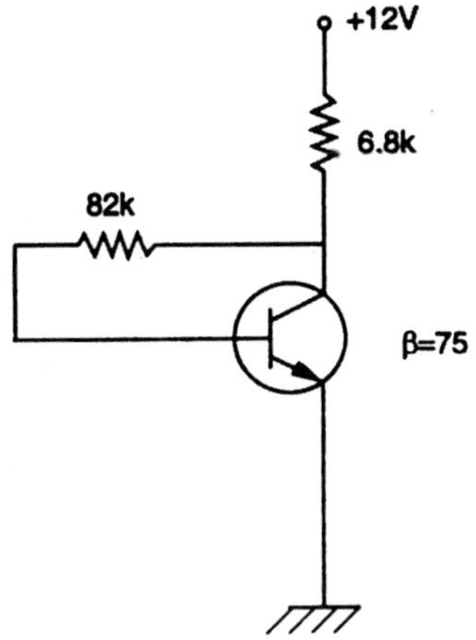

Fig 5.28 Reference example 5.14

$= 12 - (1{\cdot}43 \times 10^{-3}) \times 6{\cdot}8 \times 10^3$
$= 2{\cdot}27$ V

EXAMPLE 5.15 *Find the Q point values for the above circuit in Fig. 5.27 if* $\beta = 200$.

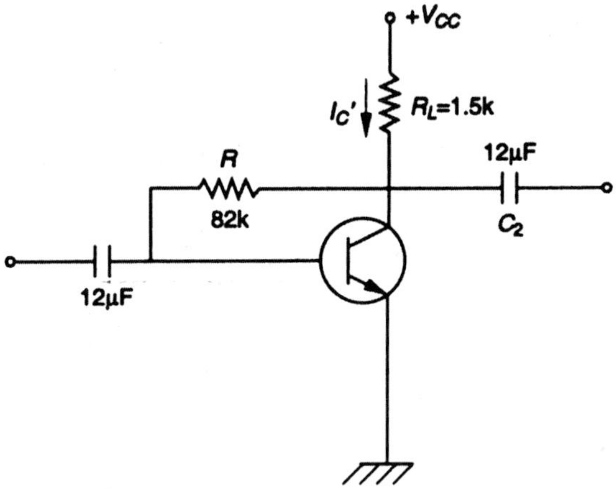

Fig E5.15.

SOLUTION

$$\text{Collector current} = I_C = \frac{V_{CC} - V_{BE}}{R_L + \dfrac{R}{\beta}}$$

$$= \frac{12 - 0{\cdot}7}{6{\cdot}8\text{ k} + \dfrac{82\text{ k}}{200}} = \frac{11{\cdot}3}{7{\cdot}21\text{ k}} = 1{\cdot}56 \text{ mA}$$

Since the voltage at the emitter is zero

$V_{CE} = V_C \simeq V_{CC} - I_C R_L$

$= 12 - (1{\cdot}56 \times 10^{-3}) \times 6{\cdot}8 \times 10^3$
$= 12 - 10{\cdot}66 = 1{\cdot}34$ V.

EXAMPLE 5.16 *A silicon bipolar junction transistor is used in the common emitter mode [Fig. E5.15]. Its* $\beta = 100$, $V_{CC} = 15$ V, $R_L = 1{\cdot}5$ kΩ, $R = 82$ kΩ. *Find the Q point analytically. Also determine the stability factors S, S′, S″.*

SOLUTION In this case

$V_{CC} = I_C' \times R_L + I_B R + V_{BE}$

or, $(I_C + I_B) \times 1{\cdot}5 \times 10^3 + I_B \times 8{\cdot}2 \times 10^3 + 0{\cdot}7 = 15$ V

$1500 I_C + 1500 I_B = 82000 I_B + 0{\cdot}7 = 15$ ...(1)

Also, $I_C = \beta I_B = 100 I_B$

Solving we get

$I_B = 61{\cdot}2$ μA

$I_C = 6{\cdot}12$ mA

Finally $V_{CE} = V_{BE} + I_B R$

$= 0{\cdot}7 + 61{\cdot}2 \times 10^{-6} \times 82 \times 10^3$

$= 5{\cdot}72$ volts

The Q point is defined by the values of I_B, I_C and V_{CE} above.

Thus $I_B = 61{\cdot}2$ μA, $I_C = 6{\cdot}12$ mA, $V_{CE} = 5{\cdot}72$ V

The stability factors are

$$S = \frac{\beta + 1}{1 + \dfrac{\beta R_L}{R_L + R}} = \frac{100 + 1}{1 + \dfrac{100 \times 1{\cdot}5}{(1{\cdot}5 + 82)}} = 36{\cdot}11$$

$$S' = \frac{-\beta}{R + R_L(1 + \beta)} = \frac{-100}{82 \times 10^3 + 1{\cdot}5 \times 10^3 \times 101}$$

$= -0{\cdot}428$ mA/V.

$$S'' = \frac{V_{CC} - V_{BE} - I_C R_L}{R + R_L/(1 + \beta)}$$

$$= \frac{15 - 0{\cdot}7 - 6{\cdot}12 \times 10^{-3} \times 1{\cdot}5 \times 10^3}{82\text{ k} + \dfrac{1{\cdot}5\text{ k}}{101}}$$

$= 64{\cdot}42$ μA

EXAMPLE 5.17 *A silicon transistor with* $\beta = 100$ *is arranged as shown in the circuit diagram of Fig. E5.17. The parameters are* $V_{CC} = 15$ V, $R_L = 1$ kΩ, $V_{CE} = 7{\cdot}5$ V, $I_C = 6$ A, $V_{BE} = 0{\cdot}7$ V *and. The stability factor has a value* $S = 12$. *Find the values of* R_1, R_2 *and* R_E.

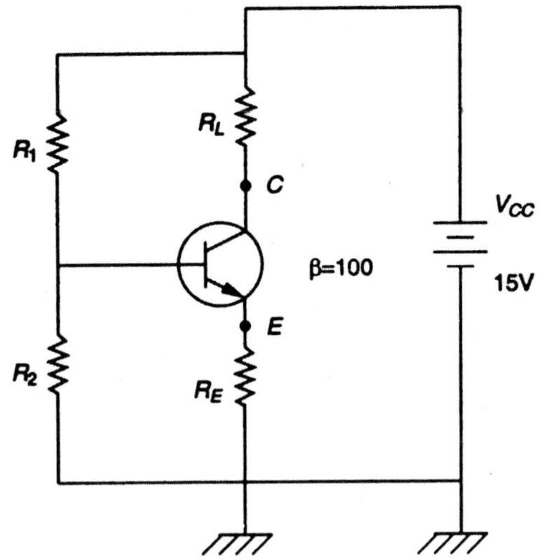

Fig E5.17.

SOLUTION

$$I_B = \frac{I_C}{\beta} = \frac{6\text{ mA}}{100} = 60\ \mu\text{A}$$

$$R_E = \frac{V_{CC} - V_{CE} - I_C R_L}{(I_B + I_C)} = \frac{15 - 7{\cdot}5 - 6 \times 10^{-3} \times 1 \times 10^3}{(6 \times 10^{-3} + 60 \times 10^{-6})}$$

$$= 0{\cdot}248\ \text{k}\Omega$$

From the equation

$$S = (1 + \beta)\,\frac{1 + R_{Th}/R_E}{1 + \beta + \dfrac{R_{th}}{R_E}}$$

$$12 = (1 + 100)\,\frac{1 + (R_{Th}/R_E)}{1 + 100 + (R_{Th}/R_E)}$$

$$\text{or,}\quad 12 \times 101 + 12\left(\frac{R_{Th}}{R_E}\right) = 101 + 101\left(\frac{R_{Th}}{R_E}\right)$$

$$\text{or,}\quad \frac{R_{Th}}{248} = \frac{11 \times 101}{89}$$

$$\text{or,}\ R_{Th} = \frac{11 \times 101 \times 248}{89} = 3096\ \Omega = \frac{R_1 R_2}{R + R_2},$$

Using the equation

$$V_{Th} = I_B R_{Th} + V_{BE} + (I_C + I_B) R_E$$

$$\text{or,}\quad \frac{15 R_2}{R_1 + R_2} = 60 \times 10^{-6} \times 3096 + 0{\cdot}7 + 6{\cdot}06 \times 10^{-3} \times 248$$

Solving the above equations, we get

$$R_1 = 26{\cdot}46\ \text{k}\Omega, \quad R_2 = 35{\cdot}07\ \Omega$$

Emitter bias with two supplies

The circuit in Fig. 5.29 uses two supplies, a positive supply and a negative supply.

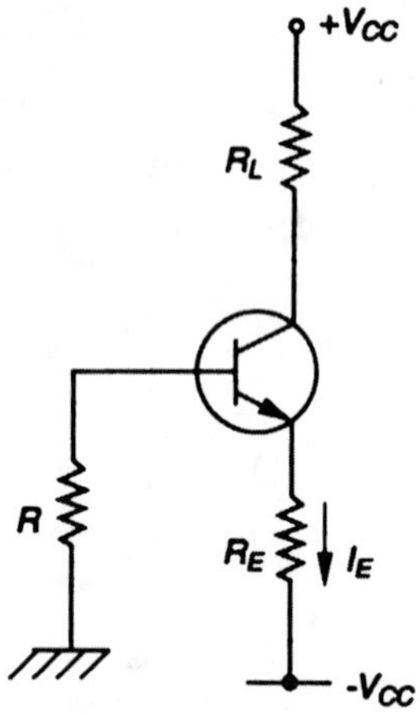

Fig 5.29. Emitter bias circuit with two suppliers

Using Kirchhoff's voltage laws (clockwise) we get

$$-I_B R - V_{BE} - I_E R_E + V_{EE} = 0 \quad ...(5.56)$$

$$I_B R + I_E R_E = V_{EE} - V_{BE} \quad ...(5.57)$$

Now $\quad I_C = \beta I_B \quad ...(5.58)$

$\therefore$
$$I_E = \frac{V_{EE} - V_{BE}}{R_E + \frac{R}{\beta}} \quad ...(5.59)$$

If $\quad V_{EE} >> V_{BE}$ and $\beta R_E >> R_B$,

$$I_E = \frac{V_{EE}}{R_E} \quad ...(5.60)$$

The emitter to ground voltage V_E is

$$V_E = -(I_B R + V_{BE}) \simeq -V_{BE} \quad ...(5.61)$$

The stability factor

$$S = \frac{(1 + (R/R_E)}{1 + (R/\beta R_E)} \quad ...(5.62)$$

EXAMPLE 5.18 *For the circuit in Fig. 5.30, find the value of (i) I_E (ii) I_C (iii) V_C (iv) V_E (v) V_{CE} and (vi) stability factor if* $\beta = 100$, $V_{BE} = 0.7$ V, $V_{CC} = 18$ V, $V_{EE} = 9$ V.

SOLUTION

(i) $I_E = \frac{V_{EE} - V_{BE}}{R_E + R/\beta} = \frac{9 - 0.7}{\left(30 + \frac{15}{100}\right)\text{k}} = \frac{8.3}{30.15\text{ k}} = .275$ mA

(ii) $I_C \simeq I_E = 0.275$ mA

(iii) $V_C = V_{CC} - I_C R_L = 18 - 0.275 \times 10^{-3} \times 9 = 18 - 2.47$

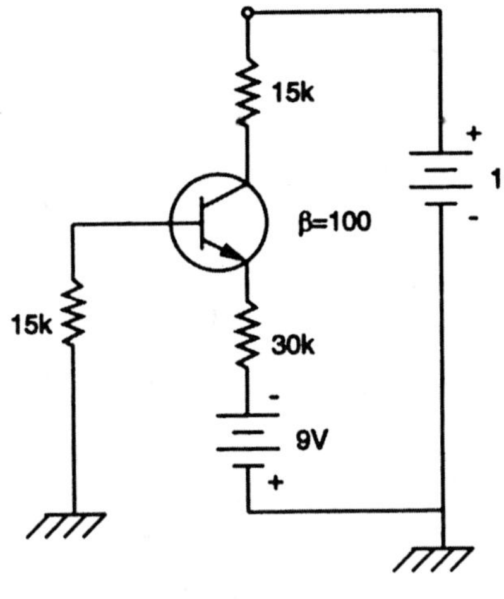

Fig 5.30. Reference Example 5.18

$= 15{\cdot}53$ V.

(iv) $V_E = -\left(\dfrac{I_C R}{\beta} + V_{BE}\right)$

$= -\left(\dfrac{0{\cdot}275 \times 10^{-3} \times 15 \times 10^3}{100} + 0{\cdot}7\right)$

$= -({\cdot}04 + 0{\cdot}7) = -0{\cdot}74$ V

(v) $V_{CE} = V_C - V_E = 15{\cdot}53 - (-0{\cdot}74)$

$= 16{\cdot}27$ V

(vi) Stability factor $S = \dfrac{1 + R/R_E}{1 + R/\beta R_E} = \dfrac{1 + \dfrac{15\text{ k}}{30\text{k}}}{1 + \dfrac{15\text{ k}}{100 \times 30\text{ k}}}$

$= \dfrac{1{\cdot}5}{1{\cdot}005} = 1{\cdot}4925$

Objective Type Questions

Select the correct alternative(s) in the following :

1. For operation of a bipolar junction transistor as an amplifier, the transistor operates in
(a) cut off region (b) saturation region
(c) active region (d) deep in saturation

2. The load line shifts parallel to itself on the *CE* output characteristics of a transistor when
(a) R_L is changed (b) V_{CC} is changed
(c) both R_L and V_{CC} is changed (d) base current is changed

3. The d.c. load line of a transistor
(a) has a positive slope (b) has a negative slope

(c) does not contain the Q point (d) is a plot of I_C versus I_B

4. A CE transistor circuit employing base bias with collector feedback possesses greater stability than the one without feedback because
(a) V_{BE} decreases (b) I_C decreases
(c) I_C is independent of β (d) of negative feedback effect

5. The self bias arrangement gives an improved Q-point stability when
(a) R_E is low (b) β is small, but R_E is large
(c) both β and R_E are large

6. The biasing method which is considered independent of transistor β_{dc} is
(a) fixed biasing (b) collecor feedback bias
(c) voltage divider bias (d) base bias with collector feedback

7. Under ideal conditions, the d.c. line is a straight line drawn on the collector characteristic curves between
(a) the Q point and cut off (b) the Q point and saturation
(c) $I_B = 0$ and $I_B = I_C/\beta$ (d) $(V_{CE})_{cut\ off}$ and $(I_C)_{sat}$

8. The voltage divider bias network
(a) can be independent of β (b) cannot be independent of β
(c) is not widely used (d) has very good stability factor

9. In a voltage divider biased npn transistor, if the upper voltage divider resistor (R_1) is an open circuit ($R_1 \to \infty$) then
(a) the transistor is destroyed (b) the transistor goes to cut off
(c) the transistor becomes saturated (d) the transistor temperature rises

10. In the saturation region of operation of a transistor
(a) the emitter-base junction is forward biased and the collector-base junction is reverse biased
(b) the emitter-base junction is reverse biased and the collector base junction is forward biased
(c) the emitter base junction is forward biased and the collector base junction is forward biased
(d) Both the emitter base and collector base junctions are reverse biased.

11. If the value of V_{CC} in a transistor bias circuit is changed without changing other parameters of components,
(a) the load line shifts parallel to itself (b) the load line slope changes
(c) the Q point remains unchanged (d) the load line slope becomes positive.

12. If the load resistance in a feedback network is increased, then
(a) the Q point remains unchanged (b) the d.c. load line shifts parallel to itself
(c) the Q point has lower $(V_{CE})_Q$ values (d) the Q point has higher $(I_C)_Q$ values

13. For a silicon n–p–n transistor
(a) β increases with increases in temperature (b) $|V_{BE}|$ increases with increase in temperature
(c) β is independent of temperature changes (d) $|V_{BE}|$ is independent of temperature changes

14. The stability factor S increases
(a) the rate of variation of I_C with $|V_{BE}|$ (b) the rate of variation of I_C with β
(c) the rate of variation of I_C with I_{CO} (d) the rate of variation of I_C with temperature

15. For the emitter-bias configuration, the stability factor S
(a) increases with increase in the ratio of (base resistance R/emitter resistance R_E)
(b) decreases with increase of R/R_E (c) reaches a value $(1 + \beta)$ when $\frac{R}{R_E} < \beta + 1$
(d) can reach a maximum value of $1 + 2\beta$.

16. For a good stabilized bias circuit design the ratio (R/R_E) or (R_{Th}/R_E) should be
(a) as high as possible (b) as small as possible
(c) constant (d) independent of temperature

Exercise Problems

1. Calculate for the circuit in Fig. 5.31, the following quantities for the Q point

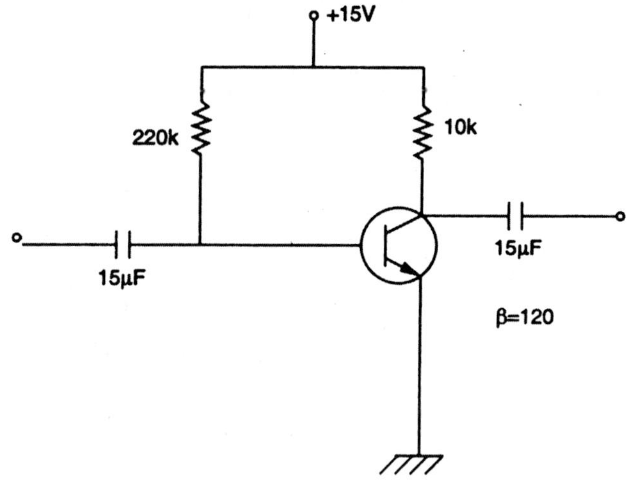

Fig 5.31. Reference Exercise Problem 5.1

(i) I_B and I_C (ii) V_{CE}

(iii) V_C and V_B (iv) V_{CE}

2. Repeat problem 1 if V_{CC} = 18 V and β = 200.

3. Find the saturation current $(I_C)_{sat}$ in Fig. 5.31 above (Prob. 1)

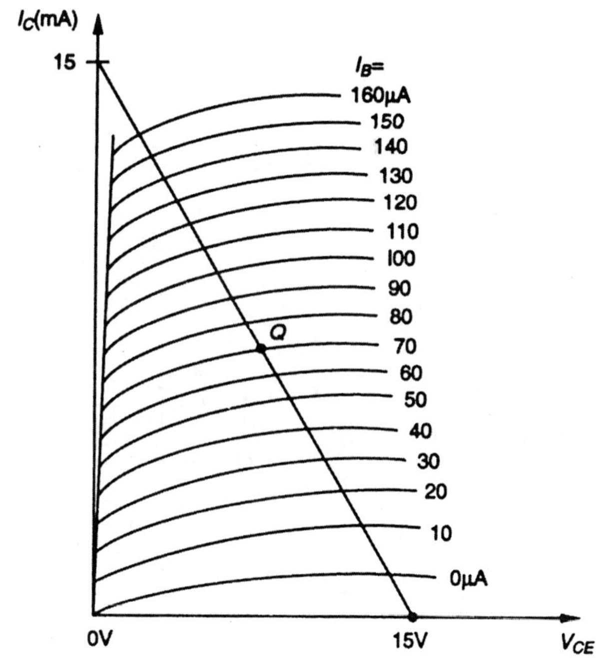

Fig 5.32. Reference Exercise Problem 5.4

4. Given the load line and the defined Q point for the Fig. 5.32, find the values of V_{CC}, R and R_L under fixed bias configuration

5. For the emitter bias stabilized circuit of Fig. 5.33 find

(i) I_B (ii) I_C

(iii) V_{CE} (iv) V_C

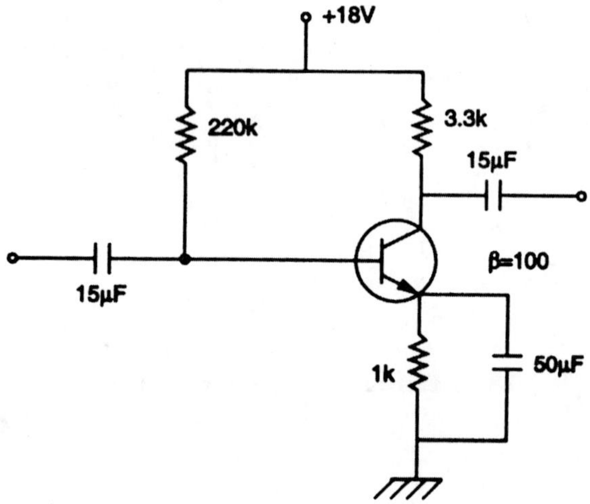

Fig 5.33. Reference Exercise Problem 5

(v) V_E (vi) V_B

(vii) V_{CB}

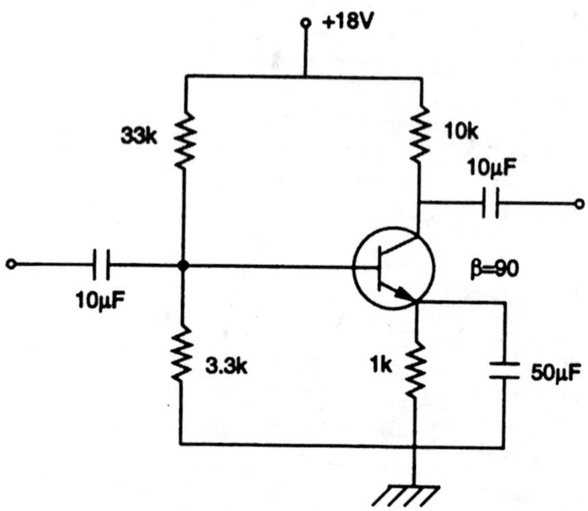

Fig 5.34. Reference Exercise Problem 6

6. Find the values of I_C and V_{CE} for the d.c. bias circuit depicted in Fig. 5.34.

7. Calculate the values of the stability factors S, S' and S'' for the circuit of Fig. 5.34. [Assume I_{CO} = 1 μA]

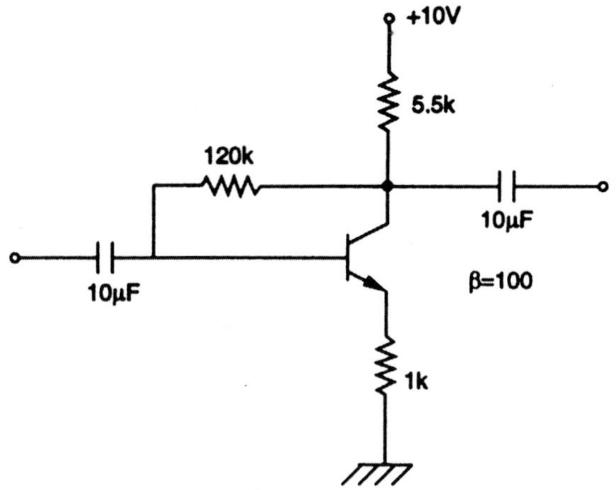

Fig 5.35. Reference Exercise Problem 8

8. Find the Quiscent values of current I_C and voltage V_{CE} for the circuit of Fig. 5.35.
9. Repeat the problem 8 above for β = 200.

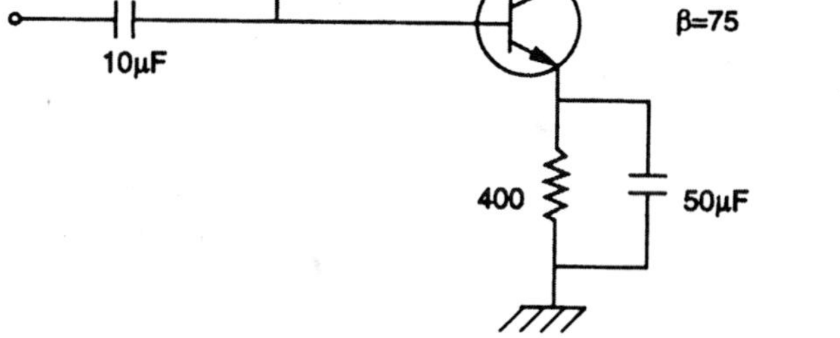

Fig 5.36. Reference Exercise Problem 10

10. Find the d.c. level of I_B and V_C for the network in Fig. 5.36.

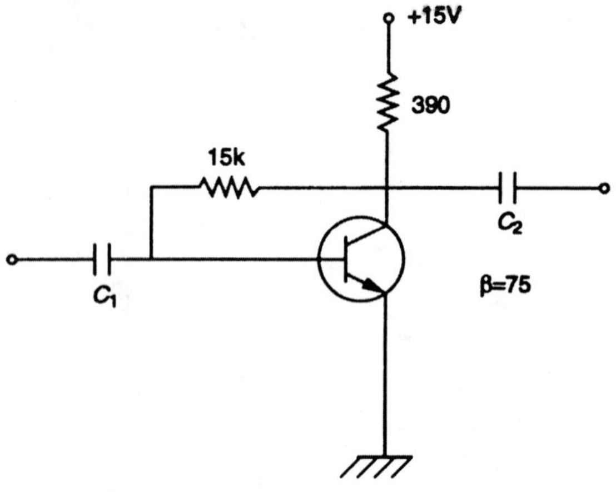

Fig 5.37. Reference Exercise Problem 11

11. Find the stability factor S [I_{CO}] for the collector-to-base bias circuit of Fig. 5.37.
12. What would be the stability factor if $R_L = 10\ \Omega$ instead of 390 Ω in Fig. 5.37.
13. Design a bias circuit for the *CE* amplifier to establish a current of $I_E = 2$ mA using a power supply of $V_{CC} = +\ 18$ V. [Assume β= 75]
14. Design a bias circuit for a silicon *npn* common emitter amplifier to establish as emitter current $I_E = 4$ mA using a power supply $V_{CC} = 20$ V. [Assume β = 100]
15. Find the *Q* point values for the circuit of Fig. 5.37.

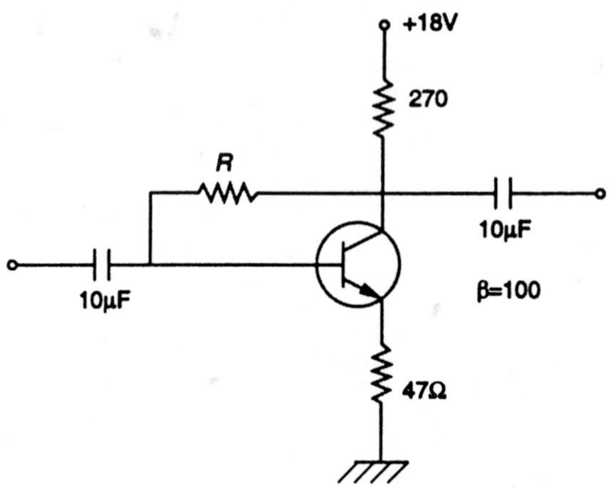

Fig 5.38. Reference Exercise Problem 16

17. Repeat the above problem 16 if $\beta = 200$ instead of 100.

18. A silicon bipolar junction transistor is used in the common emitter mode. [Fig. 5.39]. If $\beta = 150$, $V_{CC} = +18$ V, $R_L = 1{\cdot}5$ kΩ, $R = 100$ kΩ, find the Q point analytically. Also calculate the stability factors S, S' and S''.

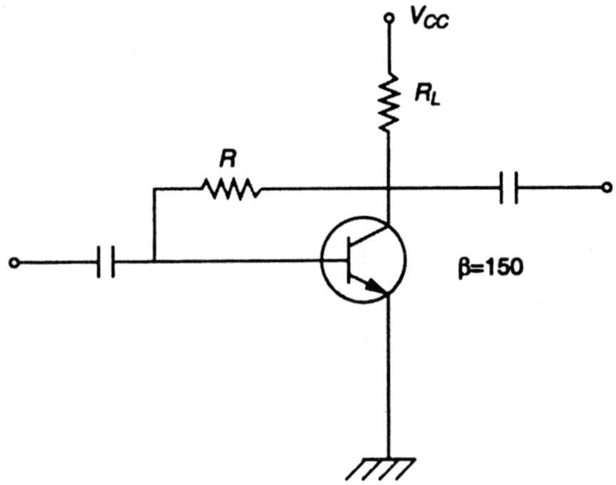

Fig 5.39. Reference Exercise Problem 16

19. A silicon transistor with $\beta = 100$ is arranged as shown in the circuit diagram of Fig. 5.40. The parameters are

$V_{CC} = 18$ V, $R_L = 1{\cdot}5$ kΩ, $V_{CE} = 9$ V, $I_C = 5$ mA and $V_{BE} = 0{\cdot}7$ V. The stability factor has a value of $S = 10$. Find the values of R_1, R_2 and R_E.

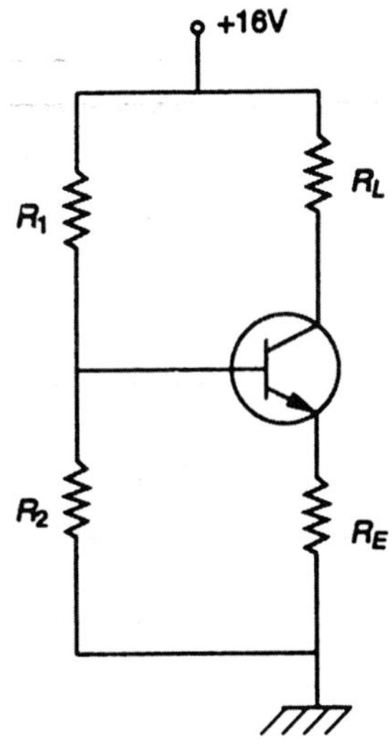

Fig 5.40. Reference Exercise Problem 18

20. Repeat the problem 19 above if the specified value of stability factor is (i) 5 (ii) 20.

Chapter 6

SINGLE STAGE BJT AMPLIFIERS

6.1 INTRODUCTION

Amplifiers are needed in electronic circuits so as to increase the magnitude of a weak signal. In many electronic equipments, amplification of weak electrical signals is a necessity. In the radio receivers, very feeble signals at the antenna are amplified so as to be audible for human hearing. In television, both the sound and picture signals should be suitably amplified. In the medical and scientific instrumentation, amplifiers are needed so as to amplify very feeble (μV and mV) order signals so as to be useful. The amplified signal may be necessary to drive various indicating instruments. Electronic amplifiers are a constant and important part of various electronic equipments and systems.

6.2 SINGLE STAGE TRANSISTOR AMPLIFIER

By the term single stage transistor amplifer, we mean one stage comprising a single transistor along with its bias circuit and auxiliary components. If only one transistor with associated circuits is used for amplifying a weak signal, the circuit is known as a single stage transistor amplifier.

Practical Transistor Amplifier Circuit

In a practical transistor amplifier circuit, when a weak signal is fed to the input or base terminals of a common emitter amplifier [Fig. 6.1], a base current i_B flows. Due to transistor action, a much larger current $\beta\, i_B$ flows in the collector circuit. As the value of R_L is large, a quite high value of voltage drop occurs in the collector circuit. Thus a small change i_B in the base current gives rise to a very large change i_C in the collector circuit. Therefore a weak a.c. signal in the base circuit appears as a large a.c. voltage in the collector circuit. A transistor can faithfully reproduce the input signal if proper biasing circuit and other components are employed.

In the circuit diagram (Fig. 6.1), the input capacitor normally is an electrolytic capacitor of magnitude about 10-15 μF. It is used to couple the weak a.c. signal to the base of the transistor. If C_{in} is not employed the source resistance R_s will fall across R_2 and disturb the bias circuit. The capacitor C_{in} permits only the a.c. signal to flow while isolating the signal source from R_2.

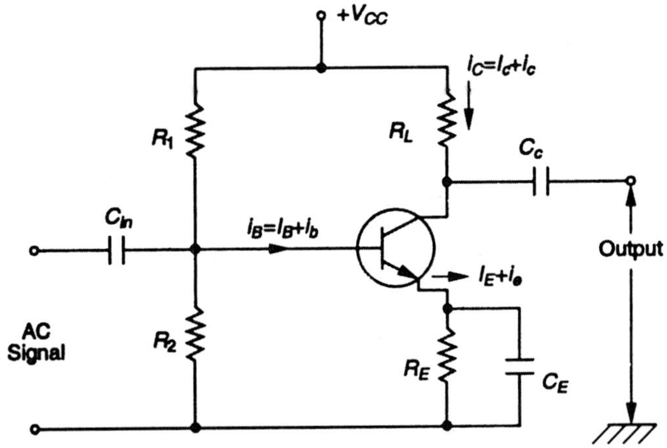

Fig 6.1. Practical transistor amplifier circuit

The emitter bypass capacitor ($\simeq$ 50 μF) is used to give a low reactance path to the amplified a.c. signal in parallel to R_E. If C_E is not connected, the amplified a.c. signal flowing through R_E will produce a voltage drop across R_E and reduce the output voltage.

The coupling capacitor (10-15 μF) couples one stage of amplification to the next stage. The coupling capacitor provides d.c. isolation of one stage from the next stage.

When no signal is applied to the base circuit, the d.c. base current is I_B, the d.c. collector current is $I_C = \beta\, I_B$ and the d.c. emitter current is I_E. The d.c. collector-to-emitter voltage drop is V_{CE}. If i_b is the base current change (or a.c. base current) then the total base current is

$$i_B = I_B + i_b \quad ...(6.1)$$

The a.c. collector current is $i_c = \beta\, i_b$ ...(6.2)

= collector current due to application of signal at the base.

The total collector current is

$$i_C = I_C + i_c \quad ...(6.3)$$

with I_C = Quiscent collector current

$$= \beta\, I_B \quad ...(6.4)$$

The total emitter current is

$$i_E = I_E + i_e \quad ...(6.5)$$

where $I_E = DC$ emitter current

i_e = a.c. magnitude of the emitter current

The base current is usually very low and with reasonable approximation we may write

$$I_E \simeq I_C \quad ...(6.6)$$

$$i_e \simeq i_c \quad ...(6.7)$$

Further, $i_e = i_b + i_c$ ...(6.8)

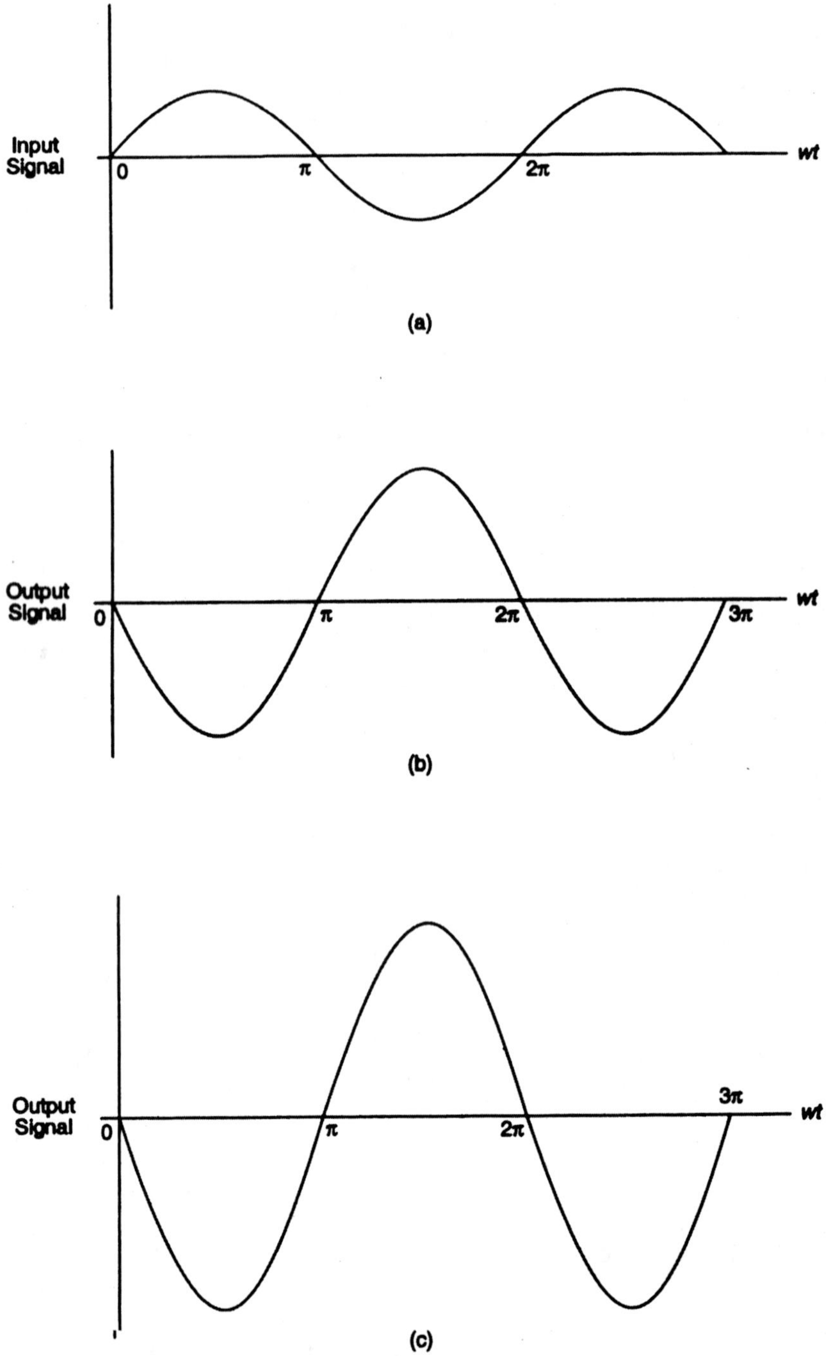

Fig 6.2. Input and output signals to a single stage transistor amplifier. There is a phase difference of 180° between the output and input signals (phase reversal) Figures (b) and (c) refer to different gain values.

A phase difference of 180° ($= \pi$) occurs between the signal voltage and output voltage in a common emitter amplifier. This is also called as **phase reversal**. [vide Fig. 6.2]

Let us consider the common emitter amplifier circuit of Fig. 6.1. For the circuit, tbe total instantaneous voltage output is

$$v_{CE} = V_{cc} - i_c R_C \quad \ldots(6.9)$$

With increase in the input signal voltage v_s during the positive half cycle, the base current i_b increases. When i_b increases, i_c the collector current also increases. With increase in i_c, the voltage drop $i_c R_c$ increases. Since V_{cc} is constant, the output voltage v_{Ce} decreases. This means that as the signal voltage is increasing during the positive half cycle, the output voltage is decreasing (or increasing in the negative direction). This implies that the output is 180° out of phase with the input. Thus in a common emitter amplifier, the positive half cycle will appear as amplified negative in the output [between 0 and π in Fig. 6.2]. Similarly, the negative

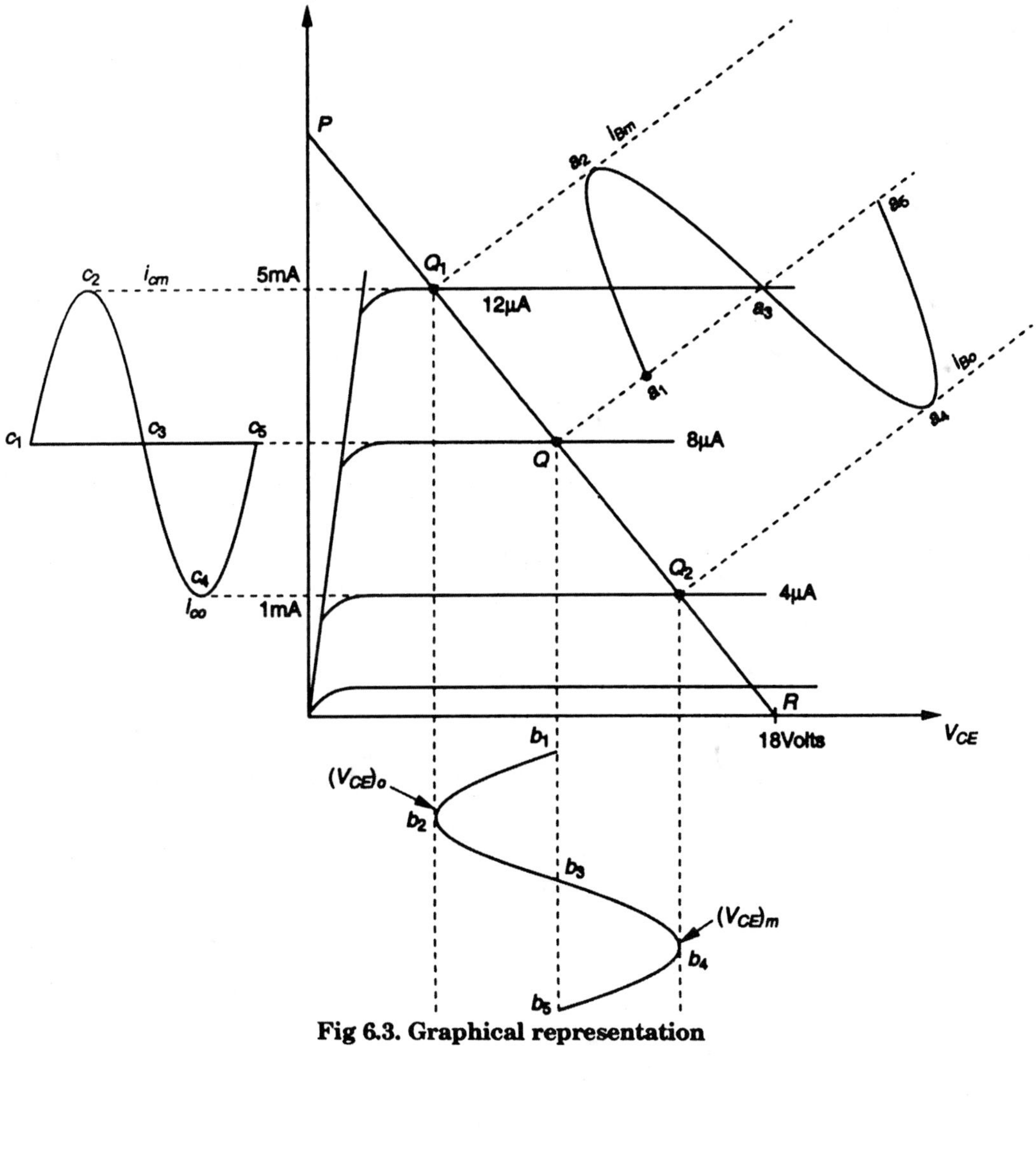

Fig 6.3. Graphical representation

half wave in the input during the negative half cycle will appear as amplified positive half wave [between π and 2π in Fig. 6.2].

A differentiation of Equation (6.9) shows that $\frac{dv_{CE}}{di_C} = 0 - R_C$ ...(6.10)

The negative sign implies that output is 180° out of phase with the input.

The collector current characteristics can be demonstrated as in Fig. 6.3. It will be seen that there is a phase difference of 180° between the output voltage and the input signal voltage. In the $I_C - V_{CE}$ characteristics of Fig. 6.3, PR is the load line and Q is the quiscent operating point. The quiescent base current is 8 μA. The base current fluctuates by ± 4 μA. It will be seen that when the value of V_{CE} is represented by point 1 ($v_{CE} = 0$, $i_C = 0$), the base current is also zero (represented by point 1). When the a.c. base current and a.c. value of V_{CE} is i_{Bm} and V_{CEO} respectively (point 2), the corresponding collector current is i_{Cm} (point 2). When the a.c. base current returns back to point 3, the quiscent operating point is Q. When the base current corresponds to point a_4 ($i_B = i_{Bo}$), the a.c. collector to emitter voltage is $(V_{CE})_m$ corresponding to point b_4 and the a.c. collector current is i_{CO} (corresponding to point c_4). The corresponding operating point is Q_2 in the load line. When the a.c. base current is zero again (point a_5), the a.c. magnitude of v_{CE} is (point b_5) and the a.c. collector current corresponds to point c_5 (operating point Q in the load line. Thus as i_B swings between i_{Bo} and i_{Bm}, the Q point moves between Q_2 to Q_1. In the plot of Fig. 6.3, the value of i_B swings around 8 μA by ± 4 μA. It should also be noticed from the graph that the input and output voltages are in phase opposition.

6.3 EQUIVALENT CIRCUITS FOR DC AND AC

In order to analyse a common emitter amplifier for example, we have to analyse both d.c. and a.c. currents and voltages. It will be seen that d.c. sources estabilsh d.c. voltages and currents. Similarly a.c. sources produces changes or variations of the current and voltages.

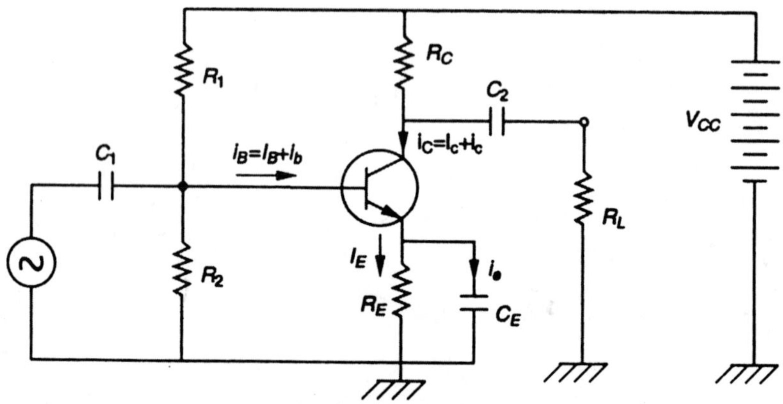

Fig 6.4. Common emitter amplifier circuit

We can then think of the circuit analysis into two parts namely d.c. analysis and a.c. analysis. In the d.c. analysis d.c. sources establish the d.c. voltages and currents into the circuit. For a.c. analysis, we consider all a.c. components at the same time and find the a.c. currents and voltages. By addition of the d.c. and a.c.

components of currents and voltages we then get the total currents and voltages in the circuit. The circuit can be easily analysed by considering seperately the d.c. equivalent circuit and a.c. equivalent circuit.

(a) Electrical equivalent circuit for d.c.

The d.c. electrical equivalent circuit can be drawn from the circuit diagram by considering that all a.c. sources are absent and by considering the capacitors to be looking like open circuits $\left(\frac{1}{\omega C} \to \infty \text{ as } \omega \text{ is zero}\right)$

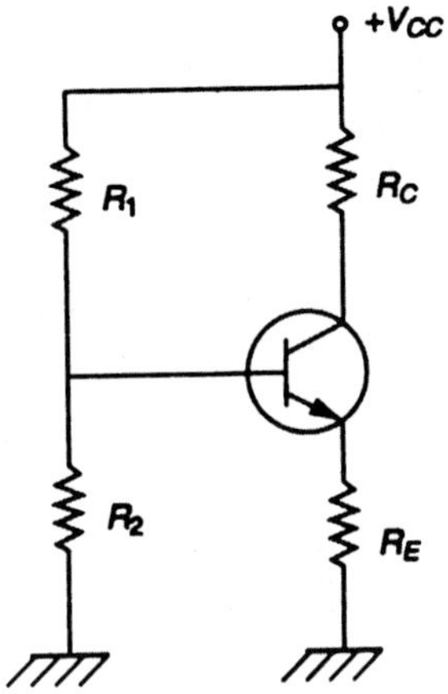

Fig 6.5. Electrical equivalent circuit for d.c.

(b) Electrical equivalent circuit for a.c.

For determining the a.c. equivalent circuit, all d.c. sources are assumed to be zero ($V_{CC} = 0$) and all the capacitors are short circuited $\left(\frac{1}{\omega C} \to 0\right)$

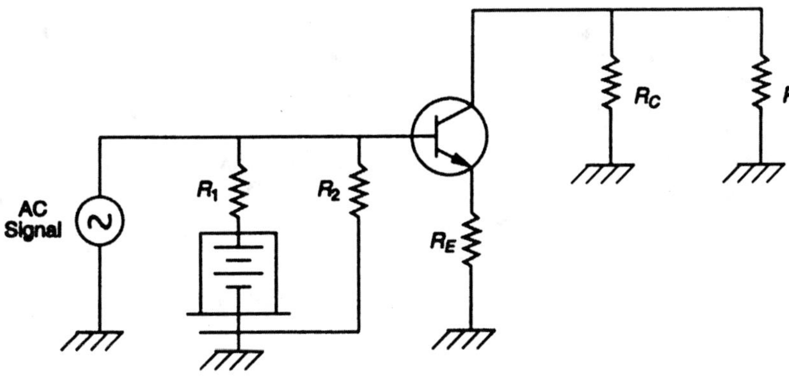

Fig 6.6. Electrical equivalent circuit for a.c.

It will be seen that the total current in any branch is the algebraic sum of the d.c. and a.c. currents in the branch while the total voltage across two points is the algebraic sum of the d.c. and a.c. voltages across the points.

EXAMPLE 6.1 *The common emitter circuit has* $V_{CC} = 20$ V, $R_C = 1{\cdot}5$ kΩ, $R_E = 1{\cdot}8$ kΩ, $R_1 = 8{\cdot}2$ kΩ *and* $R_2 = 3{\cdot}9$ kΩ. R_E *is bypassed to a.c. signals by the large capacitor* C_E *and the transistor employed has the characteristics shown in Fig. 6.7. Draw the d.c. and a.c. load lines for the circuit.*

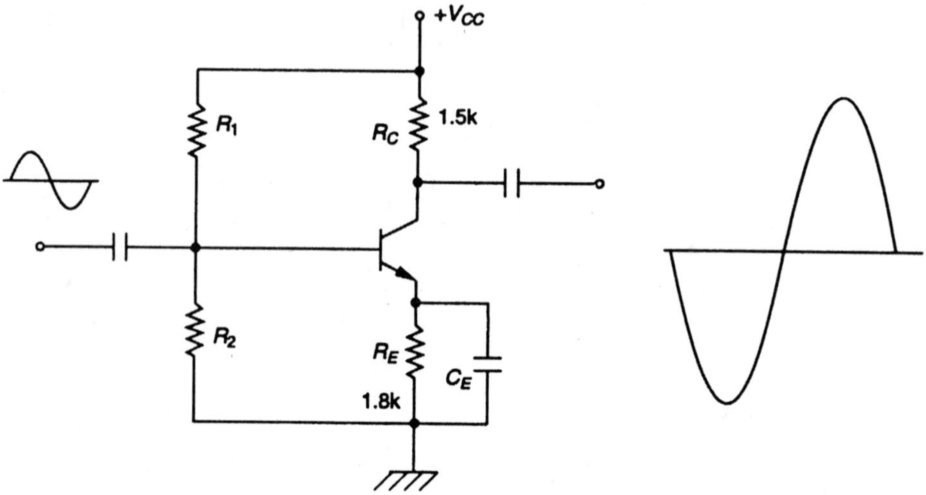

Fig 6.7. Reference example 6.1

SOLUTION **The D.C. load line**

Total d.c. load = $R_C + R_E = 1{\cdot}5\text{ k} + 1{\cdot}8\text{ k} = 3{\cdot}3\text{ k}$

$$V_{CE} = \text{Collector-to-emitter voltage}$$
$$= V_{CC} - I_C(R_C + R_E)$$
$$= 20 - I_C(R_C + R_E)$$

If $I_C = 0$, $V_{CE} = 20$ V. This gives point R in Fig.

When $V_{CE} = 0,\ I_C = \dfrac{20}{R_C + R_E} = \dfrac{20}{3{\cdot}3\text{ k}} = 6{\cdot}06\text{ mA}$

This gives point P in the $I_C - V_{CE}$ characteristics. The load line can then be drawn through points P and R.

Quiscent Operating (Q) Point

The approximate base bias conditions can be determined by assuming that the base current is very low so as to influence V_B.

Thus $$V_B = V_{CC} \times \frac{R_2}{R_1 + R_2} = \frac{20\text{ V} \times 3{\cdot}9\text{ k}}{3{\cdot}9\text{ k} + 8{\cdot}2\text{ k}} = \frac{20\text{ V} \times 3{\cdot}9\text{ k}}{12{\cdot}1\text{ k}}$$

$= 6{\cdot}45$ V.

∴ Voltage drop across

$$R_E = V_B - V_{BE}$$
$$= 6{\cdot}45 \text{ V} - 0{\cdot}7 \text{ V} \quad \text{(for silicon)}$$
$$= 5{\cdot}75 \text{ V}.$$

Thus $$I_C \approx I_E = \frac{V_B - V_{BE}}{R_E} = \frac{5{\cdot}75 \text{ V}}{1{\cdot}8 \text{ k}}$$
$$= 3{\cdot}19 \text{ mA}.$$

Accordingly, the bias point Q is marked on the d.c. load line at $I_C = 3{\cdot}19$ mA. (see Fig. 6.8).

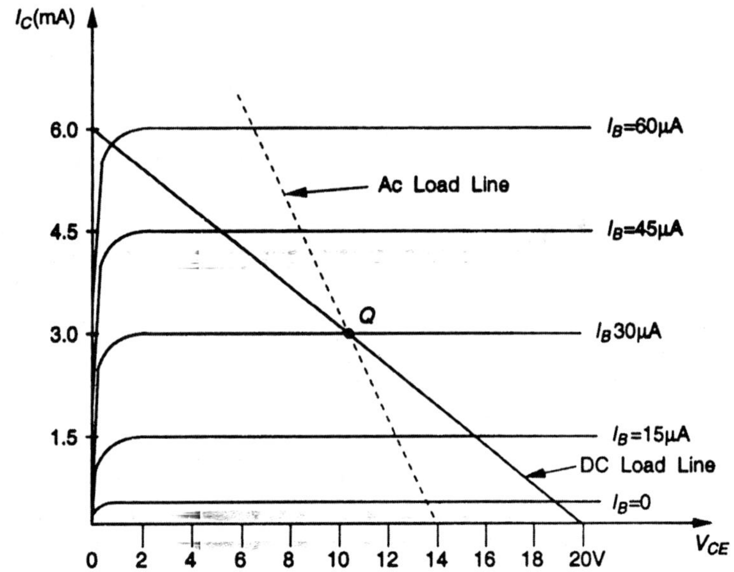

Fig 6.8.

The a.c. load line

Total a.c. load = R_C

When the collector current changes by a small amount ΔI_C,

$$\Delta V_{CE} = \Delta I_C \,.\, R_C$$

Thus, taking ΔI_C to be 1 mA

$$\Delta V_{CE} = -(1 \text{ mA} \times 1{\cdot}5 \text{ k}) = -1{\cdot}5 \text{ V}.$$

Thus when I_C increases by 1 mA (from point Q), V_{CE} decreases by 1·5 V. Plot ΔV_C and ΔI_C as another point on the a.c. load line. Draw a line through Q and C. This gives the a.c. load line.

EXAMPLE 6.2 *For the circuit in Fig. 6.9, the d.c. equivalent circuit is as follows.*

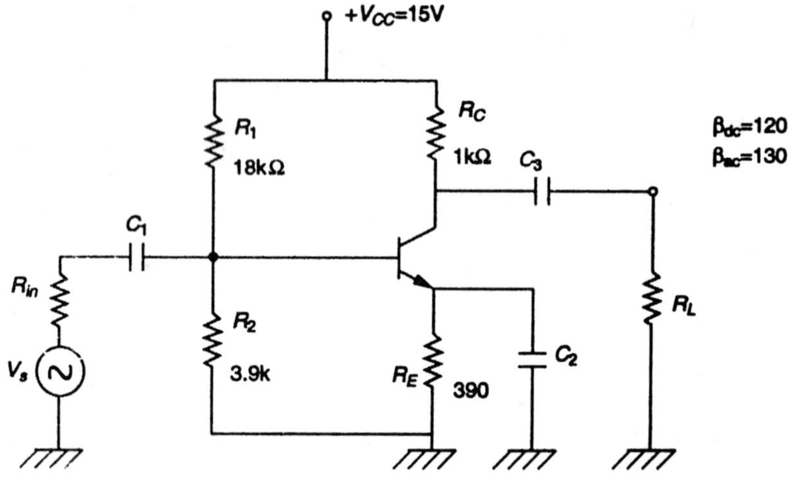

Fig 6.9. A common emitter amplifier

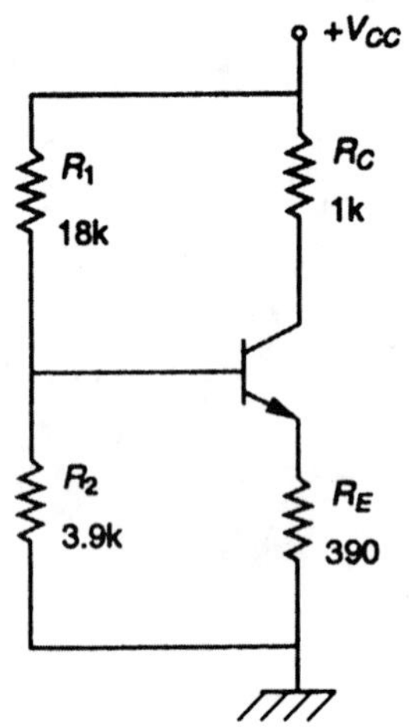

Fig 6.10. DC equivalent of circuit of fig. 6.9

SOLUTION **D.C. Analysis**

The d.c. input resistance = $R_{in} = \beta_{dc} \cdot R_E$

$= 120 \times 390 = 46{\cdot}8$ kΩ

Since this value R_{in} is more than 10 times the value of R_2, it can be neglected while finding the base voltage.

Thus $V_B = \left(\frac{R_2}{R_1 + R_2}\right) V_{CC} = \left(\frac{3{\cdot}9 \text{ k}}{18 \text{ k} + 3{\cdot}9 \text{ k}}\right) \times 15 = 2{\cdot}67$ V

Thus $V_E = 2{\cdot}67\text{ V} - V_{BE} = 2{\cdot}67 - 0{\cdot}7 = 1{\cdot}97\text{ V}.$

and $I_E = \frac{V_E}{R_E} = \frac{1{\cdot}97\text{ V}}{390} = 5{\cdot}05\text{ mA}$

The collector voltage $V_c = V_{cc} - I_c R_c$

$= 15 - 5{\cdot}05 \times 10^{-3} \times 1 \times 10^3$

$= 9{\cdot}95$ Volts.

(assuming $I_E \simeq I_C$)

The collector to emitter voltage

$$V_{CE} = V_C - V_E = 9{\cdot}95\text{ V} - 1{\cdot}97\text{ V} = 7{\cdot}98\text{ V}$$

The a.c. voltage at the base can be calculated from the circuit shown in Fig. 6.11.

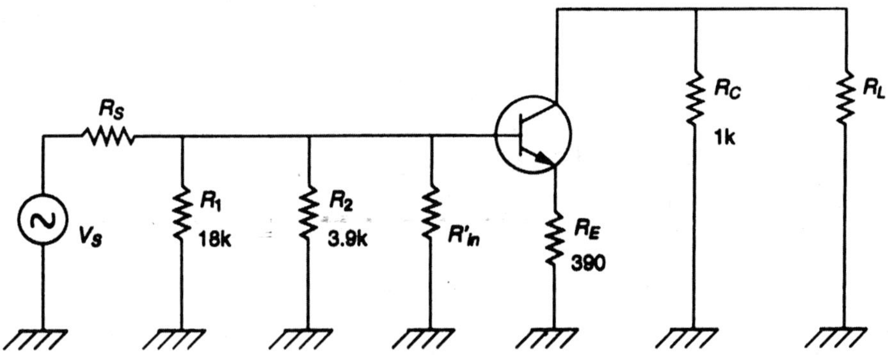

Fig 6.11. AC equivalent circuit (omitting the bypass capacitor, $X_C \simeq 0$)

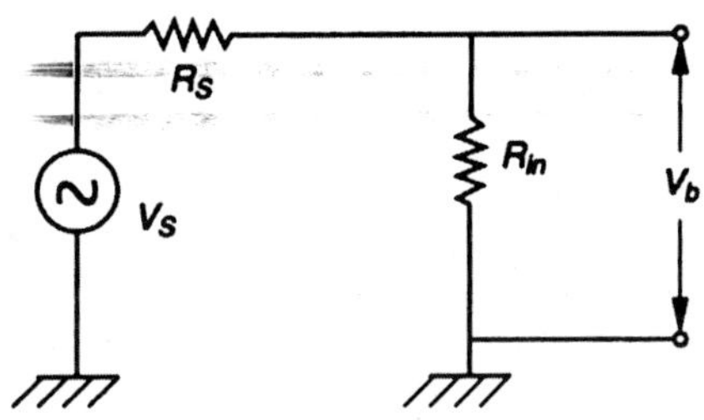

Fig 6.12. Calculation of a.c. base voltage

where R'_{in} is the parallel combination of R_1, R_2 and R_{in}. The value of R_{in} (base) can be obtained from,

$$R_{in} = \frac{V_b}{I_b} \quad ...(6.11)$$

$$V_b = I_e (r_e + R_E)$$

$$\simeq \beta_{ac} I_b (r_e + R_E) \quad ...(6.12)$$

This yields, $(R_{in})' = \beta_{ac} (r_e + R_E)$ ...(6.13)

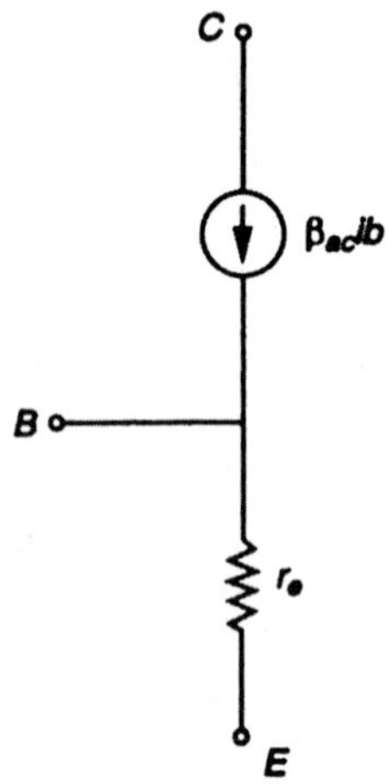

Fig 6.13. Calculation of R_{in} (base)

With the load removed, the output resistance is the parallel combination of R_C and r_C. If r_C is very much larger compared to R_C

$$R_{out} = R_C \parallel r_C \simeq R_C \quad \text{...(6.14)}$$

EXAMPLE 6.3 *Find the base signal voltage in the circuit of Fig. 6.11 if V_s= 5 mV r.m.s., 600 Ω signal source. I_E was determined as 5·05 mA in Example 6.3.*

SOLUTION In this case $I_E = 5{\cdot}05$ mA

Thus $r_e = \dfrac{25 \text{ mV}}{5{\cdot}05 \text{ mA}} \simeq 5\ \Omega$

$$R_{in}' = \beta_{ac}(R_E + r_e) = 130\,(390 + 5) = 51{\cdot}35 \text{ k}\Omega$$

Total input resistance $R_{in} = R_1 \mid R_2 \mid R_{in}'$

$$= 18 \text{ k} \parallel 3{\cdot}9 \text{ k} \parallel 51{\cdot}35 \text{ k}\Omega$$

$$\frac{1}{R_{in}} = \frac{1}{18 \text{ k}} + \frac{1}{3{\cdot}9 \text{ k}} + \frac{1}{51{\cdot}35 \text{ k}}$$

$$= (5{\cdot}555 + 25{\cdot}641 + 1{\cdot}947) \times 10^{-5}$$

or $R_{in} = 3{\cdot}02 \text{ k}\Omega$

The signal voltage at the base is thus

$$V_b = \left(\frac{R_{in}}{R_s + R_{in}}\right) V_{in}$$

$$= \frac{3{\cdot}02 \text{ k}}{(0{\cdot}6 + 3{\cdot}02) \text{ k}} \times 5 \text{ mV}$$

$$= 4{\cdot}17 \text{ mV}.$$

Voltage gain

The a.c. voltage gain expression is given by

$$G_v = \frac{\text{A.C. output voltage}}{\text{A.C. input voltage at the base}} = \frac{V_c}{V_b} \quad ...(6.15)$$

In the a.c. equivalent circuit

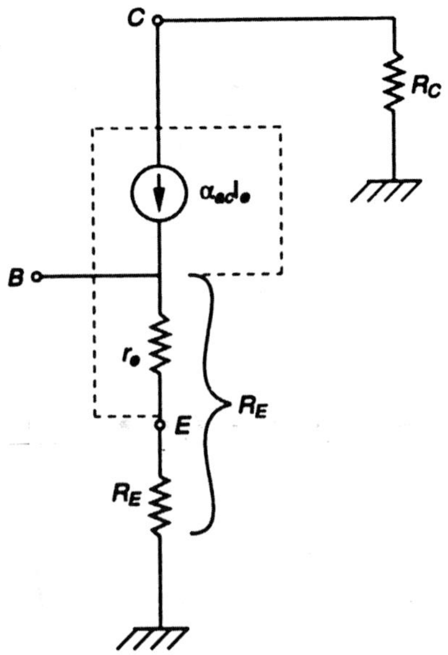

Fig 6.14. AC equivalent circuit

$$V_b = I_e R_E' \quad ...(6.16)$$

where $R_E' = R_E + r_e$ ...(6.17)

and $V_c = \alpha_{ac} I_e R_C$

$$\simeq I_e R_c \quad ...(6.18)$$

Thus, the voltage gain

$$G_v = \frac{I_e R_c}{I_e (R_E')} = \frac{R_C}{R_E'} = \left(\frac{R_C}{R_E + r_e}\right) \quad ...(6.19)$$

The voltage gain calculated in equation (6.19) is the voltage gain from base to collector. The overall voltage gain is

$$\text{Overall voltage gain } G_v' = \frac{V_c}{V_s} \quad ...(6.20)$$

Obviously, the overall voltage gain will be much lower as compared to the base to collector voltage gain by a factor f where

$$f = \frac{R_{in}}{R_s + R_{in}} \quad ...(6.21)$$

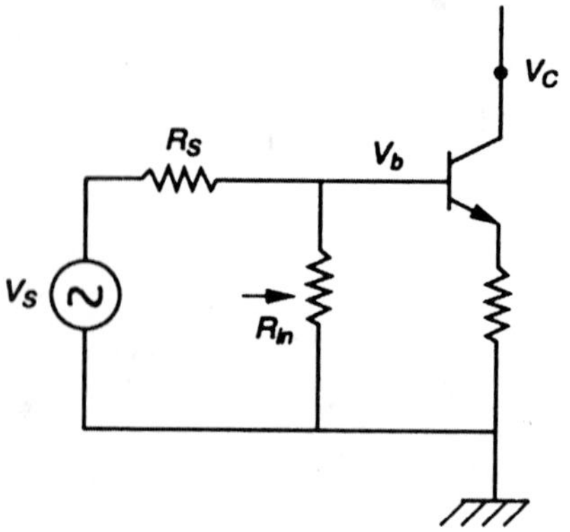

Fig 6.15. Calculation of overall voltage gain

Thus $G_v' = f \cdot G_v$. ...(6.22)

Bypass Capacitor (C_2)

The emitter bypass capacitor effectively acts as a short circuit to the signal frequency. It, however, does not disturb the d.c. bias conditions. At the signal frequency, since it effectively short circuits the emitter resistance R_E, the a.c. voltage gain expression becomes

$$G_v = \frac{R_C}{r_e} \quad ...(6.23)$$

Compare the expression (6.23) with that given by equation (6.19). **The emitter bypass capacitor thus increases gain by a large magnitude.**

EXAMPLE 6.4 *In the problem of Fig. 6.11, calculate the base to collector voltage gain assuming (i) the circuit without an emitter bypass capacitor (ii) R_E shorted.*

SOLUTION In this example,

$R_C = 1\text{ k}\,\Omega,\ r_e = 5\ \Omega,\ R_E = 390\ \Omega$

(i) Without emitter bypass capacitor, gain

$$= \frac{R_C}{R_E + r_e} = \frac{1\text{ k}\Omega}{395\ \Omega} = 2{\cdot}53$$

(ii) with R_E shorted

$$Gain = \frac{R_C}{r_e} = \frac{1\text{ k}\Omega}{5\ \Omega} = 200$$

6.4 COMMON EMITTER AMPLIFIER WITH A.C. COUPLED LOAD

When the ac coupled load R_L is present as in Fig. 6.16 (a) the effective load resistance is ($R_L' = R_L \| R_C$) the parallel combination of R_L and R_C because the capacitor C_3 is effectively a short circuit ($X_{C_3} = 0$) and the upper terminal of R_C is effective a.c. ground.

Fig. 6.16 (b) shows the effective a.c. equivalent circuit for this case.

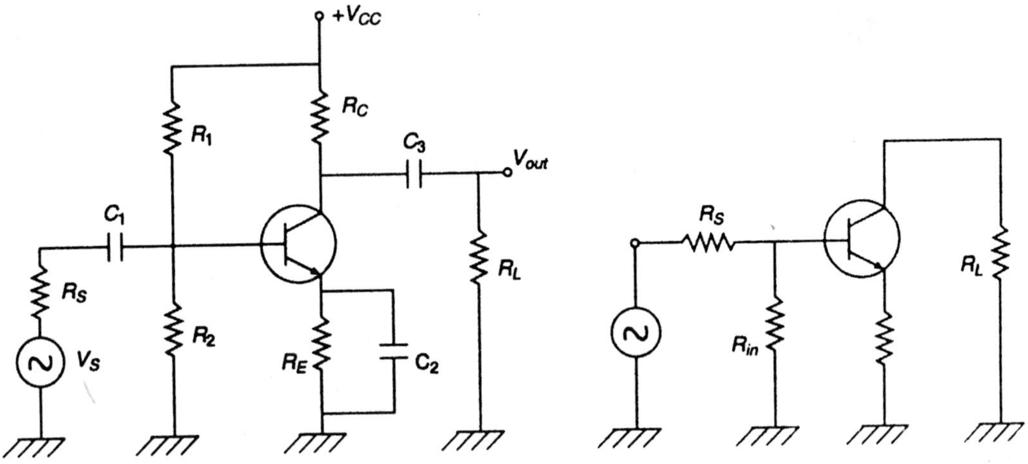

(a) Common emitter amplifier (b) AC equivalent circuit

Fig 6.16. Complete circuit diagram of a common emitter amplifier

The total a.c. collector resistance is

$$R_L' = R_L \parallel R_C = \frac{R_L R_C}{R_L + R_C} \qquad \ldots(6.24)$$

Replacing R_C with R_L', the effective equation for the voltage gain becomes

$$G_v = \frac{R_L'}{r_e} \qquad \ldots(6.25)$$

Since R_L' is the parallel combination of R_C and R_L its value is less compared to R_C alone. **The voltage gain is thus reduced due to the AC coupled load.** However, if $R_L >> R_C$, the reduction in gain is not appreciable.

EXAMPLE 6.5 *Find (i) the voltage gain of the base collector region (ii) overall voltage gain the amplifier in Fig. 6.9 if* $R_L = 6{\cdot}8$ kΩ, $C_2 = 50$ μF, $r_e = 5$ Ω, $v_s = 5$ mV *and* $R_s = 600$ Ω.

SOLUTION (i) The a.c. value of collector resistance

$$R_L' = \frac{R_C R_L}{R_C + R_L} = \frac{1\text{ k}\Omega \times 6{\cdot}8\text{ k}}{7{\cdot}8\text{ k}} = 872\ \Omega$$

The a.c. voltage gain is thus

$$G_v = \frac{R_L'}{r_e} = \frac{872\ \Omega}{5\ \Omega} = 174{\cdot}4$$

(ii) The a.c. voltage gain of the overall amplifier system is

$$G_v' = f G_v$$

In this case

$$f = \frac{R_{in}}{R_s + R_{in}} = \frac{3{\cdot}02\ \text{k}}{(0{\cdot}6 + 3{\cdot}02)\ \text{k}}$$

$$= 0{\cdot}834$$

Thus $\quad G_v' = 0{\cdot}834 \times 174{\cdot}4$

$$= 145{\cdot}5$$

6.5 STABILITY OF GAIN

The internal emitter resistance of a transistor r_e depends on temperature and since

$$G_v = \frac{R_L'}{r_e} \qquad \text{...(6.26)}$$

the voltage gain G_v also depends on temperature.

In order to overcome this problem, a circuit for partially bypassing the emitter resistance is considered in Fig. 6.17. In this circuit, the emitter resistance R_E, is bypassed by capacitor C_2. In the circuit diagram, an attempt has been made to make the gain independent of the changes in r_e. In this the gain is lower compared to what it was when the emitter resistor was fully bypassed.

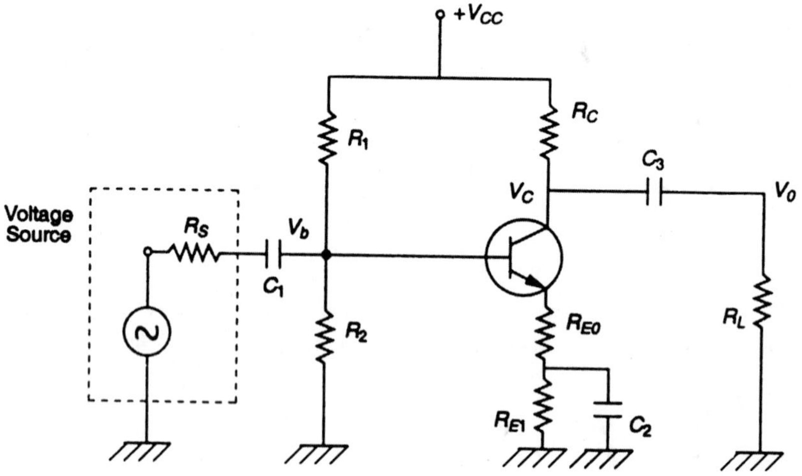

Fig 6.17. A common emitter amplifier circuit modified for gain stability

The voltage gain for the base collector circuit is

$$G_v = \frac{v_0}{v_b} = \frac{R_L'}{r_e + R_{E_0}} \qquad \text{...(6.27)}$$

Thus, in this circuit operation, the total emitter resistance $R_{E_0} + R_{E_1}$ is to be considered for d.c. biasing. However, for a.c. signals R_{E_0} is used to swamp out r_e. Therefore the effect of r_e on the voltage is negligible.

EXAMPLE 6.6 *Calculate the output voltage for the circuit in Fig. 6.18.*

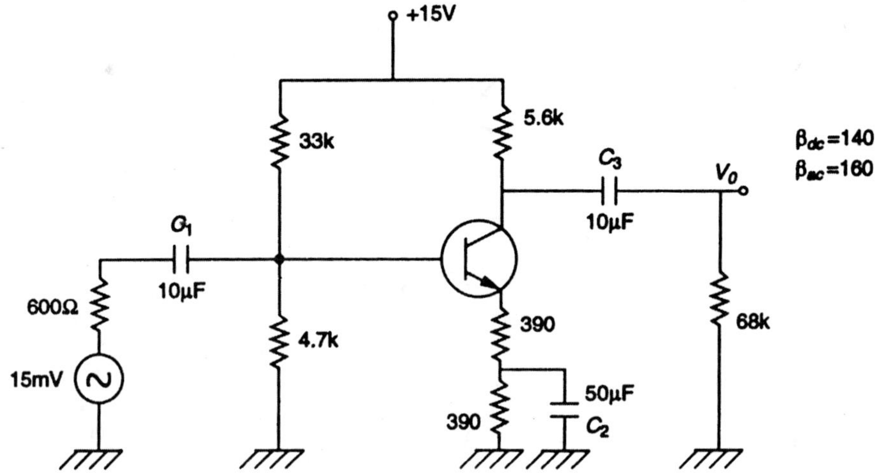

Fig 6.18. Common emitter amplifier circuit in example 6.6

SOLUTION **DC. Calculations**

For the d.c. equivalent circuit in Fig. 6.19

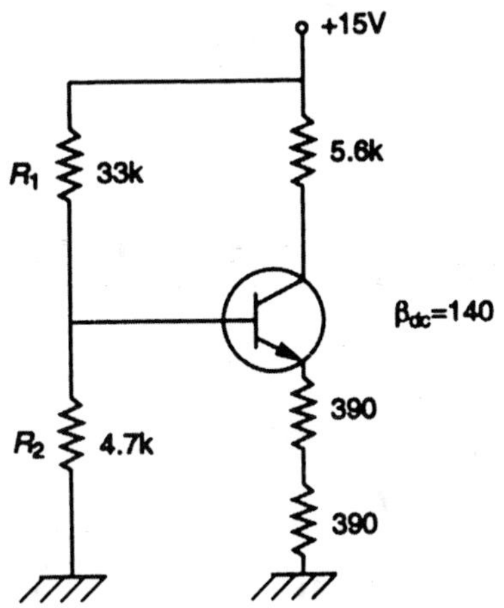

Fig 6.19. DC equivalent circuit

$R_{in}' = \beta_{dc} (R_{E_0} + R_{E_1}) = 140\ (780\ \Omega) = 109{\cdot}2\ \text{k}\Omega$

Now $R_{in}' >> R_2$, so we may ignore its effect on base voltage calculation. Thus

$$V_B = \left(\frac{4{\cdot}7\ \text{k}}{33\ \text{k} + 4{\cdot}7\ \text{k}}\right) 15 = 1{\cdot}87\ \text{V}.$$

$V_E = V_B - V_{BE} = 1{\cdot}17\ \text{V}$

$$I_E = \frac{V_E}{V_{E_0} + R_{E_1}} = \frac{1{\cdot}17}{780} = 1{\cdot}5 \text{ mA}$$

$$V_C = V_{CC} - I_C R_C = 15 \text{ V} - (1{\cdot}5 \text{ mA})(5{\cdot}6 \text{ k})$$
$$= 6{\cdot}6 \text{ V}$$

AC calculations

From the Figure 6.20

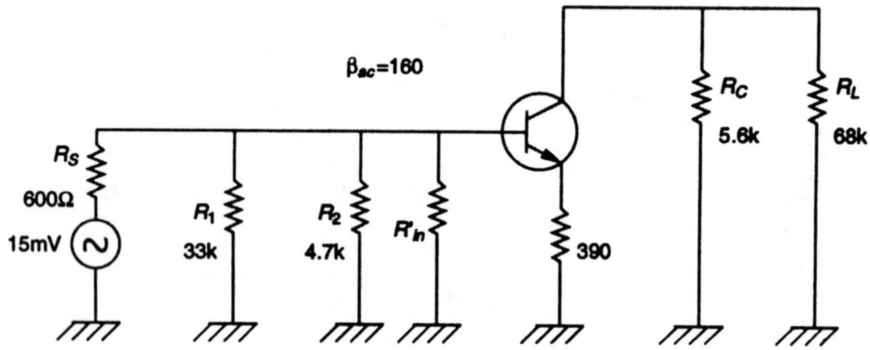

Fig 6.20. AC equivalent circuit

$$r_e = \frac{25 \text{ mV}}{I_E} = \frac{25 \text{ mV}}{1{\cdot}5 \text{ mA}} = 16{\cdot}67\ \Omega$$

$$R_{in}' = \beta_{ac}(r_e + R_{E_0}) = 160\,(16{\cdot}67 + 390)$$
$$= 160 \times 406{\cdot}67 = 65{\cdot}07 \text{ k}\Omega$$

$$\therefore \quad \frac{1}{R_{in}} = \frac{1}{R_1} + \frac{1}{R_2} + \frac{1}{R_{in}'}$$
$$= \frac{1}{33 \text{ k}} + \frac{1}{4{\cdot}7 \text{ k}} + \frac{1}{65{\cdot}07 \text{ k}}$$
$$= (3{\cdot}03 + 21{\cdot}28 + 1{\cdot}537) \times 10^{-5}$$
$$= 25{\cdot}847 \times 10^{-5}$$

$$\therefore \quad R_{in} = 3868{\cdot}9\ \Omega$$
$$= 3{\cdot}87 \text{ k}\Omega$$

$$\text{Input attenuation factor} = f = \frac{V_b}{V_s} = \frac{R_{in}}{R_s + R_{in}} = \frac{3{\cdot}87 \text{ k}}{(0{\cdot}6 + 3{\cdot}87) \text{ k}}$$
$$= 0{\cdot}866$$

Now effective load resistance

$$R_L' = R_C \parallel R_L = \frac{5{\cdot}6 \text{ k} \times 68 \text{ k}}{73{\cdot}6 \text{ k}} = 5{\cdot}17 \text{ k}\Omega$$

Thus, voltage gain from base to collector is

$$G_v = \frac{R_L'}{r_e + R_{E_0}} = \frac{5{\cdot}17\ \text{k}}{16{\cdot}67 + 390} = 12{\cdot}71$$

Overall gain $G_v' = f \,.\, G_v = 12{\cdot}71 \times 0{\cdot}866 = 11$

Since $\quad V_s = 15$ mA, the a.c. voltage at the collector is

$$v_C = G_s' = (11) \times 15\ \text{mV} = 165\ \text{mV}.$$

The collector voltage is thus an a.c. signal of 165 mV riding on a d.c. level of 6·6 V, as depicted in Fig. 6.21 (a). The capacitor C_3 will block the d.c. and the output voltage will be as shown in Fig. 6.21 (b).

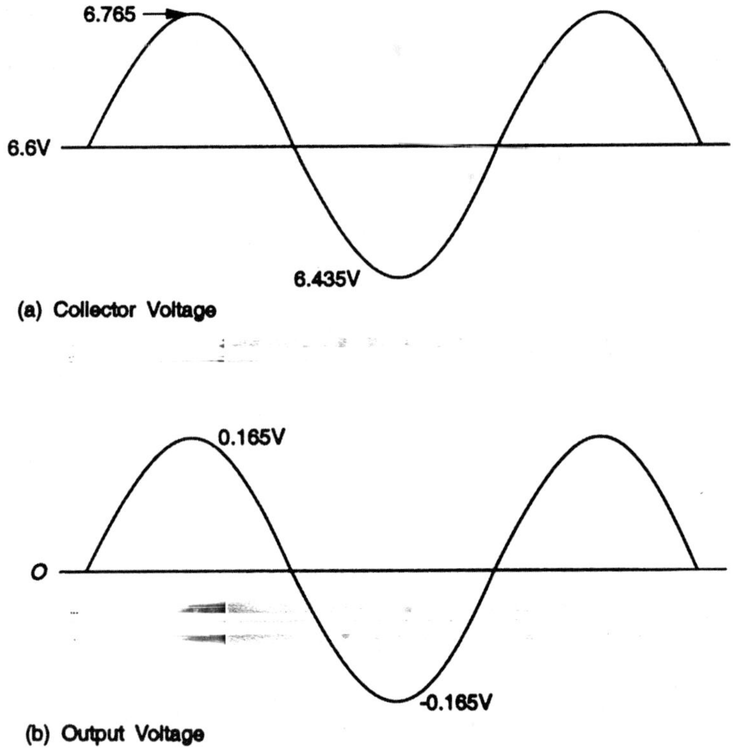

Fig 6.21. Collector voltage and output voltage in example 6.6

EXAMPLE 6.7 *Calculate the output voltage if R_L is (i) 3·3 kΩ (ii) 10 kΩ (iii) 33 kΩ (iv) 100 kΩ (v) 500 kΩ (vi) ∞*

SOLUTION The effective load resistances are

(i) $R_L' = 5{\cdot}6\ k \parallel 3{\cdot}3\ \text{k} = 2{\cdot}076\ \text{k}$

$$v_C = G_v'\, v_s = f \cdot \frac{R_L'}{r_e + R_{E_0}} \cdot v_s = \frac{2076}{406{\cdot}67} \times 15\ \text{mV} \times 0{\cdot}866 = 65{\cdot}82\ \text{mV}$$

(ii) $R_L' = 10\ \text{k} \parallel 5{\cdot}6\ \text{k} = 3{\cdot}589\ \text{k}$

$$v_c = G_v'\, v_s = f\, \frac{R_L'}{r_e + R_{E_0}} = \frac{3589}{406.67} \times 15 . f = 132 \times {\cdot}866 = 114{\cdot}312\ \text{mV}$$

(iii) $R_L' = 33\text{ k} \parallel 5{\cdot}6\text{ k} = 4{\cdot}788\text{ k}$

$$v_c = G_v' v_s = f \cdot \frac{R_L'}{r_e + R_{E_0}} = \frac{4788}{406{\cdot}67} \times 15 \times {\cdot}866$$

$= 152{\cdot}416$ mV

(iv) $R_L' = 5{\cdot}5\text{ k} \parallel 100\text{ k} = 5{\cdot}303\text{ k}$

$$v_c = G_v' v_s = f \cdot \frac{R_L' v_s}{r_e + R_{E_0}} = 0{\cdot}866 \times \frac{5303}{406{\cdot}67} \times 15$$

$= 169$ mV

(v) $R_L' = 5{\cdot}6\text{ k} \parallel 500\text{ k} = 5{\cdot}538\text{ k}$

$$v_c = G_v' v_s = f \cdot \frac{R_L' . v_s}{r_e + R_{E_0}} = 0{\cdot}866 \times \frac{5538}{406{\cdot}67} \times 15$$

$= 176{\cdot}89$ mV.

(vi) $R_L' = R_C = 5{\cdot}6\text{ k}$

$$v_c = f \cdot \frac{R_L' . v_s}{r_e + R_{E_0}} = 0{\cdot}866 \times \frac{5600}{406{\cdot}67} \times 15$$

$= 178{\cdot}87$ mV.

The values are listed in Table 6.1 below.

R_L	R_L'	v_c
3·3 kΩ	2·076 kΩ	65·82 mV
10 kΩ	3·589 kΩ	114·312 mV
33 kΩ	4·788 kΩ	152·416 mV
100 kΩ	5·303 kΩ	169 mV
500 kΩ	5·538 kΩ	176·89 mV
∞	5·6 kΩ	178·87 mV

It will be seen that with increasingly larger values of R_L, the shunting effect of R_L decreases and v_0, the output voltage increases.

Current Gain

The base to collector current gain is

$$\beta_{ac} = \frac{I_c}{I_b} \qquad \text{...(6.28)}$$

The overall current gain is

$$G_i = \frac{I_c}{I_s} \qquad \text{...(6.29)}$$

The total input current is

$$I_s = \frac{V_s}{R_{in}} \qquad \text{...(6.30)}$$

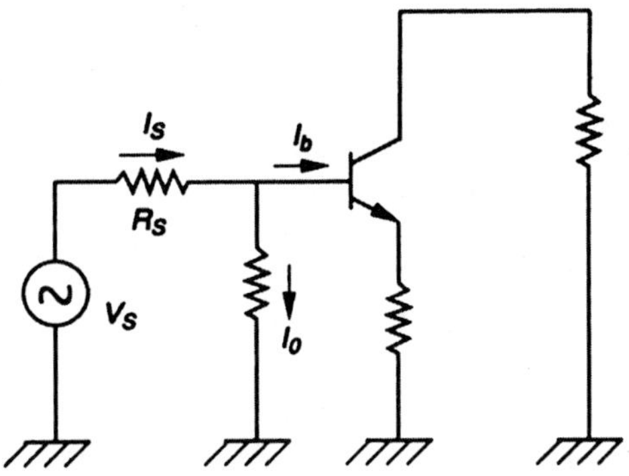

Fig 6.22. Calculation of current gain

The power gain is obtained by multiplying the current gain and the voltage gain.

$$G_p = G_i \, G_v' \qquad \ldots(6.31)$$

EXAMPLE 6.8 *Find the value of the current gain and power gain in the circuit of Fig. 6.18.*

SOLUTION

$$I_s = \frac{V_s}{R_{in}}$$

$$= \frac{15 \text{ mV}}{3{\cdot}87 \text{ k}\Omega + 600} = 3{\cdot}875 \times 10^{-6} \text{ Amps.}$$

$$I_c = \beta_{ac} \,.\, I_b = 160 \times 0{\cdot}23 \text{ } \mu A = 36{\cdot}865 \text{ } \mu A$$

$$G_i' = \frac{I_c}{I_s} = \frac{36{\cdot}865}{3{\cdot}875} = 9{\cdot}514$$

6.6 COMMON COLLECTOR AMPLIFIER OR EMITTER FOLLOWER

In the common collector amplifier, the input is applied to the base and the output is taken from the emitter. There is no collector resistor. The voltage gain of the amplifier is

$$G_v = \frac{v_{out}}{v_{in}} \qquad \ldots(6.32)$$

For the emitter follower circuit

$$v_{out} = I_e \, R_{eq} \qquad \ldots(6.33)$$

$$v_{in} = I_e \, (R_{eq} + r_e) \qquad \ldots(6.34)$$

The voltage gain is

$$G_v = \frac{R_{eq}}{R_{eq} + r_e} = \frac{1}{1 + \left(\frac{r_e}{R_{eq}}\right)} \qquad \ldots(6.35)$$

where $R_{eq} = R_E \,||\, R_L$ ...(6.36)

(considering C_{out} to be a short circuit for a.c. frequencies).

The value of G is $\simeq 1$ as $r_e << R_{eq}$.

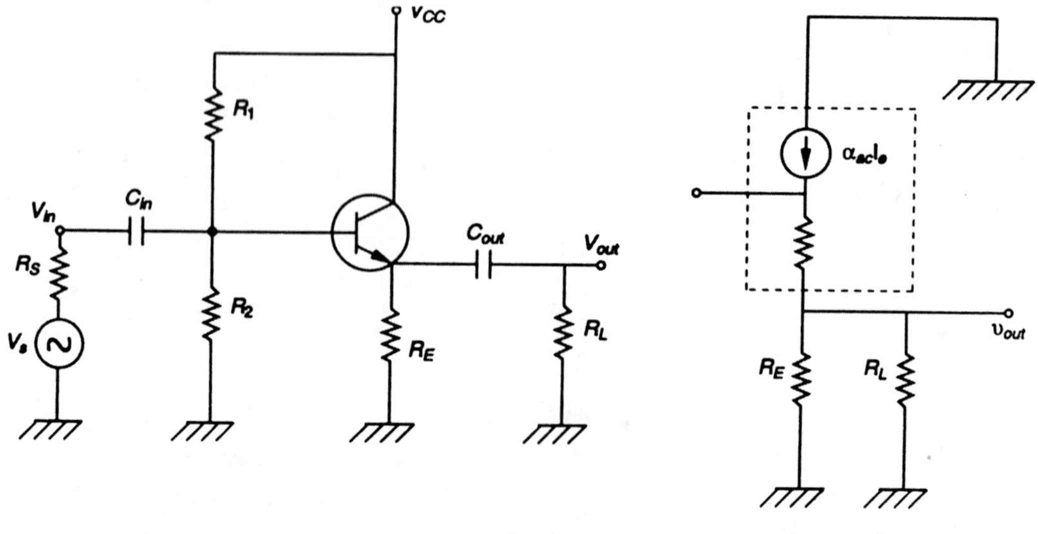

Fig 6.23. (a) Emitter follower circuit, (b) equivalent circuit for gain calculation

The output voltage is taken from the emitter and there is no phase difference between the emitter and base currents. Thus the output voltage closely follows the input voltage and the circuit is called an emitter follower.

The input resistance is

$$R_{in}' = \frac{v_b}{I_b} = \frac{I_e(r_e + R_{eq})}{I_b}$$

$$\simeq \frac{\beta_{ac} I_b (r_e + R_{eq})}{I_b} \quad \ldots(6.37)$$

or, $$R_{in}' \simeq \beta_{ac}\left(r_e + \frac{R_E R_L}{R_E + R_L}\right) \quad \ldots(6.38)$$

If $r_e << R_{eq}$, then equation (6.38) simplifies further to

$$R_{in}' = \beta_{ac} \frac{R_E R_L}{R_E + R_L} \quad \ldots(6.39)$$

The input resistance as it appears to the source is

$$R_{in} = R_1 \,||\, R_2 \,||\, R_{in}' \quad \ldots(6.40)$$

Output Resistance Calculation

In Fig. 6.24 (a), the emitter follower circuit is redrawn using the *r*-parameter a.c. equivalent circuit for the transistor. Fig. 6.24 (b) depicts the simplified Thevenin equivalent circuit. In the circuit diagram,

$$\left.\begin{aligned} V_{out} &= V_e \\ I_{Out} &= I_e \\ I_{in} &= I_b \end{aligned}\right\} \quad \ldots(6.41)$$

By definition R_{out} = Output Resistance

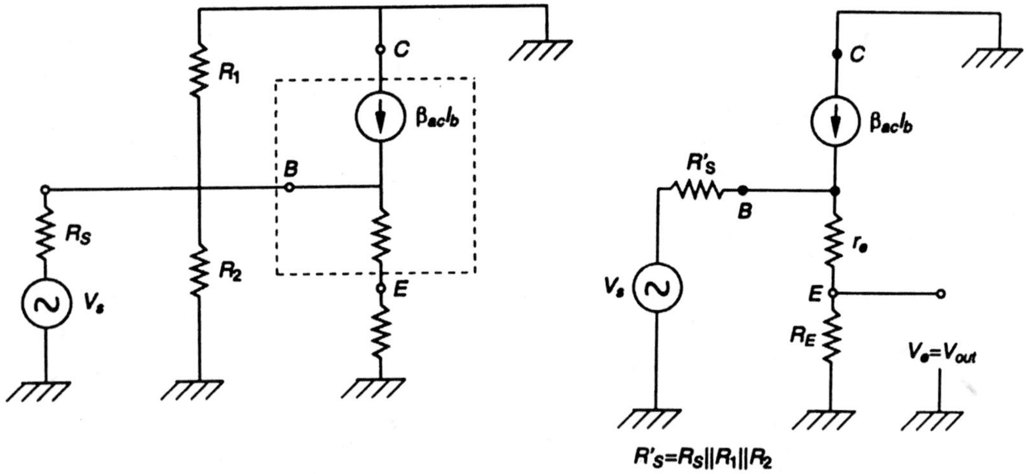

Fig 6.24. (a) Emitter follower equivalent circuit with the transistor replaced by its *r* parameter a.c. equivalent circuit (b) Simplified circuit using Thevenin equivalence

$$= \frac{V_e}{I_e} \qquad \ldots(6.42)$$

Also $$\frac{I_e}{I_b} \approx \beta_{ac} \qquad \ldots(6.43)$$

If v_s is assumed to be zero and I_b is produced by the output voltage v_O and also V_{BE} is considered negligible,

$$I_b \simeq \frac{V_e}{R_s'} \qquad \ldots(6.44)$$

Further if R_1 and R_2 both are much larger compared to R_s

$$R_s' = R_1 \parallel R_2 \parallel R_s$$

$$\approx R_s \qquad \ldots(6.45)$$

and from equation (6.44), we get

$$I_b \simeq \frac{V_e}{R_s} \qquad \ldots(6.46)$$

Further $$I_e + I_{out} = \frac{V_e\,\beta_{ac}}{R_s} \qquad \ldots(6.47)$$

so that $$R_{out} = \frac{V_e}{I_e} = \frac{V_e}{\beta_{ac}\,V_e/R_s} = \frac{R_s}{\beta_{ac}} \qquad \ldots(6.48)$$

Looking from the emitter side, R_E will be in parallel with $\frac{R_s}{\beta_{ac}}$ · Thus

$$R_{out} = R_E \parallel \left(\frac{R_s}{\beta_{ac}}\right) = \frac{R_E\,R_s}{\beta_{ac}\left(R_e + \frac{R_s}{\beta_{ac}}\right)}$$

$$= \frac{R_E R_s}{R_s + \beta_{ac} + R_E} \qquad \text{...(6.49)}$$

Current gain G_i for the common collector amplifier is

$$G_i = \frac{I_e}{I_{in}} \qquad \text{...(6.50)}$$

If $R_1 \parallel R_2 >> R_{in}'$ (the base input resistance), then most of the input current enters the base.

Thus, $$G_i \simeq \beta_{ac} \qquad \text{...(6.51)}$$

It will thus be found that the maximum achievable current gain of common collector amplifiers is β_{ac}.

The power gain for the emitter follower is

$$G_p = G_i G_v \qquad \text{...(6.52)}$$

$$\simeq G_i$$

$$\simeq \beta_{ac} \qquad \text{...(6.53)}$$

Since $$G_v \simeq 1$$

EXAMPLE 6.9 *Find the voltage gain, current gain, power gain and total input resistance for the common collector amplifier circuit in Fig. 6.25.*

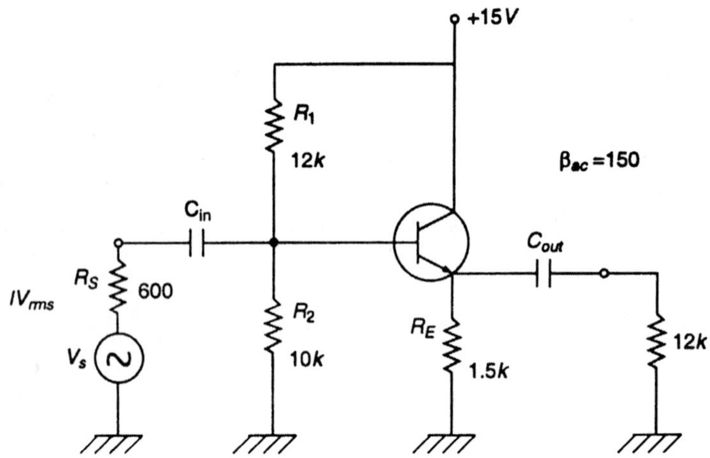

Fig 6.25. Common collector equivalent circuit

SOLUTION In the present problem

$R_E = 1{\cdot}5$ kΩ, $R_L = 12$ kΩ, $R_1 = 12$ kΩ, $R_2 = 10$ kΩ, $R_s = 600$ Ω,

$V_s = 1$ V, $\beta_{ac} = 150$ and $V_{cc} = +15$V.

$$R_{eq} = \frac{R_E R_L}{R_E + R_L} = \frac{1{\cdot}5 \text{ k} \times 12 \text{ k}}{13{\cdot}5 \text{ k}} = 1{\cdot}333 \text{ k}$$

The base input resistance $R_{in}' = \beta_{ac} \cdot R_{eq} = 150 \times 1333$

$= 200$ kΩ

The total input resistance $R_{in} = R_1 \parallel R_2 \parallel R_{in}' = 12\text{ k} \parallel 10\text{ k} \parallel 200\text{ k}$

$$\frac{1}{R_{in}} = \frac{1}{12\text{ k}} + \frac{1}{10\text{ k}} + \frac{1}{200\text{ k}} = (8{\cdot}333 + 10 + 0{\cdot}50) \times 10^{-5}$$

$$= 18{\cdot}833 \times 10^{-5}$$

$$R_{in} = 5{\cdot}31\text{ k}\Omega$$

Voltage gain $G_v \simeq 1{\cdot}0$

and $$r_e \simeq \frac{25\text{ mV}}{I_E}$$

When $I_E = v_E/R_E = \dfrac{6{\cdot}11}{1\text{k}\Omega} = 6{\cdot}11\text{ mA}$

$$r_e \simeq \frac{25\text{mV}}{6{\cdot}11\text{ mA}} = 4{\cdot}09\ \Omega$$

$$G_v = \frac{R_{eq}}{r_e + R_{eq}} = \frac{1333}{4{\cdot}09 + 1333} = \frac{1333}{1337{\cdot}09} = 0{\cdot}997$$

The current gain is $$A_i = \frac{I_e}{I_{in}}$$

where $$I_e = \frac{V_e}{R_{eq}} = \frac{A_v V_b}{R_{eq}} \simeq \frac{0{\cdot}997}{1333} = 7{\cdot}48 \times 10^{-4} = 0{\cdot}748\text{ mA}$$

$$I_{in} = \frac{V_{in}}{R_{in}} = \frac{1\text{ V}}{5{\cdot}31\text{ k}\Omega} = 0{\cdot}188\text{ mA}$$

$$G_i = \frac{I_e}{I_{in}} = \frac{0{\cdot}748\text{ mA}}{0{\cdot}188\text{ mA}} = 3{\cdot}97$$

Power gain $G_p \approx G_i = 3{\cdot}97$

6.7 COMMON BASE AMPLIFIERS

A common base amplifier circuit is employed to provide a high voltage gain. The circuit also provides a low input resistance. Fig. 6.26 provides the circuit diagram for a common base amplifer circuit. The base is the common terminal for the input and output circuits.

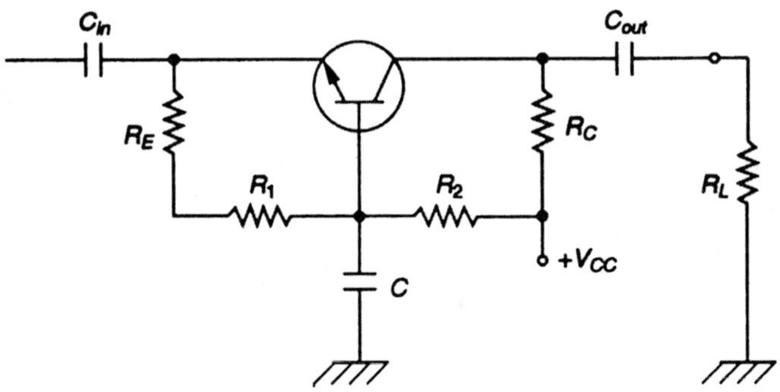

Fig 6.26. A common base amplifier circuit

The voltage gain from emitter-to-collector is

$$G_v = \frac{\text{Collector voltage}}{\text{Emitter voltage}}$$

$$\simeq \frac{R_L'}{r_e} \qquad \text{...(6.54)}$$

where $$R_L' = \frac{R_C R_L}{R_C + R_L} \qquad \text{...(6.55)}$$

The input resistance is the parallel combination of R_E and r_e, where r_e is the resistance looking at the emitter terminals. Since r_e is much smaller compared to R_E, the overall input resistance will be the same as r_e.

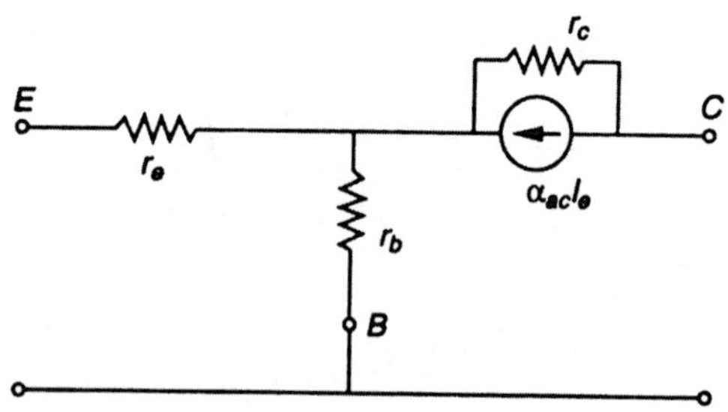

Fig 6.27. *r* parameter equivalent circuit

The output resistance R_{out} is the parallel combination of R_C and r_c. Since $r_c >> R_C$

R_{out} is approximately the same as R_C $\qquad R_{out} = R_C \qquad$ ...(6.56)

The current gain G_i for this circuit is $\qquad G_i = \frac{I_c}{I_e} \simeq 1 \qquad$ ...(6.57)

And the power gain is equal to the voltage gain $G_p \simeq G_v \qquad$ ...(6.58)

EXAMPLE 6.10 *Determine the value of the input resistance, voltage gain, current gain and power gain for the common base amplifier depicted in Fig. 6.28.*

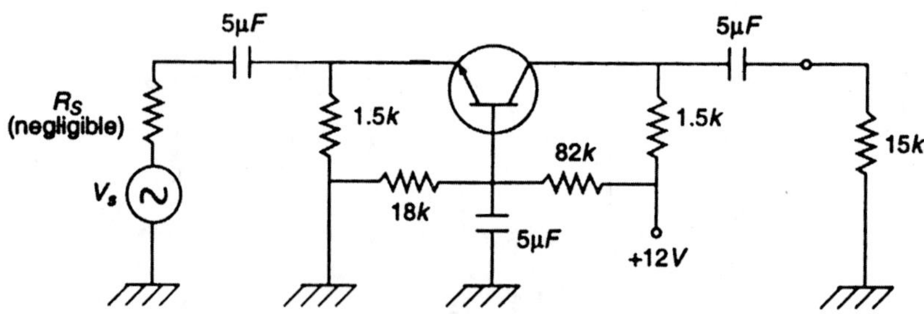

Fig 6.28. Common base amplifier circuit

SOLUTION For the circuit shown

$R_{in} \simeq r_e$

In order to calculate r_e, we have to calculate I_E.

Now $V_{base} = \frac{R_1}{R_1 + R_2} = \frac{18\text{ k}}{100\text{ k}} \times 12 = 2{\cdot}16$ V

Emitter voltage $= 2{\cdot}16 - 0{\cdot}7 = 1{\cdot}46$ V

Emitter current $= \frac{1{\cdot}46\text{ V}}{1{\cdot}5\text{ k}\Omega} = 0{\cdot}973$ mA

The input resistance $R_{in} \simeq r_e = \frac{25\text{ mV}}{0{\cdot}973\text{ mA}} = 25{\cdot}68\ \Omega$

Thus $R_{in} \approx 25{\cdot}68\ \Omega$.

Now AC voltage gain $= \frac{R_L'}{r_e} = \left(\frac{R_C R_L}{R_C + R_L}\right) \times \frac{1}{r_e}$

$= \frac{1{\cdot}5\text{ k} \times 15\text{ k}}{16{\cdot}5\text{ k}} \times \frac{1}{25{\cdot}68}$

$= 53{\cdot}1$

Also power gain $G_p = 53{\cdot}1$

The following table (Table 6.2) gives a comparative idea about the various amplifier configurations.

Table 6.2

Parameter	*Common Emitter*	*Common Collector or Emitter Follower*	*Common Base*
Input resistance	Low	High	Very low ($\approx r_e$)
Output Resistance	High	Very low	High
Voltage gain	High	Low (≈ 1)	High
Maxm. current gain	High	High	Low (≈ 1)
Power gain	Very high	High	High
Phase relationship between input and output	$\phi = 180°$	$\phi = 0°$	$\phi = 0°$

6.8 DARLINGTON CONNECTION OR SUPERBETA TRANSISTORS

In the Darlington transistor (also called superbeta transistor) connection, the composite transistor works as a single unit such that the current gain is the multiplication of the curent gains of the individual transistors. (Fig. 6.29) The collectors of two transistors are connected and the emitter of the first transistor drives the base of the second transistor. The emitter current of the first transistor is

$$I_{e_1} \simeq \beta_{ac} I_b \quad \ldots(6.59)$$

The emitter current is now the base current of the next transistor such that

$$I_{e_2} \simeq \beta_{ac_1} I_{e_1}$$

$$I_{e_2} \simeq \beta_{ac_1} \beta_{ac_2} I_b \quad \ldots(6.60)$$

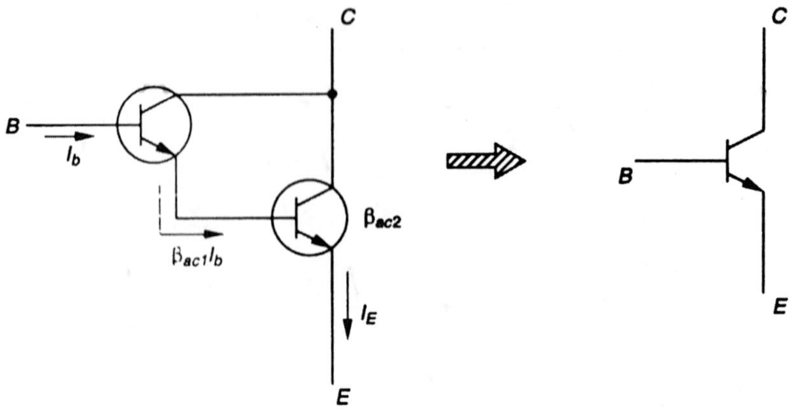

Fig 6.29. Darlington transistor connection

The effective current gain of the Darlington pair is

$$\beta_{ac} = \beta_{ac_1}\beta_{ac_2} \quad \ldots(6.61)$$

If $\beta_{ac_1} = \beta_{ac_2} = \beta$, the current gain of the Darlington pair is

$$\beta_{ac} = \beta^2 \quad \ldots(6.62)$$

In a Darlington transistor, the value of the current gain is very high, of the order of few thousands.

EXAMPLE 6.11 *Calculate the value of the current gain in a superbeta transitor if the each transistor employed possesses a current gain* $\beta = 190$.

SOLUTION $\beta_{ac} = \beta^2 = (190)^2 = 36100$

Biasing Circuit for a Darlington Transistor

For the configuration shown in Fig. 6.30, the base current I_B may be calculated as

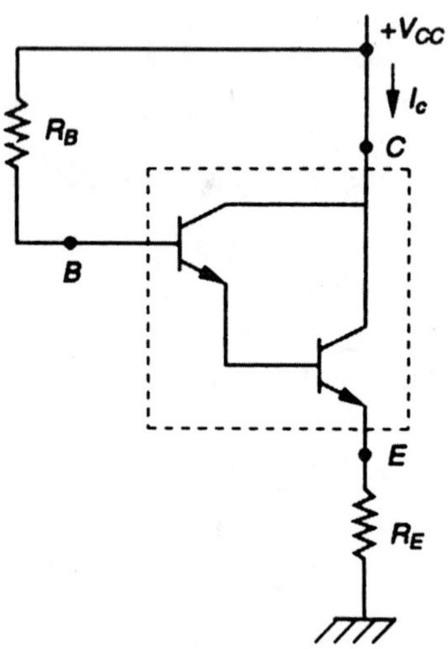

Fig 6.30. Biasing circuit

$$I_B = \frac{V_{cc} - (V_{BE})_D}{R_B + \beta_{ac} R_E} \qquad \text{...(6.63)}$$

For the Darlington or superbeta transistor, the typical value of $(V_{BE})_D$ is 1·6 volts.

Also $$V_E = I_E R_E \qquad \text{...(6.64)}$$

where I_E is given by

$$I_E = (1 + \beta_{ac}) I_B \simeq \beta_{ac} I_B \qquad \text{...(6.65)}$$

and $$V_B = V_E + (V_{BE})_D \qquad \text{...(6.66)}$$

EXAMPLE 6.12 *Find the d.c. bias voltages and currents for the circuit in Fig. 6.31.*

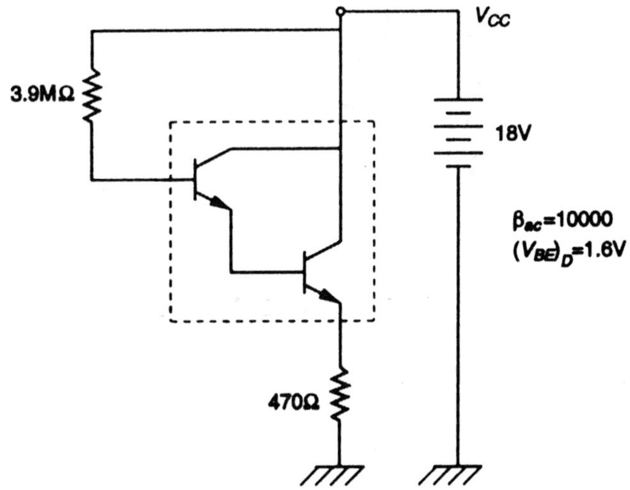

Fig 6.31. Transistor circuit in Example 6.12

SOLUTION $$I_B = \frac{V_{cc} - (V_{BE})_D}{R_B + \beta_{ac} R_E} = \frac{18 - 1{\cdot}6}{3{\cdot}9\text{ M} + 10{,}000 \times 470}$$

$$= \frac{16{\cdot}4}{8{\cdot}6\text{ M}} = 1{\cdot}91\ \mu\text{A}$$

The emitter current is

$$I_E = 10{,}000 \times (1{\cdot}91\ \mu\text{A}) = 19{\cdot}1\text{ mA}$$

Also $$I_C \simeq I_E = 19{\cdot}1\text{ mA}$$

Therefore $$V_E = 19{\cdot}1\text{ mA } (470) = 8{\cdot}96\text{ Volts}$$

The voltage at the base is

$$V_B = 8{\cdot}96 + 1{\cdot}6 = 9{\cdot}56\text{ Volts}$$

AC equivalent circuit

Let us consider the Darlington emitter follower circuit of Fig. 6.32.

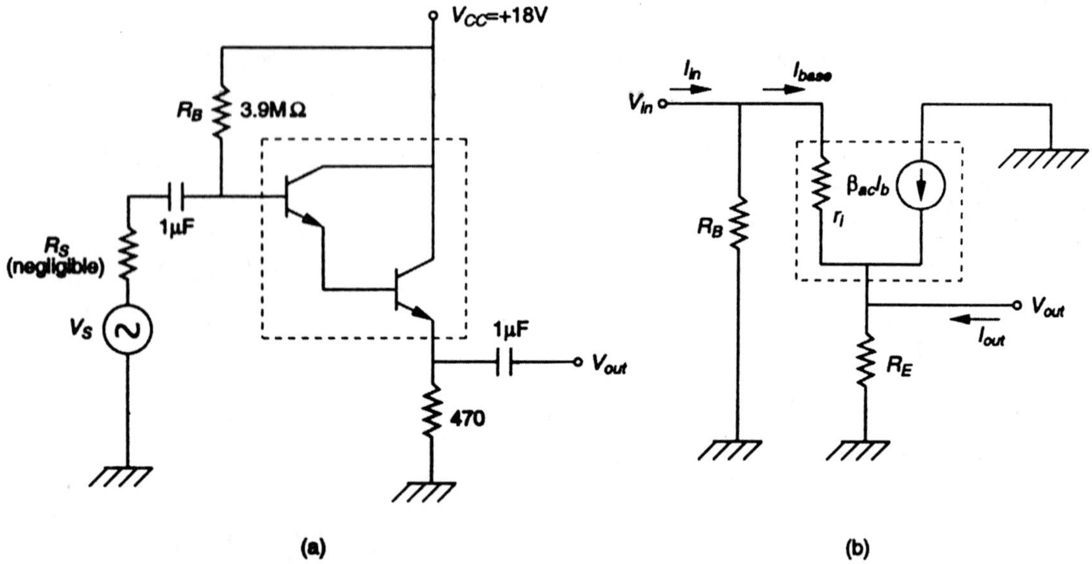

Fig 6.32. (a) Superbeta emitter follower circuit
(b) a.c. equivalent circuit of the superbeta emitter follower

Calculation of AC input impedance

The a.c. equivalent circuit of the superbeta emitter follower circuit is depicted in Fig. 6.32 (b). From the circuit we get

$$I_{base} = I_b = \frac{V_{in} - V_{out}}{r_i} \quad \ldots(6.67)$$

Also $$V_{out} = (I_b + \beta_{ac} I_b) R_E \quad \ldots(6.68)$$

From equation (6.67) we get

$$I_b r_i = V_{in} - V_{out}$$

or, $$I_b r_i = V_{in} - (I_b + \beta_{ac} I_b) R_E \quad \ldots(6.69)$$

or, $$V_{in} = I_b [(1 + \beta_{ac}) R_E + r_i] \quad \ldots(6.70)$$

The a.c. input impedance is thus (looking at the transistor base)

$$Z_{in} = \frac{V_{in}}{I_b} = r_i + (1 + \beta_{ac}) R_E$$

$$\simeq r_i + \beta_{ac} R_E \quad \ldots(6.71)$$

The a.c. input impedance looking into the transistor circuit is

$$Z_{in} = R_B \parallel (r_i + \beta_{ac} R_E)$$

$$= \frac{(r_i + \beta_{ac} R_E) R_B}{R_B + (r_i + \beta_{ac} R_E)} \quad \ldots(6.72)$$

EXAMPLE 6.13 *Find the value of the input impedance for the numerical values in the circuit of Fig. 6.31 if r_i = 6 kΩ.*

SOLUTION From equation (6.72)

$$Z_{in} = 3{\cdot}9\ \text{M}\Omega \parallel [6\ \text{k}\Omega + 10{,}000 \times 470]$$
$$= 3{\cdot}9\ \text{M}\Omega \parallel 4{\cdot}706\ \text{M}\Omega$$
$$= \frac{3{\cdot}9 \times 4{\cdot}706\ \text{M}}{8{\cdot}606} = 2{\cdot}133\ \text{M}\Omega$$

AC current gain for the Darlington emitter follower

From Fig. 6.32 (b) we get

$$I_{out} = I_b + \beta_{ac} I_b = (\beta_{ac} + 1) I_b \simeq \beta_{ac} I_b \qquad ...(6.73)$$

The a.c. current gain is

$$G_i = \frac{I_{out}}{I_{in}} = \frac{I_{out}}{I_b} \cdot \frac{I_b}{I_{in}} \qquad ...(6.74)$$

where $$\frac{I_{out}}{I_b} = \beta_{ac} \qquad ...(6.75)$$

We, thus, get from figure

$$I_B = \frac{R_B I_{in}}{R_B + (r_i + \beta_{ac} R_E)}$$
$$\simeq \frac{R_B I_{in}}{R_B + \beta_{ac} R_E} \qquad ...(6.76)$$

The a.c. current gain is

$$G_i = \beta_{ac} \frac{R_B}{\beta_{ac} R_E + R_B} = \frac{R_B}{\left(R_E + \dfrac{R_B}{\beta_{ac}}\right)} \qquad ...(6.77)$$

EXAMPLE 6.14 *Find the value of the base current and the a.c. circuit current gain for the circuit in Fig. 6.31.*

SOLUTION

$$G_i = \frac{R_B}{R_E + \dfrac{R_B}{\beta_{ac}}} = \frac{3{\cdot}9\ \text{M}}{470 + \dfrac{3{\cdot}9\ \text{M}}{10{,}000}}$$
$$= \frac{3{\cdot}9\ \text{M}}{860} = 4534$$

6.9 AC OUTPUT IMPEDANCE OF THE DARLINGTON PAIR

The output impedance for the Darlington pair emitter follower as seen by the load may be found by applying a voltage V_{out} and measuring I_{out} (when input V_s is set to zero). This is obtained from the circuit of Fig. 6.33 (b) above.

$$I_{out} = \frac{V_{out}}{R_E} + \frac{V_{out}}{r_i} - \beta_{ac} I_b$$

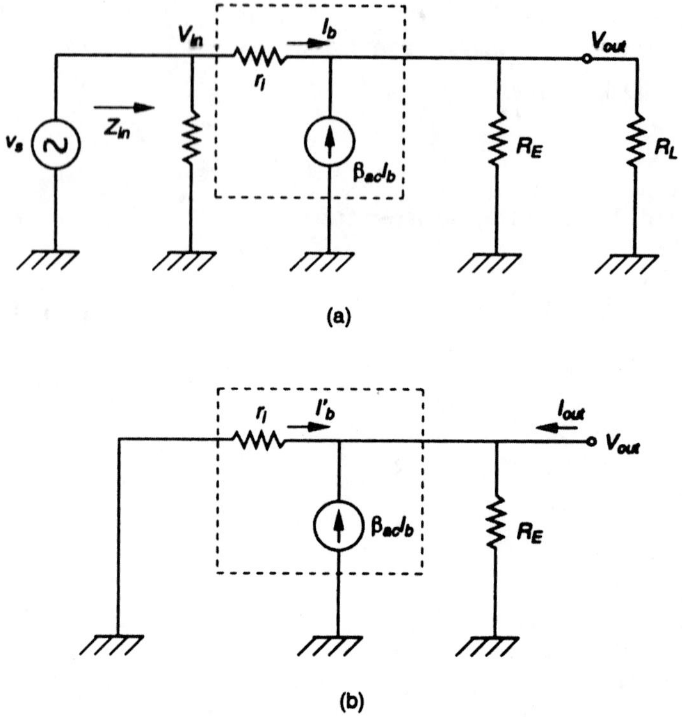

Fig 6.33. Equivalent circuits to determine Z_{out}

$$= V_{out}\left(\frac{1}{R_E} + \frac{1}{r_i} + \frac{\beta_{ac}}{r_i}\right) \qquad \ldots(6.78)$$

Thus
$$Z_{out} = \frac{V_{out}}{I_{out}} = \frac{1}{\dfrac{1}{R_E} + \dfrac{1}{r_i} + \dfrac{\beta_{ac}}{r_i}} \qquad \ldots(6.79)$$

$$\approx \frac{r_i}{\beta_{ac}} \qquad \ldots(6.80)$$

6.10 AC VOLTAGE GAIN DETERMINATION

From Fig. 6.34, we may write,

$$V_{out} = R_E\,(I_b + \beta_{ac}\, I_b) \qquad \ldots(6.81)$$

$$= I_b\, R_E\,(1 + \beta_{ac})$$

$$\simeq I_b\, R_E\, \beta_{ac} \qquad \ldots(6.82)$$

Also
$$V_{in} = I_b\, r_i + (I_b + I_b\, \beta_{ac})\, R_E \qquad \ldots(6.83)$$

$$= I_b\,(r_i + R_E + \beta_{ac}\, R_E) \qquad \ldots(6.84)$$

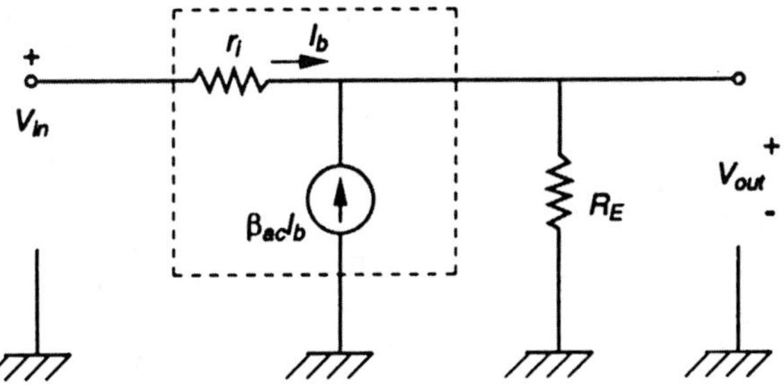

Fig 6.34 Equivalent circuit for determining voltage gain

Thus voltage gain G_v is given by

$$G_v = \frac{V_{out}}{V_{in}} = \frac{R_E(1+\beta_{ac})}{r_i + R_E(1+\beta_{ac})} \simeq 1 \qquad \ldots(6.85)$$

EXAMPLE 6.15 *Find the a.c. output impedance and a.c. voltage gain for the circuit in Fig. 6.31.*

SOLUTION

$$Z_{out} = 470\ \Omega \parallel 6\ \text{k}\Omega \parallel \frac{6\ \text{k}\Omega}{10000} \simeq \frac{6\ \text{k}\Omega}{10{,}000}$$

and

$$G_v = \frac{470 + (10{,}000) \times 470}{6\ k\Omega + [470 + (10{,}000) \times 470]}$$

$$= 0{\cdot}9987$$

Review Questions

1. What do you understand by a "single stage transistor amplifier"?
2. Show graphically how a weak voltage signal applied to the base can be amplified by a single stage transistor amplifier.
3. Draw the electrical equivalent circuit for a single stage transistor amplifier.
4. What do you understand by Quiscent operating point ?
5. Draw the a.c. and d.c. load lines on the I_c *vs.* V_{CE} characteristics of an *npn* transistor amplifier.
6. Draw the circuit diagram of a common emitter amplifier. Hence derive an expression for its voltage gain.
7. Draw the circuit diagram of a common emitter amplifier modified for gain stability. What is its voltage gain ? Derive expressions for its current gain.
8. Draw an emitter follower circuit and also find its electrical equivalent circuit.
9. Derive expressions for the voltage gain and input resistance for a common collector amplifier.
10. Derive expressions for the output resistance and current gain for a common collector amplifier.
11. What is a common base amplifier ? Derive expressions for current gain, voltage gain, input resistance and output resistance for the common base amplifier.
12. What do you understand by a Darlington amplifier ? Draw its circuit.
13. Calculate the value of the input resistance of a Darlington amplifier.
14. Calculate the value of the voltage gain for a Darlington amplifier.

Objective Type Questions

Choose (√) the correct or most appropriate answer.

1. For a transistor, the h_{fe} parameter is the same as
   (a) β_{dc} (b) β_{ac}
   (c) α_{dc} (d) r_c
2. If the d.c. emitter current in a transistor amplifier is 4 mA, the approximate value of r_e is
   (a) 4 Ω (b) 6·25 Ω
   (c) 22 Ω (d) 26 Ω
3. In a common emitter amplifier, the voltage gain = 90. If the emitter bypass capacitor is removed
   (a) the voltage gain will decrease (b) the voltage gain will increase
   (c) the Q point will shift (d) the circuit will be unstable
4. In a common emitter amplifier, $R_C = 1$ kΩ, $R_E = 200$ Ω, $r_e = 16$ Ω and $\beta_{ac} = 75$. Assuming that R_E is completely bypassed at the operating frequency, the voltage gain will be
   (a) 4 (b) 5
   (c) 63 (d) 75
5. For a common collector amplifier, $R_E = 90$ Ω, $r_e = 10$ Ω and $\beta_{ac} = 130$. The a.c. input resistance at the base will be
   (a) 100 Ω (b) 1080 Ω
   (c) 1·3 kΩ (d) 1170 Ω
6. For a common collector amplifier, $R_E = 90$ Ω, $r_e = 10$ Ω and $\beta_{ac} = 10$. The voltage gain of the amplifier will be
   (a) 1·2 (b) 120
   (c) 0·85 (d) 0·9
7. In a common emitter amplifier, the value of $R_C = 1$ kΩ, $R_L = 10$ kΩ, $r_e = 13$ Ω. The voltage gain of the transistor amplifier will be (when R_E is bypassed)
   (a) 10 (b) 69·9
   (c) 76·9 (d) 84·6
8. The input resistance of a common base amplifier is
   (a) very low (b) very high
   (c) the same as common emitter (d) the same as common collector amplifier
9. For an amplifier the current gain $\simeq 1$, power gain ≈ voltage gain, input resistance is very low ($\simeq r_e$) and output resistance $\simeq R_c$. The amplifier is
   (a) common base (b) common emitter
   (c) common collector (d) Darlington pair
10. For an amplifier, the input resistance is high, the current gain is high and voltage gain is near unity. It is a
    (a) common base (b) common emitter
    (c) common collector (d) Darlington amplifier.
11. An amplifier has good voltage, current and power gains and the input resistance is low. It is a
    (a) common base (b) common emitter
    (c) common collector amplifier.
12. In a Darlington amplifier, each transistor has a $\beta = 120$. It R_E is 500 Ω, the input resistance is
    (a) 500 Ω (b) 60 kΩ
    (c) 120 kΩ (d) 1·44 M Ω
13. The phase relationship between V_{in} and V_{out} in an amplifier is 180°. It is a
    (a) common base (b) common emitter
    (c) common collector amplifier
14. The maximum curent gain of an amplifier is unity. It is a
    (a) common base (b) common collector
    (c) common emitter amplifier.

Exercise Problems

1. A common emitter circuit has V_{cc} = 18 V, R_C = 1·2 kΩ, R_E = 1·5 kΩ, R_1 = 10 kΩ, R_2 = 2·7 kΩ. R_E is bypassed to ac signals by the large capacitor C_E and the transistor employed has the circuit shown in Figure E6.1. Draw the d.c. and a.c. load lines for the circuit.

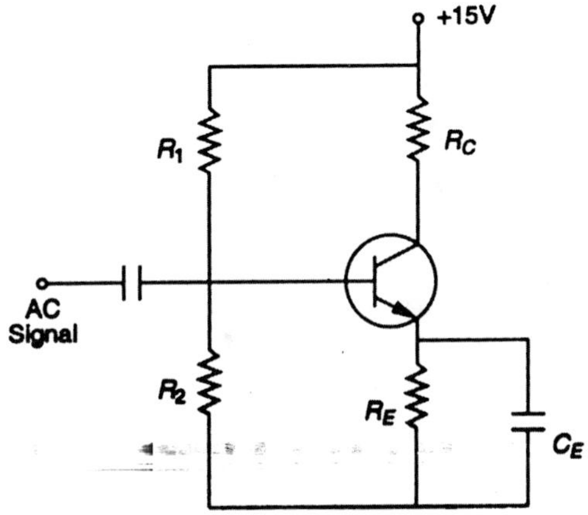

Fig E. 6.1. Reference Exercise Problem 6.1

2. Make a d.c. analysis for the circuit of (Fig. E6.2) the common emitter amplifier. Find I_C, I_E and V_{CE}.

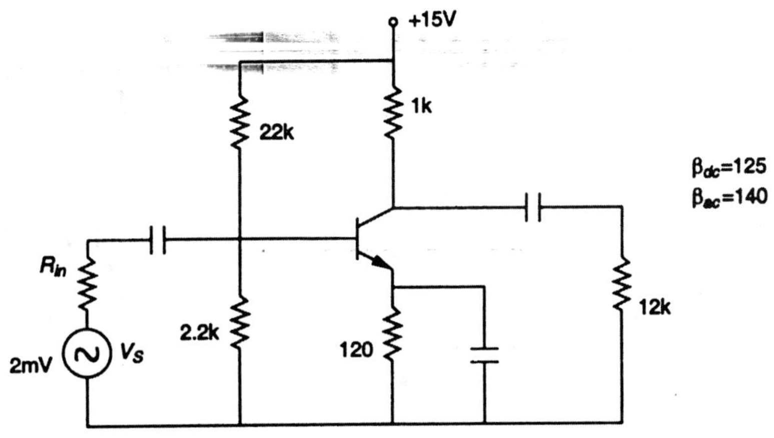

Fig E. 6.2. Reference Exercise Problem 6.2

3. Find the base signal voltage in the circuit of **Fig. E6.3** if $v_s = 2$ mV rms, 600 Ω signal source. I_E was determined as 6 mA.

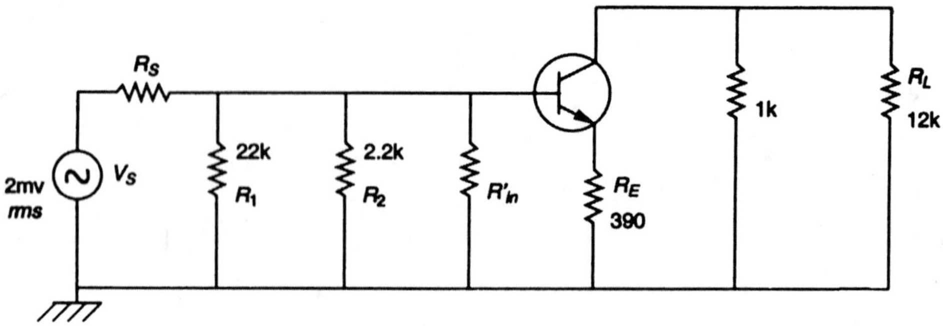

Fig E. 6.3. Reference Exercise Problem 6.3

4. In Problem 3 above, calculate the base to collector voltage gains considering
   (i) the circuit to be without an emitter bypass capacitor
   (ii) R_E shorted.

5. Find the (i) the voltage gain of the base collector region (ii) overall voltage gain of the amplifier in Fig. E6.3 if $R_L = 8{\cdot}2$ kΩ, $C_2 = 40$ μF, $r_e = 4$ Ω, $v_s = 4$ mV and $R_s = 600$ Ω.

6. Find the output voltage for the circuit in Fig. E6.6.

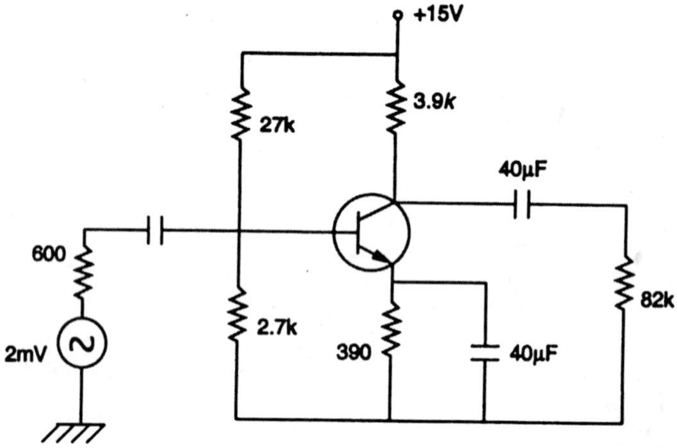

Fig E. 6.6. Reference Exercise Problem 6.6

7. Find the output voltage if R_L is
   (i) 2·7 k (ii) 10 k
   (iii) 56 k (iv) 100 k
   (v) 220 k (vi) ∞ in the problem 6 above.

8. Repeat problem 6 above if $R_s = 50$ Ω, $V_s = 5$ mV and R_C is
   (i) 1 k Ω (ii) 2·2 kΩ (iii) 4·7 kΩ (iv) 10 k Ω (vi) 47 kΩ

9. Find the value of current gain and power in the circuit of problem 6 (Fig. E6.6) above.

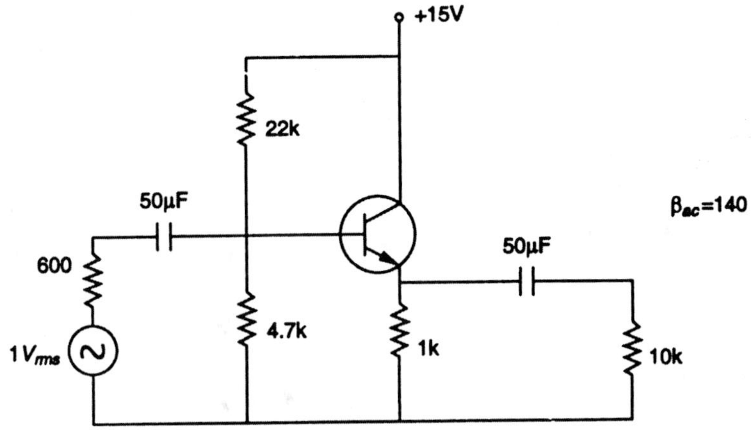

Fig E. 6.10. Reference Exercise Problem 6.10

10. Find the voltage gain, current gain, power gain, total input resistance for the common collector amplifier circuit of Fig. E6.10.

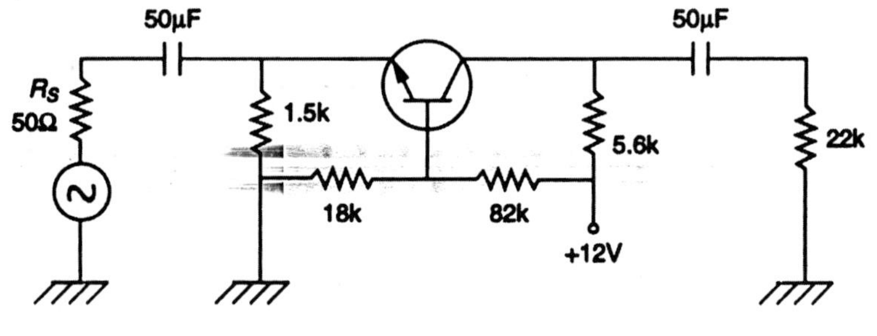

Fig E. 6.11. Reference Exercise Problem 6.11

11. Find the value of the input resistance, voltage gain, current gain, power gain for the common base amplifier depicted below in Fig. E6.11.

12. Calculate the value of the current gain in a superbeta transistor if each of the transistors employed possesses a current gain of $\beta = 150$. (or $\beta_{ac} = \beta^2 = 22500$)

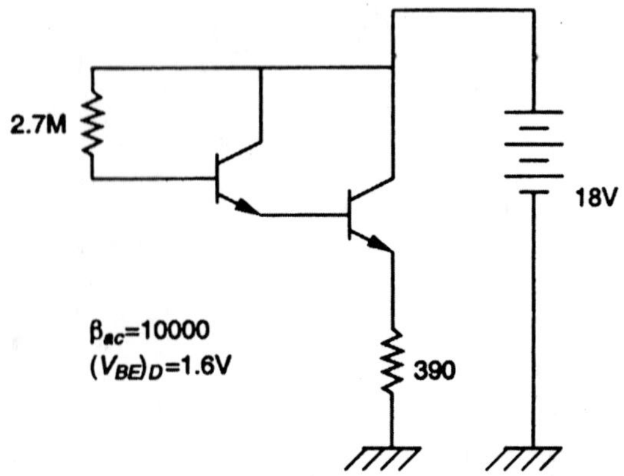

Fig E. 6.13. Reference Exercise Problem 6.13

13. Find the d.c. bias voltage and currents for the circuit in Fig. E.6.13.
14. Find the value of Z_i for the circuit above if r_i = 5 kΩ [Fig. E6.13]
15. Find the value of the base current and a.c. current gain for the circuit of problem 13 [Fig. E6.13]. Also determine the a.c. output impedance and a.c. voltage gain for the above circuit.

Chapter 7

FIELD EFFECT TRANSISTORS

7.1 INTRODUCTION

The field effect transistor (*FET*) is different from a bipolar junction transistor in that is a unipolar device, *i.e.* here only one type of charge carriers (the majority carriers) take part in the conduction process. In effect the field effect transistor is a three terminal unipolar solid state device in which the current flow is controlled by an electric field as in vacuum tubes. There are two classes of *FET*s namely

(1) Junction Field Effect Transistor (*JFET*)

(2) Insulated Gate Field Effect Transistor (*IGFET*) or Metal-Oxide-Semiconductor Field Effect Transistor (*MOSFET*)

The *MOSFET*s can be either depletion type or enhancement type.

Besides each of the *FET*s can be either *N*-channel or *P*-channel devices. The *FET* family can be symbolised as depicted in the Table 7.1.

7.2 JUNCTION FET

The three terminals for a field effect transistor are called **Source *S*, Drain *D*** and **Gate *G***. In the case of a *P*-channel *FET*, a narrow slab of *P*-type semiconductor and two *N*-type materials are diffused on opposite sides of it. The *P*-region is the channel and the two *N*-regions are electrically connected together and form the gate *G*. The source is the terminal through which the majority carriers enter into the channel.

Drain The drain is the terminal through which the electrons leave the bar. They are drained out of the *FET*.

Gate The *P* region is the channel and the two *N*-type regions are electrically connected together to form the gate. The operation of the *JFET* depends on the reverse biasing of the *p–n* junction between the gate and the channel. The gate thus controls the channel width and the current flow from drain to source.

In the *N*-channel *FET*, the channel is of *N*-type and two *P*-type regions form the gate regions (Fig. 7.2). In this *N*-type *JFET*, the electrons are the majority carriers and they flow from the source to the drain.

Once again the two *P*-type regions are electrically connected together and they form the gate. The reverse biasing of the *P–N* junction between the gate and the channel determines the device operation. The gate once again controls the current flow from source to drain.

Table 7.1 : Classification of the FETs

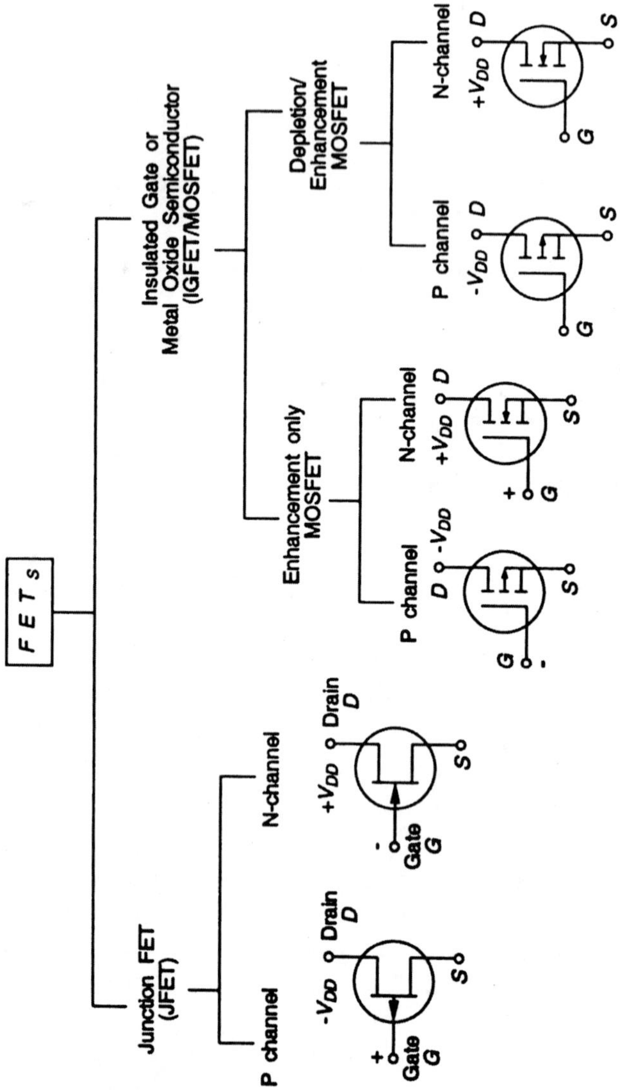

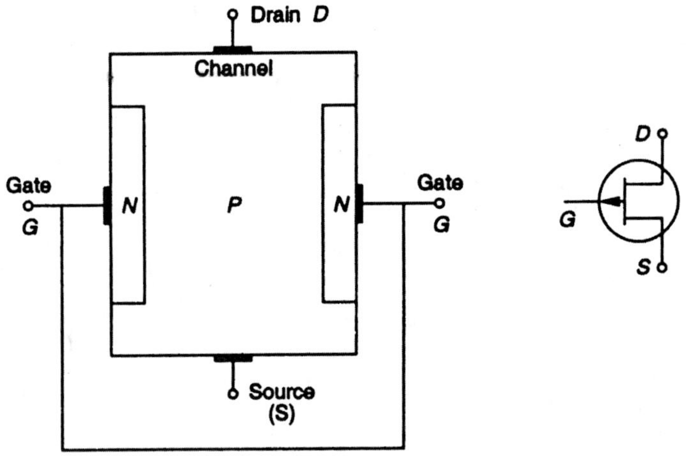

Fig 7.1. (a) Basic structure of a *P*-channel *JFET* (b) Circuit symbol

Device Operation

Let us consider the *N*-channel *JFET* of Fig. 7.3. When gate-to-source voltage $V_{GS} = 0$, the application of a drain-to-source voltage V_{DS} causes current to flow from drain to source. When V_{GS} is negative, the space charge or depletion region becomes large and the channel width is restricted. The channel resistance increases and the current I_D decreases. Since V_{DS} is small, the channel is of uniform width. The *JFET* behaves as a resistor whose value is determined by V_{GS}. When V_{GS} is increased in the negative direction, a value is reached at which the depletion region encompasses the entire channel.

It is clear that the reverse bias voltage between gate and channel varies at different points along the channel and is highest at the drain end. Thus the channel becomes tapered and the i_D vs. V_{DS} plot becomes nonlinear. When the reverse voltage at the drain end v_{GD} becomes less than the **thereshold voltage (V_t), the channel is pinched off at the drain end and the drain current becomes saturated.**

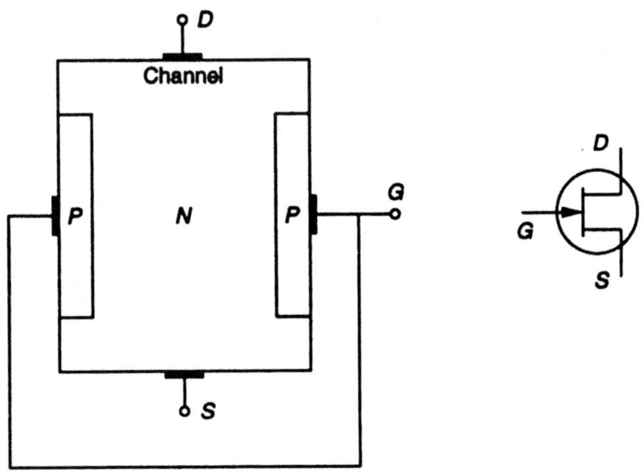

Fig 7.2. Basic structure of the *N*-channel *JFET* and its circuit symbol

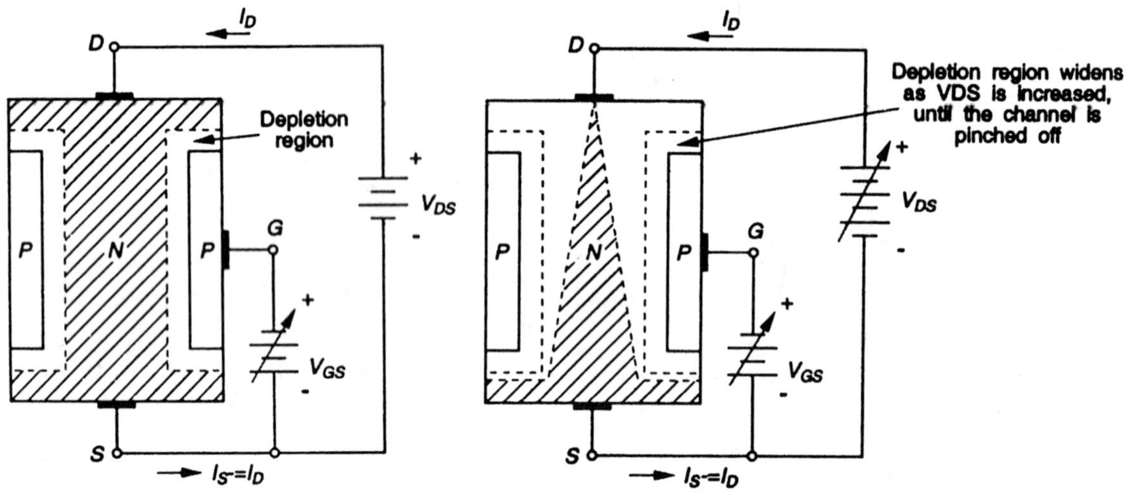

Fig 7.3. N channel JFET operation (a) for small V_{DS} and (b) for higher V_{DS}

7.2.1 *JFET* Characteristics

The $i_D - V_{DS}$ characteristics curves (Fig. 7.4) of a *JFET* indicate that there are three distinct regions of operation, **the cut-off region, the triode region and the pinch-off region.**

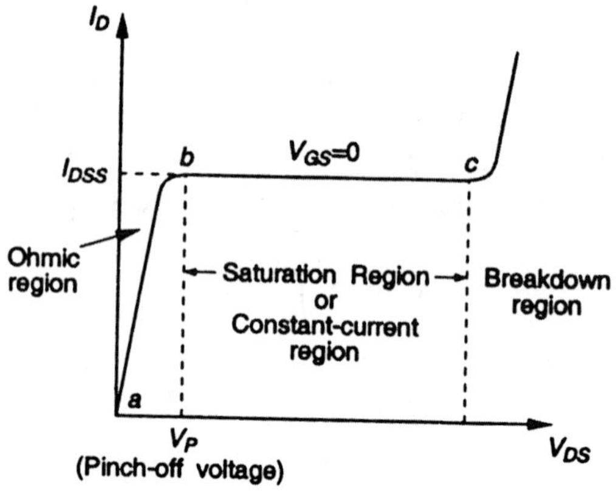

Fig 7.4. Drain characteristics (I_D *vs* V_{DS}) for a JFET at V_{GS} = OV

Fig. 7.5 depicts the drain characteristics for a *JFET* for $V_{GS} = 0$ (gate shorted to source). Initially as V_{DS} is increased, the current I_D rises linearly (portion *ab*). When the current I_D reaches a value I_{DSS} (point *b*), the current becomes constant (portion *bc*). This portion *bc* of the curve is known as the saturation region or the constant current region. For $V_{GS} = 0$ V, the value of V_{DS} at which I_D becomes nearly constant is called the **pinch-off voltage V_P**. Every *JFET* has a fixed value of V_P which is its characteristics. As V_{DS} is increased above V_P, the value of current remains fixed at I_{DSS}. I_{DSS} is also a characteristic for every *JFET* and its value is specified by the manufacturer. The value of I_{DSS} is thus the maximum value of drain current for the

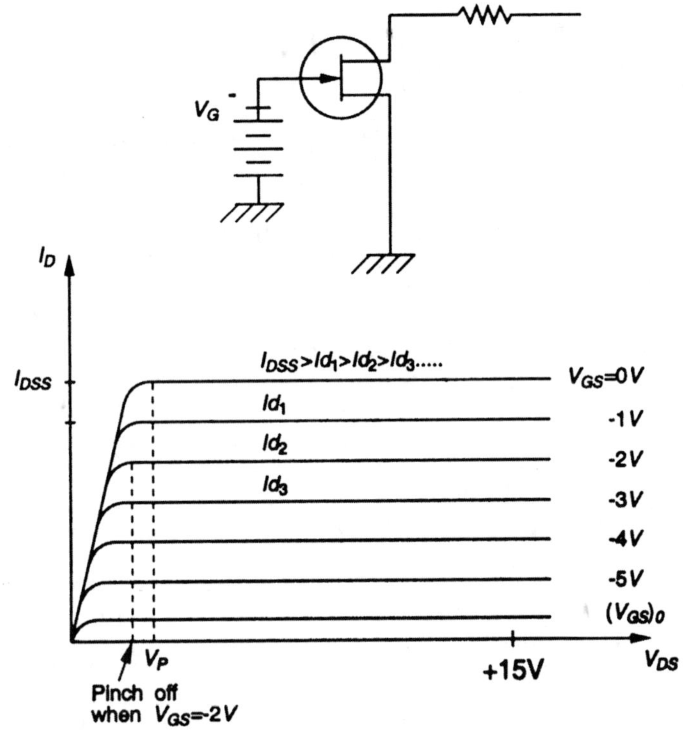

Fig 7.5. Drain characteristic curves for an *n*-channel *JFET*

JFET for $V_{GS} = 0$. As V_{DS} is increased beyond *C*, the current rises rapidly with increase in V_{DS}. The breakdown region is the region in which the operation of the *JFET* may produce an irreversible damage. *JFET*s are not operated in this region.

Let us now examine as to what happens if the gate-channel junction is reverse biased. In an *n*-channel junction, the gate is *p* type and the *p–n* junction is reverse biased by connecting the negative of the battery voltage V_G to the gate while the positive terminal is connected to the source. The $I_D - V_{DS}$ characteristics is now similar to what was obtained for $V_{GS} = 0$ expect for that the current I_D is lower and it further decreases as V_{GS} assumes increasing negative values. Thus with increasing negative values of V_{GS}, the depletion region widens in the channel and I_{DS} decreases. Further, it will be seen that as V_{GS} is decreased, the pinch off now occurs at lower values of V_{DS} (vide Fig. 7.5). It will be seen that pinch off corresponding to $V_{GS} = -2$ Volt occurs at a lower voltage than the pinch off at $V_{GS} = 0$ V.

The value of V_{GS} at which the current becomes almost zero is called cut off. That is, at the cut off value $V_{GS} = V_{GS0}$, the current is almost zero. The *JFET* is normally operated between $V_{GS} = 0$ and $V_{GS} = V_{GS0}$. **At $V_{GS} = 0$ volts, the current is $I_D = I_{DSS}$ and at $V_{GS} = V_{G0}$, the current is almost zero $I_D \simeq 0$.**

It should be noted that $(V_{GS})_{\text{cut off}} = V_{GS0}$ and V_P are always equal in magnitude but opposite in sign. For example if $V_P = +7$ V, the value of $(V_{GS})_{\text{cut off}}$ or V_{G0} is -7 V.

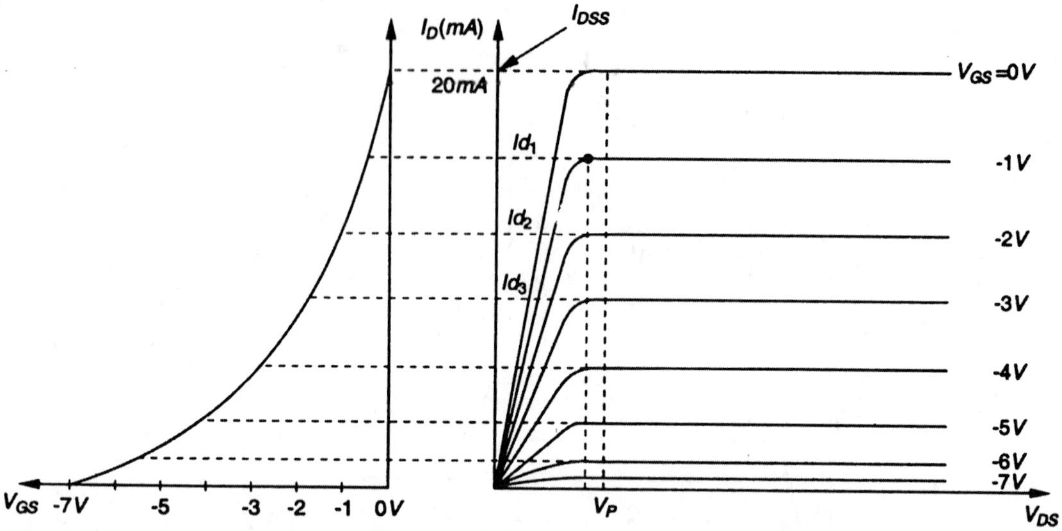

Fig 7.6. Finding the *JFET* transfer characteristic curves from *JFET* drain characteristics

EXAMPLE 7.1 *For the JFET in Fig. 7.7 if I_{DSS} = 15 mA and $(V_{GS})_{cut\ off} = -5$ V, then find the minimum value of V_D' which will place the JFET in the constant current region. What will be the drain current if V_D' is increased to 15 V.*

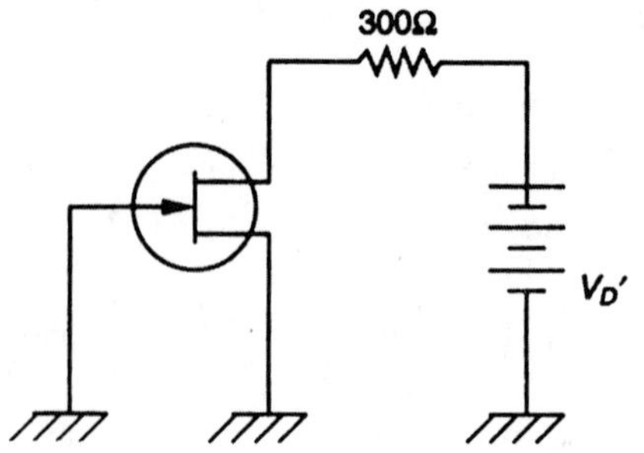

Fig 7.7. Circuit for Example 7.1

SOLUTION

$(V_{GS})_{cut\ off} = -5$ V

So $V_P = +5\text{ V} = V_{DS}$

Now for $V_{GS} = 0$ V, $I_D = I_{DSS} = 15$ mA

The voltage drop across the 500 Ω resistor is $V = 300 \times 15\text{ mA} = 4{\cdot}5$ V

Thus minimum value of V_D' battery required is $V_D' = 4{\cdot}5\text{V} + V_{DS} = 4{\cdot}5\text{V} + 5\text{V} = 9{\cdot}5\text{V}$. If V_D' is increased to 15 V, the drain current will remain constant at 15 mA and the increased voltage will appear as drop in the drain source terminals.

Transfer Characteristics

By the term *JFET* transfer characteristics we mean the variation of drain current I_D with the gate to source voltage V_{GS}. The transfer characteristics of an *n*-channel *JFET* is depicted in Fig. 7.6.

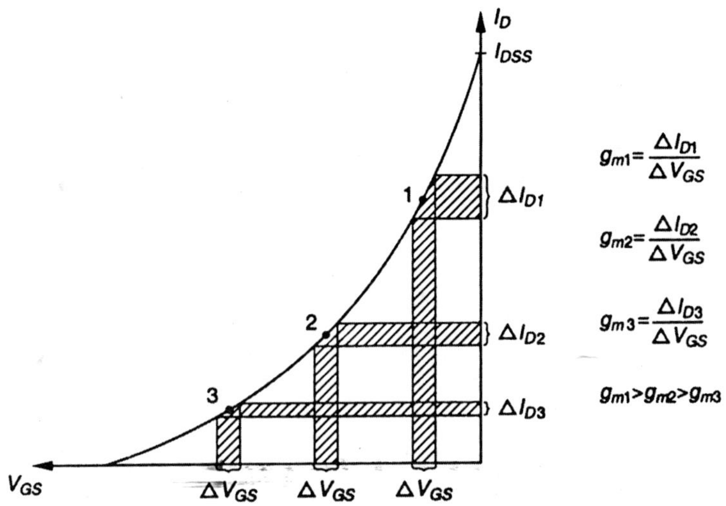

Fig 7.8. Transfer characteristics for an *n*-channel *JFET* and the dependence of g_m on V_{GS} (the bias point)

It will be seen in Fig. 7.6 that the transfer characteristics starts at $V_{GS} = (V_{GS})_{\text{cut off}}$ when $I_{DS} = 0$ and with less and less values of $|V_{GS}|$ the drain current increases until $I_D = I_{DSS}$ at $V_{GS} = 0$ Volts. The *JFET* transfer characteristics curves are parabolic in nature and can be described by the relation, called the **Shockley equation.**

$$I_D = I_{DSS}\left[1 - \frac{V_{GS}}{V_{GS0}}\right]^2 \qquad \text{...(7.1)}$$

The transfer characteristics curves can be obtained from the drain characteristics curves by plotting the values of the drain current for various values of V_{GS} taken from the drain characteristics curves in the pinch off region as depicted in Fig. 7.6.

Transconductance

One very important parameter of a *JFET* is the forward transconductance g_m. The forward transconductance g_m is given by

$$g_m = \frac{\Delta I_D}{\Delta V_{GS}} \qquad \text{...(7.2)}$$

where ΔI_D is the change in drain current for a change in gate to source voltage of ΔV_{GS}.

The transconductance g_m is also sometimes denoted by g_{fs} or y_{fs} (forward transfer admittance). The forward transconductance (g_m) parameter is very important in designing *FET* amplifiers as it determines the voltage gain. Since the transfer characterics of a *JFET* is nonlinear, the value of g_m (the slope of the transfer

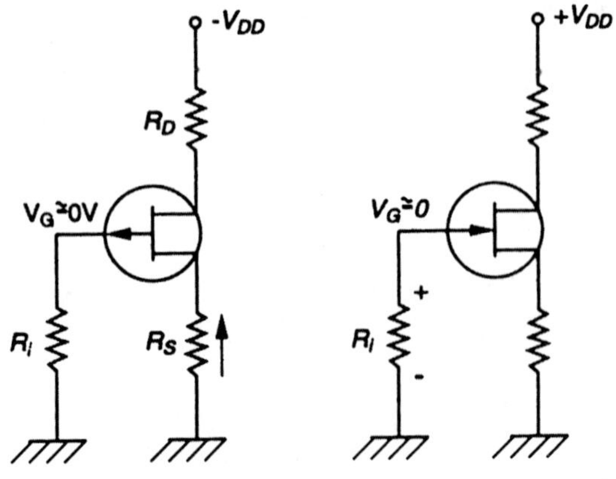

Fig 7.9. Calculation of g_m at $V_{GS} = 0$ V

characteristics) changes depending on the location of operating point. Thus, it will be apparent from the curve of Fig. 7.8 that g_{m_1} is different from g_{m_2} and g_{m_2} is different from g_{m_3} in the characteristics curves. The *JFET* manufacturers normally specifies the value of g_m at $V_{GS} = 0$ Volts (g_{m_0}). For a typical *n*-channel (2N3796) *JFET* $|y_{fs}| = 2000$ μmhos at $V_{DS} = 10$ V, ambient temperature = 25°C, $f = 1$ kHz for a drain current of 5 mA. Typically y_{fs} lies between 1000 to 6000 μS.

For the *JFET*, if g_{m0} is known, the value of g_m at $V_{GS} \neq -$ V can be obtained from the equation

$$g_m = g_{m0}\left[1 - \frac{V_{GS}}{(V_{GS})_{\text{cut off}}}\right] \qquad \text{...(7.3)}$$

This equation can be proved very easily by taking the derivative of drain current I_D with respect to V_{GS}, the gate-to-source voltage, using Shockley's equation as

$$g_m = \left.\frac{\Delta I_D}{\Delta V_{GS}}\right|_Q \qquad \text{...(7.4)}$$

$$= \left.\frac{dI_D}{dV_{GS}}\right|_Q \qquad \text{...(7.5)}$$

$$= \frac{d}{dV_{GS}}\left[I_{DSS}\left(1 - \frac{V_{GS}}{V_P}\right)^2\right] \qquad \text{...(7.6)}$$

$$= I_{DSS}\frac{d}{dV_{GS}}\left(1 - \frac{V_{GS}}{V_P}\right)^2 = 2I_{DSS}\left[1 - \frac{V_{GS}}{V_P}\right]\frac{d}{dV_{GS}}\left(1 - \frac{V_{GS}}{V_P}\right)$$

$$= 2I_{DSS}\left[1 - \frac{V_{GS}}{V_P}\right]\left[\frac{d}{dV_{GS}}(1) - \frac{1}{V_P}\frac{dV_{GS}}{dV_{GS}}\right]$$

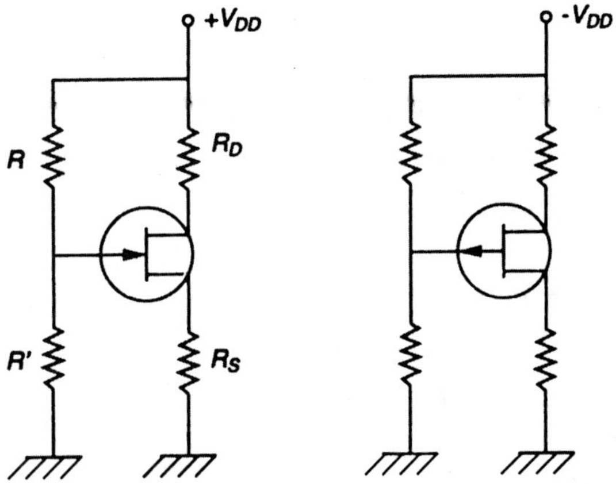

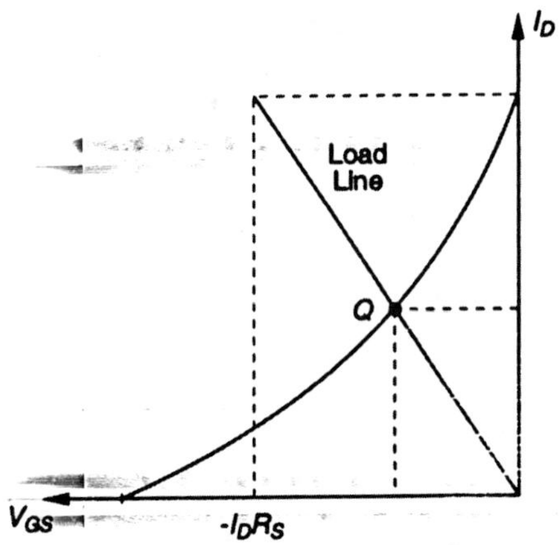

Fig 7.10. Location of *Q* (Quiscent) point

$$= 2I_{DSS}\left[1 - \frac{V_{GS}}{V_P}\right]\left[-\frac{1}{V_P}\right]$$

$$= \frac{2I_{DSS}}{|V_P|}\left[1 - \frac{V_{GS}}{V_P}\right] \quad \text{...(7.7)}$$

Since $$g_{m0} = g_m\Big|_{V_{GS}=0} = \frac{2I_{DSS}}{|V_P|}\left[1 - \frac{0}{V_P}\right] \quad \text{...(7.8)}$$

We may express g_m as

$$\mathbf{g_m = g_{m0}\left[1 - \frac{V_{GS}}{V_P}\right]} \quad \text{...(7.9)}$$

This can also be written as

$$g_m = g_{m0}\sqrt{\frac{I_D}{I_{DSS}}} \qquad ...(7.10)$$

The amplification factor (μ) of a *JFET* is defined as the change ΔV_{DS} in V_{DS} for unit change in V_{GS} while I_D is kept constant. This is written as

$$\mu = \left.\frac{\Delta V_{DS}}{\Delta V_{GS}}\right|_{I_D = \text{constant}} \qquad ...(7.11)$$

The a.c. drain resistance (r_d) is given by

$$r_d = \left.\frac{\Delta V_{DS}}{\Delta I_D}\right|_{V_{GS} = \text{constant}} \qquad ...(7.12)$$

where ΔV_{DS} is the change in V_{DS} for a change ΔI_D in the drain current while V_{GS} = constant.

It can be shown very easily that

$$\mu = g_m \times r_d \qquad ...(7.13)$$

or

$$\left\{\begin{matrix}\textbf{Amplification}\\ \textbf{factor}\end{matrix}\right\} = \textbf{Transconductance} \times \left(\begin{matrix}\textbf{AC drain}\\ \textbf{resistance}\end{matrix}\right) \qquad ...(7.14)$$

The DC drain resistance is given as

$$R_{DS} = \frac{V_{DS}}{I_D} \qquad ...(7.15)$$

EXAMPLE 7.2 *For an n-channel JFET, I_{DSS} = 30 mA and $(V_{GS})_{\text{cut off}}$ = – 10 V. The value of g_{m0} is 5000 μS. Find g_m and the drain current I_D at V_{GS} = –5V.*

SOLUTION

$$g_m = g_{m0}\left[1 - \frac{V_{GS}}{(V_{GS})_{\text{cut off}}}\right]$$

$$= 5000\ \mu\text{S}\left[1 - \frac{-5\text{ V}}{-10\text{ V}}\right] = 250\ \mu\text{S}$$

$$\text{Also } I_D = I_{DSS}\left[1 - \frac{-5\text{ V}}{-10\text{ V}}\right]^2 = 7{\cdot}5\text{ mA.}$$

EXAMPLE 7.3 *In an n-channel JFET, I_{DSS} = 10 mA, V_P = 6V, V_{GS} = – 3 V. Find the values of (i) I_D (ii) g_{m0} and (iii) g_m.*

SOLUTION

(i) $$I_D = I_{DSS}\left(1 - \frac{V_{GS}}{V_P}\right)^2 = 10\left(1 - \frac{3}{6}\right)^2$$

$$= 2{\cdot}5\text{ mA}$$

(ii) $$g_{m0} = -\frac{2I_{DSS}}{V_P} = \frac{-2 \times 10}{-6} = 3{\cdot}33\text{ mS}$$

(iii) $g_m = g_{m0}\left(1 - \dfrac{V_{GS}}{V_P}\right) = 3{\cdot}33\left(1 - \dfrac{-3}{-6}\right) = 1{\cdot}67$ mS

EXAMPLE 7.4 *For an n-channel JFET, g_{m0} = 5 mS, g_m = 2·5 mS and I_{DSS} = 15 mA. Find the drain current.*

SOLUTION From equation (7.10)

$$g_m = g_{m0}\sqrt{\frac{I_D}{I_{DSS}}}$$

$$\therefore\ I_D = I_{DSS}\cdot\left(\frac{g_m}{g_{m0}}\right)^2 = (15\text{ mA})\times\left(\frac{2{\cdot}5}{5}\right)^2$$

$$= 3{\cdot}75\text{ mA.}$$

7.3 DC BIASING OF A *JFET*

In Fig. 7.11 is shown is d.c. biasing circuits for a number of different n-channel *JFETs*. In Fig 7.11 (a) is shown a circuit where V_{GG} is used as a separate gate bias.

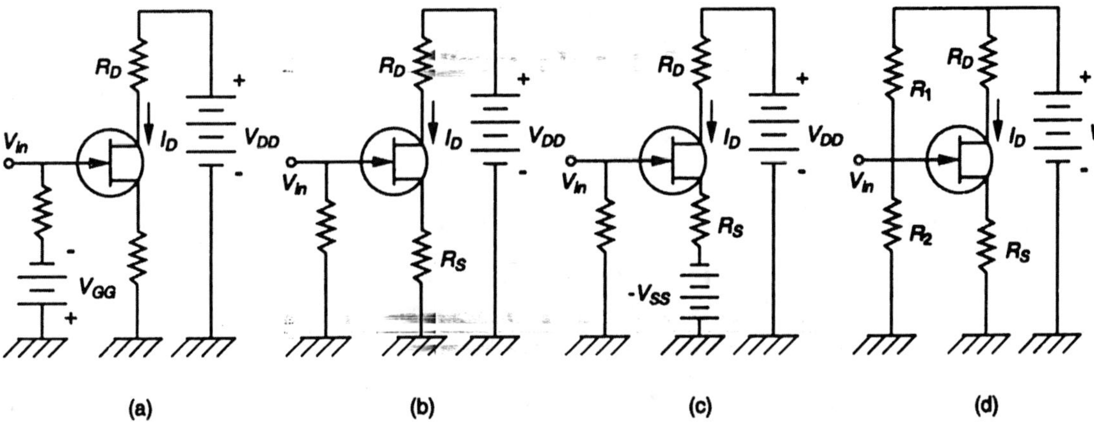

Fig 7.11. DC biasing circuits for *n*-channel *JFET* (a) seperate gate bias (b) self bias (c) source bias circuit and (d) voltage divider bias

The circuit of Fig. 7.11 (b) depicts the familiar self bias circuit. Here a seperate gate power source is not employed. Instead, the voltage drop across the resistor R_S due to current I_D flowing through it is used for gate biasing. In this case

$$V_S = I_D R_S \qquad \text{...(7.16)}$$

and so $$V_{GS} = V_G - V_S = 0 - I_D R_S = -I_D R_S \qquad \text{...(7.17)}$$

The reason why R_G has been employed in the circuit is that without it, the gate would be floating in nature, which could collect charge and ultimately cut off the *JFET*. The introduction of R_G does not, however, alter the d.c. bias since no gate current flows through it (the gate leakage current is almost zero). Thus the gate is essentially at d.c. ground. Besides R_G provides a path for the leakage current. Without R_G, the leakage current may build up static charge (voltage) at the gate which may change the bias or even destroy the

JFET.

In the source bias circuit of Fig. 7.11 (c), the voltage between gate and source is

$$V_{GS} = V_{SS} - I_D R_S \quad ...(7.18)$$

The voltage divider bias circuit of Fig. 7.9 (d) is quite popular. In this circuit

$$V_G = V_{DD} \cdot \frac{R_2}{R_1 + R_2} \quad ...(7.19)$$

and

$$V_{GS} = V_G - I_D R_S \quad ...(7.20)$$

or

$$V_{GS} = V_{DD} \cdot \frac{R_2}{R_1 + R_2} - I_D R_S \quad ...(7.21)$$

Fig. 7.12 depicts biasing circuits for *p*-channel *JFETs*.

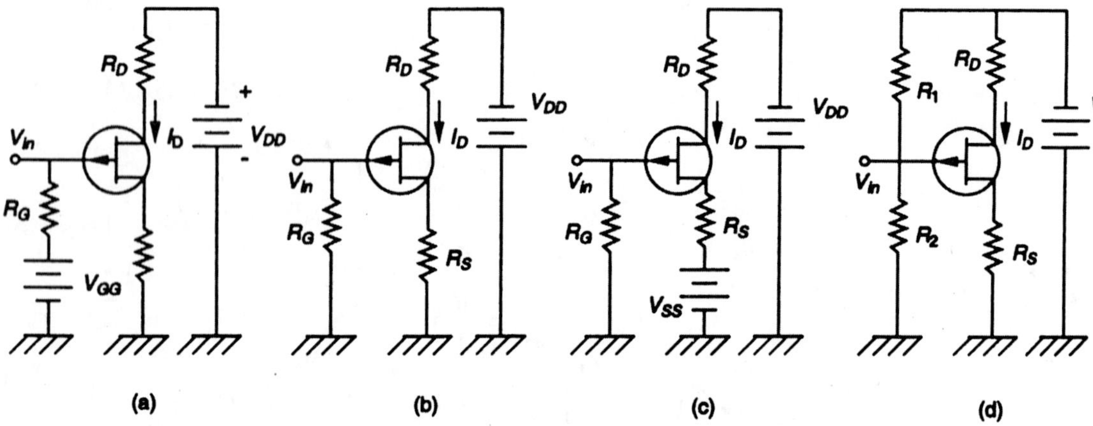

Fig 7.12. Biasing circuits for *p*-channel *JFETs* (a) fixed bias (b) self bias (c) source bias (d) voltage divider bias

(a) Fixed Bias

For the fixed bias circuit of Fig. 7.11 (a), the transfer characteristics is depicted in Fig. 7.13.

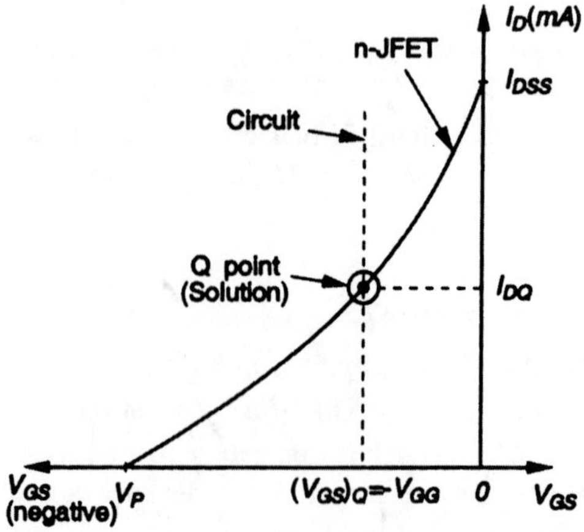

Fig 7.13. Location of *Q* point (Quiscent operating point)

The Q (or quiscent) point can be conveniently located on the transfer characteristics of the *JFET* by finding the intersection of the two curves as shown in Fig. 7.13. In this case $(V_{GS})_Q = -V_{GG}$.

$$\left.\begin{aligned} \text{Also} \quad & V_{DS} = V_{DD} - I_D R_D \\ & V_S = 0 \\ \text{and} \quad & V_G = V_{GS} \end{aligned}\right\} \qquad \ldots(7.22)$$

The corresponding drain current is $I_D = I_{DQ}$.

EXAMPLE 7.5 *Find (i) $(V_{GS})_Q$ (ii) $(I_D)_Q$, (iii) V_{DS} (iv) V_D (v) V_G and (vi) V_S for the circuit in Fig. 7.14.*

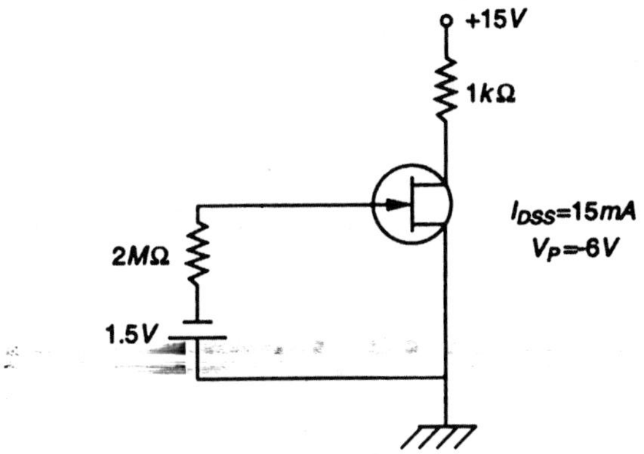

Fig 7.14. Circuit of Example 7.5

SOLUTION **Mathematical Analysis**

(i) $(V_{GS})_Q = -V_{GG} = -1{\cdot}5$ V

(ii) $I_{DQ} = I_{DSS}\left(1 - \dfrac{V_{GS}}{V_P}\right)^2$

$= 15 \text{ mA}\left(\dfrac{-1{\cdot}5}{-6}\right)^2$

$= 15 \text{ mA}\,(1 - 0{\cdot}25)^2$

$= 8{\cdot}437$ mA

(iii) $V_{DS} = V_{DD} - I_D R_D$

$= 15 \text{ V} - (8{\cdot}437 \text{ mA})(1 \text{ k}\Omega)$

$= 15 - 8{\cdot}437$

$= 6{\cdot}563$ V

(iv) $V_D = V_{DS} = 6{\cdot}563$ V

(v) $V_G = V_{GS} = -1{\cdot}5$ V

(vi) $V_S = 0$ V.

7.3.1. Self Bias Arrangement

For the self bias arrangement of Fig. 7.11 (b), the Q point can be found bv locating the interesction of the transfer characteristic curves of the n-chnnel *JFET* as defined by the Shockley equation

$$I_D = I_{DSS}\left(1 - \frac{V_{GS}}{V_P}\right)^2 \qquad \text{...(7.23)}$$

and the self bias line for the network

$$V_{GS} = - I_D R_S \qquad \text{...(7.24)}$$

This is depicted in Fig. 7.15

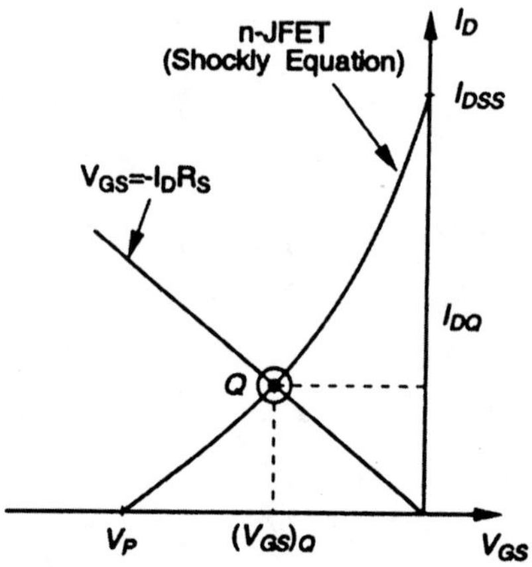

Fig 7.15. Location of Q point on the transfer characteristics for an n-channel *JFET*

EXAMPLE 7.6 *Calculate the following quantities for the circuit in Fig. 7.16*
(i) V_{GSQ} (ii) I_{DQ} (iii) V_{DS} (iv) V_S (v) V_G, (vi) V_D

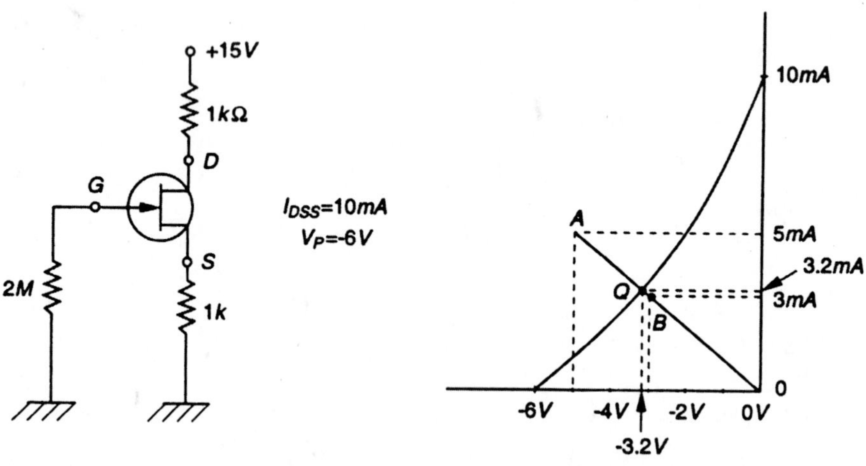

Fig 7.16. Circuit for Example 7.16

SOLUTION (i) $V_{GS} = -I_D R_S$

For $I_D = 5$ mA, $V_{GS} = -5$ V (point A)

$= 3$ mA, $V_{GS} = -3$ V (point B)

In the graph the bias line is first drawn (OA). The intersection with the transfer characteristics takes place at Q whose coordinates are (3·2mA, – 3·2 V). Thus,

$V_{GSQ} = -3{\cdot}2$ V.

(ii) $I_{DQ} = 3{\cdot}2$ mA.

(iii) $V_{DS} = V_{DD} - I_D (R_S + R_D)$

$= 16 - 1\ (3{\cdot}2 \text{ mA})\ (2\text{ k} + 1\text{ k})$

$= 16 - 3{\cdot}2 \text{ mA} \times 3\text{ k}$

$= 16 - 9{\cdot}6 = 6{\cdot}4$ V.

(iv) $V_S = I_D R_S = 3{\cdot}2 \text{ mA} \times 1\text{ k} = 3{\cdot}2$ V

(v) $V_G = 0$ V

(vi) $V_D = V_{DD} - I_D R_D = 16 - 3{\cdot}2 \text{ mA} \times 2\text{ k} = 16 - 6{\cdot}4$

$= 9{\cdot}6$ V

Also $V_D = V_{DS} + V_S = 6{\cdot}4\text{ V} + 3{\cdot}2\text{ V} = 9{\cdot}6$ V

Source Biasing Configuration

In the source biasing configuration of Fig. 7.17, we may write from the Kirchhoff's voltage laws that

$$-V_{GS} - I_S R_S + V_{SS} = 0 \qquad \text{...(7.25)}$$

or,

$$V_{GS} = V_{SS} - I_D R_S \qquad \text{...(7.26)}$$

where $I_S = I_D$

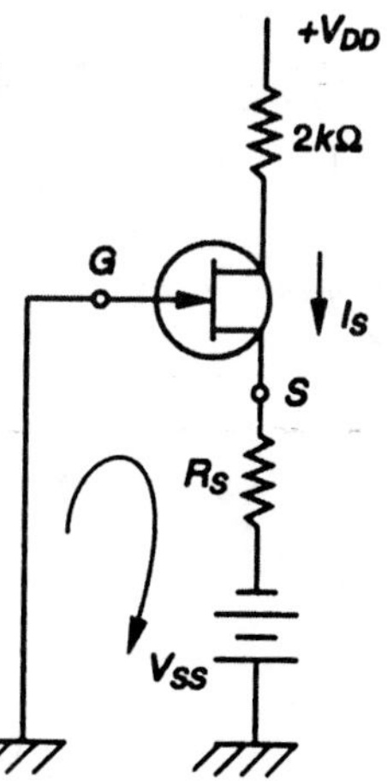

Fig 7.17. Finding the network equation for the source biasing configuration

EXAMPLE 7.7 *Find the values of the following quantiteis in the circuit of Fig. 7.18.*
(i) I_{DQ} (ii) V_{GSQ} (iii) V_{DS} (iv) V_D (v) V_S

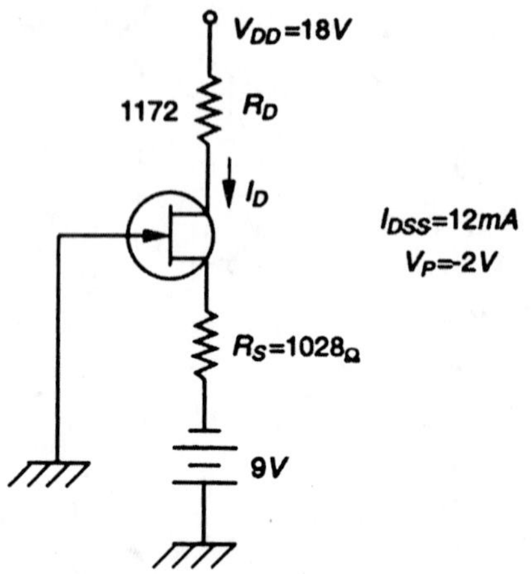

Fig 7.18. Circuit in Example 7.7

SOLUTION Using the result from equation (7.26) we get

$$V_{GS} = 9\text{ V} - I_D\,(1\text{ k})$$

For $I_D = 0$ mA, $V_{GS} = V_{SS} = 9$ V

For $V_{GS} = 0$ V, $I_D = \dfrac{9\text{ V}}{1\text{ k}} = 9$ mA

The transfer characteristics are obtained by plotting the point

$$V_{GS} = \frac{V_P}{2} = -\frac{2\text{ V}}{2} = -1\text{ V and } I_D = \frac{I_{DSS}}{4} = \frac{12\text{ mA}}{4} = 3\text{ mA}.$$

From the graph, the operating point is obtained as

$$I_{DQ} = 9\text{ mA}$$

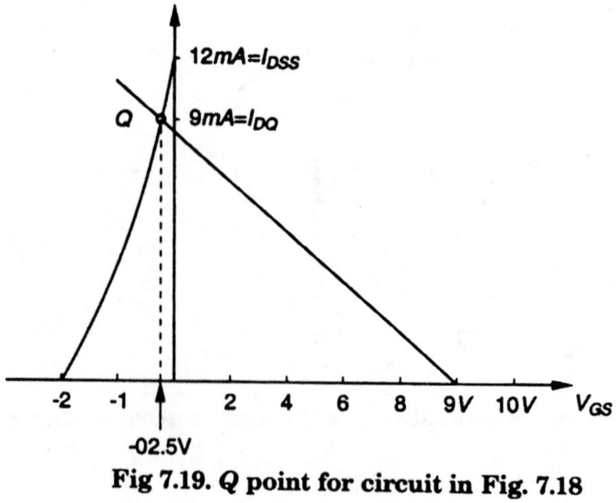

Fig 7.19. *Q* point for circuit in Fig. 7.18

$V_{GSQ} = -0{\cdot}25$ V.

(iii) Applying Kirchhoff's voltage laws to the drain circuit, we get

$$-V_{DD} + I_D R_D + V_{DS} + I_S R_S - V_{SS} = 0 \quad \text{...(7.27)}$$

or, $$V_{DS} = V_{DD} + V_{SS} - I_D (R_D + R_S) \quad \text{...(7.28)}$$

since $I_D = I_S$

Thus $V_{DS} = 18 \text{ V} + 9 \text{ V} - 9 \text{ mA } (1172 + 1028)$

$= 27 \text{ V} - 9 \text{ mA} \times 2{\cdot}2 \text{ k} = 27 - 19{\cdot}8 \text{ V} = 7{\cdot}2 \text{ V}$

(iv) $V_D = V_{DD} - I_D R_D = 18 \text{ V} - 9 \text{ mA } (1172)$

$= 18\text{V} - 10{\cdot}55\text{V} = 7{\cdot}45\text{V}$

(v) $V_{DS} = V_D - V_S = V_D - V_{DS} = 7{\cdot}45 - 4{\cdot}2 = 0{\cdot}25$ V.

7.3.2 Voltage-Divider Biasing

The voltage divider biasing configuration Fig. (7.20) which is employed in the case of *BJTs* is also applied to the case of *FETs*. However, there is an important difference. In the case of *BJTs*, the base current I_B provides the link between output and input while for *FET*, the voltage V_{GS} provides the link. For *FETs*, the gate current $I_G = 0$

Fig. 7.20 (a) depicts the voltage divider biasing configuration while Fig. 7.20 (b) shows the redrawn circuit for ease of analysis. From the circuit of Fig. 7.20(b), one gets,

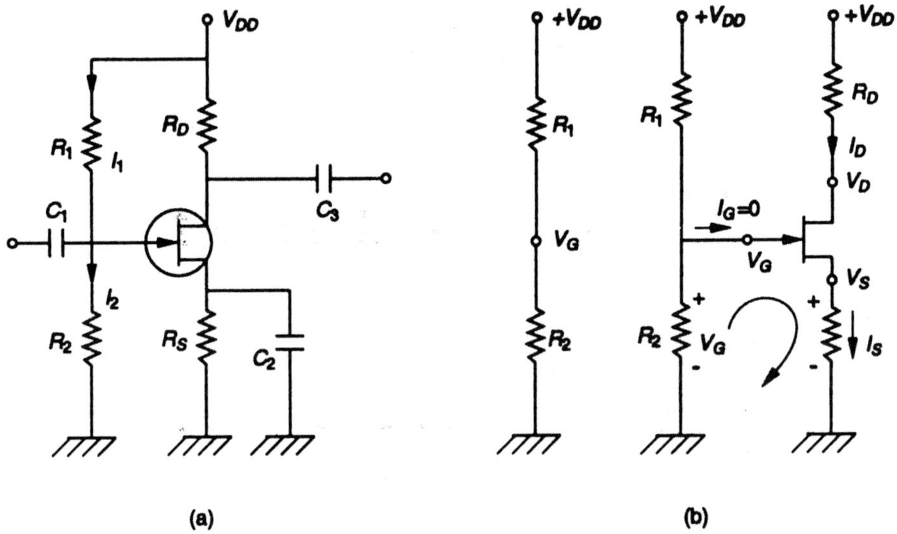

Fig 7.20. (a) Voltage divider bias configuration and (b) redrawn network

$$V_G = V_{DD} \cdot \frac{R_2}{R_1 + R_2} \quad \text{...(7.29)}$$

and $$V_{GS} = V_G - I_D R_S \quad \text{...(7.30)}$$

From equation (7.30) one can plot the voltage divider bias line. Two convenient points are $V_{GS} = 0$ V, $I_D = V_G/R_S$ and $I_D = 0$ mA, $V_{GS} = V_G$.

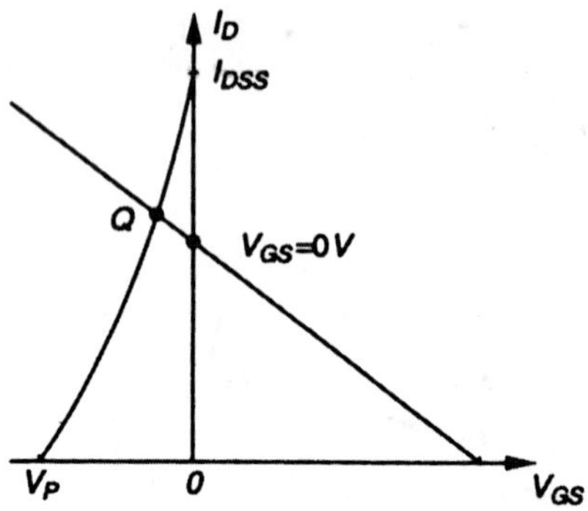

Fig 7.21. Determining the *Q* point in the voltage divider configuration

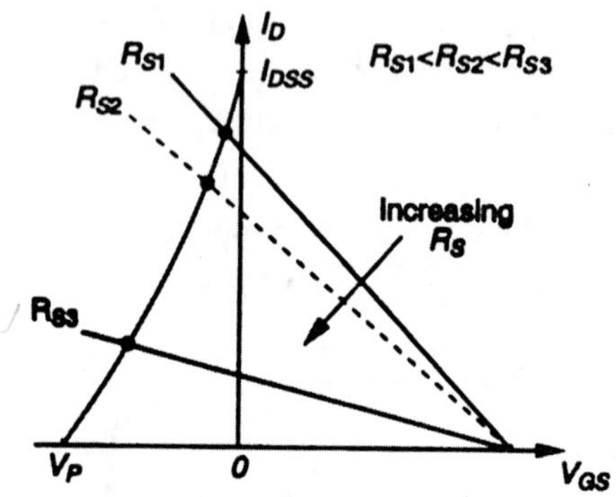

Fig 7.22. Dependence of *Q* point of the value of R_S
(higher values of R_S leads to smaller I_{DQ} and more negative V_{GSQ})

Putting these values the bias line is drawn (Fig. 7.21) on the transfer characteristics of a *JFET*. The intersection of the transfer characteristics and the bias line provides the *Q* point where coordinate (V_{GSQ} and I_{DQ}) can then be found. Fig. 7.22 depicts the dependance of the *Q* point on the values of R_S. As R_S increases, the *Q* point shows lower values of I_D and more negative V_{GS} values.

When the values of V_{GSQ} and I_{DQ} are obtained, the remaining network quantites may be obtained very easily namely

$$V_S = I_D R_S \qquad \text{...(7.31)}$$

$$V_D = V_{DD} - I_D R_D \qquad \text{...(7.32)}$$

$$V_{DS} = V_{DD} - I_D (R_D + R_S) \qquad \text{...(7.33)}$$

and

$$I_1 = I_2 = \frac{V_{DD}}{R_1 + R_2} \qquad \text{...(7.34)}$$

EXAMPLE 7.8 *Find the following parameters for the network in Fig. 7.2 (i) I_{DQ} and V_{GSQ} (ii) V_D (iii) V_S (iv) V_{DS}.*

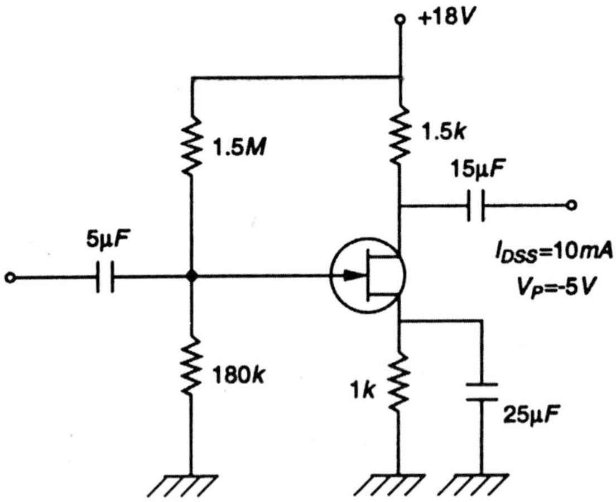

Fig 7.23. A typical voltage divider configuration

SOLUTION

For the transfer characteristics when $I_D = \frac{I_{DSS}}{4} = \frac{10}{4}$ $= 2.5$ mA, $V_{GS} = \frac{V_P}{2} = -\frac{5\text{ V}}{2} = -2.5$ V. The curve representing the Shockley equation is drawn in Fig. 7.24. For the circuit configuration,

$$V_G = \frac{V_{DD} R_2}{(R_1 + R_2)} = \frac{18\ (180\text{ k})}{(1500 + 180)\text{ k}}$$

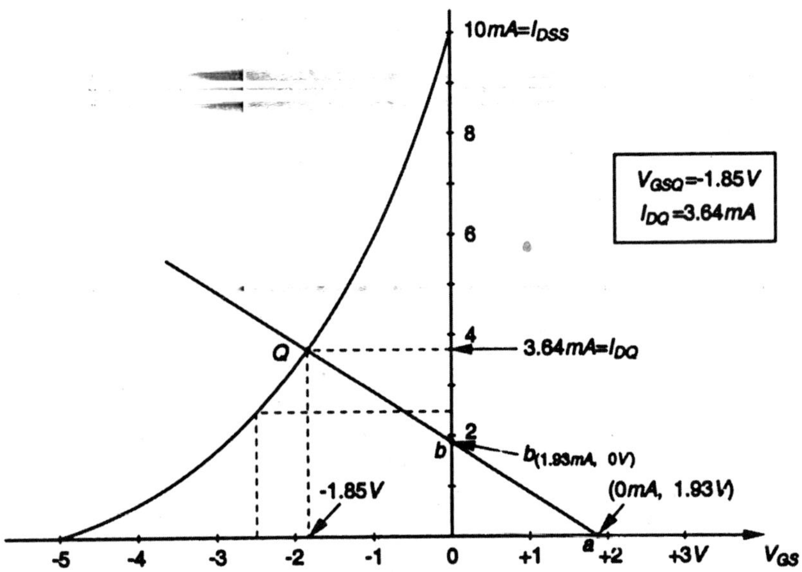

Fig 7.24. Determination of the Q point

$$= \frac{18 \times 180\text{ k}}{1680\text{ k}} = 1{\cdot}93\text{ V}$$

and $V_{GS} = V_G - I_D R_S$

$= 1{\cdot}93\text{ V} - I_D(1\text{ k})$

If $I_D = 0$, $V_{GS} = 1{\cdot}93$ V point '*a*' in figure

If $V_{GS} = 0$, $I_D = 1{\cdot}93\text{ V}/1\text{ k} = 1{\cdot}93$ mA point '*b*' in figure

The curve passing through *a* and *b* defines the bias line and when extrapolated cuts the Shockely equation curve at point *Q*, the Quiscent or operating point. The co-ordinates of *Q* are

$I_{DQ} = 3{\cdot}64$ mA

$V_{GSQ} = -\ 1{\cdot}85$ V.

(ii) $V_D = V_{DD} - I_D R_D = 18 - 3{\cdot}64\text{ mA }(1{\cdot}5\text{ k})$

$= 18 - 5{\cdot}46\text{ V} = 12{\cdot}54$ V.

(iii) $V_s = I_D R_S = (3{\cdot}64\text{ mA})\ (1{\cdot}0\text{ k}) = 3{\cdot}64$ V.

(iv) $V_{DS} = V_{DD} - I_D(R_D + R_s)$

$= 18 - 5{\cdot}46\text{ V} - 3{\cdot}64\text{ V}$

$= 18 - 9{\cdot}1\text{ V} = 8{\cdot}9$ V

7.4 COMMON SOURCE JFET AMPLIFIER

A common source *JFET* amplifier circuit is depicted in Fig. 7.25. In the circuit, the resistor R_S provides the self bias, R_D acts as the load. Also R_G provides a path for leakage current and C_S provides the a.c. ground path for the input signal.

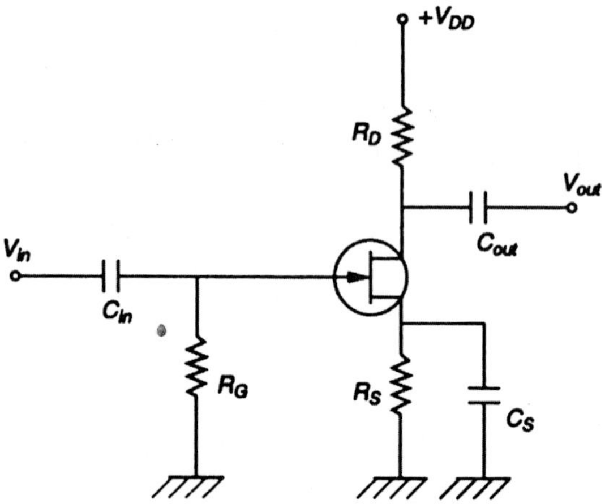

Fig 7.25. A common source *JFET* amplifier

In order to understand the working principle of this circuit, let us consider that a negative going signal is applied at the input. The capacitor C_{in} provides a low resistance path for the input signal. The input signal when applied to the gate makes it more negative with respect to the source. As a result, the depletion or space charge regions are widened, the channel resistance increases and the drain current I_D decreases. A lower I_D means a lower drop across R_D and the drain becomes more positive. Thus a **positive going signal** appears at the output terminals (considering C_{out} as a short circuit). Therefore, a negative gate signal gives rise to a positive output signal, meaning phase inversion of 180° between input and output. Similarly, a positive input signal gives a negative a.c. signal at V_{out}.

7.4.1 FET AC Equivalent Circuit

The a.c. equivalent circuit of a *JFET* is depicted in Fig. 7.26. In the circuit the current source $g_m V_{gs}$ has been shown with the arrow pointed downward meaning that there is a phase shift of 180° between the input and the output voltages.

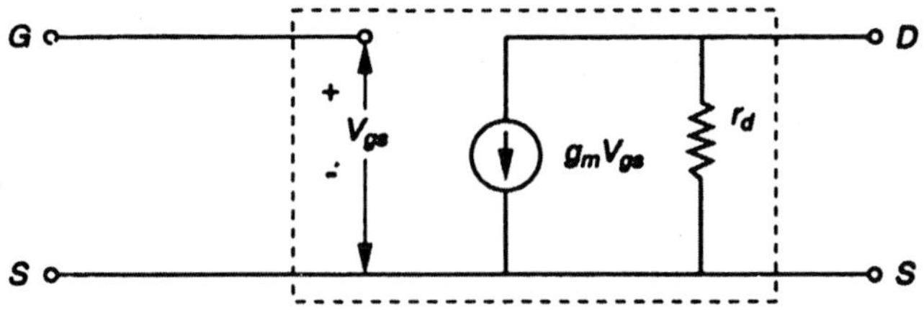

Fig 7.26. *FET* AC equivalent circuit

The input impedance is represented by an open circuit at the input terminals and the output impedance by the resistor r_d from drain to source. The gate to source voltage is represented by lower case letters since a.c. voltages are considered.

EXAMPLE 7.9 *Draw the a.c. equivalent circuit of a FET when y_{fs} = 3 mS and y_{os} = 25 μS.*

SOLUTION

$$g_m = y_{fs} = 3 \text{ mS}$$

$$r_d = \frac{1}{y_{os}} = \frac{1}{25 \ \mu\text{S}} = 40 \text{ k}\Omega$$

Thus the circuit in Fig. 7.25 (a) shows a *JFET* amplifier and Fig. 7.25 (b) shows the *JFET* amplifier when the *JFET* is replaced by its a.c. equivalent circuit, the battery V_{DD} replaced by short circuit, and $X_{C_{in}}$ and $X_{C_{out}}$ are considered as short circuits. Further, C_S is used for bypassing the a.c. signal and $X_{C_S} \cong 0 \ \Omega$. From the redrawn circuit of Fig. 7.28 (c),

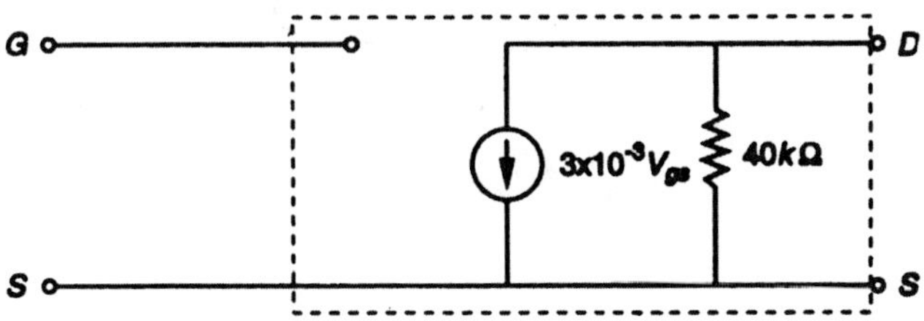

Fig 7.27. Circuit in Example 7.9

we get,

Input resistance $Z_{in} = R_G$ ...(7.35)

Output resistance $Z_{out} = R_D \parallel r_d$ ...(7.36)

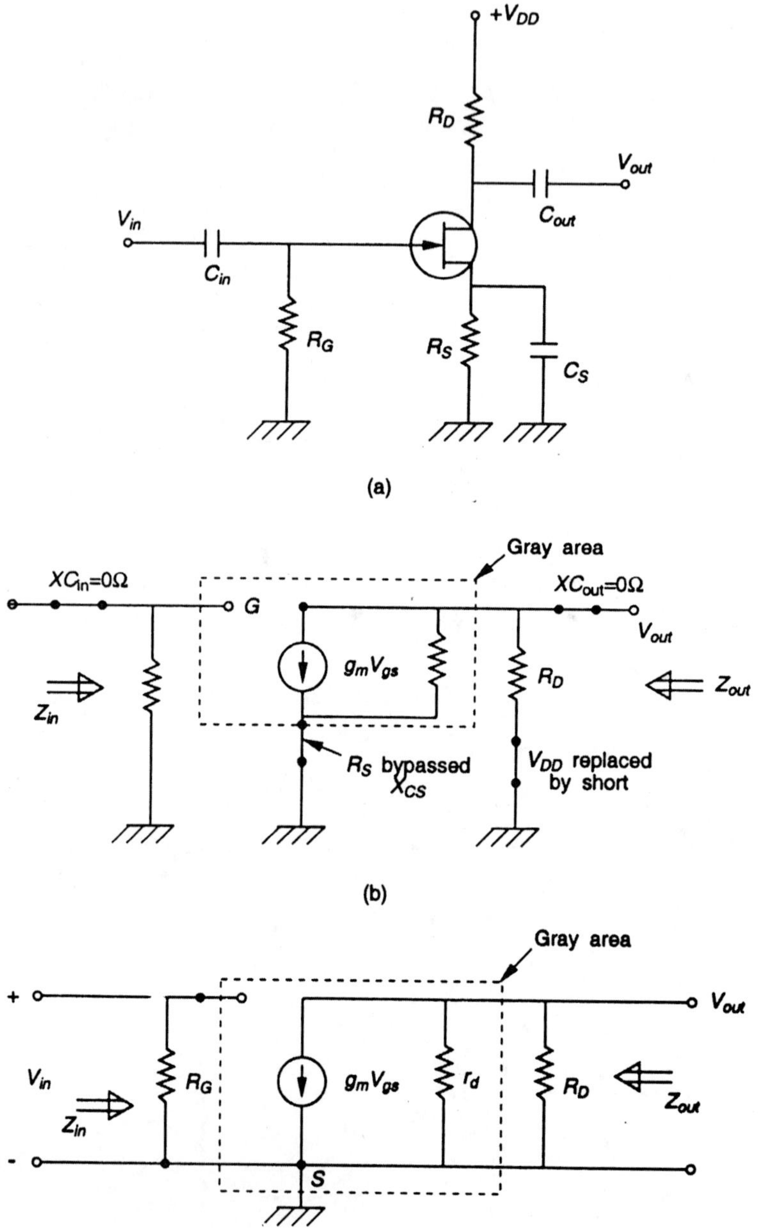

Fig 7.28. (a) *JFET* amplifier (b) common source *JFET* amplifier using *JFET* a.c. equivalent circuit and (c) redrawn network of Fig. (b)

If $\quad r_d \geq 10R_D$ then

$$Z_{out} \simeq R_D \qquad ...(7.37)$$

The expression for the voltage gain is

$$G_V = \frac{V_{out}}{V_{gs}} \qquad ...(7.38)$$

Now $V_{ds} = V_{out} = I_d\,(R \parallel r_d)$

$$= I_d \cdot \frac{R_D \,.\, r_d}{R_D + r_d} \qquad ...(7.39)$$

Also $$V_{gs} = \frac{I_d}{g_m} \qquad ...(7.40)$$

Thus $$G_V = I_d\left(\frac{R_D \,.\, r_d}{R_D + r_d}\right) \cdot \frac{g_m}{I_d}$$

$$= g_m \cdot \left(\frac{R_D \,.\, r_d}{R_d + r_d}\right) \qquad ...(7.41)$$

If $\quad r_d >> 10R_D$ then

$$(R_D \parallel r_d) \simeq R_D' \qquad ...(7.42)$$

and the voltage gain is

$$G_V = g_m \,.\, R_D \qquad ...(7.43)$$

EXAMPLE 7.9 *For a certain JFET,* $g_m = 5$ mS, $R_D = 1$ kΩ *and* $r_d = 7$ kΩ. *Find the voltage gain.*

SOLUTION

$$G_V = g_m\left(\frac{R_D \,.\, r_d}{R_D + r_d}\right)$$

$$= (5 \text{ mS} \times \left[\frac{1 \text{ k}\Omega \times 7\text{k}\Omega}{(1+7)\text{ k}\Omega}\right]$$

$$= 4{\cdot}375$$

EXAMPLE 7.10 *In problem 7.9 above, what would be the gain if the effect of* r_d *could be neglected ?*

SOLUTION

In this case

$$G_V = g_m \,.\, R_D$$

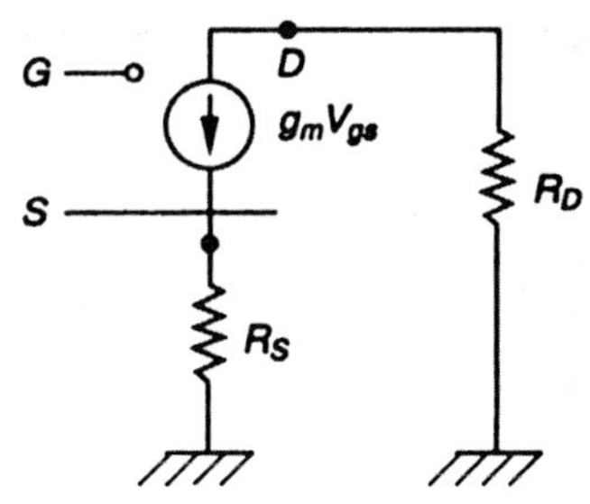

Fig 7.29. Equivalent circuit in which the effect of R_S is included

$= (5 \text{ mS}) \times 1 \text{ k}\Omega$

$= 5$

If the effect of the source resistance R_S is included, we may write from Fig. 7.29.

$$V_{\text{in}} = V_{gs} + I_d \cdot R_S \qquad \ldots(7.44)$$

$$V_{\text{out}} = I_d R_D \qquad \ldots(7.45)$$

$$\text{And gain } G_V = \frac{V_{\text{out}}}{V_{\text{in}}} = \frac{I_D R_D}{V_{gs} + I_d R_S}$$

$$= \frac{g_m V_{gs} \cdot R_D}{V_{gs} + g_m V_{gs} \cdot R_s} = \frac{G_m R_D V_{gs}}{V_{gs}(1 + g_m R_S)}$$

$$= \frac{G_m R_D}{(1 + g_m R_S)} \qquad \ldots(7.46)$$

EXAMPLE 7.11 *A FET equivalent circuit is shown in Fig. 7.30. Find the voltage gain when the output is obtained across R_D. Neglect r_d.*

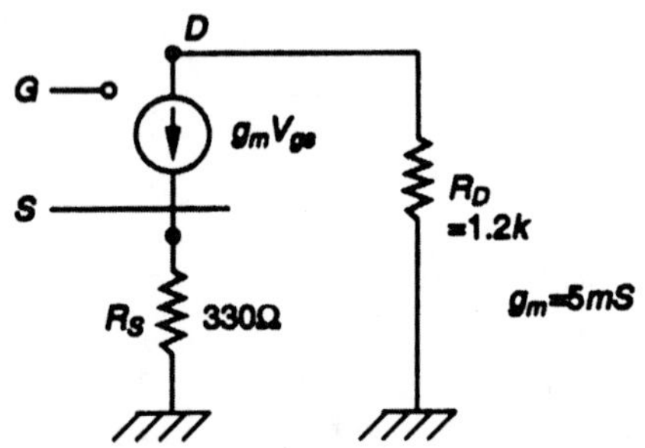

Fig 7.30. Circuit in Example 7.11

SOLUTION From equation (7.46), the voltage gain is

$$G_V = \frac{g_m R_D}{1 + g_m R_S} = \frac{(5 \text{ mS})(1{\cdot}2 \text{ k}\Omega)}{1 + (5 \text{ mS})(330 \ \Omega)}$$

$$= \frac{6}{1 + 1{\cdot}65} = \frac{6}{2{\cdot}65} = 2{\cdot}26$$

7.4.2 Voltage Gain when an External Load Resistance R_L is Connected

DC Analysis

To analyse the amplifier circuit of Fig. 7.31, d.c. analysis is first carried out. The Q point is normally taken to be at the midpoint of the transfer characteristic curves such that

$$I_D = \frac{I_{DSS}}{2} \qquad \ldots(7.47)$$

where I_{DSS} is obtained from the manufacturer's specification sheets. If I_D is not known, it should be obtained from the equation

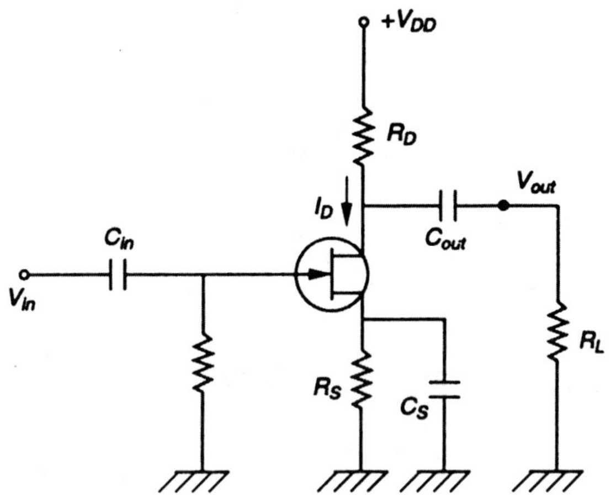

Fig 7.31. *JFET* common source amplifier with external R_L load

$$I_D = I_{DSS}\left[1 - \frac{I_D R_S}{V_{GS0}}\right]^2 \qquad \ldots(7.48)$$

$$= I_{DSS}\left[1 - \frac{I_D R_S}{V_{GS0}}\right]\left[1 - \frac{I_D R_S}{V_{GS0}}\right]$$

$$= I_{DSS}\left[1 - \frac{2I_D R_S}{V_{GS0}} + \frac{I_D^2 R_S^2}{V_{GS0}^2}\right]$$

$$= I_{DSS} - \frac{2R_S I_{DSS}}{V_{GS0}} I_D + \frac{I_{DSS} R_S^2}{V_{GS0}^2} \cdot I_D^2 \qquad \ldots(7.49)$$

which can be arranged as

$$\frac{I_{DSS} R_S^2}{V_{GS0}^2} \cdot I_D^2 - \left(1 + \frac{2I_{DSS} \cdot R_S}{V_{GS0}}\right) I_D + I_{DSS} = 0 \qquad \ldots(7.50)$$

The equation can be rewritten as

$$A_1 I_D^2 + A_2 I_D + A_3 = 0 \qquad \ldots(7.51)$$

where
$$\left.\begin{aligned} A_1 &= \frac{I_{DSS} R_S^2}{V_{GS0}^2} \\ A_2 &= -\left(1 + \frac{2I_{DSS} R_S}{V_{GS0}}\right) \\ A_3 &= I_{DSS} \end{aligned}\right\} \qquad \ldots(7.52)$$

The solution for I_D is

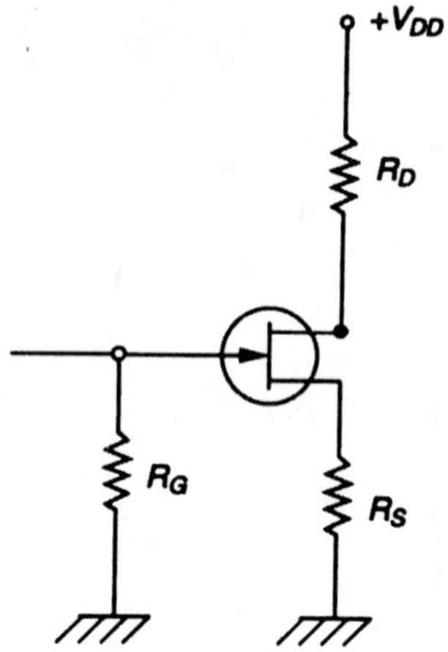

Fig 7.32. DC equivalent circuit

$$I_D = \frac{-A_2 \pm \sqrt{A_2^2 - 4A_1A_3}}{2A_1} \qquad ...(7.53)$$

when I_D is found, the other quantities may be obtained from the circuit of Fig. 7.29, namely

$$V_S = I_D R_S \qquad ...(7.54)$$

$$V_D = V_{DD} - I_D R_D \qquad ...(7.55)$$

and $$V_{DS} = V_D - V_S \qquad ...(7.56)$$

The AC equivalent circuit of Fig. 7.28 will be as follows (Fig. 7.33)

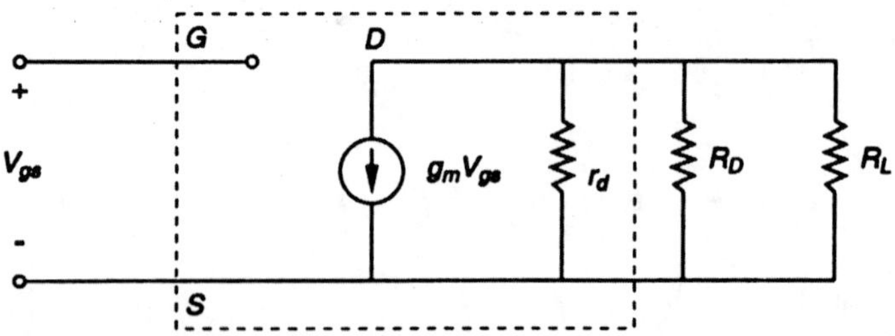

Fig 7.33. AC equivalent circuit for the electronic circuit of Fig. 7.31

Neglecting r_d, the total a.c. drain resistance will be,

$$R_L' = \frac{R_D \cdot R_L}{R_D + R_L} \qquad ...(7.57)$$

and the a.c. voltage gain will be

$$G_V = \frac{V_{out}}{V_{in}} = g_m R_L' \qquad ...(7.58)$$

It should be remembered here that the output voltage (at the drain) is 180° out of phase with the input voltage. The gain is sometimes written as $-G_V$.

EXAMPLE 7.12 *If $R_D = 2{\cdot}7$ kΩ, $R_L = 3{\cdot}3$ kΩ and $g_m = 5$ mS in the circuit of Fig. 7.31, find the voltage gain.*

SOLUTION In this case, the equivalent load resistance

$$= 2{\cdot}7\ \text{k}\Omega \parallel 3{\cdot}3\ \text{k}\Omega = \frac{2{\cdot}7\ \text{k} \times 3{\cdot}3\ \text{k}}{6{\cdot}0\ \text{k}}$$

$$= 1{\cdot}485\ \text{k}\Omega.$$

The voltage gain $G_V' = g_m R_L' = (5000\ \mu\text{S}) \times 1{\cdot}485\ \text{k}\Omega$

$$= 7{\cdot}43$$

The unloaded a.c. voltage gain is

$$G_V = g_m R_D = (5000\ \mu\text{S})\ (2{\cdot}7\ \text{k}\Omega)$$

$$= 13{\cdot}5$$

The effect of R_L is thus to reduce the gain.

7.5 JFET COMMON DRAIN AMPLIFIERS (SOURCE FOLLOWER CIRCUIT)

The common drain amplifier is also sometimes called as source follower circuit because the voltage at the source is approximately of the same amplitude as the input (gate) voltage and is in phase with it. The source thus follows the input (gate) voltage and the circuit is called source follower. The circuit is thus comparable to the *BJT* common collector (or emitter follower) circuit. A common drain *JFET* amplifier is depicted in Fig. 7.34.

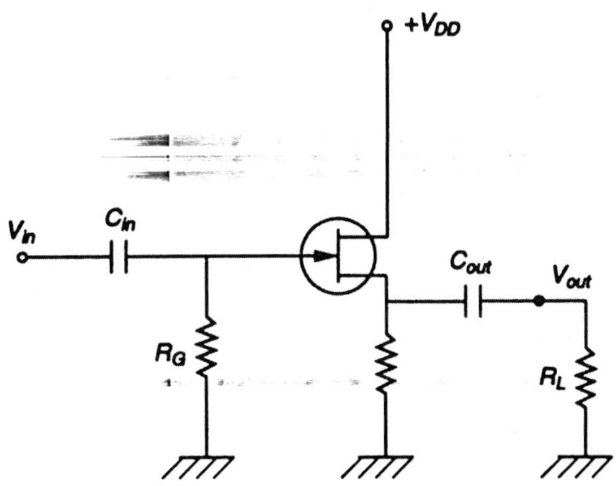

Fig 7.34. *JFET* common drain amplifier (source follower)

For this circuit, the total input resistance is the parallel combination of R_G and R_{gs} (the gate to source resistance). Since R_{gs} is very high $Z_{in} \simeq R_G$.

The a.c. equivalent circuit for the common drain amplifier is depicted in Fig. 7.35.

For the source follower circuit,

$$V_{out} = I_d R_S' \qquad \ldots(7.59)$$

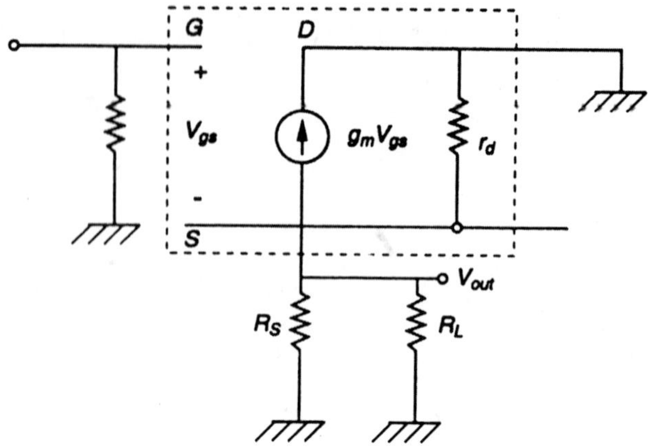

Fig 7.35. AC equivalent circuit for the source follower

where $$R_S' = R_S \parallel R_L = \frac{R_S \cdot R_L}{R_S + R_L} \qquad ...(7.60)$$

The gate to source voltage is

$$V_{gs} = \frac{I_D \cdot R_S'}{V_{gs} + I_d R_S'} \qquad ...(7.61)$$

Since $$I_d = g_m V_{gs}. \qquad ...(7.62)$$

$$G_V = \frac{g_m V_{gs} \cdot R_S'}{V_{gs} + g_m V_{gs} R_S'} \qquad ...(7.63)$$

or, Voltage gain $$G_V = \frac{g_m R_S'}{1 + g_m R_S'} \qquad ...(7.64)$$

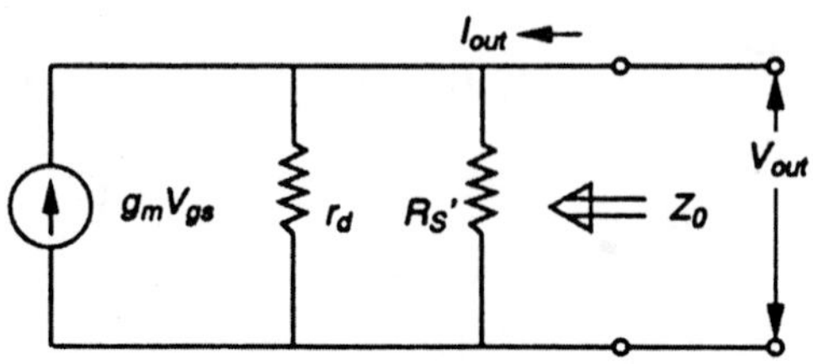

Fig 7.36. Determination of Z_{out}

$$Z_{out} = \frac{V_{out}}{I_{out}} = r_d \parallel R_S' \parallel \frac{1}{g_m} \qquad ...(7.65)$$

Since G_V in equation (7.64) is positive, V_{out} and V_{in} are **in phase**

EXAMPLE 7.13 *For the circuit of Fig. 7.37, find (a) g_m (b) r_d (c) Z_{in} (d) Z_{out} with and without r_d (e) G_v with and without r_d. Assume $V_{GSQ} = -2.5$ V and $I_{DQ} = 5$ mA.*

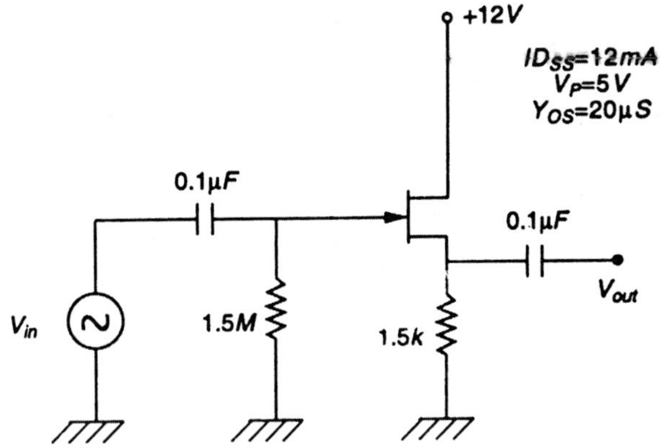

Fig 7.37. Circuit in Example 7.13

SOLUTION (a) $g_{mo} = \dfrac{2I_{DSS}}{|V_P|} = \dfrac{2 \times 12\text{ mA}}{5\text{ V}} = 4{\cdot}8\text{ mS}$

$$g_m = g_{mo}\left(1 - \frac{V_{GSQ}}{V_P}\right) = (4{\cdot}8\text{ mS}) \times \left(1 - \frac{-2{\cdot}5\text{ V}}{-5\text{ V}}\right)$$

$$= -2{\cdot}4\text{ mS}$$

(b) $r_d = \dfrac{1}{y_{os}} = \dfrac{1}{20\ \mu\text{S}} = 50\text{ k}\Omega$

(c) $Z_{in} = R_G = 1{\cdot}5\text{ M}\Omega$

(d) **Z_{out} with r_d**

$$Z_{out} = r_d \,||\, R_S \,||\, \frac{1}{g_m} = 50\text{ k}\Omega \,||\, 1{\cdot}5\text{ k}\Omega \,||\, \frac{1}{2{\cdot}4\text{ mS}}$$

$$= 50\text{ k}\Omega \,||\, 1{\cdot}5\text{ k}\Omega \,||\, 416{\cdot}66\ \Omega$$

$$\frac{1}{Z_{out}} = \frac{10}{50\text{ k}} + \frac{1}{416{\cdot}66} + \frac{1}{1{\cdot}5\ k}$$

$$= 20 \times 10^{-6} + 6{\cdot}66 \times 10^{-4} + 2{\cdot}4 \times 10^{-3}$$

$$= 3{\cdot}086 \times 10^{-3}$$

$$Z_{out} = 324\ \Omega$$

Z_{out} without r_d

$$Z_{out}' = 1{\cdot}5\text{ k}\Omega \,|\, 416{\cdot}66\ \Omega$$

$$= 326{\cdot}15\ \Omega$$

The result shows that Z_{out} is almost independent of r_d.

(e) Gain G_V with r_d.

$$G_V = \frac{g_m (r_d \| R_S)}{1 + g_m (r_d \| R_s)}$$

Now $(r_d \| R_s) = 50\ k\Omega \| 1{\cdot}5\ k\Omega$

$$= \frac{50 \times 1{\cdot}5}{51{\cdot}5} = 1{\cdot}456\ k\Omega$$

$$G_V = \frac{2{\cdot}4\ mS\ (1{\cdot}456\ k\Omega)}{1 + (2{\cdot}4\ mS)\ (1{\cdot}456\ k\Omega)} = \frac{3{\cdot}495}{1 + 3{\cdot}495}$$

$$= \frac{3{\cdot}495}{4{\cdot}495} = 0{\cdot}78$$

which is less than 1.

G_V without r_d.

$$G_V = \frac{g_m R_S}{1 + g_m R_s} = \frac{(2{\cdot}4\ mS)\ (1{\cdot}5\ k\Omega)}{1 + (2{\cdot}4\ mS)\ (1{\cdot}5\ k\Omega)}$$

$$= \frac{3{\cdot}6}{4{\cdot}6} = 0{\cdot}782$$

which implies that r_d has very small influence on the voltage gain.

7.6 COMMON GATE AMPLIFIERS

The common gate amplifier is similar to the common base *BJT* amplifier. The common gate (*CG*) amplifier has a low input resistance.

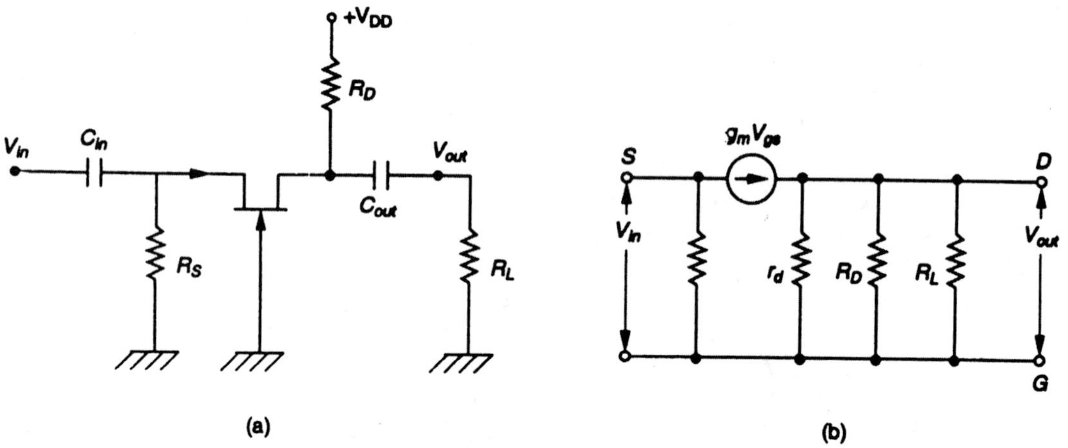

Fig 7.38. (a) *JFET* Common gate amplifier and (b) Electrical equivalent circuit

Fig. 7.38 (a) depicts a *JFET* (*n*-channel) common gate (*CG*) amplifier configuration. The gate is connected to the ground. The input signal is connected to the source through C_{in} while the output is coupled through C_{out} from the drain terminal.

The voltage gain of the *CG* amplifier is

$$G_V = \frac{V_{out}}{V_{in}} = \frac{V_d}{V_{gs}} = \frac{I_d R_L'}{V_{gs}}$$

$$= \frac{g_m V_{gs} R}{V_{gs}} = g_m R_L' \quad \text{...(7.66)}$$

where R_L' is the parallel combination of R_D and R_L *i.e.*,

$$R_L' = \frac{R_D \cdot R_L}{R_D + R_L} \quad \text{...(7.67)}$$

The input resistance of the *CG* amplifer is low. This may be desired as follows.

$$I_{in} = I_s = I_d = g_m V_{gs} \quad \text{...(7.68)}$$

and $$V_{in} = V_{gs} \quad \text{...(7.69)}$$

$$\therefore R_{in} \text{ (source)} = \frac{V_{in}}{I_{in}} = \frac{V_{gs}}{g_m V_{gs}} = \frac{1}{g_m} \quad \text{...(7.70)}$$

If g_m has a value of 5000 μS,

$$R_{in} \text{ (source)} = \frac{1}{5000 \ \mu S} = 200 \ \Omega$$

The output resistance is $r_d \parallel R_D \parallel R_L = R_{out}$ ...(7.71)

EXAMPLE 7.13 *Find the voltage gain and input resistance for the amplifier in Fig. E7.13 (given g_m = 4000 μS and y_{os} = 20 μS).*

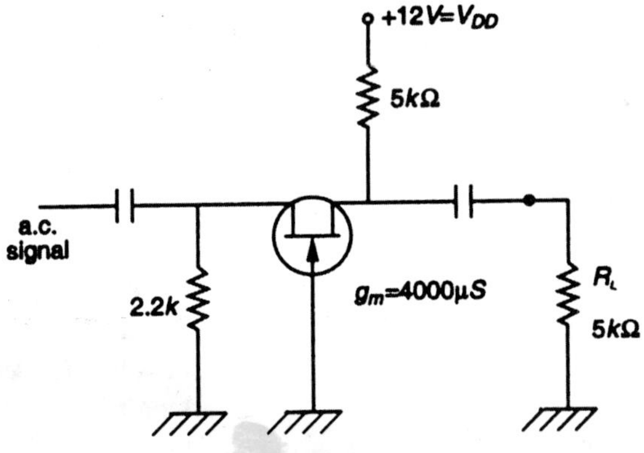

Fig E7.13 Circuit in Example 7.13

SOLUTION G_V = voltage gain = $g_m \cdot R_L'$

Now $R_L' = R_D \parallel R_L = 5 \text{ k}\Omega \parallel 5 \text{ k}\Omega$

$= 2{\cdot}5 \text{ k}\Omega$

$\therefore \quad G_V = g_m (R_L') + (4000 \ \mu S)(2{\cdot}5 \text{ k}\Omega)$

$= 10$

The input resistance at the source terminal is

$$R_{in} \text{ (source)} = \frac{1}{g_m} = \frac{1}{4000 \ \mu S} = 250 \ \Omega.$$

The signal source finds R_S in parallel with R_{in} (souce). So the net input resistance is the parallel combination of R_S and R_{in} (source)

$\therefore R_{in}$ (net) = 250 Ω ||2·2 kΩ

$$= \frac{250 \times 2200}{250 + 2200} = 224{\cdot}5\ \Omega$$

The output insistance is $r_d \parallel R_D \parallel R_L = R_{out}$ = 50 kΩ || 5 kΩ || 5 kΩ = 2·38 kΩ

The *FET* amplifier voltage gains and input and output resistance formulas for *CS*, *CD* and *CG* amplifiers is given in Table 7.2

Table 7.2

G_V, Z_{in} and Z_{out} for CS, CD and CG amplifiers

	Common Source (**CS**)	*Common Drain* (**CD**) *or souce follower*	*Common Gate* (**CG**)
Voltage gain	$g_m R_D'$	$\frac{g_m R_S}{1 + g_m R_S}$	$g_m R_L'$
Input Resistance	$\left(\frac{V_{GS}}{I_{GSS}}\right) \parallel R_G$	$\frac{V_{GS}}{I_{GSS}} \parallel R_G$	$\frac{1}{g_m} \parallel R_S$
Output Resistance	$r_d \parallel R_D \parallel R_L$	$\frac{1}{g_m} \parallel R_L$	$r_d \parallel R_L \parallel R_D$

7.7 THE INSULATED GATE FIELD EFFECT TRANSFOR (IGFET) [ALSO METAL OXIDE SEMICONDUCTOR FET (MOSFET)]

The metal oxide semiconductor *FET* (*MOSFET*) is another type of *FET* in which the gate is insulated from the channel by a silicon dioxide (SiO_2) layer. It is also called the insulated gate *FET* or *IGFET*. The two basic types of *MOSFETs* are depletion (*D*) and enhancement (*E*) types.

7.7.1 D–MOSFETs

Fig. 7.39 depicts a basic depletion type *MOSFET*. (*D-MOSFET*). Both *n* channel (in which majority carrier are electrons) and *p*-channel (in which the current carriers are holes) are possible. However for our discussion we would concentrate only on the *n*-channel *FETs*. The *p*-channel working principle is also similiar except for the fact that the voltage polarities are opposite to those of the *n*-channel.

The three terminals are gate, source and drain. The gate is insulated from the channel by the thin SiO_2 (oxide) layer. The drain and source are diffused onto the substrate material. It will be found that a channel is inbuilt in the depletion *MOSFET*. The symbols for the *n*-channel and *p*-channel depletion type *MOS* is depicted in Fig. 7.39 (b) and (d) respectively.

The *D–MOSFET* can be operated in either the depletion mode or the enhancement mode. This is the reason why the depletion *MOS* is also sometimes called as the depletion/enhancement *MOS* (or *D/E MOS*). Since the gate is insulated by the oxide layer (dielectric) either a positive or negative voltage can be applied. The *N*-type *MOSFET* operates in the depletion (enhancement) mode depending on whether the gate-to-source voltage is negative (positive).

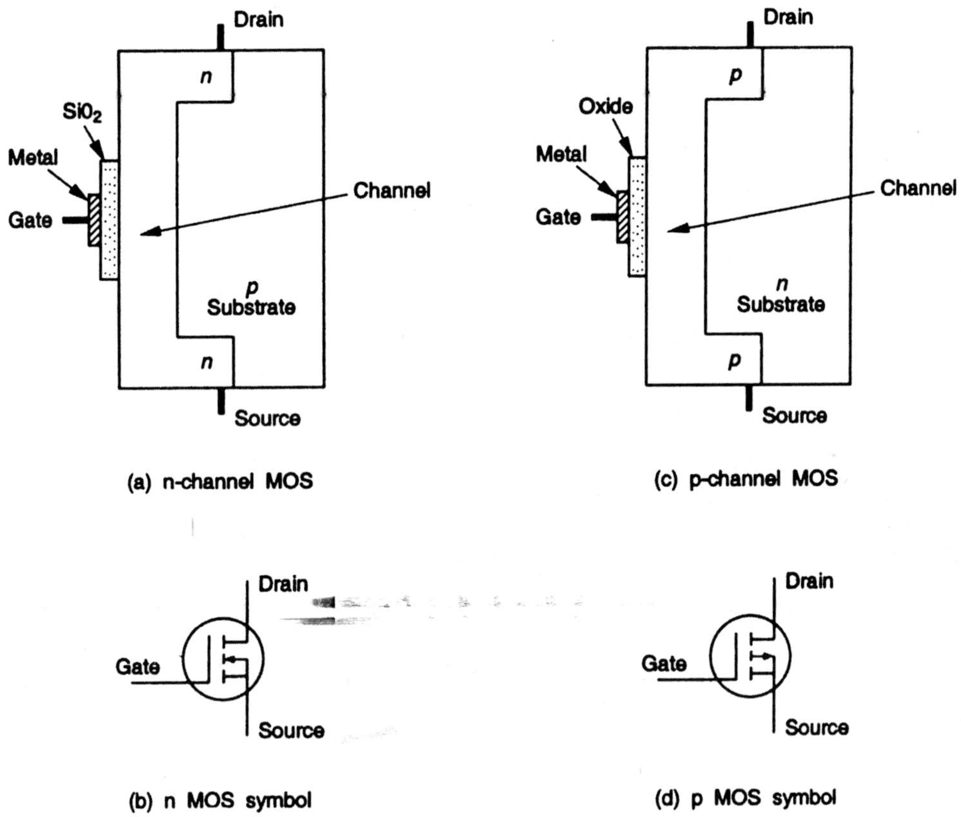

Fig 7.39. Depletion *MOSFEET* Structures (a) n-channel, (b) Symbol for n MOS, (c) p-MOS and (d) p-MOS symbol

Depletion Mode Operation

In the depletion mode operation of *n*-channel *D MOSFET* a negative voltage is applied to the gate [vide Fig. 7.37 (a)] like that of parallel plate capacitor separated by a dielectric. The electrons on the gate repel electrons in the channel leaving positive ions in their place. The channel conductivity is thus reduced. With increase in magnitude of the negative gate voltage, more and more electrons are repelled and the channel conductivity is reduced. At a sufficiently large negative gate voltage, the channel is totally depleted of electrons and the drain current is reduced to zero. This magnitude of the gate source voltage is called $(V_{GS})_{\text{cut off}}$ or V_{GS0}. The *n*-channel *DMOS* thus conducts current when V_{GS} is between zero and V_{GS0}.

When the gate voltage is made positive (enhancement mode) as depicted in Fig. 7.37 (b), the positive charges on the gate attract electrons on the channel side and the channel conductivity is increased. Thus the current increases.

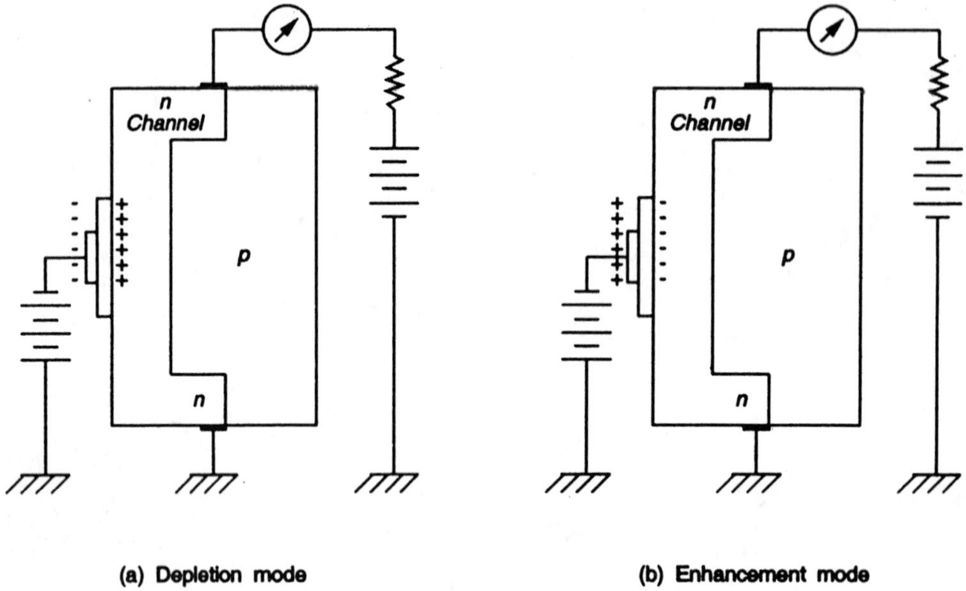

Fig 7.40 Operation of n-channel *DMOSFETs*.

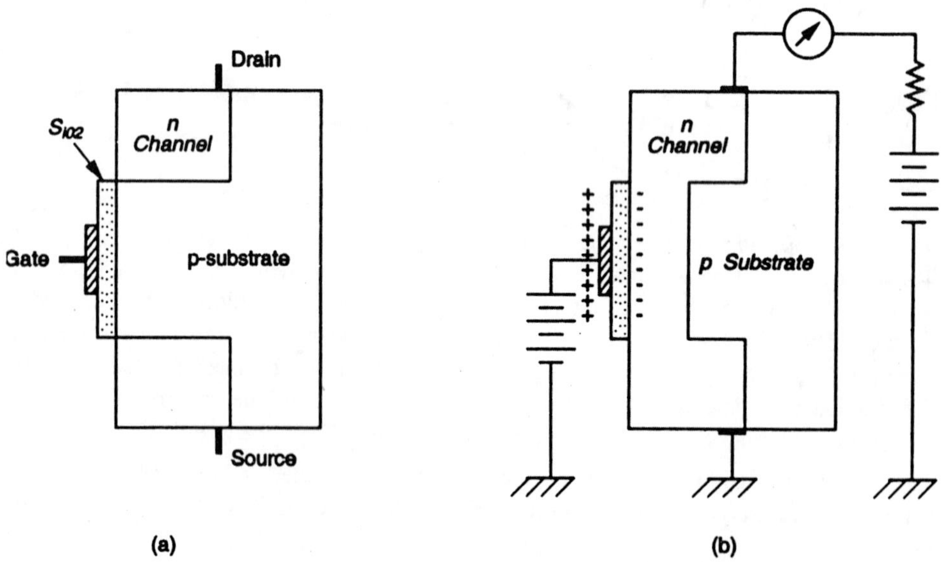

Fig 7.41 (a)n-channel *EMOSFET* structure and (b) Channel is induced in the *EMOSFET* with a positive gate voltage $V_{GS} > (V_{GS})$ threshold.

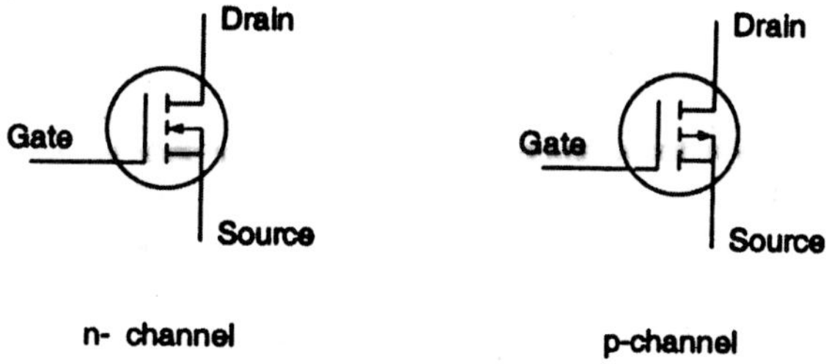

Fig 7.42 *EMOSFET* symbols.

7.7.2 E–MOSFET or Enhancement Mode MOSFET

The construction of the *EMOSFET* is different from *D MOS* in sense the that there is no physical channel present. The *E–MOSFET* thus operates only in the enhancement mode (depletion mode is not possible). Fig. 7.41 (a) depicts on *n*-channel *E–MOSFET*. It will be noticed that the substrate extends completely to the SiO_2 layer and there is no connection between the drain and the source. In an *n*-channel *EMOS* a positive value of gate voltage about a threshold value induces (or creates) a channel formed by the accumulation of a thin layer of electrons in the substrate adjacent to the oxide layer as depicted in Fig. 7.41 (b). The conductivity of channel can be increased by increasing the positive gate-to source voltage. The value $(V_{gs})_{threshold}$ or $(V_{gs})_{th}$ is the threshold value. Below $(V_{gs})_{th}$ the cahnnel is absent.

The schematic symbols for the *E–MOSFET* is depicted in Fig. 7.42.

7.7.3 V-MOSFET

The *V-MOSFET* (V grooued *MOSFET*) permits high power operation. It has a physical structure such that it can be operated at higher currents and greater drain-to-source voltages. The *V-MOSFET* is also an enhanement type device and has no in built physical channel. The channel can be induced only by application

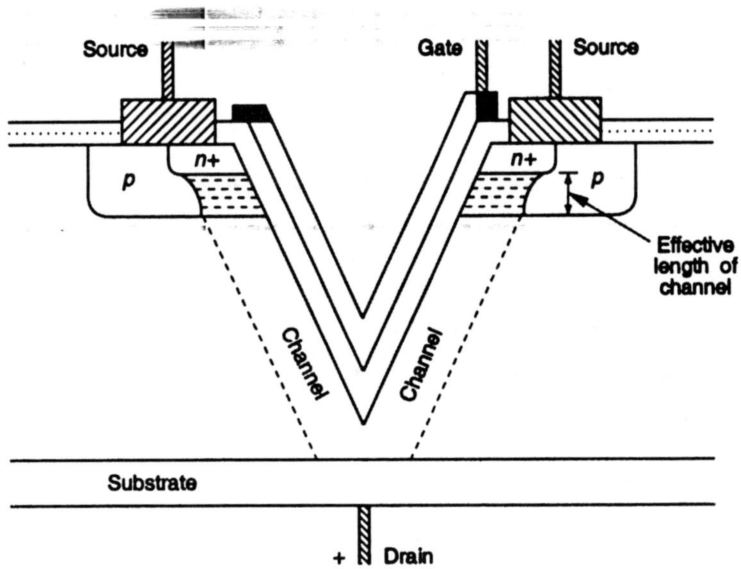

Fig 7.43 Structure of a *V-MOSFET*.

of a positive gate to source voltage for n channel type devices. The structure of the *V-MOSFET* is depicted in Fig. 7.43.

With a positive gate voltage applied, an n-channel is induced very close to the gate in the *V–MOSFET* parallel with the surface. Since small channel lengths gives rise to reduced resistance, the power dissipation of the device will be small. Besides, the contact area between the channel and the n^+ region is increased by the vertical nature of construction which leads to a further decrease of resistance and increased area for current between the doping layers.

The presence of two conduction paths between the drain and the source also gives rise to a larger current rating. The device can handle drain currents which may be a few ampere and power level may have a value exceeding 10W. **The *V–MOSFETs*, also possess a positive temperature co-efficient that will reduce of thermal runway. Besides the *V MOS* possess faster switching times.**

7.8 MOSFET CHARACTERISTICS

The *D–MOSFET* transfer characteristic curves are depicted in Fig. 7.44. It will be seen from Fig. 7.44 that $I_D = 0$ when V_{GS} reaches that cutoff value V_{GS0}. The current $I_D = I_{DSS}$ when $V_{GS} = 0$. Further it will be seen that the *MOSFET* can work with either positive or negative gate voltages.

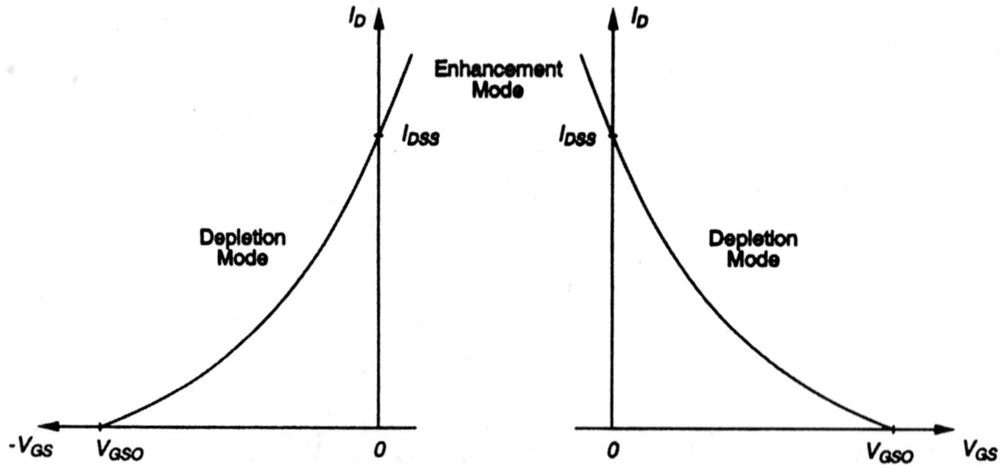

Fig 7.44 *DMOSFET* transfer characteristics Curves.

The drain characteristic curves are also depicted for a common source n-channel *D–MOSFET* for V_{GS} varying from + 2 V to V_{GS0}.

From the static drain charcteristic curves of n-channel *D–MOSFET* (Fig. 7.45 (a)) it is apparent that the device can operate either in the depletion mode (V_{GS} = negative between 0 and V_{GS0}) or the enhancement mode (V_{GS} = positive). The spacing between the different curves for various V_{GS} values indicate how much the current changes for a certain change in V_{GS}. The transfer characteristics follow the Shockley equation namely

$$I_D = I_{DSS}\left[1 - \frac{V_{GS}}{V_P}\right]^2 \qquad \ldots(7.71)$$

both in the depletion and enhancement modes. Thus the transfer characteristics can be easily plotted from equation 7.71.

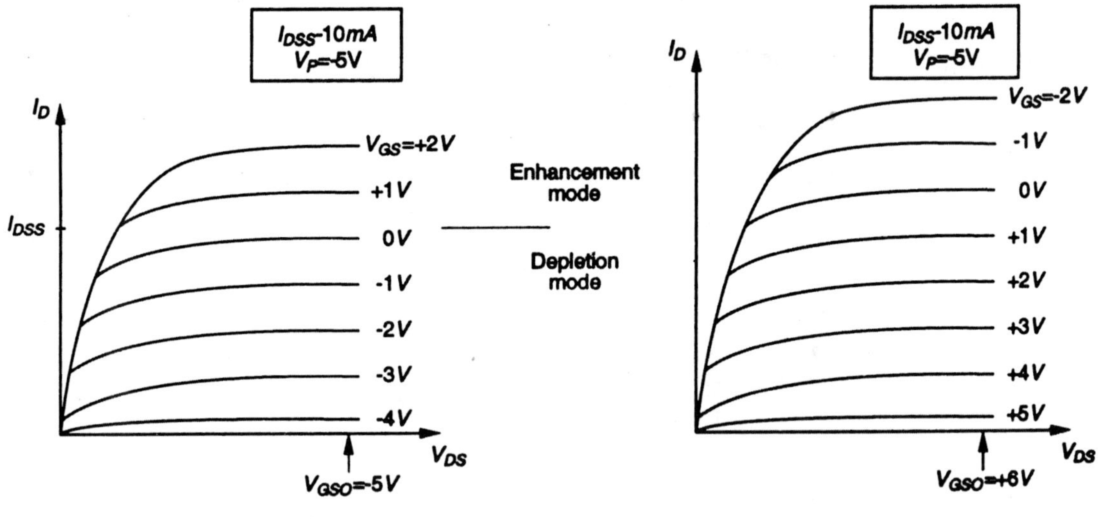

(a) n-channel DMOSFET (B) p-channel DMOSFET

Fig 7.45 Drain Characteristic Curves.

EXAMPLE 7.14 *In a D MOSFET, I_{DSS} = 15 mA and V_{GS0} = – 6 V. Find the drain current for V_{GS} = – 2 V and V_{GS} = + 2 V.*

SOLUTION

$$I_D = I_{DSS}\left[1 - \frac{V_{GS}}{V_{GS0}}\right]^2$$

At $V_{GS} = -2$ V.

$$I_D = (15 \text{ mA})\left[1 - \frac{-2}{-6}\right]^2$$

$$= 15 \text{ mA}\left[1 - \frac{1}{3}\right]^2$$

$$= 6{\cdot}66 \text{ mA.}$$

At $V_{GS} = +2$ V

$$I_D = 15 \text{ mA}\left[1 - \frac{+2 \text{ V}}{-6 \text{ V}}\right]^2$$

$$= 15 \text{ mA}\left[1 + \frac{1}{3}\right]^2$$

$$= 26{\cdot}66 \text{ mA.}$$

EXAMPLE 7.15 *In a D MOSFET, I_{DSS} = 20 mA and V_{GS0} = + 6 V. Find I_D at V_{GS} = + 3 V and – 3 V. Is this an n-channel or p-channel device ?*

SOLUTION At $V_{GS} = +3$ V

$$I_D = I_{DSS}\left[1 - \frac{V_{GS}}{V_{GS0}}\right]^2$$

$$= 20 \text{ mA}\left[1 - \frac{3}{6}\right]^2$$

$$= 5 \text{ mA}$$

At $V_{GS} = -3$ V

$$I_D = I_{DSS}\left[1 - \frac{V_{GS}}{V_{GSO}}\right]^2$$

$$= 20 \text{ mA}\left[1 - \frac{-3}{6}\right]^2$$

$$= 45 \text{ mA}.$$

Since V_{GSO} is positive, this is an *p*-channel *MOSFET*.

N-Channel *EMOSFET* Characteristics

The minimum gate to source voltage which forms the *n*-type inversion layer (and the drain current flows) is

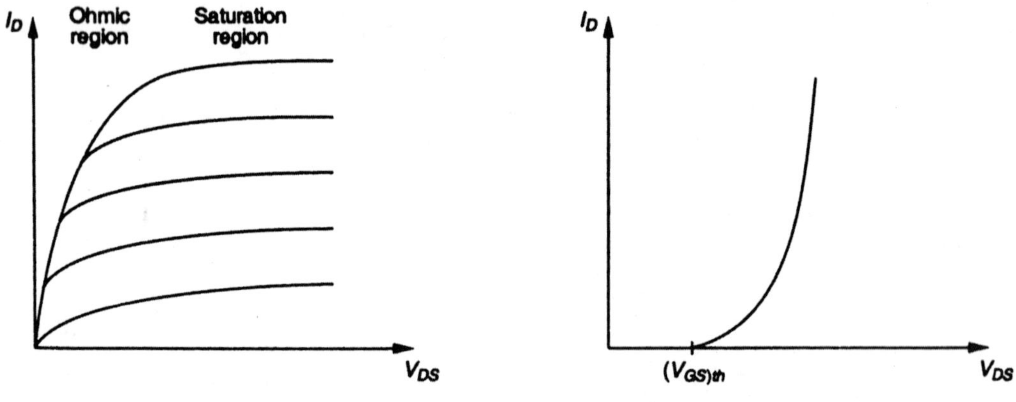

Fig 7.46 Characteristics of an *n*-channel *EMOSFET*, (a) Drain characteristic (b)Transfer characteristics.

called the threshold voltage V_{GS} (th). The drain current I_D flows only when $V_{GS} > V_{GS}$ (th). For values of $V_{GS} < V_{GS}$ (th), $I_D = 0$. For a specified V_{DS} value, with increase in V_{GS}, the channel widens and I_D increases. The value of I_D can be calculated from

$$I_D = C\left(V_{GS} - V_{GS\,(\text{th})}\right)^2 \qquad \ldots(7.73)$$

where C is a constant for the *MOSFET*. The value of the constant C can be found from the data sheet by using a specified value of the drain current called $I_{D\,(ON)}$ at the value of V_{GS} and then using equation (7.73). The value of g_m can be obtained as

$$g_m = \frac{\Delta I_D}{\Delta V_{GS}} = \frac{dI_D}{dV_{GS}} = 2C\left(V_{GS} - V_{GS\,(\text{th})}\right) \quad ...(7.74)$$

EXAMPLE 7.16 *The manufacturer's data sheet for an E–MOSFET shows $I_{D\,(ON)} = 5$ mA at $V_{GS} = 9$ V and $V_{GS\,(th)} = 5$ V. Find V_{GS} and V_{DS} for the circuit (Fig. 7.47).*

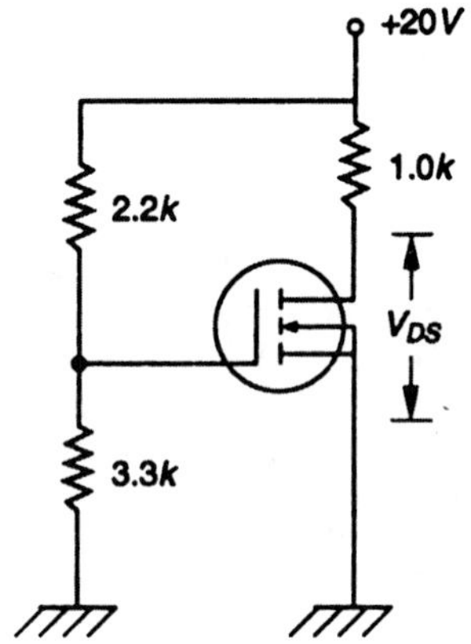

Fig 7.47 Circuit in Example 7.16.

SOLUTION

$$V_{GS} = V_G = 20 \times \frac{3{\cdot}3\text{ k}}{2{\cdot}2\text{ k} + 3{\cdot}3\text{ k}} = 12\text{ V}$$

Now C in equation. (7.73) is

$$C = \frac{I_{D\,(ON)}}{[V_{GS} - V_{GS\,(\text{th})}]^2} = \frac{5\text{ mA}}{(9-5)^2}$$

$$= \frac{5}{16}\frac{\text{mA}}{\text{V}^2}$$

Value of I_D when $V_{GS} = 12$ V is

$$I_D = C\left(V_{GS} - V_{GS\,(\text{th})}\right)^2$$

$$= \frac{5}{16}(12-5)^2$$

$$= 15{\cdot}31\text{ mA.}$$

Also $V_{DS} = V_{DD} - I_D R_D$

$$= 20 - 1\text{ k} \times 15{\cdot}31\text{ mA}$$

$$= 20 - 15{\cdot}31\text{ V} = 4{\cdot}69\text{ V}$$

7.9 MOSFET BIASING

Three different types of biasing circuits are given for *D–MOSFETs* in Fig. 7.48. The corresponding currents and voltages are also shown.

Fig. 7.45 depicts two different types of *E–MOSFET* biasing circuits. The resulting equations are also given.

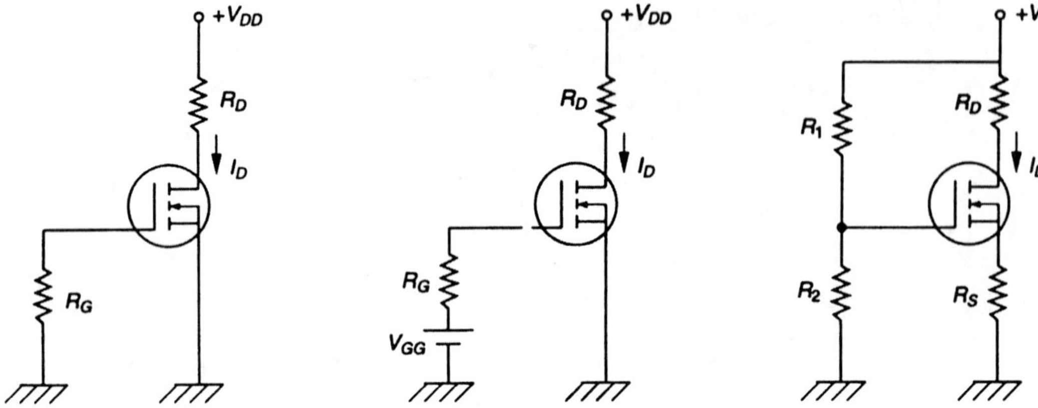

(a) $V_{DS} = V_{DD} - I_D R_D$

(b) $V_{GS} = V_{GG}$

(c) $V_G = \dfrac{R_2 V_{DD}}{R_1 + R_2}$

$V_{GS} = V_G - I_S R_S$

$V_{DS} = V_{DD} - I_D (R_D + R_S)$

Fig 7.48 *D–MOSFET* biasing circuits.

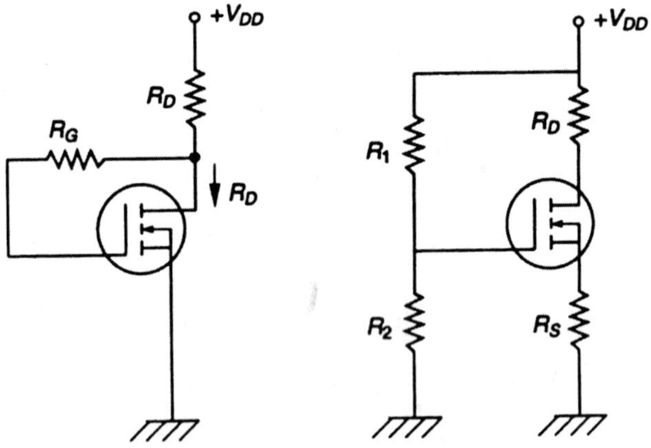

(a) $V_{GS} = V_{DS}$

$V_{GS} = V_{DD} - I_D R_D$

(b) $V_G = \dfrac{R_2 V_{DD}}{R_1 + R_2}$

$V_{GS} = V_G - I_D R_S$

$V_{DS} = V_{DD} - I_D R_D - I_D R_S$

Fig 7.49 *E–MOSFET* biasing circuits along with voltages and currents.

EXAMPLE 7.17 *Find the drain to source voltage for the D–MOSFET circuit in Fig. 7.50 (given I_{DSS}= 15 mA and V_{GS0} = – 6 V)*

SOLUTION $I_D = I_{DSS} = 15$ mA.

$\therefore\ V_{DS} = V_{DD} - I_{DSS} R_D = 20 - 15 \text{ mA} \times 470\ \Omega$

$= 20 - 7{\cdot}05 = 12{\cdot}95$ V.

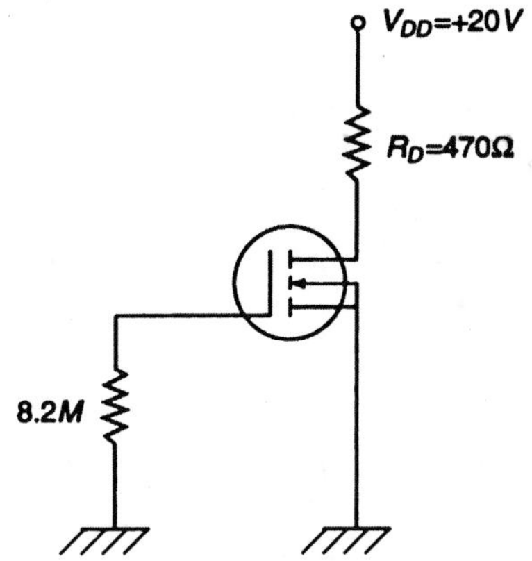

Fig 7.50 Circuit in Example 7.17.

EXAMPLE 7.18 *Find the drain to source voltage for the EMOSFET of Fig. 7.51. Given* $V_{GS\,(th)} = 4$ *V,* $I_D = 3$ *mA.*

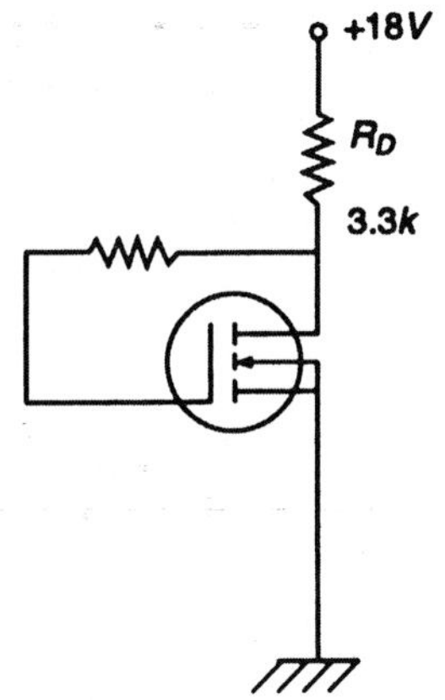

Fig 7.51 Circuit in Example 7.18.

SOLUTION $V_{DS} = V_{DD} - I_D R_D = 18 - 3 \text{ mA} \times 3{\cdot}3\ k = 8{\cdot}1 \text{ V}$

EXAMPLE 7.19 *In Fig. 7.49 (b) if* $R_1 = 8{\cdot}2$ k, $R_2 = 15$ k, $R_D = 680\ \Omega$ *and* $R_S = 0$ *and* $V_{DD} = +\,20$ V *find* V_{DS}. *The data sheet gives* $I_{D\,(ON)} = 2$ mA *at* $V_{GS} = 10$ *V and* $V_{GS\,(th)} = 5$ *V.*

SOLUTION $$V_{GS} = \left(\frac{R_2}{R_1 + R_2}\right) V_{DD} = \frac{15 \text{ k}}{(8{\cdot}2 + 15) \text{ k}} \times 20 = 12{\cdot}93 \text{ V}$$

Now C is eqn. (7.73) is

$$C = \frac{I_{D\,(ON)}}{[V_{GS} - V_{GS\,\text{th})}]^2}$$

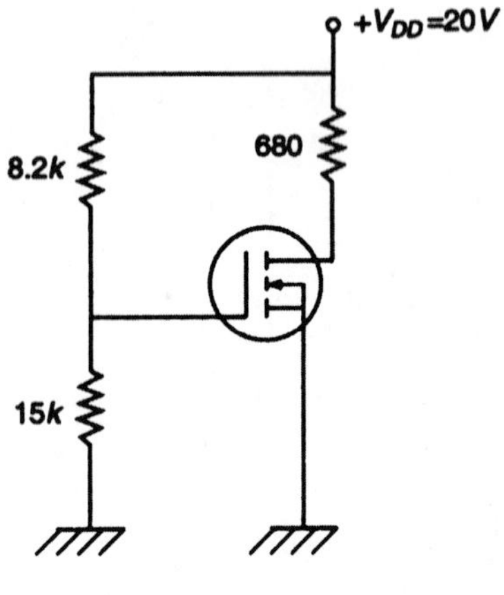

Fig 7.52 Circuit in Example 7.19.

$$= \frac{2 \text{ mA}}{[10-5]^2} = \frac{2}{25} \text{ mA/V}^2$$

Now I_D at $V_{GS} = 12{\cdot}93$ V is $I_D = \frac{2}{25} \frac{\text{mA}}{\text{V}^2} (12{\cdot}93 - 5)^2$

$I_D = 5{\cdot}03$ mA.

$\therefore\ V_{DS} = V_{DD} - I_D R_D = 20 - (5{\cdot}03) \times 680 = 16{\cdot}58$ V.

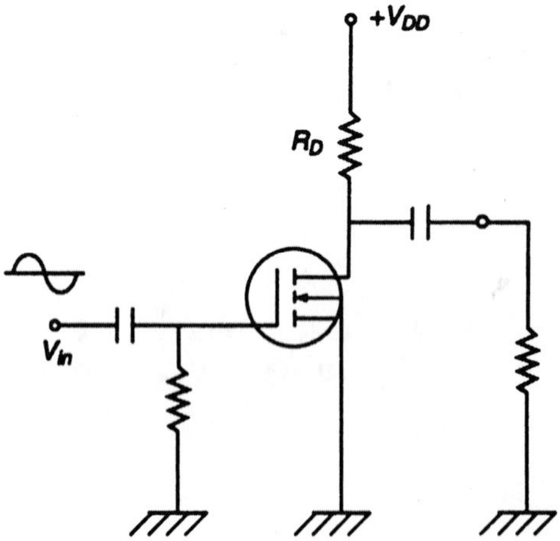

Fig 7.53. D channel MOSFET common source amplifier

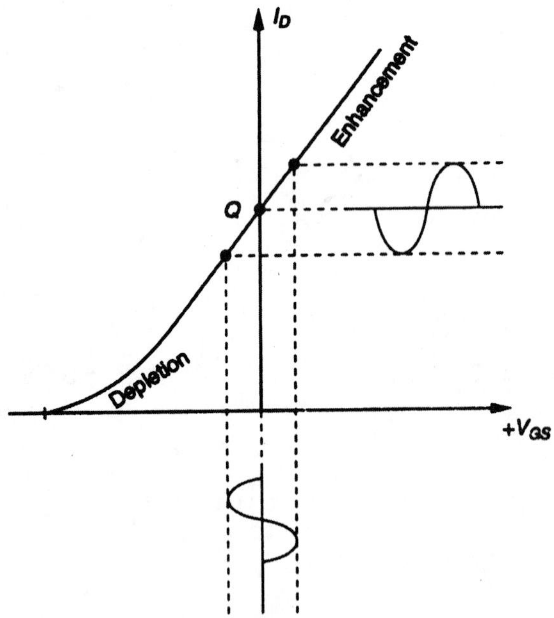

Fig 7.54 Operation of the *DMOSFET* Amplifier

7.10 MOSFET AMPLIFIERS

D MOSFET

Fig. 7.53 depicts a *D* channel *MOSFET* amplifier. In the circuit the gate to source bias voltage is zero. The signal voltage V_{in} swings the gate above and below the zero gate voltage $V_{GS} = 0$. When V_{gs} is negative, the *FET* operates in the depletion mode while the *FET* operates in the enhancement mode when V_{gs} is positive. The *D-MOSFET* amplifier operation is depicted in Fig. 7.54.

Fig. 7.55 shows the *E-MOSFET* amplifier circuit in the common source mode. The input signal voltage swings the V_{GS} above and below the *Q* (quiscent) point. The operating principle is depicted in the transfer characteristics curve as shown in Fig. 7.56.

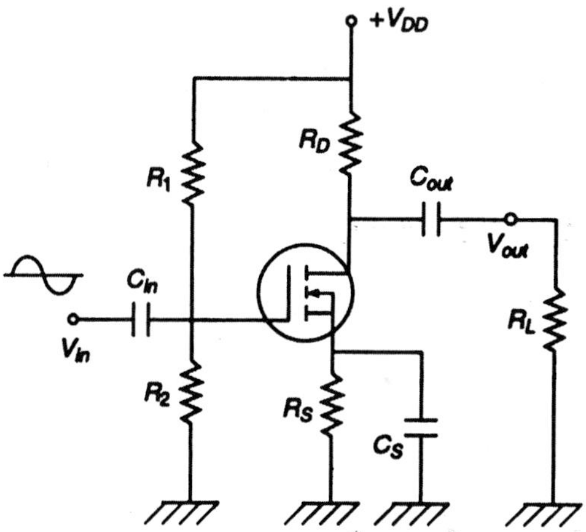

Fig 7.55 *E-MOSFET* amplifier common source operation

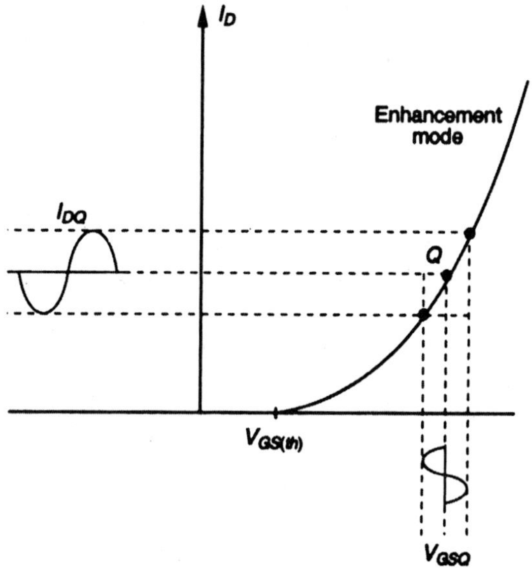

Fig 7.56.*E-MOSFET* amplifier operation shown on the transfer characteristics

AC Equivalent Circuit

The AC equivalent circuit for the *E–MOSFET* voltage divider configuration is depicted in Fig. 7.57

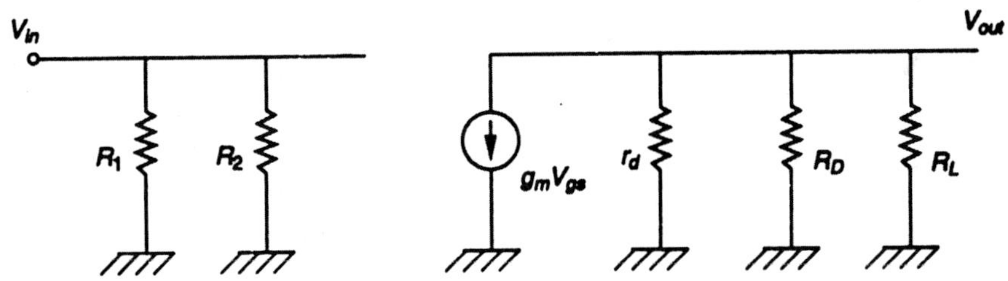

Fig 7.57. AC equivalent circuit for the common source amplifier

From Fig. 7.57, one can very easily obtain.

Input impedance $Z_{in} = R_1 \parallel R_2 = \frac{R_1 R_2}{R_1 + R_2}$ ...(7.75)

Output impedance $Z_{out} = r_d \parallel R_D \parallel R_L$ ...(7.76)

or $\frac{1}{Z_{out}} = \frac{1}{r_d} + \frac{1}{R_D} + \frac{1}{R_L}$ ...(7.77)

If $r_d >> 10 R_D$

$Z_{out} \simeq R_D \parallel R_L$ ...(7.78)

The voltage gain

$$G_V = \frac{V_{out}}{V_{in}} = -g_m (r_d \parallel R_D \parallel R_L) \quad ...(7.79)$$

If $r_d >> 10 R_L'$, the

$$G_V = \frac{V_{out}}{V_{in}} = -g_m R_L' \quad ...(7.80)$$

where $R_L' = R_D \parallel R_L = \frac{R_D \cdot R_L}{R_D + R_L}$ ...(7.81)

7.11 MOSFET HANDLING

In the case of a *MOSFET*, the input resistance is very high ($\simeq$ 1000 MΩ). The leakage current is low. Also the input capacitance gives rise to a high input impedance. Thus electrostatic charges may accumulate on the gate and can damage the device. To prevent a *MOSFET* from being damaged due to electrostatic discharge (*ESD*), the following procedures shold be accepted.

(1) All *MOS* devices should be stored and transported in conductive form.

(2) All metal benches, tools and instruments used in assembly or testing, should be connected to earth ground.

(3) The person handling the *MOS* should have his wrist connected to earth ground.

(4) No *MOS* device should be removed from the circuit while electrical power is *ON*.

(5) When the DC power supply is switched off, no signals should be applied to it.

Review Questions

1. What is the basic difference between a *FET* and *BJT* ?
2. Mention the different classes of *FETs*.
3. Distinguish between depletion type and enhancement type devices.
4. Draw the basic structure of a *n*-channel *JFET* and explain its principle of operation.
5. What is meant by "pinch off"? What is pinch off voltage ?
6. Draw the I_D vs. V_{DS} characteristics of an *n*-channel *JFET* for $V_{GS} = 0$ V and show I_{DSS}, V_P on the curve.
7. What is threshold voltage ?
8. Draw the $I_D - V_{DS}$ and I_D *vs.* V_{GS} characteristics of *n*-channel *JFETs*.

9. What do you understand by the following parameters g_m, V_P, V_{GS0}, I_{DSS}.
10. What is Shockley equation ?
11. Derive equation (7.8).
12. Show the relationship between the amplifcation factor μ, transconductance g_m and a.c. drain resistance for a *FET*.
13. Draw the various DC biasing circuits for a *JFET* (*n*-channel).
14. Mention some differences between the *n*-channel and the *p*-channel *JFETs*.
15. Draw the various DC biasing circuits for *p*-channel *JFETs*.
16. Show the location of the Quiscent operating point of an *n*-channel *JFET* on the transfer characteristics curves.
17. Draw the circuit diagram of a common source *JFET* amplifier and explain it.
18. Draw the a.c. equivalent circuit of a *JFET*.
19. Draw the equivalent circuit of a common source *JFET* amplifier.
20. Derive the expression for the voltage gain of a common source *JFET* amplifier circuit.
21. Derive equation (7.50).
22. Draw the circuit diagram for a *JFET* common drain amplifier circuit. Why it is called an emitter follower ?
23. Derive an expression for the voltage gain of a *JFET* source follower circuit.
24. What is a common gate (*CG*) amplifier circuit ? Draw its electrical equivalent circuit.
25. Derive the expression for the voltage gain of a common gate amplifier circuit.
26. Draw the structure of an *n*-channel depletion mode *MOSFET*. Why is it called a depletion mode circuit ?
27. What is a *MOSFET* ? What are the different types of *MOSFETs* ?
28. Distinguish between a *JFET* and a *MOSFET*.
29. Discuss how an *n*-channel depletion type *MOSFET* can be operated in the enhancement mode.
30. Discuss the principle of operation of an depletion mode and an enhancement mode *MOSFET*.
31. What is an enhancement mode *MOSFET* ?
32. Draw the structure of an *n*-channel enhancement mode *MOSFET*.
33. What is a *V-MOSFET* ? What are its special features ?
34. Draw the drain characteristics and transfer characteristics of an *n*-channel *MOSFET*.
35. Draw the drain characteristics and transfer characteristics of a *p*-channel *MOSFET*.
36. Draw the drain characteristics and transfer characteristics of an *n*-channel *EMOSFET*.
37. Draw the voltage divider bias type *D–MOSFET* biasing circuit. Find how I_D is related to V_{DS} in such a bias circuit.
38. Draw the voltage divider type *E–MOSFET* circuit for an *n-MOS* device.
39. Draw the electrical equivalent circuit for an *n*-channel *E MOSFET* with voltage divider type bias circuit.
40. Derive an expression for the voltage gain of *n*-channel *EMOSFET* circuit in the voltage divider type bias configuration.
41. What precautions should be followed in handling a *MOSFET* ?

Objective Type Questions

1. The *JFET* is
(a) a bipolar device (b) unipolar device

(c) current controlled device (d) voltage controlled device
(e) *b* and *d* (f) *a* and *b*

2. The channel of a *JFET* exists between
(a) gate and source (b) drain and source
(c) gate and drain (d) input and output

3. For low values of V_{DS}, the *JFET* behaves like a
(a) resistance (b) constant voltage device
(c) constant current device (d) negative resistor

4. When V_{DS} reaches the pinch off voltage V_P the drain current becomes
(a) low (b) zero
(c) constant (saturated) (d) reversed.

5. In an *n*-channel *JFET*, the drain current reaches its maximum value when V_{GS} is
(a) negative (b) zero
(c) positive (d) equal to V_P

6. In an *n*-channel *JFET*
(a) the current carriers are holes (b) the current carriers are electrons
(c) V_{GS} is positive (d) the input resistance is very low

7. In a *p*-channel *JFET*
(a) the current carriers are electrons (b) the current carriers are holes
(c) V_{GS} is negative (d) the input resistance is very small

8. For an *n*-channel *JFET*,
(a) V_{GS} can vary between zero negatively to V_{GSO} (b) V_{GS} can vary between zero positively to V_{GSO}
(c) pinch off occurs for positive V_{GS} (d) V_{DD} is negative

9. An *FET* can not operate at $V_{GS} = 0$ V. The *FET* is
(a) *JFET* (b) *D–MOSFET*
(c) *E–MOSFET* (d) both (a) and (b)

10. The transconductance g_m of *JFET* is defined as
(a) $\Delta I_D/\Delta V_{GS}$ (b) $\Delta I_D/\Delta V_{DS}$
(c) $\Delta V_{GS}/\Delta I_D$ (d) $\Delta I_{DSS}/\Delta I_D$

11. The amplification factor μ of *JFET* is given by
(a) $g_m \times r_d$ (b) g_m/r_d
(c) $g_m \cdot I_{DSS}$ (d) $g_m\left(1 - \frac{V_{GS}}{V_P}\right)$

12. A channel is induced in a *E–MOSFET* by the application of a
(a) $V_{GS} < (V_{GS})_{th}$ (b) $V_{GS} > V_P$
(c) $V_{GS} > V_{GS\,(th)}$ (d) $V_{DS} > V_P$

13. The Shockley equation is given by
(a) $I_D = I_{DSS}\left[1 - \frac{V_{GS}}{V_P}\right]^2$ (b) $I_D = I_{DSS}\left[1 - \frac{V_{GS}}{V_P}\right]$
(c) $g_m = g_{m0}\left(1 - \frac{V_{GS}}{V_P}\right)$ (d) $I_D = C\left[V_{GS} - V_{GS\,(th)}\right]^2$

14. Ideally, the equivalent circuit of an *FET*consists of
(a) a resistance between drain and source
(b) a current source between the gate and the source
(c) a current source between the drain and the source
(d) a current source between the gate and the drain

15. The magnitude of the current source in the a.c. equivalent circuit of an *FET* depends on
(a) the d.c. supply voltage (b) V_{ds}
(c) external drain resistance (d) transconductance and gate to source voltage

16. A CS amplifier has a drain load $R_D = 1$ kΩ and an external load resistance of $R_L = 1$ kΩ. If $g_m = 10$ mS, the voltage gain will be

(a) 10 (b) 5
(c) 20 (d) 1

17. If the external load resistance is removed in problem 16, the voltage gain will
(a) decrease (b) increase
(c) remain the same (d) become zero

18. In a **CD** amplifier, R_S = 500 Ω and the transconductance is 9000 μS. The voltage gain will be
(a) 0·82 (b) 1
(c) 1·22 (d) 9

19. In a common gate amplifier the transconductance is g_m = 6000 μS, R_G = 1 MΩ. The total input resistance will be
(a) 1666 Ω (b) 166·6 Ω
(c) 1 MΩ (d) 100166·6 Ω

20. In a CD amplifier, I_{DSS} = 5 nA at V_{GS} = 10 V. If the resistor R_G is 20 MΩ, the total input resistance will be
(a) 20 MΩ (b) 200 MΩ
(c) 40 MΩ (d) 100 MΩ

21. If there is an internal open between the source and the drain in a *CS* amplifier, the drain voltage will be
(a) 0 V (b) V_{Gs}
(c) V_{DD} (d) $V_{DD}/2$

Exercise Problems

1. For the *JFET* shown in Fig. 7.58, if I_{DSS} = 20 mA and $V_{GS\,(cut\,off)}$ = – 6 V, find the minimum value of V_{DD} which will place the *JFET* in the constant current region. What will be the drain current if V_{DD} is increased to 20 V ?

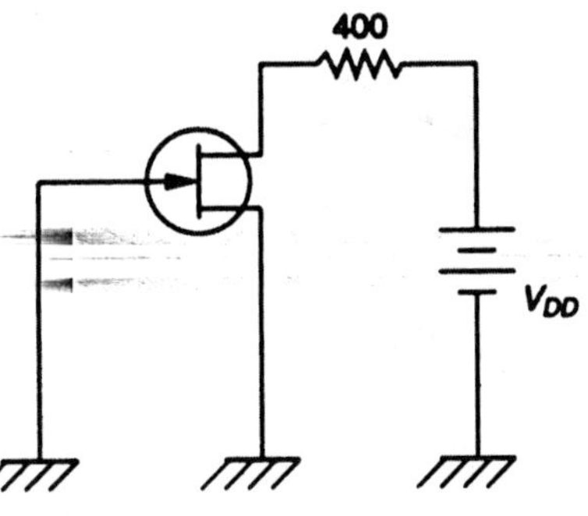

Fig 7.58. Circuit in Exercise Problem 1

[**Ans.** 13V, 20mA]

2. For an *n*-channel *JFET*, I_{DSS} = 25 mA and $V_{GS\,(cut\,off)}$ = – 10 V. The value of g_{m0} = 5000 μS. Find g_m and the drain current I_D at V_{GS} = – 5 V.

[**Ans.** 2500 μS, 12·5 mA]

3. In an *n*-channel *JFET*, I_{DSS} = 15 mA, V_P = 6 V, V_{GS} = – 3 V. Find (i) I_D (ii) g_{m0} and (iii) g_m.

[**Ans.** (i) 3·75 mA (ii) 5 mS (iii) 2·5 mS]

4. For an *n*-channel *JFET*, g_{m0} = 6 mS, g_m = 3 mS and I_{DSS} = 20 mA. Find I_D.

[**Ans.** 5 mA]

5. Find (a) V_{GSQ} (b) I_{DQ} (c) V_{DS} (d) V_D (e) V_G and (f) V_S for the circuit of Fig. 7.59.

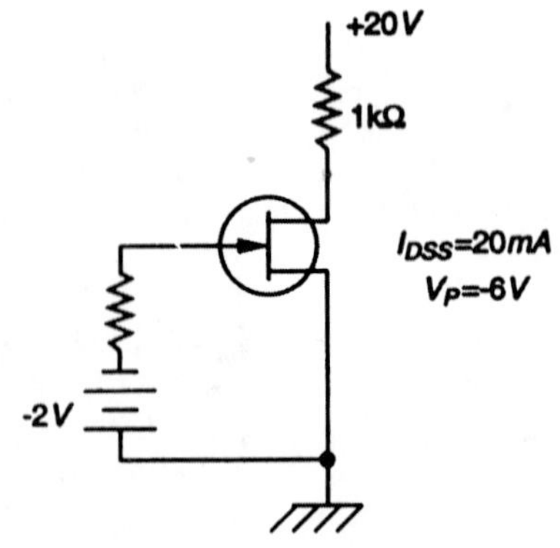

Fig 7.59. Circuit in Exercise Problem 5

[**Ans.** (a) – 2 V (b) 8·88 mA (c) 11·12 V (d) 11·12 V (e) – 2 V (f) 0 V.]

6. Find the following quantities in the Fig. 7.60, find (a) V_{GSQ} (b) I_{DQ} (c) V_{DS} (d) V_S (e) V_G and (f) V_D.

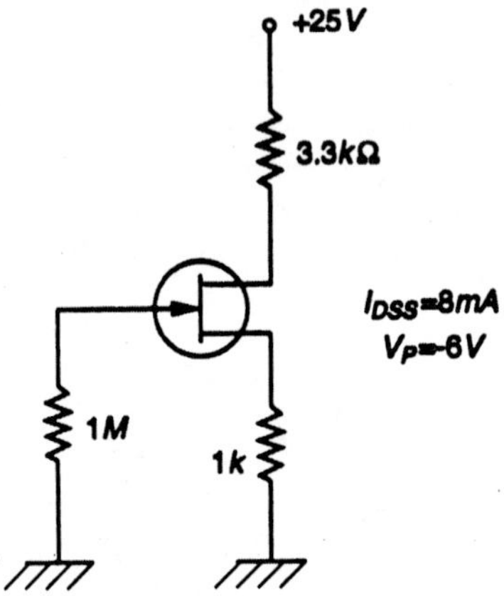

Fig 7.60. Circuit is Exercise Problem 6

[**Ans.** (a) – 2·6 V (b) 2·6 mA (c) 13·82 V (d) 2·6 V (e) 0 V (f) 11·18 V.]

7. Find the following parameters for the network in Fig. 7.61. (i) V_{GSQ} (ii) I_{DQ} (iii) V_D (iv) V_S (v) V_{DS}

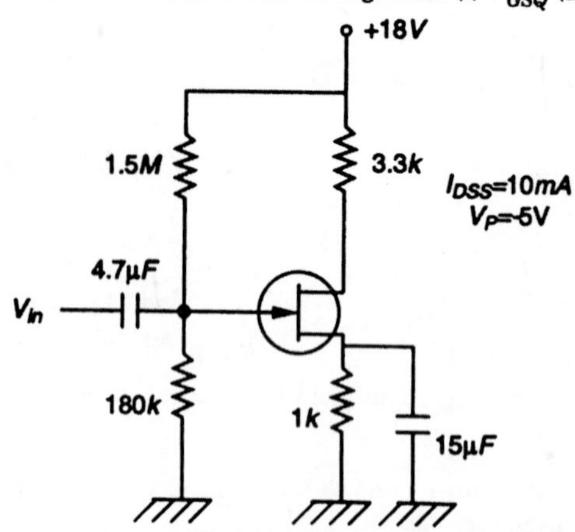

Fig 7.61. Circuit in Exercise Problem 7

[**Ans.** (i) – 1·85 V (ii) 3·64 mA (iii) 6 V (iv) 12 V (v) 2·35 V]

8. Draw the a.c. equivalent circuit of a *FET* when y_{fs} = 4 mS and y_{os} = 20 μS.

[**Hints :** g_m = 3 mS, r_d = 25 kΩ]

9. For a certain *JFET*, g_m = 6 mS, R_D = 1 kΩ, r_d= 6 kΩ. Find the voltage gain.

[**Ans.** 5·14]

10. In Problem 9. above, what would be the magnitude of the gain if the effect of r_d could be neglected.

[**Ans.** G_V = 6]

11. A *FET* equivalent circuit is depicted in Fig. 7.62. Find the voltage gain if the output is taken across R_D. Neglect r_d.

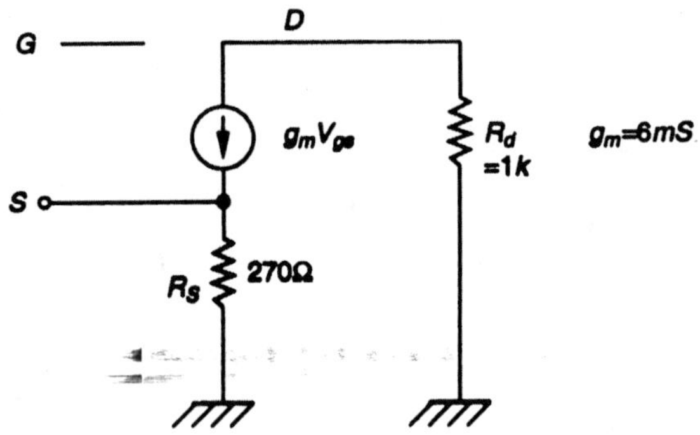

Fig 7.62. Circuit in Exercise Problem 11

[**Ans.** 2·29]

12. If R_D = 3·3 kΩ, R_L = 3·3 kΩ, g_m = 7 mS in the circuit of Fig. 7.31, find the voltage gain.

[**Ans.** 11·55]

13. What would be the voltage gain in Problem 12 above if R_L is removed from the circuit ?

[**Ans.** Voltage gain = 23]

14. In the circuit of Fig. 7.63, find (i) g_m (ii) r_d (iii) Z_{in} (iv) Z_{out} with r_d (v) G_V with r_d. Assume V_{GSQ} = – 3 V, I_{DQ} = 5 mA.

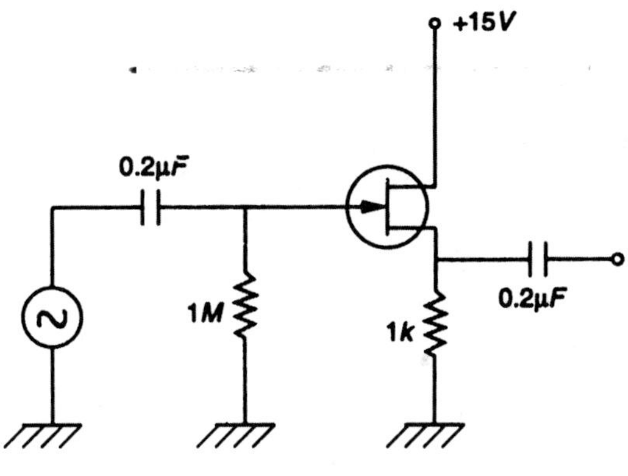

Fig 7.63. Circuit in Exercise Problem 14

I_{DSS} = 15 mA

V_P = − 6 V

y_{os} = 25 μS

[**Ans.** (i) 2·5 mS (ii) 40 kΩ (iii) 1 M (iv) 283·7 Ω (v) 0·7092]

15. In Problem 14 above find

(i) Z_{out} and (ii) voltage gain if the effect of r_d could be neglected.

[**Ans.** (i) 285·7 Ω (ii) 0·7142]

16. Find the voltage gain and the input resistance for the amplifier in Fig. 7.64. (Given g_m = 5000 μS)

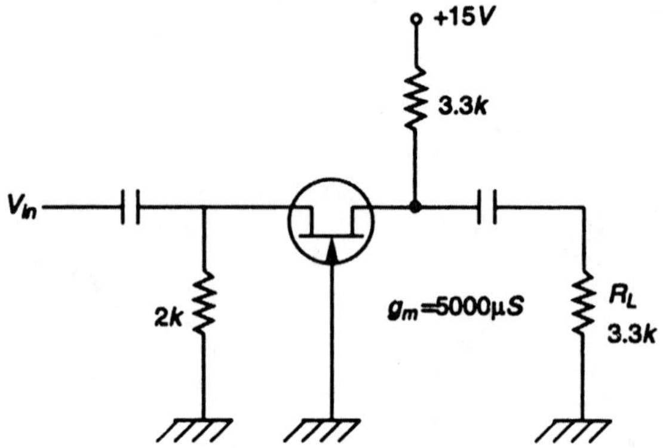

Fig 7.64. Circuit in Exercise Problem 16

[**Ans.** 8·25, 181·8 Ω]

17. In a *D-MOSFET*, I_{DSS} = 20 mA, V_{GS0} = − 6 V. Find the drain current if V_{GS} = − 3 V and V_{GS} = + 3 V.

[**Ans.** (i) 5 mA (ii) 45 mA]

18. In a *D-MOSFET*, I_{DSS} = 25 mA and V_{GS0} = + 6 V. Find I_D at V_{GS} = + 3 V and − 3 V. Is this an *n*-channel or a *p*-channel device ?

[**Ans.** (i) 6·25 mA (ii) 56·25 mA (iii) *p*-channel]

19. The manufacturer's data sheet of an *E-MOSFET* device shows $I_{D(ON)}$ = 6 mA at V_{GS} = 10 V and $V_{GS(th)}$ = 5 V. Find V_{GS}, I_D and V_{DS} for the circuit of Fig. 7.65.

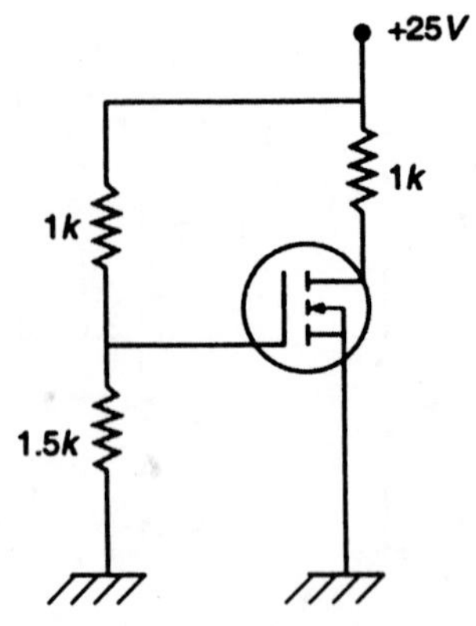

Fig 7.65. Circuit in Exercise Problem 19

[**Ans.** $V_{Gs} = 15$ V, $I_D = 24$ mA, $V_{Ds} = 1$ V]

20. Find the drain-to-source voltage for the *D-MOSFET* circuit of Fig. 7.66 (given $I_{DSS} = 20$ mA and $V_{GS} = -6$ V).

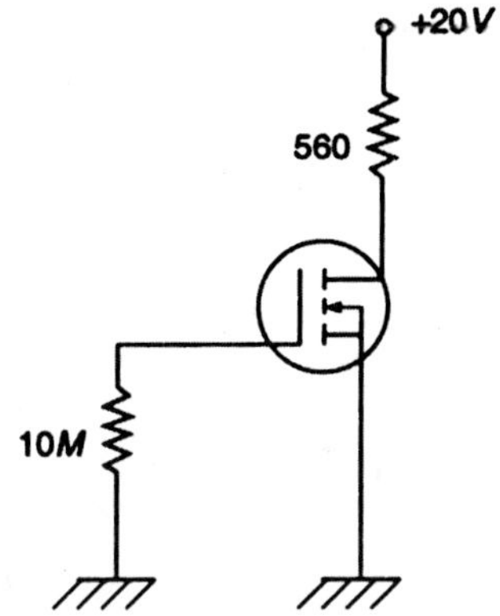

Fig 7.66. Circuit in Exercise Problem 20

[**Ans.** 8·8 V]

21. Find the drain-to source voltage for the *E–MOSFET* of Fig. 7.67. Given $V_{GS\,(th)} = 5$ V, $I_D = 4$ mA.

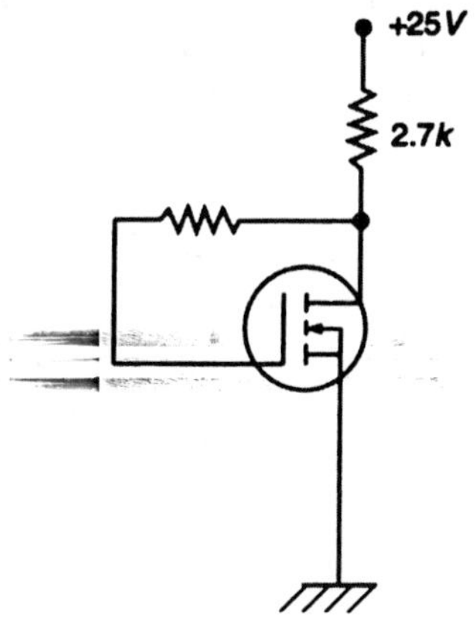

Fig 7.67. Circuit in Exercise Problem 21

[**Ans.** 9·2 V]

Chapter 8

POWER AMPLIFIERS

8.1 INTRODUCTION

In a practical amplifier, there are a number of amplifer stages whose purpose is to amplify a weak signal until sufficient power is obtained to operate the loudspeaker(s), the transmitting antenna or other output devices. A distinction between a voltage amplifier and a power amplifier is that–a voltage amplifier is designed to provide maximum voltage gain while a power amplifier is designed to achive maximum output power. To deliver a large amount of output power the amplifier is required to handle large output current. Power amplifiers are constructed using transistors that can handle large current. As a result the size of the power transistor is quite large as it is intended to dissipate the heat produced in the transistor during operation. Besides, the base is considerably thicker in a power transistor such that it can handle high currents. The β of the transistor is thus low.

Power amplifiers are also called as large signal amplifiers. Thus in a power amplifier a wide range of the load line is used during signal operation than in a small signal amplifier. Power amplifiers can be classified as class *A*, class *B* and class *C* amplifiers. The classification is based on the percentage of the input cycle of which the amplifier operates in the linear region. The common emitter, common base and common collector such that it operates in the linear region. The current thus flows for 360° of input cycle. It is thus a class *A*

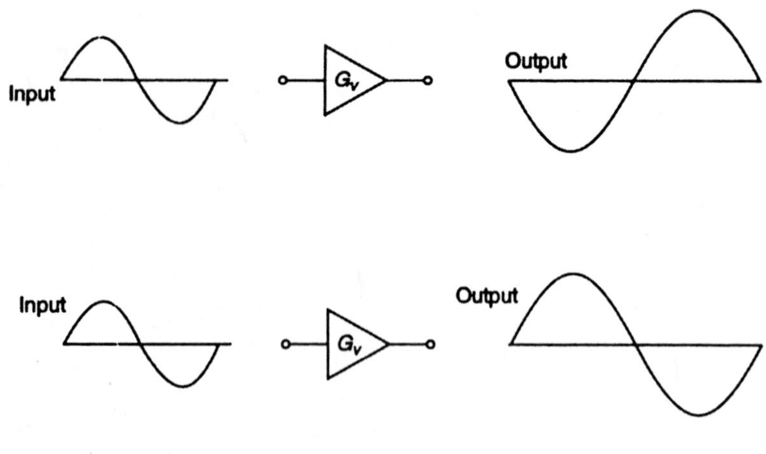

Fig 8.1. Class *A* amplifier operation (a) inverting (b) non-inverting

amplifier. The output voltage waveform is of the same shape as the input waveform. A class *A* amplifier operation is depicted in Fig. 8.1.

It can be seen from Fig. 8.1 that the output signal waveform may be either in phase or output-of-phase. For maximum output in class *A* operation, the *Q* point should be located at the centre of the a.c. load line. This will be apparent from Figs. 8.2 and 8.3.

It can be found from Fig. 8.3 that the current corresponding to *Q* point is I_{CQ} and the collector to emitter voltage is V_{CEQ} or V_{CQ}. If the peak value of the input waveform crosses the $I_{C\,sat}$ line, the transistor is driven into saturation. Similarly, if the value of the collector current falls below zero, the transistor is driven into cut off as illustrated in the Figs. of 8.2 and 8.3 respectively.

EXAMPLE 8.1 *A power amplifier is operated from a 15 V battery produces an output of 5 W. Find the maximum collector current.*

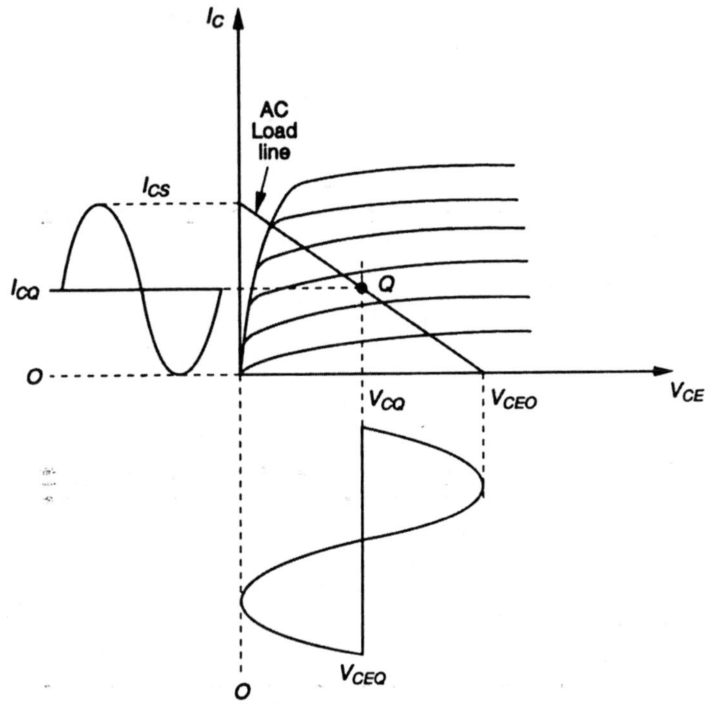

Fig 8.2. Class *A* operation for maximum output

SOLUTION Output power = Battery voltage × Collector current.

or, 5W = 15 V × I_C

or, $I_C = \frac{5}{15} = 333$ mA.

EXAMPLE 8.2 *A power amplifier produces 32 W in an 8 Ω speaker. Find (i) the a.c. output voltage (ii) the a.c. output current.*

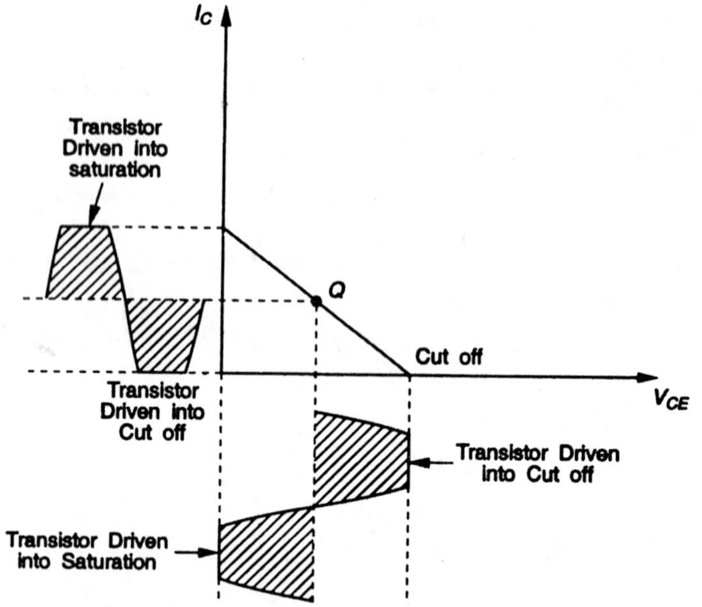

Fig 8.3. Overdriven amplifier

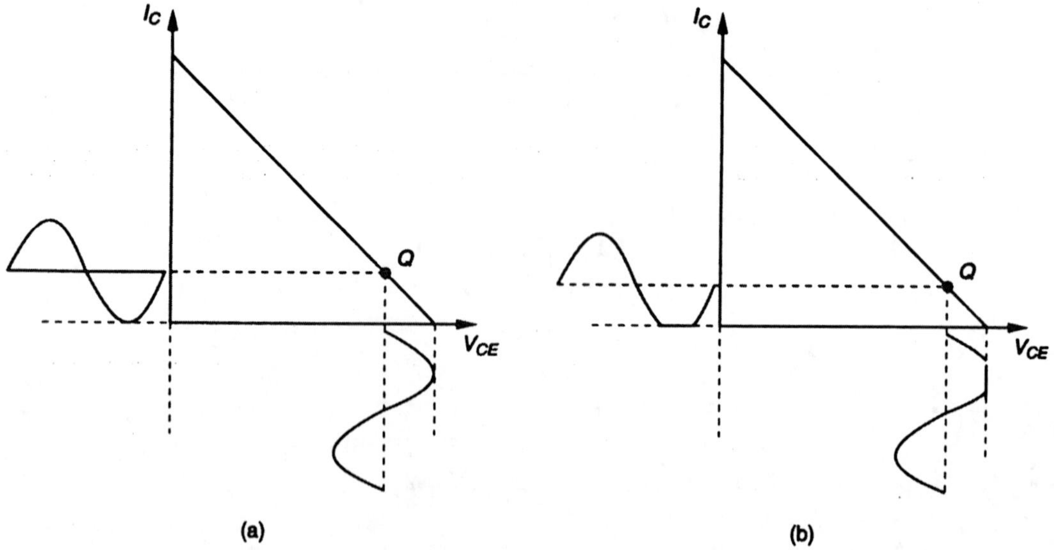

Fig 8.4. *Q* point located near cut off

SOLUTION (i) Power $= \dfrac{(\text{Voltage})^2}{\text{Resistance}}$

$$P = \frac{V^2}{R}$$

Thus, a.c. output voltage $= V = \sqrt{P \cdot R} = \sqrt{32 \times 8} = 16$ V

a.c. output-current $= I = \frac{V}{R} = \frac{16}{8} = 2$ Amps.

Fig 8.4 (a) depicts the location of the Q point near cut off. The output swing is limited by cut off in this case. The operation of an amplifier where the Q point is shifted near cut off is shown in Fig. 8.4(a). However, in this case, the output is a clipped waveform at one peak. Fig 8.5 (a) depicts the operation of the amplifier where the Q point is shifted near saturation. Fig 8.5 (a) however depicts an unclipped waveform. Fig 8.5 (b) shows clipping at one of the peaks due to saturation.

It should be remembered in this connection that a power amplifier does not effect amplifing power. What it does, is that it takes power from the d.c. supply source and converts it into a.c. signal power. A power amplifier can, in fact, be termed as d.c.-to-a.c. power converter,

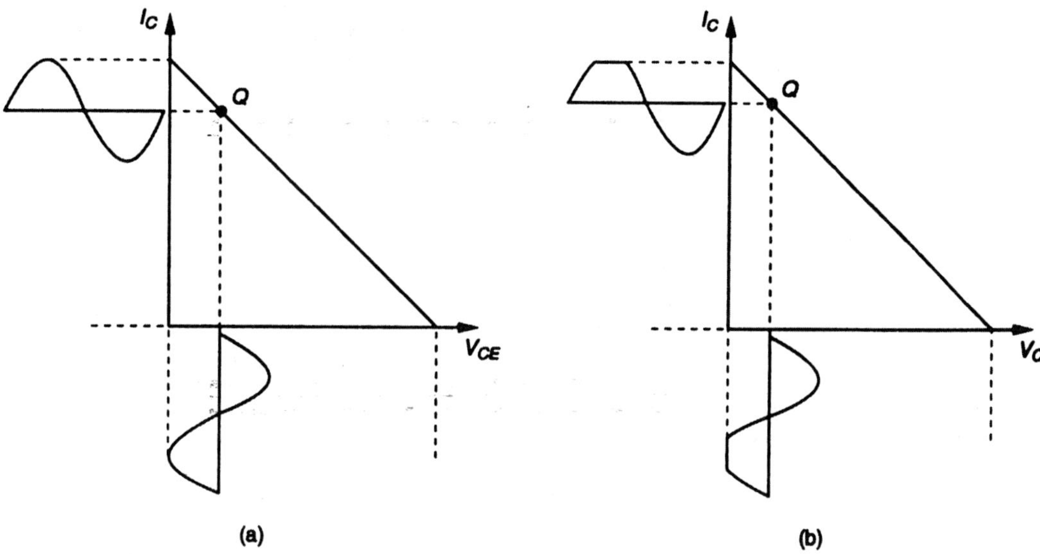

Fig 8.5. Quiscent (Q) operating point near saturation

whose action is controlled by the input signal.

8.2 POWER DISSIPATION IN A VOLTAGE AMPLIFIER

Let us consider a simple voltage divider bias type small signal *BJT* amplifier which is widely employed as a voltage amplifier [vide Fig. 8.6]

A voltage amplifier is not required to handle large power. So the power dissipating rating of the tansistor Q is small. The transistor must possess a large power dissipation rating, if it is to work in a power amplifier circuit.

Let us consider whether this circuit is useful as a power amplifier. In the circuit of Fig. 8.6, the voltage that appears across the transistor under quiscent operating condition is

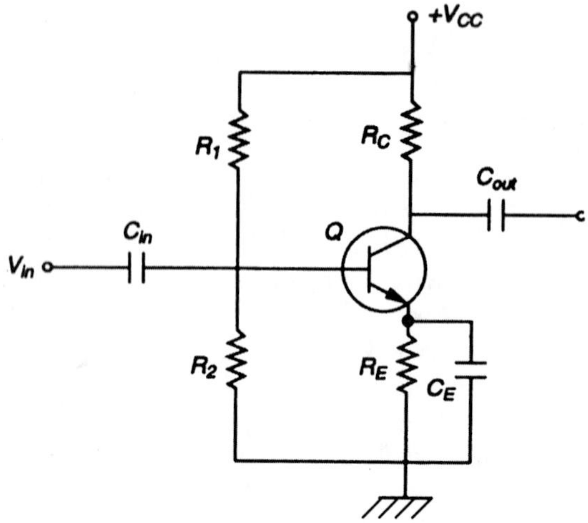

Fig 8.6. A *BJT* amplifier

$$(V_{CE})_Q = V_{CQ} = V_{CC} - I_{CQ}\,(R_C + R_E) \quad \text{...(8.1)}$$

i.e. the voltage between the collector-to-emitter at Q is less than the supply voltage by means of the voltage drops across R_C and R_E. The d.c. power that enters the transistor is

$$P_{DQ} = V_{CQ}\, I_{CQ}. \quad \text{...(8.2)}$$

This may also be called as the effective d.c. input power to the amplifier. A fraction of P_{DQ} defined by equation (8.2) is actually converted to a.c. power. It should be kept in mind that the total d.c. power supply drain is

$$P_{CC} = V_{CC} \cdot I_{CQ} \quad \text{...(8.3)}$$

Since $V_{CQ} < V_{CC}$, the question is where does the remaining power goes. The answer to this is that the power is lost in heating the resistors R_C and R_E. This can also be seen from the following equation

$$P_R = P_{CC} - P_{DQ} \quad \text{...(8.4)}$$

$$= V_{CC}\, I_{CQ} - V_{CQ}\, I_{CQ}$$

$$= I_{CQ}\,(V_{CC} - V_{CQ})$$

$$= I_{CQ} \cdot I_{CQ}\,(R_C + R_E)$$

$$= I_{CQ}^2\,(R_C + R_E) \quad \text{...(8.5)}$$

where P_R is the power loss in the resistors R_C and R_E. Thus, this circuit will not act as a good power amplifier unless the drop in the resistors R_C and R_E is reduced.

The resistor R_E is a part of the bias circuit and removing it altogether may result in poor stabilization of the circuit. R_L may, however, be replaced by a component which has a low (or zero) d.c. resistance and high a.c. impedance. A choke or the primary of a transformer can be used for this purpose.

Fig. 8.7 shows a power amplifier in which R_C has been replaced by the primary of a transformer while the secondary is connected to the load (the loudspeaker).

For a transformer, if the primary voltage is V_1, and the secondary voltage V_2, the voltage transformation ratio is

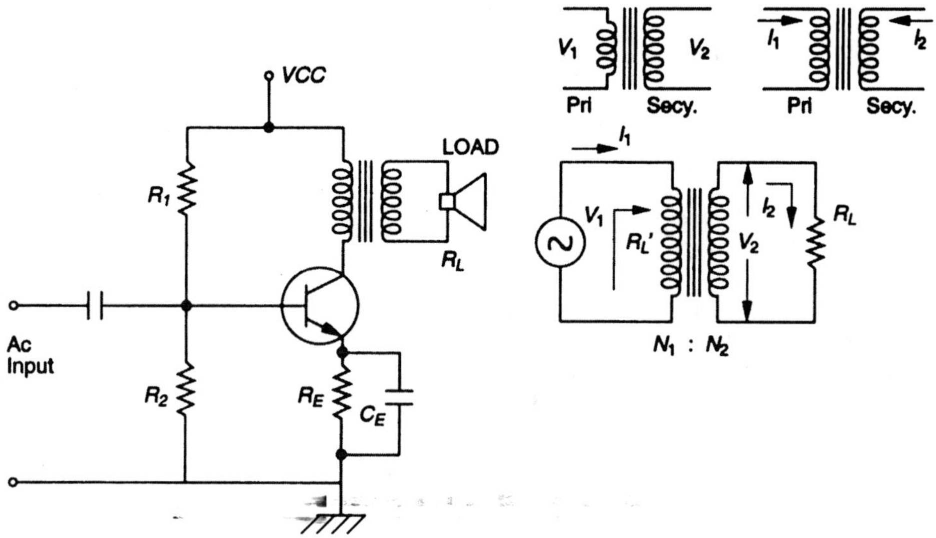

Fig 8.7. (a) A power amplifier circuit (b) impedance transformation

$$\frac{V_2}{V_1} = \frac{N_2}{N_1} \qquad \text{...(8.6)}$$

where N_1 and N_2 are the number of turns in the primary and secondary of the transformer respectively.

For current transformation, the ratio is

$$\frac{I_2}{I_1} = \frac{N_1}{N_2} \qquad \text{...(8.7)}$$

Thus the effective load R_L' on the primary side of the transformer due to a load R_L on the secondary side is

$$\frac{R_L'}{R_L} = \frac{V_1/I_1}{V_2/I_2} = \left(\frac{V_1}{V_2}\right) \cdot \left(\frac{I_2}{I_1}\right)$$

$$= \left(\frac{N_1}{N_2}\right)^2 \qquad \text{...(8.8)}$$

or,
$$R_L' = \left(\frac{N_1}{N_2}\right)^2 \cdot R_L \qquad \text{...(8.9)}$$

Thus a load of R_L at the secondary of the transformer will present an a.c. load impedance of $(N_1/N_2)^2 R_L$ to the transistor. Note also that the d.c. resistance of the primary winding is zero (or near zero).

EXAMPLE 8.3 *Find the effective resistance at the primary of a 10 : 1 transformer when connected to a 8 Ω load.*

SOLUTION
$$R_L' = \left(\frac{N_1}{N_2}\right)^2 \cdot R_L$$
$$= (10)^2 \times 8$$
$$= 800 \text{ ohms.}$$

It should also be kept in mind that maximum power will be transferred to the load when the effective resistance at the primary of the transformer equals the output impedance of the amplifier *i.e.* the transofmer provides impedance matching *i.e.* when

$$R_L' = R_{out} \qquad \ldots(8.10)$$

EXAMPLE 8.4 *An 8 Ω speaker load is to be matched to an power amplifier having an effective resistance of 5 kΩ. Find the transformer turns ratio.*

SOLUTION
$$R_L' = \left(\frac{N_1}{N_2}\right)^2 \cdot R_L$$

In this case

$$R_L = 8\ \Omega$$
$$R_L' = 5\ \text{k}\Omega$$

or,
$$\left(\frac{N_1}{N_2}\right)^2 = \frac{R_L'}{R_L} = \frac{5000\ \Omega}{8\ \Omega}$$
$$= 625$$

or,
$$\frac{N_1}{N_2} = \frac{25}{1}$$

Another power amplifier circuit using transformer coupling is depicted in Fig. 8.8.

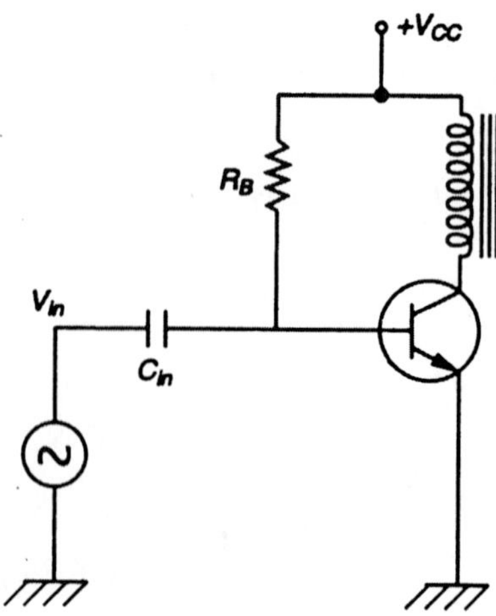

Fig 8.8. Transformer coupled power amplifier

8.3 CLASSIFICATION OF AMPLIFIERS.

The power amplifiers can be classified as class A, class B, class AB, class C, class D depending on the output signal swing.

Table 8.1
Classification of Power Amplifiers

Class A	Class AB	Class B	Class C	Class D
(360°)	(180° to 360°)	(180°)	(< 180°)	Pulse operation

In a class A amplifier the output signal swings for full 360° of the cycle. The Q point is selected such that the signal swing can vary up and down without being limited by cut off or saturation.

Class *B* Operation : In the class B operation, the reproduction of the input occurs over 180° of the output signal swing. In this operation, the d.c. operating point is chosen at 0 V. It should be noted, however, that by combining two class B operations such that one of them operates over the positive output half cycle and the another operates over the negative output half cycle, it is possible to provide output for full 360° of operation. This is also known as push pull operation.

Class *AB* Operation : In the class AB operation, the amplifier is biased at a quiscent operating point that is above the zero base curent level of class B and above one half the supply voltage level of class A. The class AB operation also requires push pull connection to for amplification over the full output cycle.

Class *C* operation : In class C operation, the amplifier is biased at less than 180° of the cycle and will operate with a tuned (resonant) circuit.

Class *D* Operation : In class D operation, the amplifier is operated with pulses or digital signals that operate for small *ON* periods and long *OFF* periods.

8.4 SERIES FED CLASS A AMPLIFIER

For the single ended class A amplifier operation (Fig. 8.9), the d.c. bias conditions are specified by

$$\text{Base current} = I_B = \frac{V_{CC} - 0{\cdot}7\ \text{V}}{R_B} \qquad \text{...(8.11)}$$

$$\text{Collector current} = I_C = \beta I_B \qquad \text{... (8.12)}$$

and collector-to-emitter voltage

$$V_{CE} = V_{CC} - I_C R_C \qquad \text{...(8.13)}$$

The $I_C - V_{CE}$ characteristics is depicted in Fig. 8.10.

In the collector characteristics is shown the load line. The load line is drawn by connecting the two points a and b having coordinates a (V_{CC}, 0) and $b\left(0, \dfrac{V_{CC}}{R_L}\right)$. The intersection of the d.c. load line with the d.c. base current line (I_B) determines the Q (Quiscent operating point). The co-ordinates of Q are maked as I_{BQ}, I_{CQ} and V_{CQ} which possess the values defined by equation (8.11) to (8.13). As discussed earlier, the maximum collector current swing occurs when I_{CQ} has a value at the midpoint (*i.e.*

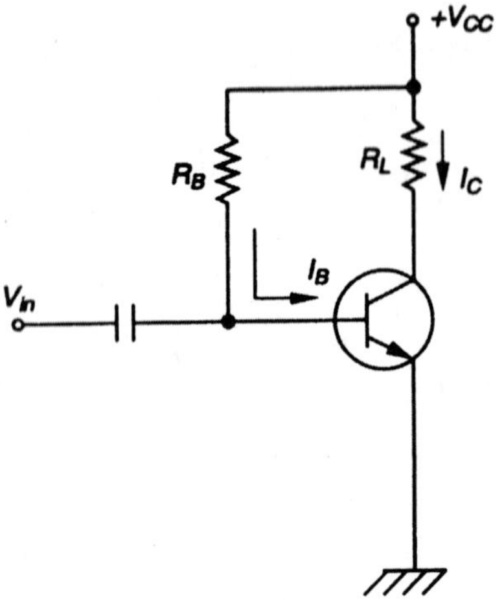

Fig 8.9. Circuit diagram of a single ended class *A* amplifier

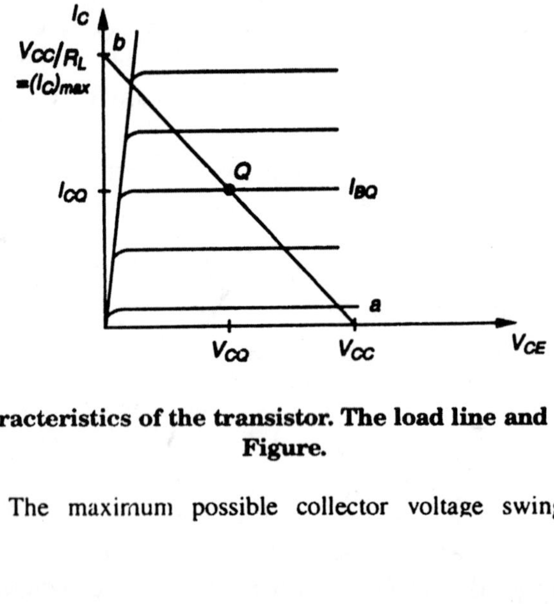

Fig 8.10. I_C vs. V_{CE} characteristics of the transistor. The load line and Q point is also shown in Figure.

$I_{CQ} = \frac{1}{2}(I_C)_{\max} = \frac{1}{2}\frac{V_{CC}}{R_L}\bigg)$. The maximum possible collector voltage swing also takes place when $V_{CQ} = \frac{1}{2}V_{CC}$.

When an a.c. signal is superimposed at the base of the amplifier, the base current changes. In Fig. 8.11 it will be apparent that when the base current swings from I_{BQ} to I_{B_1} (a to b), the Q point shifts from Q to Q_1, the collector current shifts from I_{CQ} to I_{C_1} (a' to b') and the collector to emitter voltage shifts from V_{CQ} to V_{C_1} (a'' to b''). When the base current now swings from I_{B_1} to I_{BQ} again (from point b to c), the Q-point changes along the load line from Q_1 to Q, I_c changes from I_{c1} to I_{CQ} (b' to c'), V_{CE} changes from V_{C_1} to V_{CQ} (b'' to c''). Now when the input base signal goes negative in the negative half cycle, I_B changes from I_{BQ} to I_{B_2} (c to d), the Q point changes along the load time from Q to Q_2, the collector current changes from I_{CQ} to I_{C_2} (c' to d') and the collector to emitter voltage changes from V_{CQ} to V_{C_2} (C'' to d''). Now when the base current changes from I_{B_2} to I_{BQ} again (completing the full cycle) (from d to e), the Q point moves along the load line from Q_2 to Q, and I_C changes from I_{C_2} to I_{CQ} (d' to e'). At any point of time during the full cycle, it is a simple matter to find the values of I_C and V_{CE} from Fig. 8.11 by following the load line variations.

A variation of Fig. 8.11 is Fig. 8.12 where the characteristics and the load line has been plotted for large

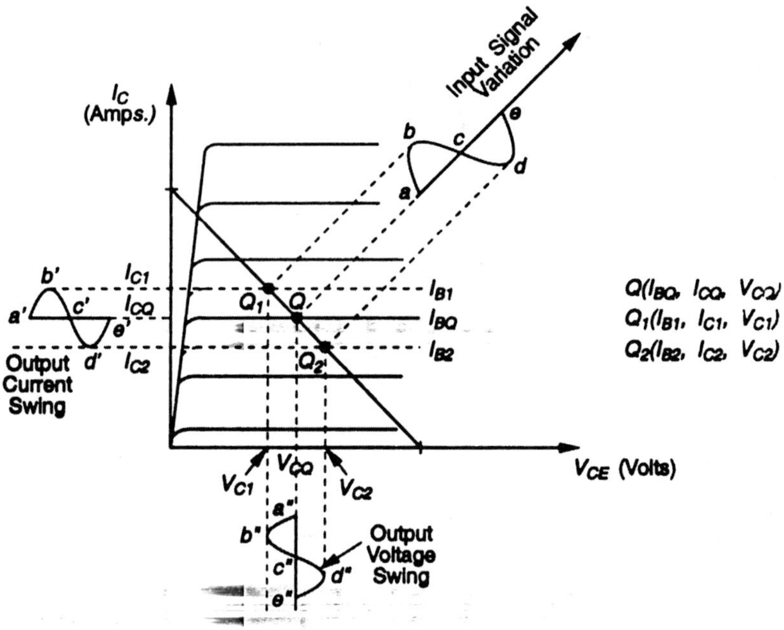

Fig 8.11. Input and output signal variations in a power amplifier

input voltage signal swings.

8.4.1 Input and Output Powers

The input power supplied to the class A amplifier of Fig. 8.9 when no input signal is applied is

$$P_{in}\,(dc) = V_{CC} \cdot I_{CQ} \qquad \ldots(8.14)$$

The quiscent power is the maximum power that the amplifier will handle. When an a.c. signal is applied, the average current drawn from the supply remains the same so that the power drain for the supply is once again $V_{CC}\,I_{CQ}$.

The a.c. power delivered to the load (output power) is

$$P_{out}(\text{a.c.}) = V_{cer} \,.\, I_{cr} \qquad \ldots(8.15)$$

$$= I_{cr}^2 \,.\, R_L \qquad \ldots(8.16)$$

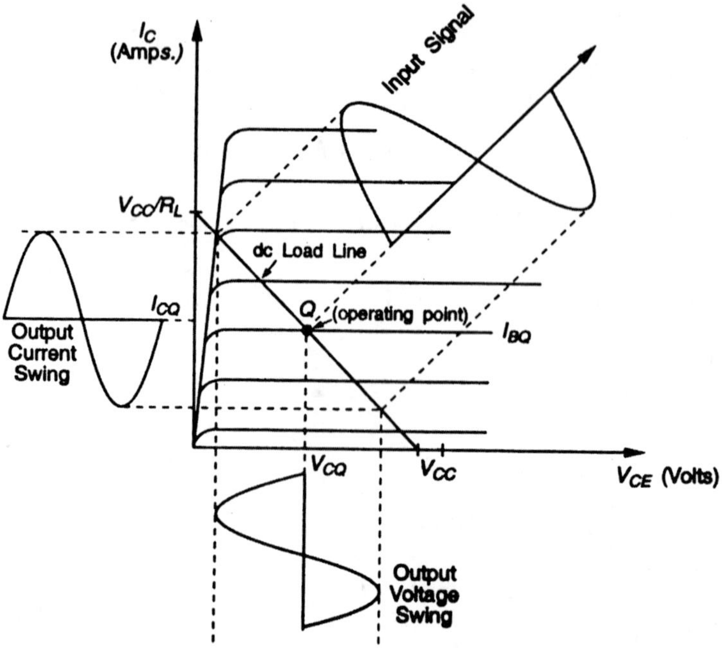

Fig 8.12. Input and output signal variations in a power amplifier (large swings)

$$= \frac{V_{cer}^2}{R_L} \quad \ldots(8.17)$$

where I_{cr} and V_{cer} are the r.m.s values for I_C and V_{CE} respectively.

The efficiency of the amplifier is defined as

$$\% \text{ Efficiency} = \frac{P_{\text{out}}\,(\text{ac})}{P_{\text{in}}\,(\text{dc})} \times 100\% \quad \ldots(8.18)$$

The maximum power input will occur when the Q point is set at the mid point (the d.c. bias current is set to half the maximum value)

Maximum P_{in} (dc) = $V_{cc} \cdot (I_C)_{max}$

$$= \frac{V_{cc} \cdot V_{cc}/R_L}{2} = \frac{V_{cc}^2}{2R_L} \quad \ldots(8.19)$$

The maximum a.c. output will occur, when the voltage and current swings are maximum. From Fig. (8.12)

$$\text{Maximum current swing (peak-to-peak)} = \frac{V_{cc}}{R_L} \quad \ldots(8.20)$$

$$\text{Maximum } V_{Ce} \text{ swing (peak-to-peak)} = V_{cc} \quad \ldots(8.21)$$

Thus maximum current swing (peak) $= \dfrac{V_{cc}}{2R_L}$ ...(8.22)

maximum current swing (r.m.s.) $= 0{\cdot}707 \dfrac{V_{cc}}{2R_L}$ ...(8.23)

maximum V_{CE} swing (peak) $= \dfrac{V_{cc}}{2}$ ...(8.24)

maximum V_{CE} swing (r.m.s.) $= 0{\cdot}707 \dfrac{V_{cc}}{2}$ ...(8.25)

Further maximum a.c. power putput (r.m.s.) =

maximum current swing (r.m.s.) × maximum V_{CE} swing (r.m.s.)

$$= 0{\cdot}707 \frac{V_{cc}}{2R_L} \times 0{\cdot}707 \frac{V_{cc}}{2} \quad ...(8.26)$$

$$= \frac{(0{\cdot}707)^2 V_{cc}^2}{4R_L}$$

$$= \frac{0{\cdot}5 V_{cc}^2}{4R_L} = \frac{V_{cc}^2}{8R_L} \quad ...(8.27)$$

Thus, maximum percentage efficiency

$$\eta = \frac{\text{Maximum } P_{out} \text{ (ac)}}{P_{in} \text{ (dc)}}$$

$$= \frac{V_{cc}^2/8R_L}{V_{cc}^2/2R_L} \times 100\% = 25\%$$

Thus the maximum efficiency of a class *A* series fed amplifier is 25%. As the maximum efficiency will occur in an amplifier when the current and voltage variation are ideal, the typical class *A* amplifier efficiency is less than 25%.

EXAMPLE 8.5 *Find (a) the maximum transistor power rating (b) AC output power and (c) efficiency for the amplifier depicted in Fig. 8.13.*

SOLUTION

$$V_B = \left(\frac{R_2}{R_1 + R_2}\right) V_{CC}$$

$$= \frac{1{\cdot}5 \text{ k}}{1{\cdot}5 \text{ k} + 3{\cdot}3 \text{ k}} \times 20 \text{ V} = 6{\cdot}25 \text{ V}$$

$$V_E = V_B - V_{BE} = 6{\cdot}25 \text{ V} - 0{\cdot}7 \text{ V} = 5{\cdot}55 \text{ V}$$

$$I_E = \frac{V_E}{R_E} = \frac{5{\cdot}55 \text{ V}}{150} = 37 \text{ mA}$$

$$I_{CQ} \simeq 37 \text{ mA}$$

$$V_C = V_{CC} - I_{CQ} R_C = 20 \text{ V} - 37 \times 10^{-3} \times 270$$

$$= 10{\cdot}01 \text{ V}.$$

$$V_{CEQ} = V_{CQ} = V_C - V_E = 10{\cdot}01 \text{ V} - 5{\cdot}55 \text{ V} = 4{\cdot}46 \text{ V}.$$

The transistor power rating must be greater than

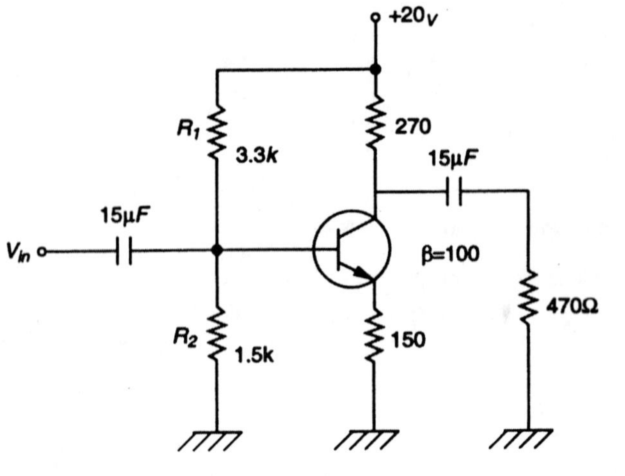

Fig 8.13. Power amplifier circuit (reference Example 8.5)

$P_D = V_{CEQ}\, I_{CQ} = (4{\cdot}46\ \text{V}) \times 37\ \text{mA} = 0{\cdot}165\ \text{W}.$

To find out the a.c. output power under maximum signal, the Q point position should be known.

Now $R_L' = R_C \parallel R_L = 270\ \Omega \parallel 470\ \Omega$

$$= 171{\cdot}5\ \Omega$$

$$(I_C)_{\text{sat}} = I_{CQ} + \frac{V_{CQ}}{R_L'}$$

$$= 37\ \text{mA} + \frac{4{\cdot}46}{171{\cdot}5}$$

$$= 37\ \text{mA} + 26\ \text{mA} = 63\ \text{mA}$$

$$(V_{CE})_{\text{cut off}} = V_{CQ} + I_{CQ} \cdot R_L'$$

$$= 4{\cdot}46 + (37\ \text{mA}) \times 171{\cdot}5\ \Omega$$

$$= 4{\cdot}46 + 6{\cdot}35 = 10{\cdot}81\ \text{V}$$

If the Q point is at the mid point, its co-ordinates would have been

$$I_{CQ} = \frac{63\ \text{mA}}{2} = 31{\cdot}5\ \text{mA}$$

$$V_{CQ} = \frac{10{\cdot}81\ \text{V}}{2} = 5{\cdot}40\ \text{V}.$$

Thus the Q point is shifted towards cut off.

The maximum collector current swing will be

$$P_{\text{out}} = 0{\cdot}5\, I_{CQ}^2 \cdot R_L'$$

$$= 0{\cdot}5\,(37\ \text{mA})^2 \times 171{\cdot}5$$

$$= {\cdot}117\ \text{W} \quad \text{or} \quad 117\ \text{mW}$$

∴ Efficiency is

$$\eta = \frac{P_{out}}{P_{in}\,(dc)} = \frac{117\ \text{mW}}{(20\ \text{V}) \times 37\ \text{mA}} = 0{\cdot}15$$

So the efficiency is only 15%. This is much below the 25% maximum efficiency possible for class *A* amplifiers. The fall in efficiency is due to the shift of the *Q* point from the centre position.

EXAMPLE 8.6 *Find the maximum value of the load power for the amplifier of Example 8.5.*

SOLUTION If the *Q* point is located at the central position,

$V_{CEQ} = V_{CQ} = 5{\cdot}4$ V.

$$\text{or, } P_{out}\,(\text{max}) = \frac{0{\cdot}5 \times (5{\cdot}4)^2}{171{\cdot}5}$$

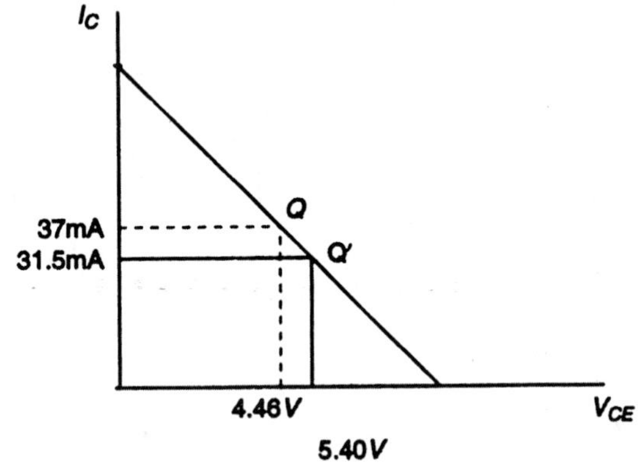

Fig 8.14. I_C vs. V_{CE} characteristics (reference Fig. 8.6)

EXAMPLE 8.7 *Find (i) Input power (ii) Output power (iii) efficiency for the amplifier circuit depicted in Fig. 8.15.*

SOLUTION ***Q* point**

$$I_{BQ} = \frac{V_{CC} - V_{BE}}{R_B}$$

$$= \frac{25 - 0{\cdot}7}{1500} = 16{\cdot}2\ \text{mA}$$

$I_{CQ} = \beta I_{BQ} = 50\,(16{\cdot}2\ \text{mA}) = 0{\cdot}81$ A

$V_{CEQ} = V_{CQ} = V_{CC} - I_C R_L = 25 - (0{\cdot}81\ \text{A}) \times 25$

$= 25 - 20{\cdot}25 = 4{\cdot}75$ V.

The peak collector current swing

$= I_C\,(p) = \beta I_B\,(p) = 50\,(12\ \text{mA})$

$= 600$ mA peak.

$$P_{out} = \frac{I_C\,(\text{peak})^2 \times R_L}{2}$$

$$= \frac{(600\ \text{mA})^2}{2} \times 25$$

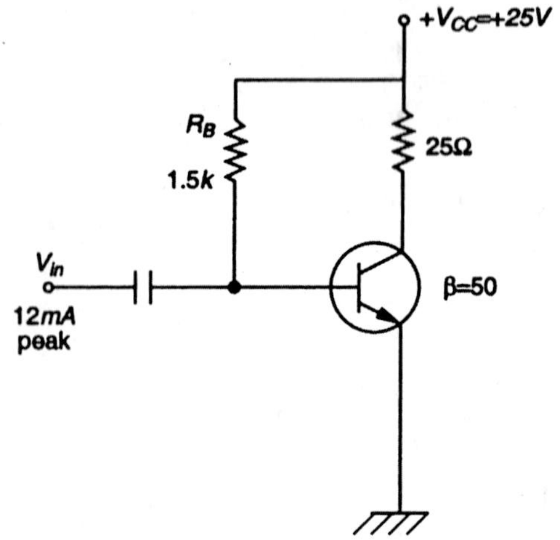

Fig 8.15. Amplifier circuit (reference Example 8.7)

$$= 4{\cdot}5 \text{ W}$$

$$P_{\text{in}} \text{ (dc)} = V_{CC} I_{CQ} = (25 \text{ V}) \times (0{\cdot}81 \text{ A})$$

$$= 20{\cdot}25 \text{ W}.$$

$$\text{So, efficiency} = \eta = \frac{P_{\text{out}} \text{ (ac)}}{P_{\text{in}} \text{ (dc)}} \times 100\%$$

$$= \frac{4{\cdot}5 \text{ W}}{20{\cdot}25 \text{ W}} \times 100\%$$

$$= 22{\cdot}22\%$$

8.5 OPERATION WITH A TRANSFORMER COUPLED CLASS A AMPLIFIER

For the transformer coupled class A power amplifier of Fig. 8.16, the d.c. load resistance is very small (typically 0·1 to 1 Ω). In the ideal case, the d.c. load resistance is zero. There will be zero d.c. voltage drop across the primary of the transformer and the d.c. load line will be a straight vertical line (shown dotted) in Fig. 8.17. The a.c. load line is also depicted in the above figure. The a.c. load line will have a slope of $-\frac{1}{R_L'}$ where $R_L' = \left(\frac{N_1}{N_2}\right)^2 \cdot R_L$.

The quiscent operating point is obtained from the intersection of the d.c. load line with the d.c. base current line. If the input signal has a voltage swing less than the maximum possible value, the efficiency of the circuit will fall below 25%.

The power developed across the load R_L is

$$P_L = \frac{V_L^2 \text{ (rms)}}{RL} \qquad \text{...(8.28)}$$

where $$V_L = V_2 = \frac{N_2}{N_1} V_1 \qquad \text{...(8.29)}$$

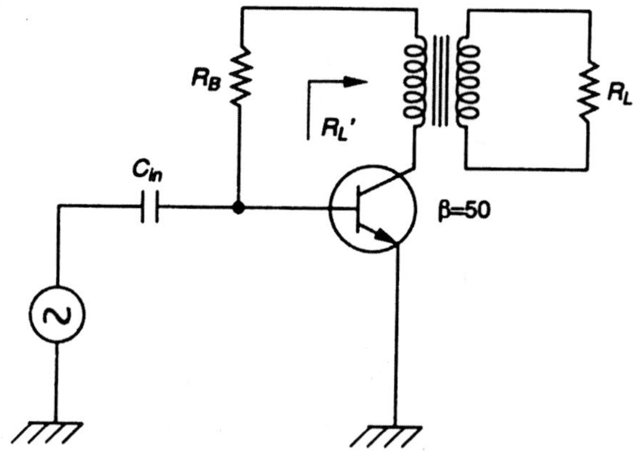

Fig 8.16. Transformer coupled class *A* amplifier

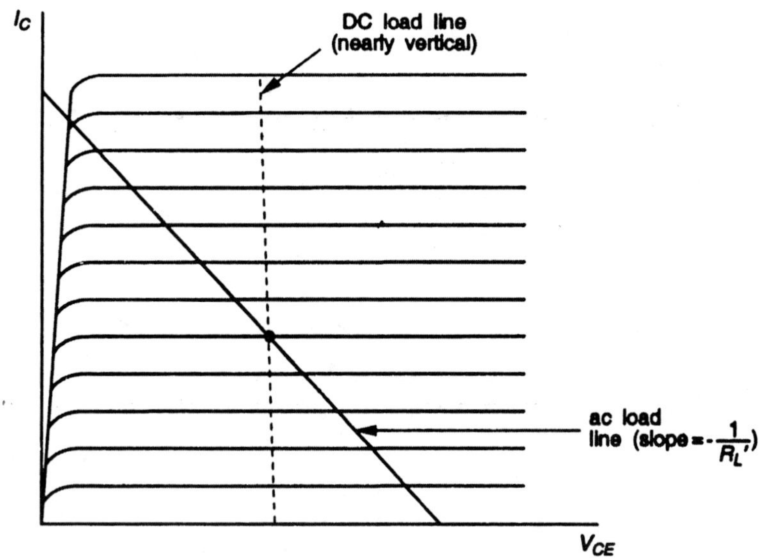

Fig 8.17. AC and DC load lines

8.6 CLASS *B* AMPLIFIER OPERATION

If the bias voltage is zero, the transistor is biased just off. The transistor turns on when the a.c. voltage signal is applied. For getting the output over a full cycle, two transistors are necessary, they conduct in opposite half cycles and the combined working of the two transistors provide a full cycle output signal. In the class *B* operation, one half of the circuit pushes the signal high during one half cycle and the other half-pulls the signal low during the other half cycle and the circuit is called a pushpull circuit. The class *B* push-pull operation is explained by the circuit. The class *B* push-pull operation is explained by the circuit of Fig. 8.18. In the push-pull connection, the two transistors are so connected, that when the collector current in one transistor increases then in the other transistor the collector current decreases.

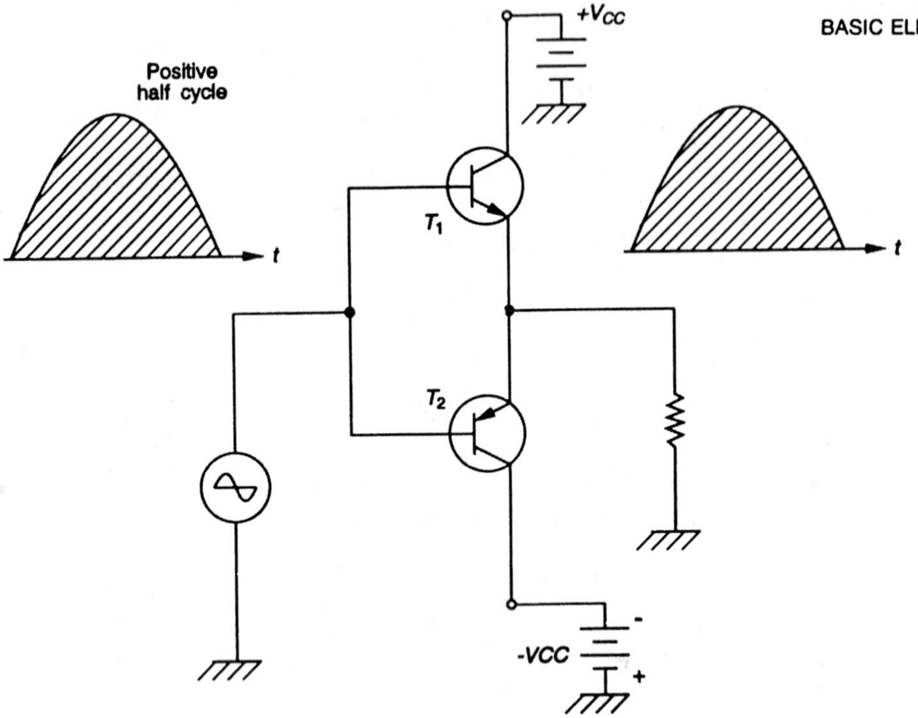

(a)

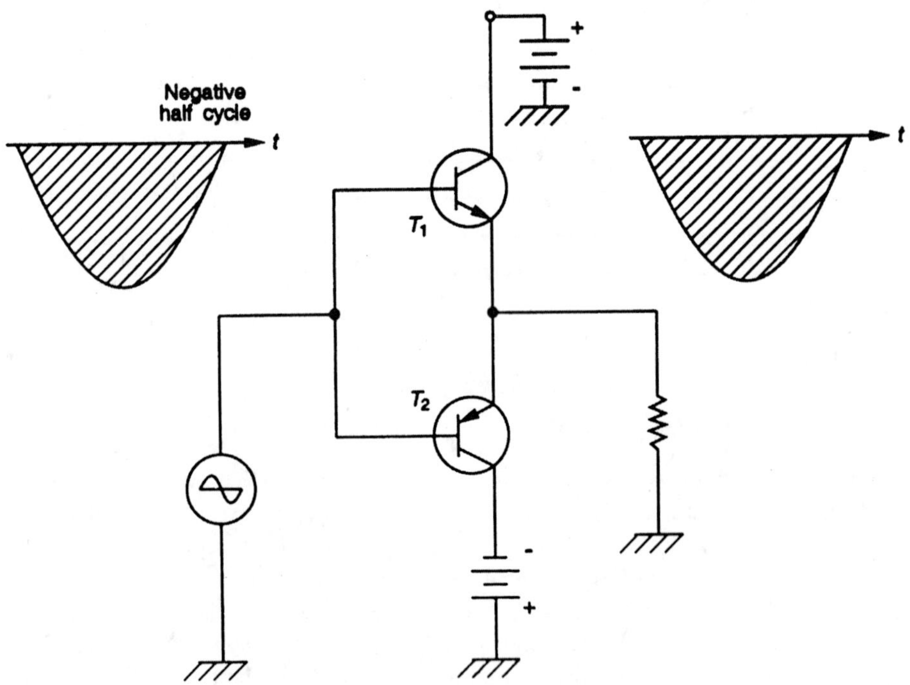

(b)

Fig 8.18. Push-pull class *B* amplifier working principles

In Fig. 8.18, it with be observed that the push-pull class *B* amplifier uses two emitter followers. Here one of the transistors is *npn* while the other is *pnp*. They conduct on opposite alternation of the input voltage. The d.c. base bias voltage is zero. Therefore, the signal voltage drives the transistors into conduction. The transistor T_1 conducts (on) during the positive half of the input cycle while the transistor T_2 is on during the negative half cycle.

It should, however, the kept in mind that when the d.c. base voltage is zero, the input signal voltage must exceed V_{BE} before the transistor conducts. So, there is a time interval between the positive and negative alternation during which both the transistors T_1 and T_2 are non-conducting. This results in a distortion called the cross-over distortion. This is shown clearly in Fig. 8.19.

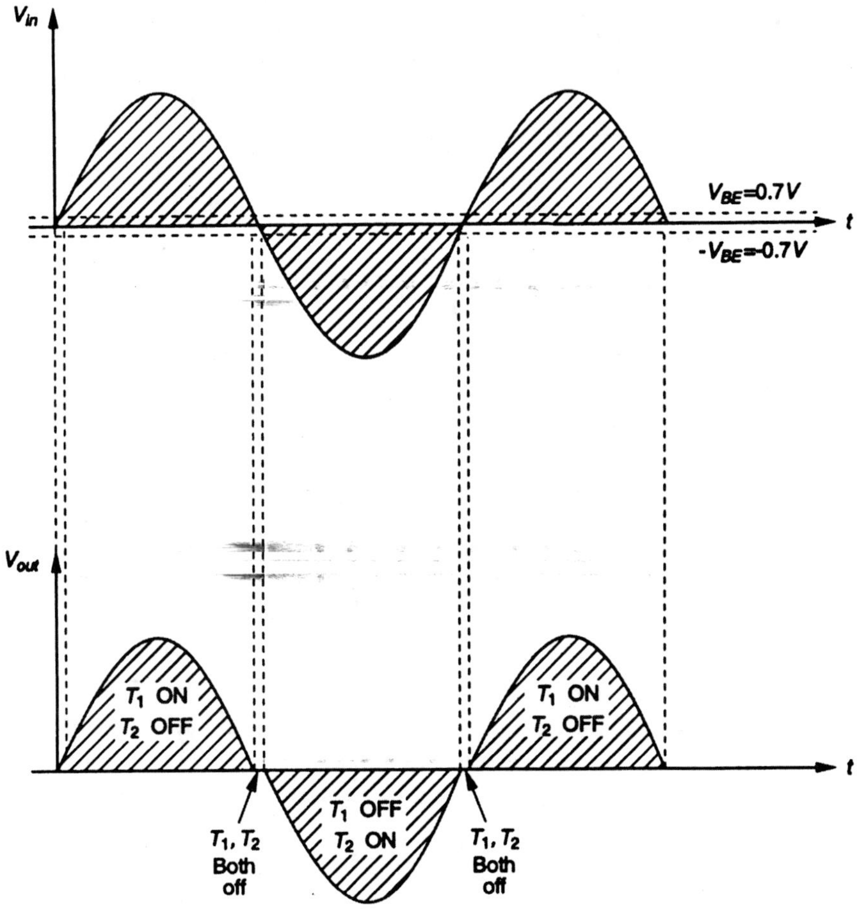

Fig 8.19. Cross over distortion in a class *B* push-pull amplifier

8.6.1 Transformer Coupled Push Pull Circuit

A practical push pull amplifier circuit using two transistors is depicted in Fig. 8.20.

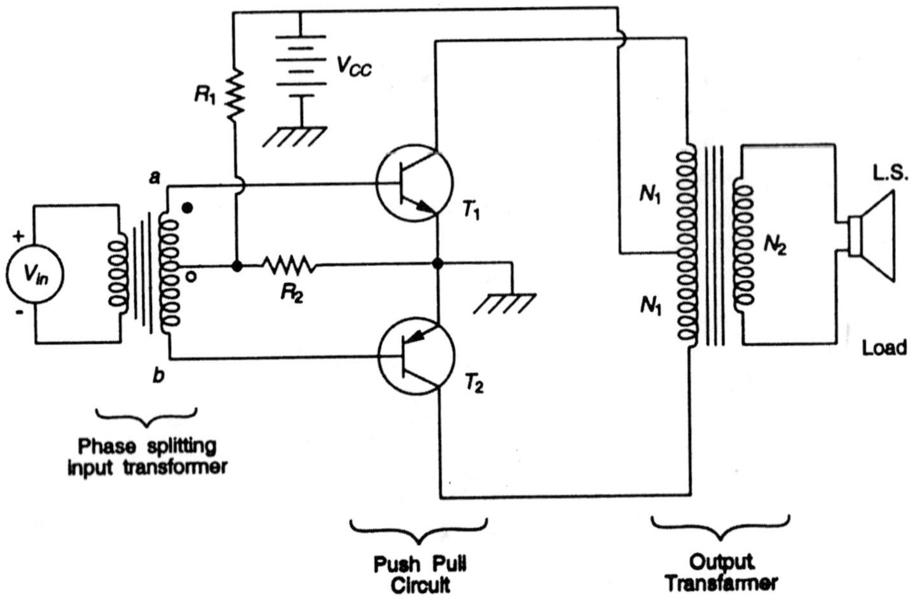

Fig 8.20. Push pull amplifier circuit

Let us examine the circuit of Fig. 8.20. Let us consider an instant of time when the upper end *a* of the secondary of the input transformer is positive with respect to the centre point O of the transformer secondary. The lower end *b* is negative. So at this instant (in the half cycle), the emitter base junction of the transistor T_1 is forward biased. Transistor T_1 thus conducts and T_2 becomes non conducting (*OFF*).

In the next half cycle, end *b* of the transformer becomes positive with respect to the centre point O in the secondary of transformer and transistor T_2 becomes conducting (*ON*). However the end *a* is negative with respect to centre point O of the transformer secondary and T_1 remains cut off. The collector currents i_{c1} and i_{c2} of the transistors T_1 and T_2 are half sinusoidal voltages. As the collector current of each transistor flows through each half of the primary of the output transformer, the total load current is a full sinusoidal waveform. The load resistance for each transistor appearing between one end of primary and the centre tap is

$$R_L' = \left(\frac{N_1}{N_2}\right)^2 R_L = \frac{N_1^2 R_L}{N_2^2} \qquad \ldots(8.30)$$

Since the collector current and collector voltage are half sinusoids, the r.m.s. collector voltage

$$V_{ce}\,(\text{rms}) = \frac{V_{cm}}{2} \qquad \ldots(8.31)$$

and r.m.s. collector current is

$$I_{ce}\,(\text{r.m.s.}) = \frac{I_{cm}}{2} \qquad \text{...(8.32)}$$

where V_{cm} is the maximum collector voltage and I_{cm} is the maximum collector current.

The a.c. power output from the transistor is

$$P_{ac} = V_{ce}\, I_{ce} = \frac{V_{cm}\, I_{cm}}{4} \qquad \text{...(8.33)}$$

The maximum swing of the collector voltage is

$$V_{cm} = V_{cc} \qquad \text{...(8.34)}$$

Therefore, the maximum value of the output power is

$$(P_{\text{out}})_{\max} = \frac{V_{cc}\, I_{cm}}{2} = \frac{V_{cc}^2}{2R_L'} \qquad \text{...(8.35)}$$

The d.c. power drain for each transistor is

$$P_s = V_{cc} \cdot I_{c\,(\text{av})} \qquad \text{...(8.36)}$$

where $I_{c\,(\text{av})}$ is the average value of the half cycle collector current.

The total d.c. power supplied by the source for two transistors is

$$(P_s)_{\text{dc}} = 2P_s = 2V_{cc} \cdot \frac{I_{cm}}{2} = \frac{2}{2}\frac{V_{cc}^2}{R_L'} \qquad \text{...(8.37)}$$

Thus the maximum efficiency for push-pull class *B* amplifier is

$$\eta_{\max} = \frac{P_{\text{out (max)}}}{(P_s)_{\text{dc}}} \qquad \text{...(8.38)}$$

$$= \frac{\pi}{4}$$

$$= 78{\cdot}5\% \qquad \text{...(8.39)}$$

8.6.2 Complementary Symmetry Push Pull Class *B* Power Amplifier

A *p–n–p* transistor and an *n–p–n* transistor are termed as complementary symmetry if they possess identical characteristics and the currrent conduction in the transistors in the circuit are complementary to each other. The advantage of the complemantry symmetry circuit is that it can provide a push pull output without a centre tapped output transformer. The input transformer is also not necessary.

Fig. 8.21 shows the circuit of a complementary symmetry class *B* push-pull amplifier.

One disadvantage of a complementary symmetry amplifier is that it is not easy to get a pair of transistors having identical characteristics. Some imbalances in the transistor characteristics give rise to distortion.

8.7 HARMONIC DISTORTION IN POWER AMPLIFIERS

The quality of sound provided by a loudspeaker depends very much upon the nature of the output signal. The quality of sound will be poor if the output is distorted. Because the power amplifiers are intended to handle large signals, distortion will always be preset. If output signal is sinusoidal, harmonic distortion occurs. One reason for the occurrence of distortion in an amplifier is the presence of non-linearity in the amplifier characteristics which is emphasized by the large signal levels involved. Because of this nonlinearity, the signals are distorted in waveshape. If a sinusoidal signal

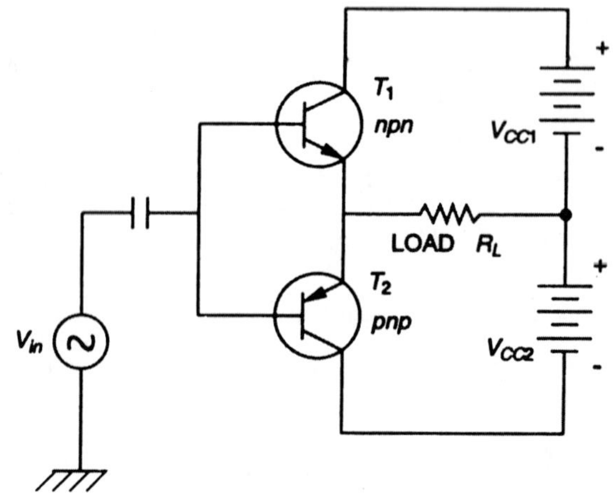

Fig 8.21. Complementary symmetry class *B* push pull amplifier

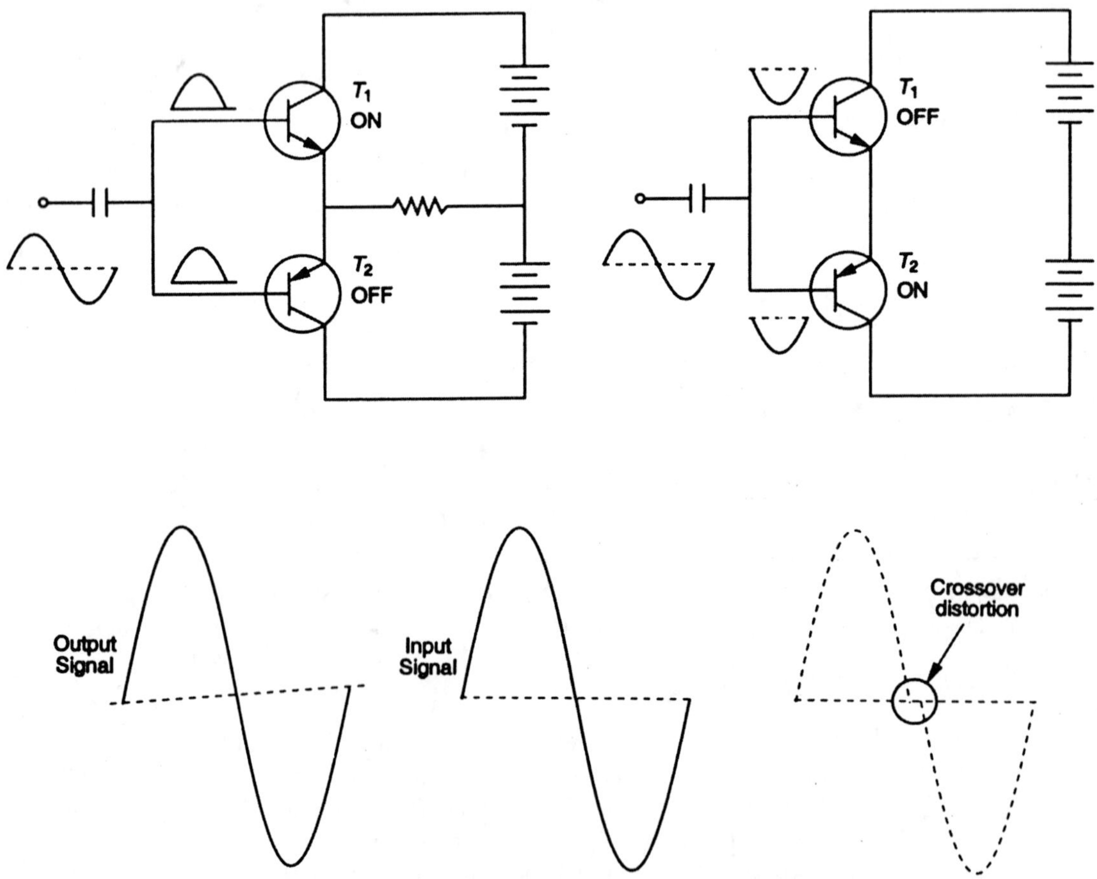

Fig 8.22. Complementary symmetry push pull circuit

$$v_{in} = V_{in} \sin \omega t \qquad ...(8.40)$$

is applied to the input of the power amplifier, the waveform of the output signal can be written as

$$i_0 = I_0 + I_1 \sin \omega t + I_2 \sin 2\omega t + I_3 \sin 3\omega t + ... \qquad ...(8.41)$$

where I_0 = d.c. component

I_1 = peak value of the first harmonic

I_2 = peak value of the second harmonic

The harmonic distortion is given as

$$\text{Second harmonic distortion} = D_2 = \frac{I_2}{I_1} \qquad ...(8.42)$$

$$\text{Third harmonic distortion} = D_3 = \frac{I_3}{I_1} \qquad ...(8.43)$$

$$\text{Fourth harmonic distortion} = D_4 = \frac{I_4}{I_1}, \text{ etc.} \qquad ...(8.44)$$

If distortion is present in the output signal, the power output calculated will be different from the value of the power output with no distortion. If distortion occurs in a circuit, the output power due to the fundamental component of the distorted signal is

$$P_1 = \frac{I_1^2 R_L'}{2} \qquad ...(8.45)$$

The total power due to all harmonic components of the output is

$$P = (I_1^2 + I_2^2 + I_3^2 + I_4^2 + ...) \frac{R_L'}{2} \qquad ...(8.46)$$

$$= \left[1 + \left(\frac{I_2}{I_1}\right)^2 + \left(\frac{I_3}{I_1}\right)^2 + \left(\frac{I_4}{I_1}\right)^2 + ...\right] \frac{I_1^2 R_L'}{2}$$

$$= [1 + D_2^2 + D_3^2 + D_4^2 + ...] \frac{I_1^2 R_L'}{2} \qquad ...(8.47)$$

The total distortion is given as

$$D = \sqrt{D_2^2 + D_3^2 + D_4^2 + ...} \qquad ...(8.48)$$

The total power can be written as

$$P = (1 + D^2)\, P \qquad ...(8.49)$$

If the total disortion is 20%, the total power will be

$$P = [1 + (0{\cdot}2)^2]\, P_1$$

$$= 1{\cdot}04\, P_1 \qquad ...(8.50)$$

or if the total distortion is 10%, the total power will be

$$P = [1 + (0{\cdot}1)^2]\, P = 1{\cdot}01\, P_1 \qquad ...(8.51)$$

This means that a 20% distortion represent a power which is only 4% of the fundamental. Similarly a 10% distortion represent a power which is only 1% of the fundamental. Thus, a very small error is introduced in using only the fundamental term P_1 in calculating the total power.

EXAMPLE 8.8 *A sinusoidal signal $v_{in} = 2 \sin 100\pi t$ is fed to a large signal amplifier. The resulting output current is*

$i_0 = 10 \sin 314t + 1{\cdot}5 \sin 628t + 0{\cdot}7 \sin 942t + 0{\cdot}3 \sin 1256t + \ldots$

Find (a) second, third and fourth harmonic distortions

(b) percentage increase in power due to distortion.

SOLUTION The percentage harmonic distortion of each component is

$$D_2 = \frac{I_2}{I_1} \times 100 = \frac{1{\cdot}5}{10} \times 100 = 15\%$$

$$D_3 = \frac{I_3}{I_1} \times 100\% = \frac{0{\cdot}7}{10} \times 10\% = 7\%$$

$$D_4 = \frac{0{\cdot}3}{10} \times 100\% = 3\%$$

The distortion factor is given as

$$D = \sqrt{D_2^2 + D_3^2 + D_4^2} = \sqrt{(\cdot 15)^2 + (\cdot 07)^2 + (\cdot 03)^2}$$

$$= \sqrt{\cdot 0225 + \cdot 0049 + \cdot 0009} = \sqrt{\cdot 0283} = \cdot 1682 = 16{\cdot}82\%$$

Now, net output power $= P(1 + D^2)\,P_1 = [1 + (\cdot 1682)^2]\,P_1 = 1{\cdot}0283P_1$

Therefore, percentage increase in power

$$= \frac{(1{\cdot}0283P_1 - P_1)}{P_1} \times 100\% = 2{\cdot}83\%$$

EXAMPLE 8.9 *Find the value of the output power delivered in the amplifier circuit of Fig. 8.23.*

SOLUTION Power developed $= P_{CC} = V_{CEQ}\,I_{CQ} = V_{CC} \cdot \dfrac{V_{CC}}{R_L}$

$$= \frac{V_{CC}^2}{R_L} = \frac{(25)^2}{220} = 2{\cdot}84 \text{ Watts}$$

8.8 CLASS *C* AND CLASS *D* AMPLIFIERS

The class *A*, class *AB* and class *B* amplifiers are used in power amplfiers while class *D* amplifiers are very attractive for their high efficiency. Class *C* amplifiers are very useful in tuned circuits as in communications.

A class *C* amplifier is depicted in Fig. 8.24.

EXAMPLE 8.10 *Find the maximum value of efficiency of an inductor coupled amplifier in Fig.8.23 if $R_E/R_L << 1$ and the amplifier is biased for maximum symmetrical swing.*

SOLUTION Maximum effciency $\eta_m = \dfrac{(P_L)_{max}}{\text{Power delivered to the amplifier}} \times 100\%$

$$= \frac{(V_{CC}/R_L)^2\,R_L/2}{V_{CC} \cdot \dfrac{V_{CC}}{R_L}} \times 100\% = 50\%$$

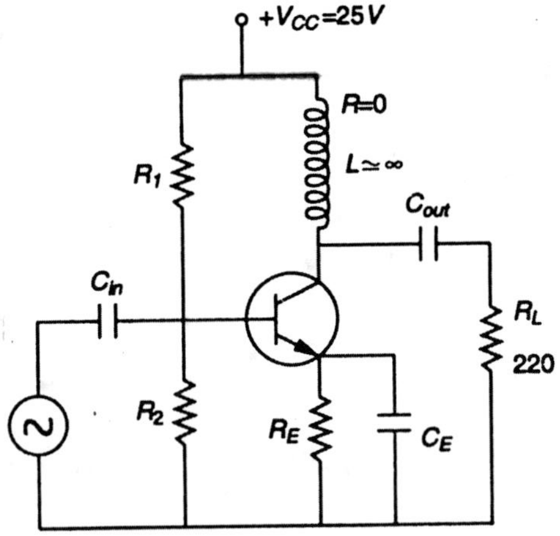

Fig 8.23. Power amplifier circuit (reference Example 8.9)

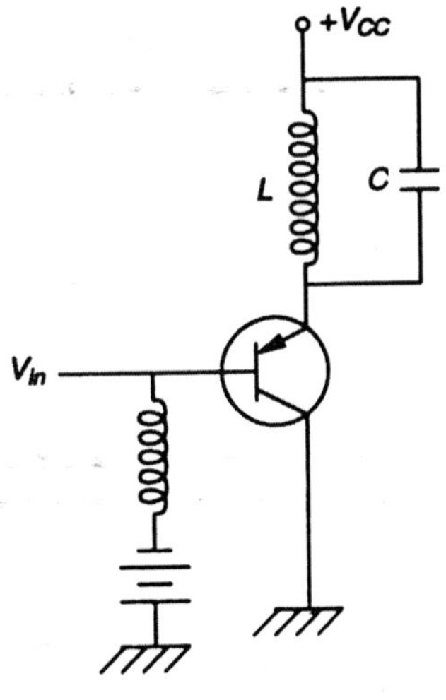

Fig 8.24. Circuit of a class C amplifier

EXAMPLE 8.11 *Find the maximum value of load power in Example 8.10 if* V_{CC} *is chosen to be 12 V and* $R_L = 12\ \Omega$.

SOLUTION $(P_L)_{max} = \dfrac{V_{CC}^2}{2R_L} = \dfrac{(12)^2}{2 \times 12} = 6$ Watts

EXAMPLE 8.12 *A loudspeaker with a 16 Ω input resistance requires a power of 1 watt. It is driven by the push pull amplifier of Fig. 8.20. If* $V_{CC} = 12$ *Volts and the transistors have* $(V_{CE})_{sat} = 0{\cdot}7\ V$ *and* $I_{BE_0} = 0$*, find the turns ratio of the transformer.*

SOLUTION $$\eta = \frac{V_{CC} - V_{CEsat}}{\sqrt{2R_L P_{max}}} = \frac{12 - 0{\cdot}7}{\sqrt{2 \times 16 \times 1}} = \frac{11{\cdot}3}{\sqrt{32}} = 1{\cdot}997$$

or 2 turns.

EXAMPLE 8.13 *Calculate the value of the supplied power in example 8.12 above. Find also the value of collector dissipated power.*

SOLUTION The average power supplied by V_{CC} is

$$P_1 = \frac{1}{\pi}\int_0^{\pi} V_{CC} I_{C\,max} \sin \omega t \, d(\omega t) = \frac{2}{\pi} V_{CC} I_{C\,max}$$

$$= \frac{4}{\pi} \times 1 = 1{\cdot}272 \text{ W}$$

Collector dissipated power $= \frac{1}{2}(P_{CC} - P_L) = \frac{1}{2}(1{\cdot}272 - 1) = {\cdot}136$ W

EXAMPLE 8.14 *Find the efficiency of the complementary symmetry amplifier of Fig. 8.21 when a maximum load power of 15 W is delivered. Assume $R_L = 4\ \Omega$ and $I_{C\,max} = 2{\cdot}5$ Amps and $V_{CC} = 12$ V.*

SOLUTION Power supplied $= P_1 = \frac{2}{\pi} = V_{CC} I_{C\,max} = \frac{2}{\pi} \times 12 \times 2{\cdot}5 = 19{\cdot}09$ W

$$\eta_{max} = \frac{\text{Maximum load power}}{P_1} \times 100\% = \frac{15}{19{\cdot}09} \times 100\% = 78{\cdot}54\%$$

Review Questions

1. What is a "power amplifier" ?
2. What do you understand by a class *A* amplifier ?
3. Locate the Quiscent (*Q*) point on the $I_C - V_{CE}$ characteristics of a transistor for maximum class *A* output operations.
4. Locate the Quiscent (*Q*) point on the $I_C - V_{CE}$ characteristics of a transistor for an overdriven amplifier.
5. On the I_C *vs.* V_{CE} characteristics, locate the *Q* point such that the amplifier is near cut off or saturation.
6. Draw the circuit diagram of a power amplifier and explain its principle of operation.
7. What do you understand by class *B*, class *AB*, class *C* and class *D* amplifiers ?
8. Draw the circuit diagram of a series fed class *A* amplifier (single ended) and explain its principle of operation
9. Derive expressions for the maximum percentage efficiency of class *A* amplifiers.
10. Explain the principle of operation of a class *A* transformer coupled amplifier.
11. Explain the principle of operation of a class *B* amplifier.
12. What is cross over distortion ? Discuss the cross over distortion with reference to a class *B* push-pull amplifier ?
13. Explain the principle of operation of transformer coupled push-pull amplifier circuit.
    What is the maximum efficiency for a push-pull class *B* amplifier ?
14. Explain the principle of operation of a push-pull class *B* amplifier.
15. What is harmonic distortion ?

Objective Type Questions

1. 'When the Q point in an inverting class A amplifier is closer to saturation than cut off, if the input sine wave is gradually decreased, clipping on the output will be first noticed on
   (a) the negative peaks (b) the positive peaks
   (c) simultaneously in both peaks
2. The efficiency of a class A amplifier can be increased by using
   (a) directly coupled load (b) high d.c. power input
   (c) a transistor with high rating (d) transformer coupled load
3. In a class A amplifier, the worst case conditions takes place with
   (a) maximum signal input (b) low load resistance
   (c) high load resistance (d) zero signal input
4. The maximum efficiency of a class A series fed amplfier is
   (a) 25% (b) 50%
   (c) 78·6% (d) 100%
5. In the transfermer coupled class A power amplifier the d.c. load line will be
   (a) inclined with 45° slope (b) inclined with a slope of 135°
   (c) a straight vertical line (d) a line with a slope of $1/R_L'$
6. In a transformer coupled class A power amplifier, the a.c. load line will have a slope of
   (a) 45° (b) 135°
   (c) straight vertical line (d) $-\frac{1}{R_L'}$
7. The maximum efficiency of a transformer coupled class A amplifier is
   (a) 25% (b) 50%
   (c) 78·5% (d) 100%
8. The main advantage of a class B push pull amplifier is that it possesses
   (a) no dc magnetic saturation (b) no even order harmonic distortion
   (c) no noise (d) no circuit imbalances
9. The power dissipation of class C amplifiers is
   (a) very small (b) very large
   (c) the same as class A amplifier (d) the same as class B amplifier
10. The maximum efficiency for a push-pull class B amplifier is
    (a) 50% (b) 78·6%
    (c) 87% (d) 100%
11. A class A amplifier has a current gain of 60 and a voltage gain of 50. The power gain is
    (a) 60 (b) 110
    (c) 3000 (d) 50
12. Class C amplifiers are mainly used as
    (a) audio power amplification (b) stereo amplifier
    (c) rf power amplifier

Exercise Problems

1. A power amplifier operated from a 18 V battery produces an output of 8 W. Find the maximum collector current.
2. A power amplifier produces 40 Watts in a 8 Ω speaker. Find (i) ac output voltage (ii) ac output current.
3. Find the effective resistance at the primary of a 7 : 1 transformer when connected to a 8 Ω load.

4. A 6 Ω speaker load is to be matched to a power amplifier having an effective resistance of 7404 Ω. Find the transformation ratio.
5. Find (a) the maximum transistor power rating (b) ac output power (c) efficiency for the amplifier shown in Fig. E8.5.

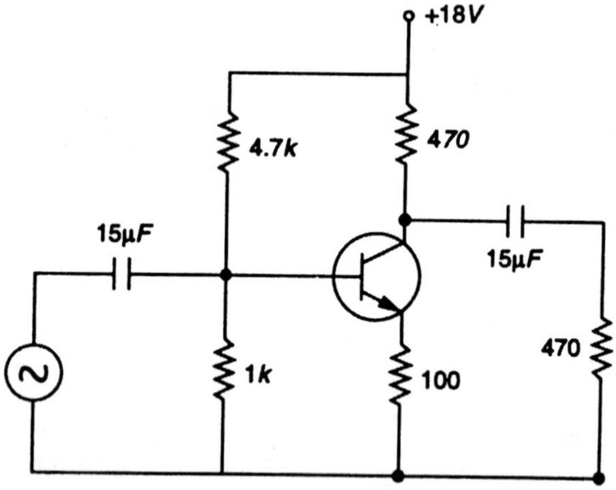

Fig E.8.5. Reference Exercise Problem 8.5

6. Find the maximum value of the load power for the amplifier of Problem 8.5 (Fig. E8.5).
7. Find the (a) input power (b) output power (iii) efficiency for the amplifier depicted in Fig. E8.7.

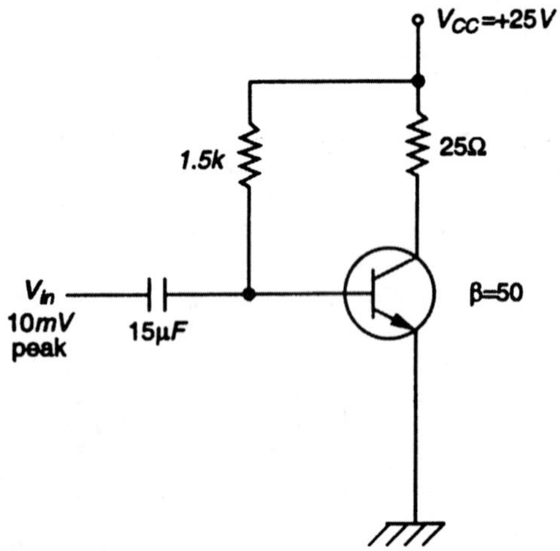

Fig E.8.7. Reference Exercise Problem 8.7

8. A sinusoidal signal $v_{in} = 3 \sin 100\pi t$ is fed to a large signal amplifier. The resulting output current is

   $i_0 = 10 \sin 314t + 2{\cdot}5 \sin 628t + 0{\cdot}5 \sin 942t + 0{\cdot}02 \sin 1256t + \ldots$

   Find second, third and fourth harmonic distortion.
9. Find the value of the output power delivered in the amplifier circuit of Fig. E8.9.

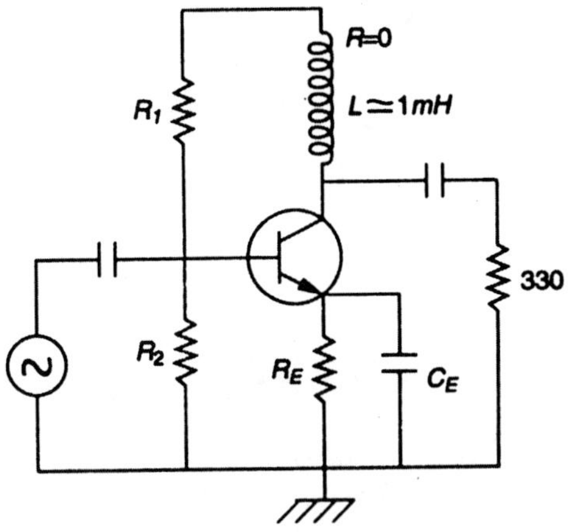

Fig E.8.9. Reference Exercise Problem 8.9

10. A class C amplifier has at its input a 100 kHz signal. The transistor is *ON* for 2μ sec and the amplifier is operating over 100 percent of its load line. If $V_{CE(sat)} = 0.2$ V and $I_{C\,sat} = 200$ mA, what is the average power dissipation.

    [**Ans.** 8 mW]

11. Find the efficiency of the amplifier in Problem 10 above if V_{CC} is 20 V and R_C is 200 ohms.

    [**Ans.** 96.15%]

Chapter 9

FREQUENCY RESPONSE OF AMPLIFIERS

9.1 INTRODUCTION

In the introductory chapters on amplifiers, the coupling capacitor reactances were assumed to the small so as to be negligible. Thus $X_C \simeq 0$. However X_C may have a non zero value. Besides, the internal capacitances of the transistor (C_{be}, C_{bc}, C_{ce} for a *BJT* and C_{gs}, C_{ds}, C_{gd} for *FET*) were also considered to be small at the signal frequency. These simplifying assumptions are extremely useful in understanding the theory of amplifier operation. However, for a proper understanding of the amplifier, the effect of these capacitance at various frequencies need to be investigated.

Role of Coupling Capacitors

Fig. 9.1 shows the circuit diagram of typical resistance - capacitance coupled amplifiers using (a) *BJT* (b) *JFET* and (c) *MOSFET*. In order to understand the role of the coupling capacitor, we should keep in mind that the reactances of a capacitor C is

$$X_C = \frac{1}{2\pi fC} \qquad \ldots(9.1)$$

At low frequencies, f is small and as a result X_C has a large value. As frequency f increases, the value of X_C decreases and ultimately at high frequencies $X_C \to 0$. Thus, in the simplifying assumptions made by us earlier, we assumed that $X_C = 0$ at the signal frequencies. However, since this assumption is not fully correct at low frequencies (< 10 Hz), there will be voltage drops across C_{in} and C_{out} and as a result the gain will be decreased. Also, the emitter voltage will not be at signal ground at these low frequencies because f is small and X_{C_E} or X_{C_S} is not zero. Further the coupling capacitor C_{in} form **phase lead** network with the input resistance R_{in} of the amplifier and C_{out} form another **phase lead** network with R_L' where $R_L' = R_C \| R_L$ or $R_D \| R_L$ as the case may be **(*A* lead network is a network where the voltage across the resistance leads the input voltage in phase)**. The effect of these lead networks in the circuit is that they introduce phase shifts into the amplifier circuit.

Role of Bypass Capacitors (C_E or C_S)

Since $X_C \neq 0$ at low frequencies (f < 10 Hz), the parallel combination of X_{C_E} and R_E (or X_{C_S} and R_S)

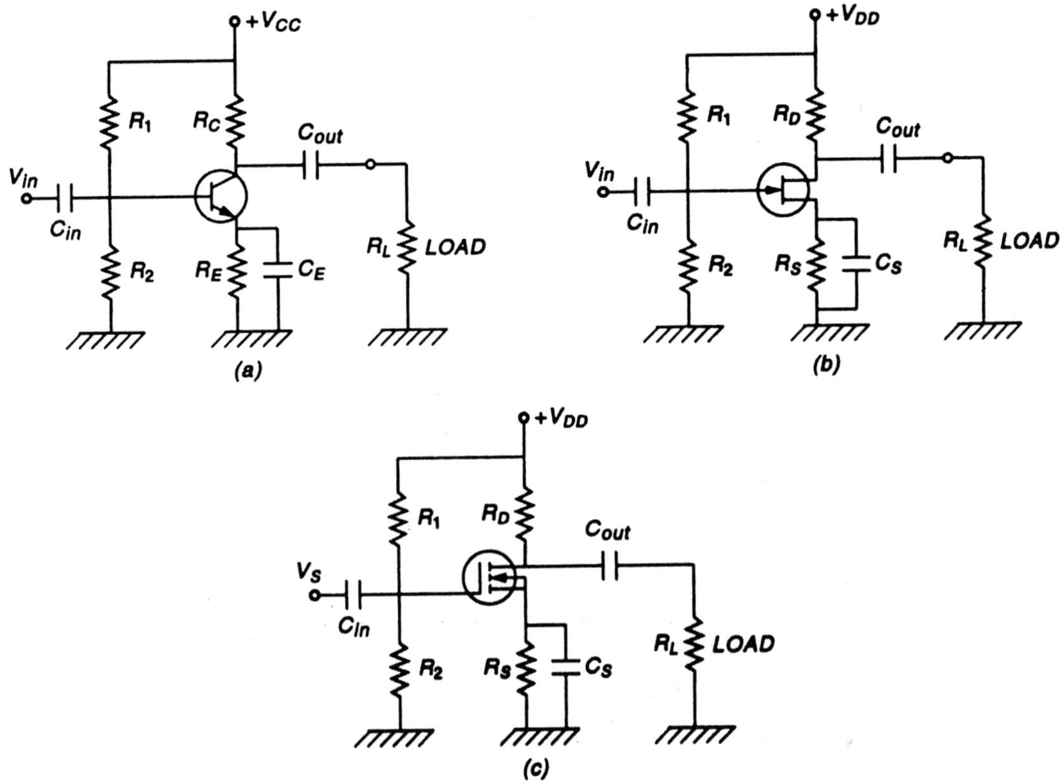

Fig 9.1 Capacitatively coupled amplifier using (a) BJT (b) JFET (c) MOSFET.

form a nonzero impedance Z. Since $Z \neq 0$, the voltage gain is reduced.

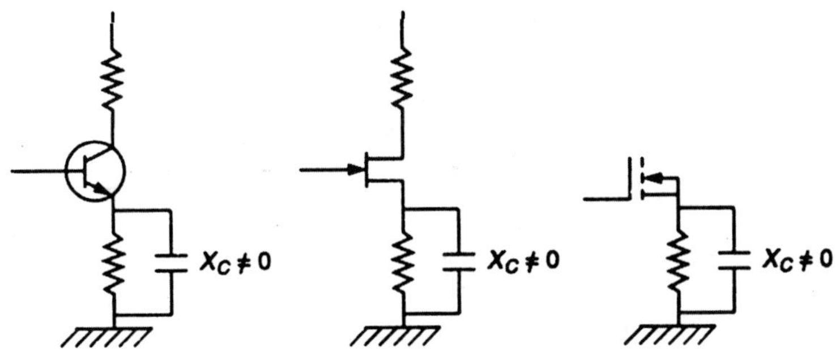

Fig 9.2 Non zero values of bypass capacitors.

Role of transistor capacitances

The interelectrode capacitances within the transistor reduces the gain of the amplifier and introduce

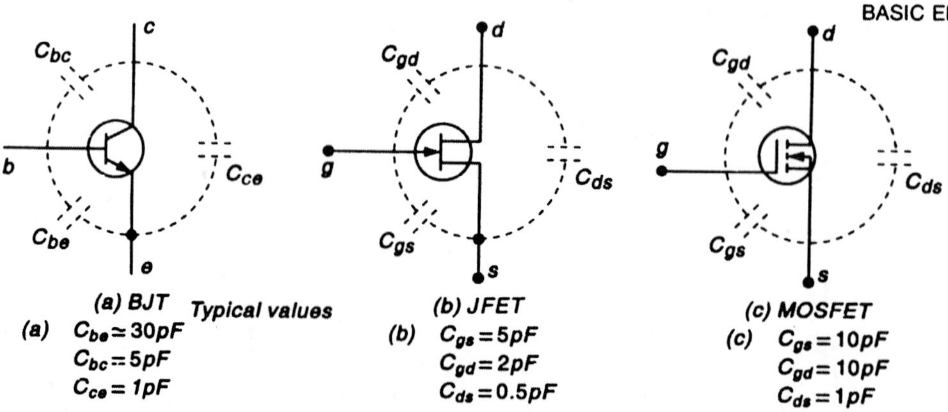

Fig 9.3 Interelectrode capacitances.

phase shift as the signal frequency is increased. In the data sheets for BJTs, the capacitance C_{be} is often mentioned as C_{ib}, the input capacitance and C_{bc} as the output capacitance. For *FETs*, the data sheets C_{iss} gives the input capacitance and C_{rss} as the reverse transfer capacitance. For example a typical *JFET* mentions C_{iss} as 25 pF at f = 1 MHz and C_{rss} = 6 pF at f = 1 MHz. The interelectrode capacitance thus play a major role at high signal frequencies while at low frequency, they behave as **OPEN** circuits. As frequency increases, X_C decreases for the internal device capacitances (C_{be} or C_{gs}) and the voltage gain is decreased.

9.2 MILLER'S THEOREM

Let us consider the circuit diagram of Fig. 9.4. If C is the interelectrode capacitance between the input and the output terminals (C_{bc} or C_{gd}) then for an inverting amplifier, the Miller effect input capacitance will be

$$C_{Mi} = (1 - G_v)\,C \qquad \text{...(9.2)}$$

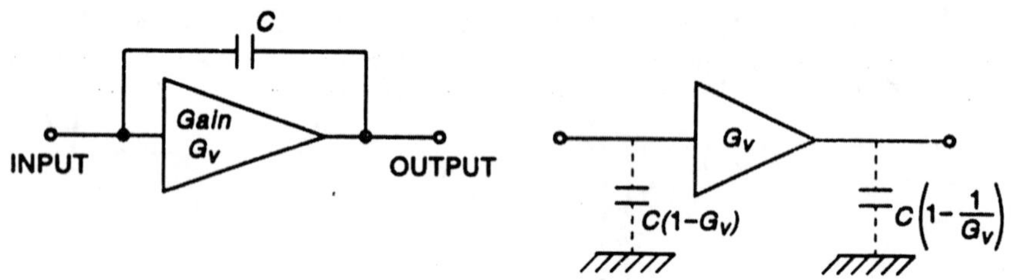

Fig 9.4 Miller capacitances.

and the Miller effect output capacitance will be

$$C_{M_0} = \left(1 - \frac{1}{G_v}\right) C \qquad \text{...(9.3)}$$

where G_v is the negative voltage gain for an inverting amplifier

If $G_v >> 1$, then

$$C_{M_0} = C \qquad \text{...(9.4)}$$

At high frequencies, the amplifier response will depend (a) the interelectrode and wiring capacitances as also the (b) h_{fe} (β) variation with frequency.

Fig. 9.6 shows the circuit diagram of a *BJT* amplifier wherein all capacitances including the parasitic

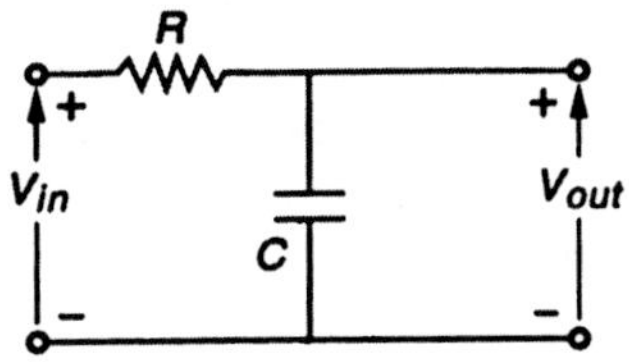

Fig 9.5 The general nature of RC combination for high frequency cut off.

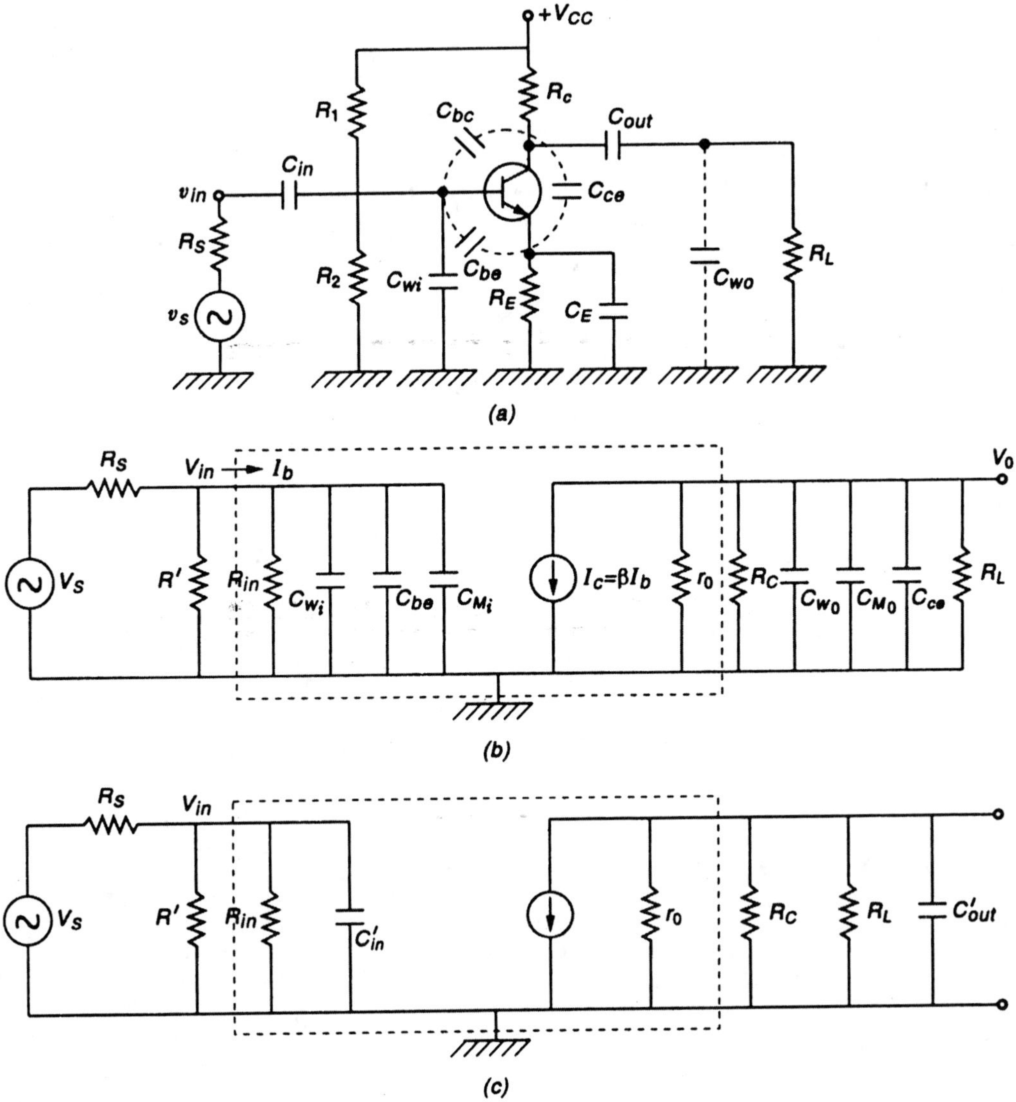

Fig 9.6. (a) Amplifier circuit (b) High frequency response electrical equivalent circuit (c) Simplified equivalent circuit

(In Fig. 9.6(c), $C_{in}' = C_{wi}+C_{be}+C_{Mi}$ and $C_{out}' = C_{wo}+C_{ce}+C_{Mo}$

capacitance and the wiring capacitances have been taken into account. The corresponding electrical equivalent circuit of the amplifier is depicted in Fig. 9.6 (b) and 9.6 (c). The Giacoletto model (hybrid π) high frequency transistor small signal model of a *BJT* is shown in Fig. 9.7.

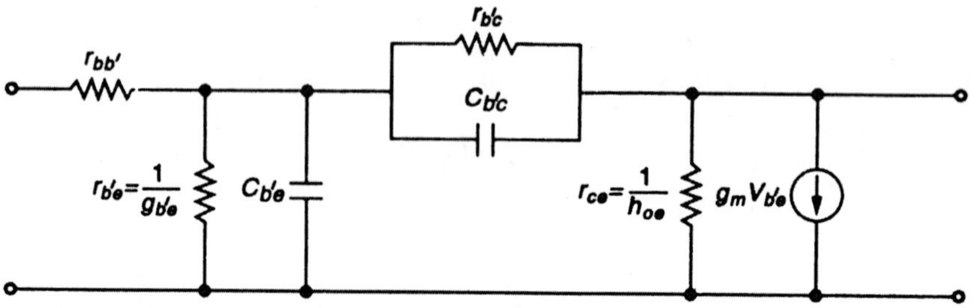

Fig 9.7 Giacoletto (or hybrid π) high frequency transistor small signal model of a BJT.

Figure 9.8 (a) depicts the circuit diagram of an *FET* amplifier wherein all capacitances (the coupling capacitors, by pass capacitors, wiring capacitances and parasitic capacitances) have been considered. The corresponding electrical equivalent circuit is shown in Fig. 9.8 (b).

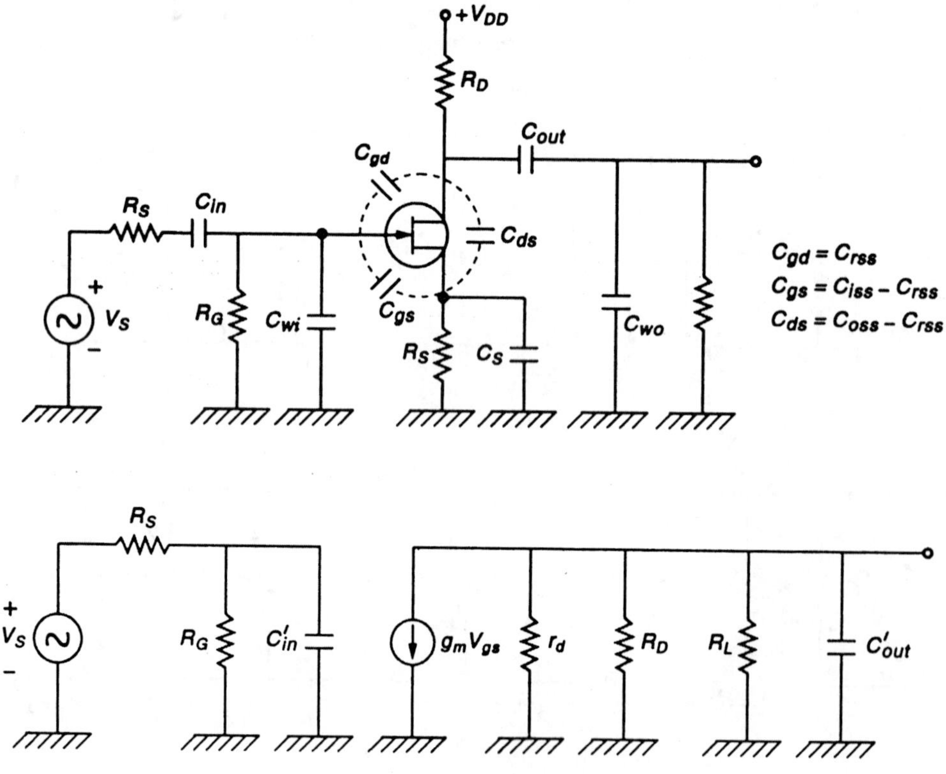

(b)

Fig 9.8 FET amplifier circuit along with interelectrode capacitances (b) Electrical equivalent circuit .

9.3 DECIBELS

The importance of the decibel unit arises from the logarithmic response of the human ear to sound intensity. The decibel unit is specified for measuring the ratio of two power units or two voltage units. The power gain is written in decibel (dB) units by the following well known formula

$$G_p\,(\text{dB}) = 10\log_{10} G_p = 20\log_{10}\frac{\textit{output voltage}}{\textit{input voltage}} \quad \ldots(9.5)$$

where G_p is the power gain G_p is given by

$$G_p = \frac{P_{\text{out}}}{P_{\text{in}}} = \frac{\text{Output power}}{\text{Input power}} \quad \ldots(9.6)$$

The voltage gain is expressed in decibel units by the following formula

$$G_v\,(\text{dB}) = 20\log_{10} G_v = 20\log_{10}\frac{\text{output voltage}}{\text{input voltage}} \quad \ldots(9.7)$$

If the gain is greater than unity, the decibel gain is positive. But if the gain is less than unit (alternation) the dB gain is expressed as negatives.

EXAMPLE 9.1 *Calculate the following quantities is* dB *units.*

(i) $\frac{P_{\text{out}}}{P_{\text{in}}} = 200$ (ii) $\frac{P_{\text{out}}}{P_{\text{in}}} = 100$ (iii) $\frac{P_{\text{out}}}{P_{\text{in}}} = 50$

(iv) $\frac{P_{\text{out}}}{P_{\text{in}}} = 10$ (v) $G_v = \frac{V_{\text{out}}}{V_{\text{in}}} = 20$ (vi) $G_v = 0{\cdot}707$

(vii) $G_p = 0{\cdot}5$ (viii) $G_v = 0{\cdot}25$ (ix) $G_v = 0{\cdot}125$ (x) $G_v = {\cdot}0625$

(xi) $G_v = 2$ (xii) $G_v = 4$ (xiii) $G_v = 8$ (xiv) $G_v = 16$

SOLUTION

(i) $G_p\,(\text{dB}) = 10\log_{10}(200) = 23{\cdot}01$ dB

(ii) $G_p\,(\text{dB}) = 10\log_{10}(100) = 20$ dB

(iii) $G_p\,(\text{dB}) = 10\log_{10}(50) = 16{\cdot}99$ dB

(iv) $G_p\,(\text{dB}) = 10\log_{10}(10) = 10$ dB

(v) $G_v\,(\text{dB}) = 20\log_{10}(20) = 26$ dB

(vi) $G_v\,(\text{dB}) = 20\log_{10}(0{\cdot}707) = -3$ dB

(vii) $G_p\,(\text{db}) = 10\log_{10}(0{\cdot}5) = -3$ dB

(viii) $G_v\,(\text{dB}) = 10\log_{10}(0{\cdot}25) = -12$ dB

(ix) $G_v\,(\text{dB}) = 10\log_{10}(0{\cdot}125) = -18$ dB

(x) $G_v\,(\text{dB}) = 10\log_{10}({\cdot}0625) = -24$ dB

(xi) $G_v\,(\text{dB}) = 10\log_{10}(2) = 6$ dB

(xii) $G_v\,(\text{dB}) = 10\log_{10}(4) = 12$ dB

(xiii) $G_v\,(\text{dB}) = 10\log_{10}(8) = 18$ dB

(xiv) $G_v\,(\text{dB}) = 10\log_{10}(16) = 24$ dB

EXAMPLE 9.2 *The voltage gain of a BJT amplifier in Fig. 9.4 is calculated to be 48. Apply Miller's theorem to find out (i) the input Miller capacitance and (ii) the output Miller capacitance if $C_{bc} = 2$ pF and $C_{be} = 0{\cdot}5$ pF.*

SOLUTION C_{in} (Miller) $= C_{bc}(1 - G_v) = 2$ pF $(1 + 48) = 98$ pF

$$C_{out}\text{ (Miller)} = C_{be}\left(\frac{1 + G_v}{G_v}\right) \simeq 2 \text{ pF}$$

EXAMPLE 9.3 *Find the input and out put Miller capacitance if the voltage gain is 120 instead of 48.*

SOLUTION C_{in} (MIller) = 2 pF (1 + 120) = 242 pF

$$C_{out}\text{ (Miller)} = 2 \text{ pF}\left(\frac{1 + 120}{120}\right) \simeq 2 \text{ pF}$$

In electronic communication equipment, the gain is generally related to decibels. The reference level is normally considered to be 1 mW. When the 1 mW reference level is accepted, the decibel symbol is expressed as dBm. The gain written as

$$(G_p)_{dB_m} = 10 \log_{10} \frac{P_2}{1 \text{ mW}}\bigg|_{600\,\Omega} \qquad \text{...(9.8)}$$

where the resistance associated with 1 mW power is considered to be 600 Ω. The voltage gain in dB is expressed as

$$G_{dB} = 20 \log_{10} \frac{V_2}{V_1}. \qquad \text{...(9.9)}$$

For cascaded stages, the overall voltage gain is expressed as

$$\left|G_{V_{total}}\right| = \left|G_{V_1}\right| \left|G_{V_2}\right| \cdots\cdots \left|G_{V_n}\right| \qquad \text{...(9.10)}$$

where $\left|G_{V_1}\right|$, $\left|G_{V_2}\right|$,... $\left|G_{V_n}\right|$ are the voltage gains for the first stage, second stage ... nth stage of the amplifier. Using logarithmic relations, one can write the overall voltage gain in decibels as

$$(G_V)_{total} = 20 \log_{10} \left|G_{V_1}\right| + 20 \log_{10} \left|G_{V_2}\right| + \ldots + 20 \log_{10} \left|G_{V_n}\right| \qquad \text{...(9.11)}$$

The voltage gain and dB levels are expressed in Table 9.1.

Voltage gain V_{out}/V_{in}	*dB level*
0·25	–12
0·5	–6
1·0	0
2·0	6
10	20
20	26
40	32
100	40
200	46
400	52
1000	60
2000	66
10000	80

EXAMPLE 9.4 *Find the gain magnitude if the decibel gain is 75.*

SOLUTION

$$G_{dB} = 10 \log_{10} \frac{P_2}{P_1} = 75 \text{ dB}$$

$$\log_{10}\left(\frac{P_2}{P_1}\right) = 7{\cdot}5$$

$$\therefore \quad \frac{P_2}{P_1} = 31{,}622{,}777$$

EXAMPLE 9.5 *An amplifier has a wattage rating of 50 W. It is connected to a 16 Ω speaker.*

(i) Find the input power required for full power output if the power gain is 22 dB.

(ii) Find the input voltage for rated output if the amplifier voltage gain is 37 dB.

SOLUTION

(i) $22 = 10 \log_{10} = \dfrac{37 \text{ W}}{P_i}$

or, $\log_{10} \dfrac{37}{P_i} = 2{\cdot}2$

or, $\dfrac{37}{P_i} = 158{\cdot}49$

or, $P_i = 23.35\text{mW}$

(ii) $G_V = 20 \log_{10} \dfrac{V_{out}}{V_{in}} = 37$

or, $\log_{10} \dfrac{V_{out}}{V_{in}} = \dfrac{37}{20} = 1{\cdot}85$

or, $\dfrac{V_{out}}{V_{in}} = 70{\cdot}79$

$$V_{out} = \sqrt{P \,.\, R} = \sqrt{37 \text{ W} \times 16\ \Omega} = 24{\cdot}33 \text{ V}.$$

$$V_{in} = \frac{V_{out}}{70{\cdot}79} = \frac{24{\cdot}33 \text{ V}}{70{\cdot}79} = 343{\cdot}69 \text{ mV}.$$

9.4 GENERAL FREQUENCY RESPONSE

The gain versus frequency response curves for an *RC* coupled, transformer coupled and direct coupled amplifiers are provided in Fig. 9.9 (a), (b) and (c) respectively. It will be noticed that the horizontal scale is logarithmic such that the frequency axis can be extended from low (through midfrequency) to high frequency ranges. For each of the plots, a low, high and mid frequency region has been defined. The main reasons for the fall in gain at low and high frequencies have been indicated in the diagram. The main reason for the reduced gain at low frequencies for an *RC* coupled amplifier is the effect of (i) input and output coupling capacitors and (ii) bypass capacitors. The fall of the voltage gain at high frequencies for the *RC* coupled amplifer takes place due to the internal parasitic capacitances of the network and active devices as also the frequency dependence of the gain of the transistors or *FET*. For the transformer coupled amplifier the fall in the voltage gain (compared to middle range of frequencies) occurs due to the shorting action (across the input terminals of the transformer) of the magnetizing inductive reactance at low frequencies ($X_L = \omega L$). The gain curve must pass through the origin for at $f = 0$ ($\omega = 0$), there is no changing flux to induce a secondary voltage. The high frequency drop in gain occurs due to the stray capacitance between the turns of the primary and secondary windings. In the directly coupled amplifiers, the gain does not fall at low frequencies because any coupling and bypass capacitors are absent. At the high frequencies, the gain curve drops due to the presence of the parasitic capacitances in the circuit and the active devices (*BJT/FET*) as also the gain variation with frequency for the *BJT* or *FET* etc.

In the gain *vs.* frequency response curves, it will be seen that the voltage falls to a value $0{\cdot}707\ (G_V)_{mid}$ at f_1 and f_2, where $(G_v)_{mid}$ is the mid frequency voltage gain. The frequency f_1 is called the

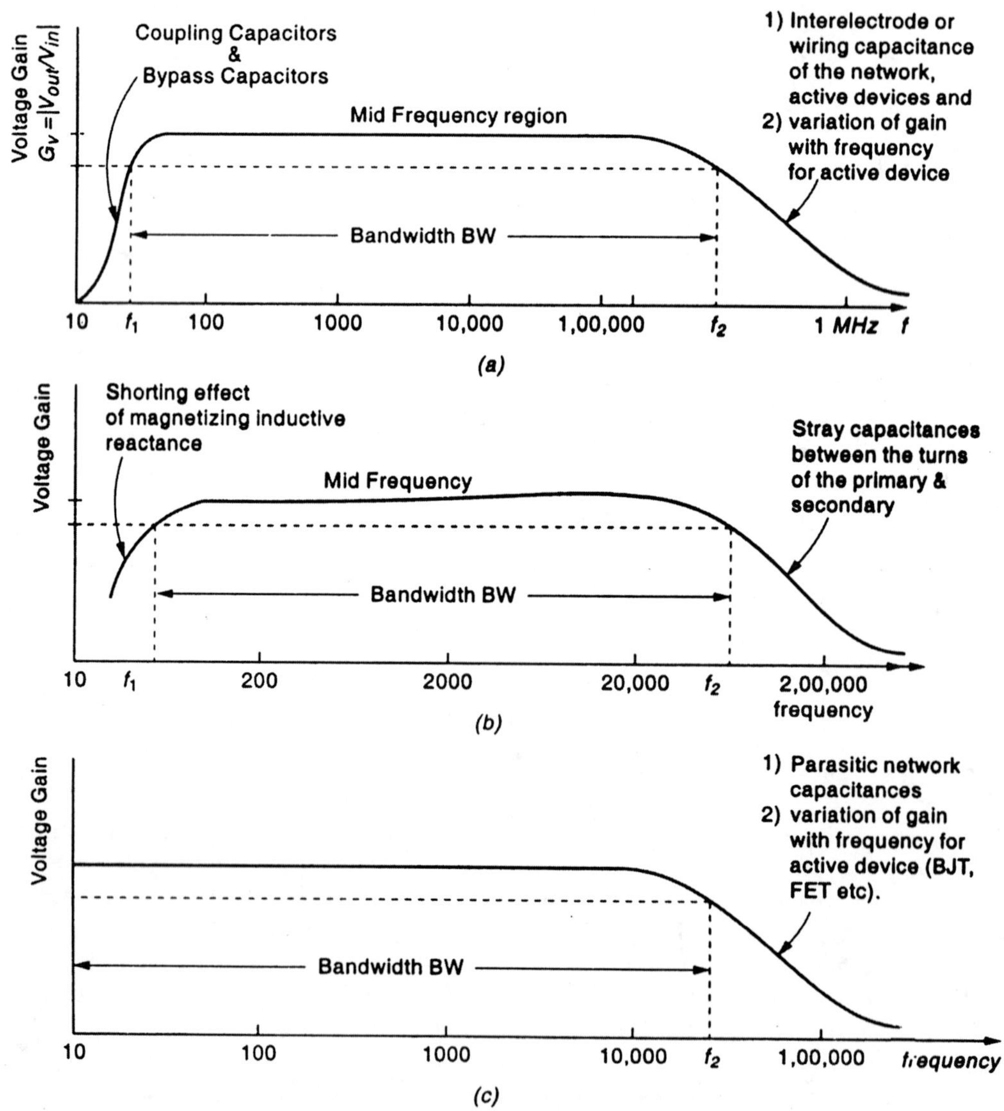

Fig 9.9 Voltage amplification (or gain) vs. frequency curves for (a) RC coupled amplifiers (b) Transformer coupled amplifiers.

lower corner, break, cut-off, band or half-power frequency. Similarly, the frequency f_2 is called the **upper corner, break, cut-off, band or half power frequency.** If $(P_{out})_{mid}$ is the middle band power output,

$$(P_{out})_{mid} = \frac{|V_{out}^2|}{R_{out}} = \frac{|G_{V\,mid}\,V_{in}|^2}{R_{out}} \quad \text{...(9.12)}$$

At f_1 and f_2, the half power or corner frequencies,

$$(P_{out})_{f_1} = (P_{out})_{f_2} = \frac{|0{\cdot}707\,G_{V\,mid}\,V_{in}|^2}{R_{out}}$$

$$= \frac{0{\cdot}5\,|G_{V\,mid}\,V_{in}|^2}{R_{out}} \quad \text{...(9.13)}$$

The bandwidth is expressed as the difference between f_1 and f_2 or

$$BW = f_2 - f_1 \quad \text{...(9.14)}$$

For various industrial and communication applications, a more useful plot of the frequency response is the decibel plot of the normalized gain. This is depicted in Fig. 9.10 for the *RC* coupled *BJT* or *FET* amplifier. The phase angle variation is depicted in Fig. 9.11.

9.5 BODE PLOT FOR LOW FREQUENCIES

Let us consider the simple circuit of Fig. 9.12 so as to study the behaviour at low frequencies.

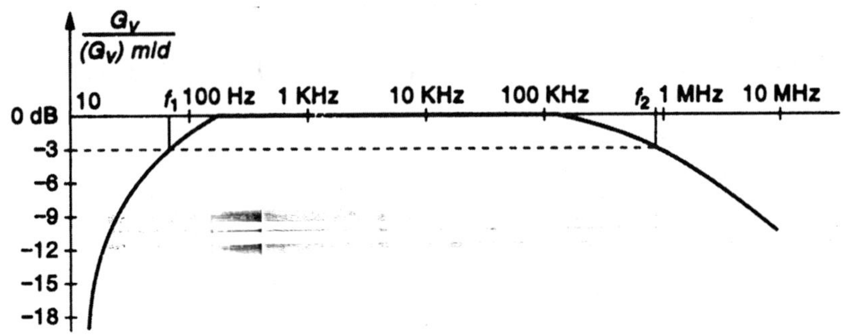

Fig 9.10 Normalised gain (dB) vs. frequency curve for an RC coupled BJT or FET amplifier.

At dc (or $f = 0$), $X_C = \dfrac{1}{2\pi f C} = \dfrac{1}{0} \simeq \infty$ ohms. and the capacitor behaves as an open circuit. Thus $V_{out} = 0$.

At a frequency which is quite high (f_x = high), the capacitative reactance $X_C = \dfrac{1}{2\pi f_x C} = 0$ ohms.

So the capacitor behaves as a short circuit and $V_{out} = V_{in}$. The output voltage would thus vary from 0 to V_{in} as frequency changes from 0 to f_x.

Mathematically, it can be shown that

$$V_{out} = \frac{R}{R + X_C} \cdot V_{in}. \quad \text{...(9.15)}$$

The magnitude part is

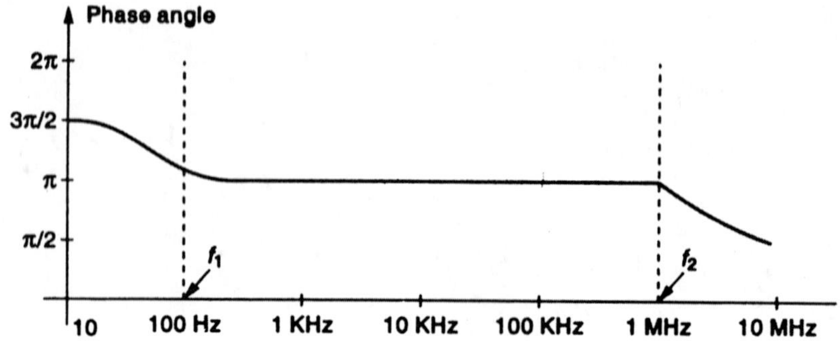

Fig 9.11 Phase angle variation with frequency for an RC coupled amplifier.

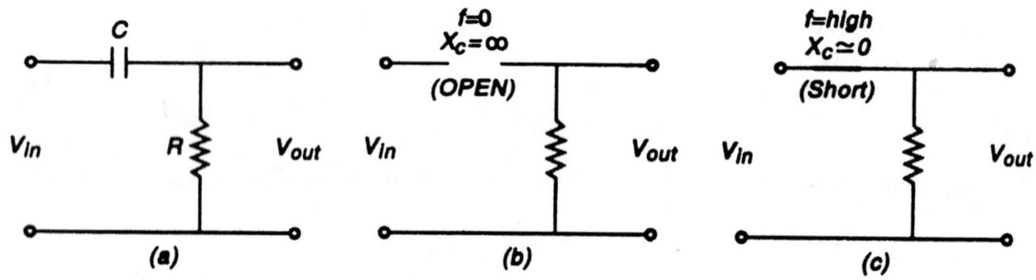

Fig 9.12 (a) A simple RC circuit, (b) RC circuit at $f = 0$ (dc) and (c) RC circuit when frequency is high.

$$|V_{out}| = \frac{R \cdot V_{in}}{\sqrt{R^2 + X_C^2}} \quad \ldots(9.16)$$

If $R = X_C = \frac{1}{2\pi fC}$ at frequency of f_1 (say)

$$f_1 = \frac{1}{2\pi RC} \quad \ldots(9.17)$$

and
$$|G_V| = \frac{V_{out}}{V_{in}} = \frac{R \cdot V_{in}}{\sqrt{R^2 + R^2}} \cdot \frac{1}{V_{in}}$$

$$= \frac{1}{\sqrt{2}} = 0{\cdot}707 \quad \ldots(9.18)$$

and
$$|G_V|_{db} = 20 \log_{10} |G_V| = 20 \log_{10} \frac{1}{\sqrt{2}} = -3 \text{ db}$$

Thus the gain falls by 3 db at $f = f_1$. This frequency f_1 is thus the lower corner frequency or cut-off frequency.

If
$$G_V = \frac{V_{out}}{V_{in}} = \frac{R}{R - jX_C} = \frac{1}{1 - j\left(\frac{1}{2\pi fCR}\right)} \quad \ldots(9.19)$$

or,
$$G_V = \frac{V_{out}}{V_{in}} = \frac{1}{\sqrt{1+\left(\frac{f_1}{f}\right)^2}} \angle \tan^{-1} f_1/f \quad \ldots(9.20)$$

In equation (9.20), the magnitude of G_V is

$$|G_V| = \frac{1}{\sqrt{1+\left(\frac{f_1}{f}\right)^2}} \quad \ldots(9.21)$$

and the phase angle is

$$\phi = \tan^{-1}\left(\frac{f_1}{f}\right) \quad \ldots(9.22)$$

Thus
$$|G_V|_{db} = 20 \log_{10} \frac{1}{\sqrt{1+\left(\frac{f_1}{f}\right)^2}}$$

$$= -20 \log_{10}\left[\sqrt{1+\left(\frac{f_1}{f}\right)^2}\right]$$

$$= -10 \log_{10}\left[1+\left(\frac{f_1}{f}\right)^2\right] \quad \ldots(9.23)$$

$$G_V\Big|_{dB} = -20 \log_{10}\left(\frac{f_1}{f}\right) \quad \ldots(9.24)$$

From the above equations, we get

$$G_V = -20 \log_{10} 1 = 0 \text{ at } f = f_1$$

$$G_V = -20 \log_{10} 2 = -6 \text{ dB at } f = \frac{f_1}{2}$$

$$G_V = -20 \log_{10} 4 = -12 \text{ dB at } f = \frac{f_1}{4}$$

$$G_V = -20 \log_{10} 8 = -18 \text{ dB at } f = \frac{f_1}{8}$$

$$G_V = -20 \log_{10} 10 = -20 \text{ dB at } f = \frac{f_1}{10}$$

$$G_V = -20 \log_{10} 20 = -26 \text{ dB at } f = \frac{f_1}{20}$$

The plot of the above values from $f_1/20$ to f_1 is shown in Fig. 9.13

By the term Bode plot we mean **a plot of dB gain versus frequency on a semilog graph paper.**

It should be noted from the graph in Fig. 9.13, that the change in frequency by a factor of 2, which is called one **Octave**, gives rise to a 6 dB change. This is seen for the values between $\frac{f_1}{2}$ to f_1 or from $\frac{f_1}{4}$ to $\frac{f_1}{2}$. The curve is said to have a slope of – 6 dB/octave.

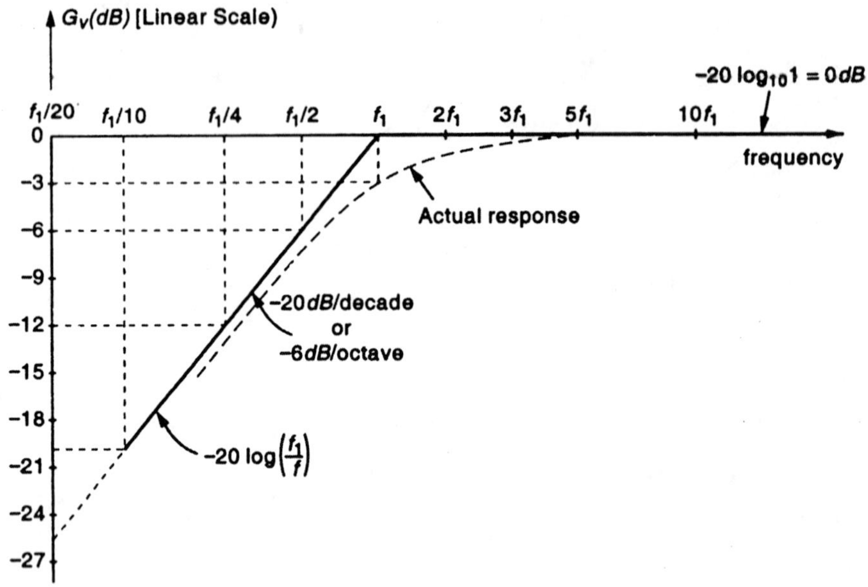

Fig 9.13. Bode plot for the amplifier in semilog scale

It should also be apparent from the plot in Fig. 9.13, that if frequency changes by a factor of 10, which is called one **decade** (say from $\frac{f_1}{10}$ to f_1) the gain changes by 20 dB. The curve is said to possess a slope of – 20 dB/decade. In fact the slope of – 6 dB/octave is the same as – 20 dB/decade. (Remember one octane is 2 to 1 change in frequency)

Thus, for the function in equation (9.24), the curve can be easily plotted by first drawing the two asymptotes (0 dB line along the *X*-axis through f_1 and the – 20 dB/decade also called – 6 dB/octave) line through f_1. It should be kept in mind that at $f = f_1$, the G_V has a value of – 3 dB. The curve can then be drawn.

For the *R-C* circuit, the phase angle is found from

$$\theta = \tan^{-1}\left(\frac{f_1}{f}\right) \qquad \text{...(9.25)}$$

If $f << f_1$,

$$\theta = \tan^{-1}\left(\frac{f_1}{f}\right) \to 90°$$

If $f_1 = 200f,\ \theta = \tan^{-1}\left(\frac{f_1}{f}\right) = \tan^{-1}(200) = 89{\cdot}7°$

$f_1 = 100f,\ \theta = \tan^{-1}\left(\frac{f_1}{f}\right) = \tan^{-1}(100) = 89{\cdot}42°$

If $f_1 = 50f,\ \theta = \tan^{-1}(50) = 88{\cdot}85°$

$f_1 = 20f,\ \theta = \tan^{-1}(20) = 87{\cdot}14°$

$f_1 = 10f,\ \theta = \tan^{-1}(10) = 84{\cdot}29°$

$$f_1 = f, \theta = \tan^{-1}(1) = 45°$$
$$f_1 = f/50, \theta = \tan^{-1}(\cdot 02) = 1\cdot 14°$$
$$f_1 = f/100, \theta = \tan^{-1}(\cdot 01) = 0\cdot 57°$$
$$f_1 = f/200, \theta = \tan^{-1}(\cdot 005) = 0\cdot 28°$$

Thus if $f_1 << f, \theta = \tan^{-1}\left(\frac{f_1}{f}\right) \to 0°$.

The phase response of the *R-C* circuit is depicted in Fig. 9.14.

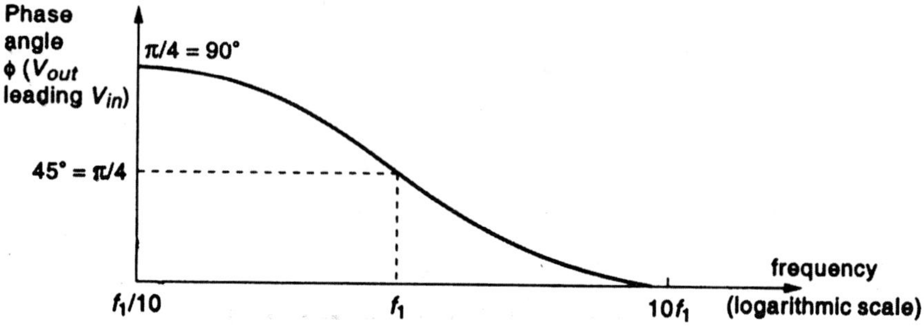

Fig 9.14. Plot of phase response for an RC coupled circuit

EXAMPLE 9.6 *Find the critical frequency for the bypass network of Fig. 9.15.*

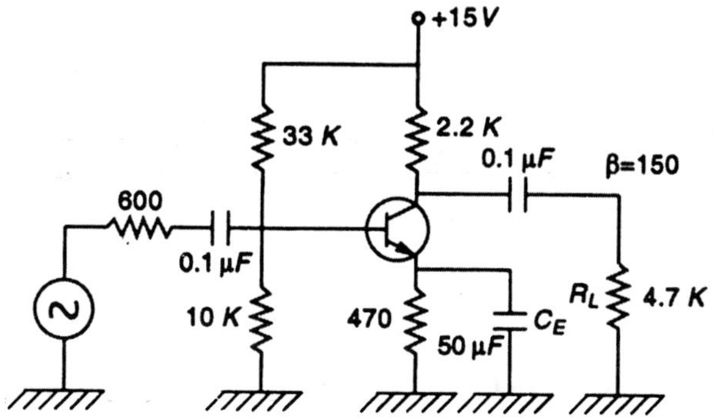

Fig 9.15. RC coupled network (Ref. Example 9.6)

SOLUTION Let us Thevenize the base circuit (looking from the base towards the input source)

$$R_{Th} = R_1 \| R_2 \| R_S = 33\text{ k} \| 10\text{ k} \| 600\ \Omega$$

or
$$\frac{1}{R_{TH}} = \frac{1}{33\text{ k}} + \frac{1}{10\text{ k}} + \frac{1}{600}$$
$$= 3\cdot 03 \times 10^{-5} + 10 \times 10^{-5} + 1\cdot 6666 \times 10^{-3} = 1\cdot 79696 \times 10^{-5}$$

or
$$R_{TH} = 556\ \Omega$$

The resistance looking into the emitter is R_{in} (emitter) $= r_e + \frac{R_{TH}}{\beta} = 4\ \Omega + \frac{556}{150} = 7\cdot 7\ \Omega$

The resistance of the equivalent *RC* network is

$$R_{in}\text{ (emitter)} \parallel R_E = 7{\cdot}7\ \Omega \parallel 470 = 7{\cdot}58\ \Omega$$

The critical frequency of the bypass network is

$$fc = \frac{1}{2\pi\,(R_{in}\text{ (emitter)} \parallel R_E)\,.\,C_E} = \frac{1}{2\pi \times 7{\cdot}58 \times 50 \times 10^{-6}}$$

$$= 419 \text{ Hz}$$

EXAMPLE 9.7 *Find the corner frequency for the bypass network of Fig. 9.16 (Assume $r_e = 11{\cdot}5\ \Omega$)*

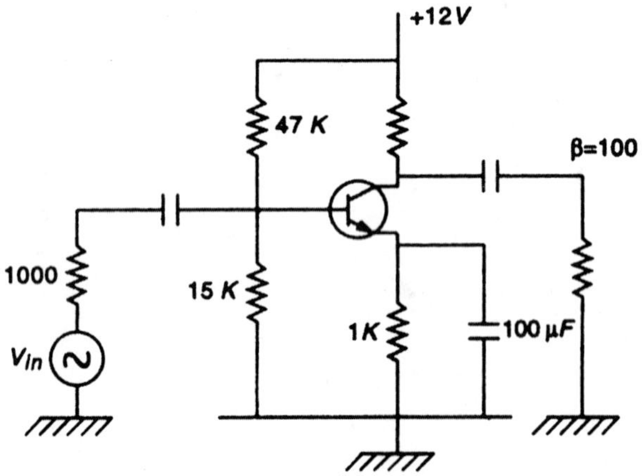

Fig 9.16. Reference Example 9.7

SOLUTION Let us Thevenize the base circuit (looking from the base towards the input source)

$$R_{TH} = R_1 \parallel R_2 \parallel R_5 = 47\text{ k} \parallel 15\text{ K} \parallel 1\text{ k}$$

or,

$$\frac{1}{R_{TH}} = \frac{1}{47\text{ k}} + \frac{1}{15\text{ k}} + \frac{1}{1\text{ k}}$$

$$= 2{\cdot}13 \times 10^{-5} + 6{\cdot}66 \times 10^{-5} + 100 \times 10^{-5}$$

$$= 108{\cdot}79 \times 10^{-5}$$

$$R_{TH} = 919{\cdot}2\ \Omega$$

The resistance looking towards the emitter is

$$R_{in}\text{ (emitter)} = r_e + \frac{R_{TH}}{\beta} = 11{\cdot}5 + \frac{919{\cdot}2}{100} = 11{\cdot}5 + 9{\cdot}19$$

$$= 20{\cdot}69\ \Omega$$

The resistance of the equivalent *R-C* network is

$$R_{in}\text{ (emitter)} \parallel R_e = 20{\cdot}69\ \Omega \parallel 1000\ \Omega = 20{\cdot}27\ \Omega$$

The critical frequency of the bypass network is

$$fc = \frac{1}{2\pi\,(R_{in}\text{ (emitter)} \parallel R_E)\,.\,C_E} = \frac{1}{2\pi \times 20{\cdot}27 \times 100 \times 10^{-6}}$$

$$= 78{\cdot}52\text{Hz.}$$

9.6 TOTAL LOW FREQUENCY RESPONSE

To determine the total frequency response, we have to consider the net effect of the three networks in Fig. 9.17.

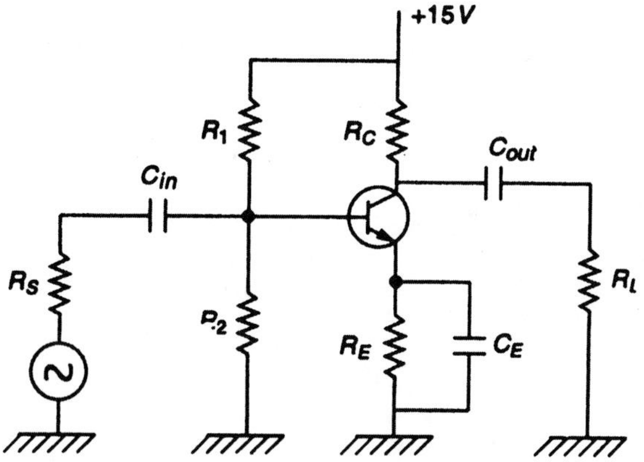

Fig 9.17. RC coupled network

Each of the networks has its own critical frequency. Further the critical frequency of the three networks may not be equal at all. The *RC* network that has a critical frequency higher than the other two, that is the dominant network. The dominant network critical frequency f_{c_1} is the frequency at which the overall gain of the amplifier begins to decrease at – 6 db/octave or – 20 db/decade roll off.

As the frequency is decreased from the mid frequency range (0 db), the first corner frequency takes place at fc_1 where the curve rolls off at – 20 db/decade. This roll off is continued till fc_2. At fc_2, the second corner frequency, the curve begins to roll off at – 40 db/decade and maintaints it till f is decreased to fc_3. fc_3 is the third corner frequency. At fc_3, the curve slope changes to – 60 db/decade. Normally fc_1 is the corner frequency corresponding to the input network, fc_2 is the corner frequency corresponding to the output network and fc_3 is the corner frequency corresponding to the bypass network.

If all the three corner (cut-off) frequencies coincide $fc_1 = fc_2 = fc_3 = fc$, the curve begins to roll off at – 60 db/decade. At fc, the actual response curve is – 9 db below the mid frequency gain of 0 db.

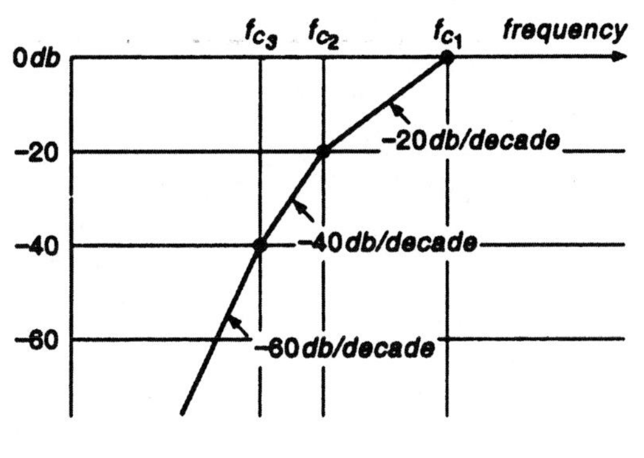

Fig 9.18. Bode plot for total frequency response at low frequencies

EXAMPLE 9.8 *Plot the total frequency response for the amplifier is Fig. 9.15.*

SOLUTION **fc (input) network**

$$R_{in} = R_1 \parallel R_2 \parallel \beta r_e = 33\text{ k} \parallel 10\text{ k} \parallel 600 = 556{\cdot}5\ \Omega$$

Thus
$$f_c\text{ (input)} = \frac{1}{2\pi (R_s + R_{in}) \cdot C_{in}} = \frac{1}{2\pi (600 + 556{\cdot}5) \times 0{\cdot}1 \times 10^{-6}}$$

$$= \frac{1}{2\pi (1156{\cdot}6) \times 0{\cdot}1 \times 10^{-6}} = 1376\text{ Hz.}$$

fc (bypass network)

$$R_{TH} = R_1 \parallel R_2 \parallel R_s = 33\text{ k} \parallel 10\text{ k} \parallel 600\ \Omega$$
$$= 556{\cdot}5\ \Omega$$

The resistance looking into the emitter

$$R_{in}\text{ (emitter)} = 7{\cdot}7\ \Omega$$

$$fc\text{ (bypass)} = \frac{1}{2\pi (R_{in(emitter)} + R_E)\, C_E}$$

$$= \frac{1}{2\pi (7{\cdot}58\ \Omega) \times 50 \times 10^{-6}}$$

$$= 458\text{ Hz.}$$

fc (output)

$$(fc)_{output} = \frac{1}{2\pi (R_C + R_L)\, C_{out}}$$

$$= \frac{1}{2\pi (2{\cdot}2\text{ k} + 4{\cdot}7\text{ k}) \times 0{\cdot}1\ \mu F}$$

$$= \frac{1}{2\pi (6{\cdot}9\text{ k}) \times 0{\cdot}1\ \mu F}$$

$$= 231\text{ Hz}$$

The composite low frequency Bode Plot is given in Fig. 9.19 below.

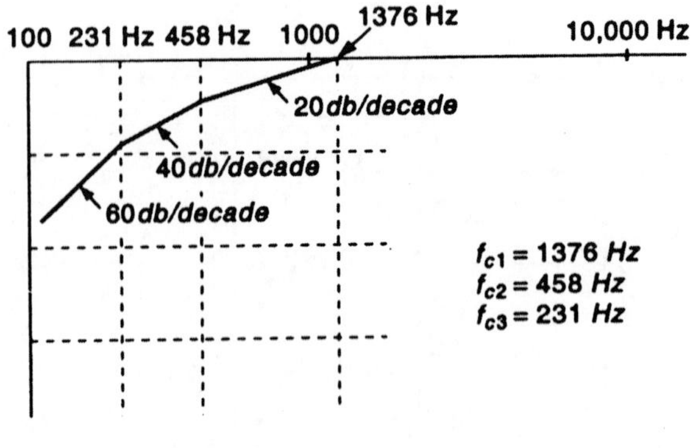

Fig 9.19. Composite low frequency Bode plot for Example 9.8

9.7 HIGH FREQUENCY RESPONSE OF AMPLIFIERS

At the low frequencies, the coupling and bypass capacitors significantly affect the voltage gain. At the mid range of frequencies, the effect of these capacitors on the voltage gain is insignificant. However as frequency is increased, a high range of frequencies is reached where the interelectrode capacitance of the active device as also the frequency dependence of the gain β of the amplifier, affect the frequency response of the amplifier. The gain rolls off at high frequencies. This is true for both *R-C* coupled and direct coupled amplifiers.

From Fig. 9.6 (c), we can develop the high frequency input *R-C* network for the *R-C* coupled amplifier of Fig. 9.6 (a). This is shown in Fig. 9.20. As frequency increases, the capacitative reactance of C_{in}' decreases. Thus, less voltage is developed across C_{in}' and more voltage appears across R_{TH}. At the cut off frequency f_{ch},

$$R_{TH} = X_{C_{in}'} \qquad \ldots(9.26)$$

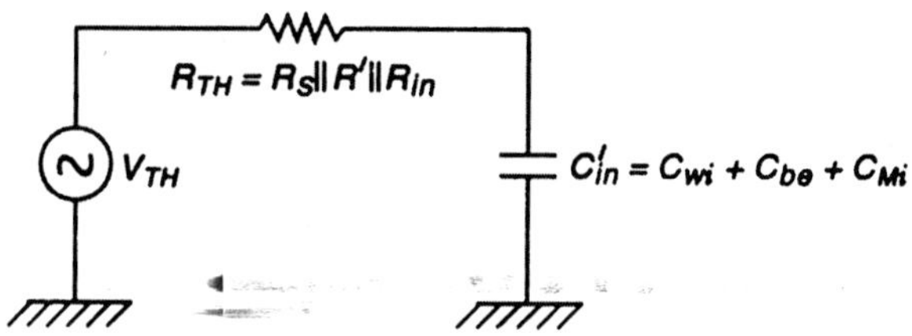

Fig 9.20. High frequency input RC network of Fig. 9.6(c)

and the gain is 3 db less than the mid band gain.

or,
$$\frac{1}{2\pi f_{ch}\, C_{in}'} = R_{TH} \qquad \ldots(9.27)$$

and
$$f_{ch} = \text{Critical frequency}$$

$$= \frac{1}{2\pi\, C_{in}'\, R_{TH}} \qquad \ldots(9.28)$$

$$= \frac{1}{2\,\pi\, C_{in}'(R_s \parallel R' \parallel R_{in}} \qquad \ldots(9.29)$$

$$= \frac{1}{2\pi(C_{wi} + C_{be} + C_{Mi})\,(R_s \parallel R' \parallel R_{in})} \qquad \ldots(9.30)$$

As the frequency f is increased above f_1, the gain rolls off at – 20 db/decade just as in the case of low frequencies. This is depicted in Fig. 9.21.

The input *R-C* network of Fig. 9.20 also acts as a lag network because the output is taken across the capacitor. The phase angle ϕ between the output and the input is given by

$$\tan\phi = \frac{R_{TH}}{X_{C_{in}'}} \qquad \ldots(9.31)$$

or,
$$\phi = \tan^{-1}\left(\frac{R_{TH}}{X_{C_{in}'}}\right)$$

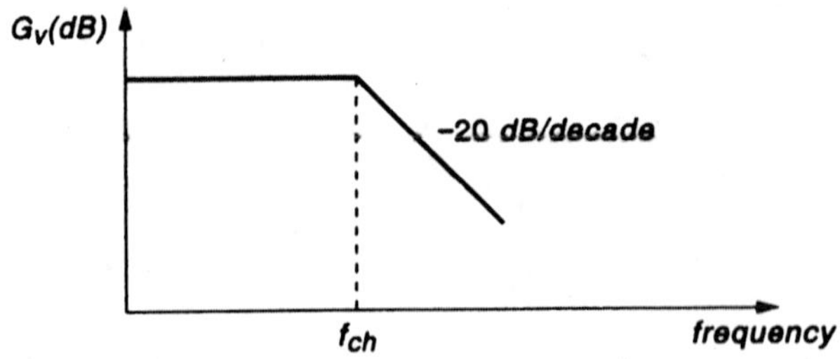

Fig 9.21. Gain roll off at high frequency

$$= \tan^{-1}\left(\frac{R_s \parallel R' \parallel R_{in}}{X_{C_{in}'}}\right) \qquad \ldots(9.32)$$

$$\text{where } X_{C_{in}'} = \frac{1}{2\pi f(C_{wi} + C_{be} + C_{Mi})} \qquad \ldots(9.33)$$

At $f = fch$, the critical frequency, the output is lagging the input voltage by 45°. The value of ϕ increases as f increases above f_{ch} and becomes 90° at f sufficiently high.

EXAMPLE 9.9 *For the circuit in Fig. 9.22, determine high frequency input R-C network and also find the critical (cut-off) frequency. For the transistor shown assume* $\beta = 100$, $C_{be} = 15\ pF$, $C_{bc} = 2\ pF$, $C_{wi} = 5\ pF$.

SOLUTION

$$V_B = \left(\frac{5{\cdot}6\text{ k}}{5{\cdot}6\text{ k} + 15\text{ k}}\right) \times 12 = 3{\cdot}11\text{ V}.$$

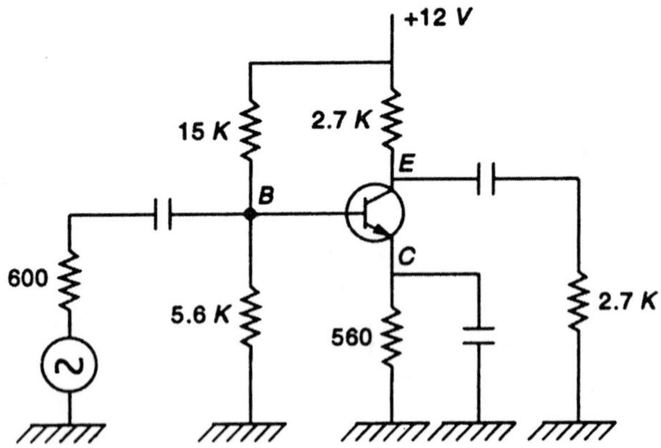

Fig 9.22. Reference Example 9.9

$$V_E = V_B - V_{BE} = 3{\cdot}11 - 0{\cdot}7 = 2{\cdot}41\text{ V}.$$

$$I_E = V_E/R_E = \frac{2{\cdot}41\text{ V}}{560} = 4{\cdot}3\text{ mA}.$$

$$r_e = \frac{25\text{ mV}}{4{\cdot}3} = 5{\cdot}8\ \Omega$$

So

$$R_{TH} = R_s \parallel R' \parallel \beta r_e$$

$$= 600 \parallel 15\text{ k} \parallel 5{\cdot}6\text{ k} \parallel 580$$

or,
$$\frac{1}{R_{TH}} = \frac{1}{600} + \frac{1}{15\text{ k}} + \frac{1}{5{\cdot}6\text{ k}} + \frac{1}{580}$$
$$= 1{\cdot}667 \times 10^{-3} + 6{\cdot}66 \times 10^{-5} + 17{\cdot}86 \times 10^{-5} + 1{\cdot}724 \times 10^{-3}$$
$$= 3{\cdot}636 \times 10^{-3}$$
$$R_{TH} = 274{\cdot}25\ \Omega$$

The midrange gain of the amplifier is
$$(G_V)_{mid} = \frac{R_L'}{r_e} = \frac{1{\cdot}35\text{ k}}{5{\cdot}8} = 232{\cdot}76$$

Using Miller's theorem,
$$C_{in}\text{ (Miller)} = C_{bc}\,(1 + 232{\cdot}76)$$
$$= 2\,(1 + 232{\cdot}76) = 467{\cdot}5\text{ pF}$$

The total input capacitance is
$$= C_{wi} + C_{be} + C_{in}\text{ (Miller)}$$
$$= 5 + 15 + 467{\cdot}5 = 487{\cdot}5\text{ pF.}$$

The high frequency input R–C network is shown in Fig. 9.23.

The critical frequency is

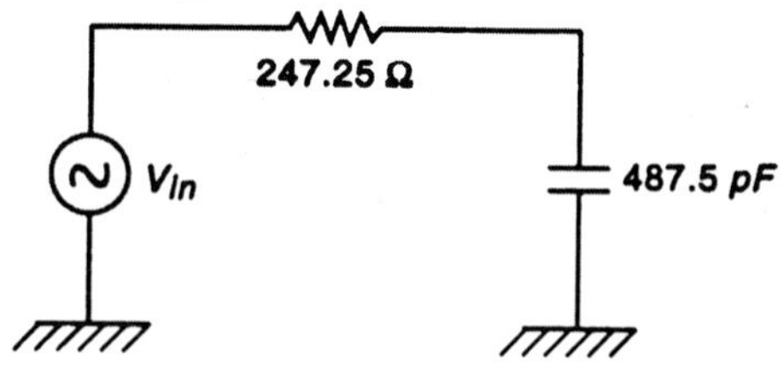

Fig 9.23. Reference Example 9.9

$$fc = \frac{1}{2\pi\,(274{\cdot}25\ \Omega)\,(487{\cdot}5\text{ pF})}$$
$$= 1{\cdot}19\text{ MHz.}$$

EXAMPLE 9.10 *Find the phase shift of the input R-C network in Example 9.9 at f = 1·19 MHz, 2·38 MHz, 4·76 MHz.*

SOLUTION From equation (9.32)
$$\phi = \tan^{-1}\left(\frac{R_{TH}}{X_{C_{in}}'}\right)$$
$$= \tan^{-1}\left(\frac{274{\cdot}25}{X_{C_{in}}'}\right)$$

At $f = 1{\cdot}19$ MHz, $X_{C_{in}}' = \dfrac{1}{2\pi \times 1{\cdot}19 \times 10^6 \times 487{\cdot}5\text{ pF}} = 274{\cdot}25\ \Omega$
$$\phi = \tan^{-1}\left(\frac{274{\cdot}25}{274{\cdot}25}\right) = \tan^{-1}(1) = 45°$$

At $f = 2{\cdot}38$ MHz,

$X_{C_{in}}' = 137{\cdot}125\ \Omega$

$$\phi = \tan^{-1}\left(\frac{274{\cdot}25}{137{\cdot}125}\right) = \tan^{-1}(2) = 63{\cdot}43°$$

At $f = 4{\cdot}76$ MHz, $X_{C_{in}}' = 68{\cdot}562\ \Omega$

$$\phi = \tan^{-1}\left(\frac{274{\cdot}25}{68{\cdot}562}\right) = \tan^{-1}(4) = 75{\cdot}96°$$

Output *R–C* network

The output $R–C$ network can also be developed from Fig. 9.6 (c) as shown in Fig. 9.24.

For the output network of Fig. 9.24, the critical frequency would occur if

$$R_L' = X_{Cout}' \quad \ldots(9.34)$$

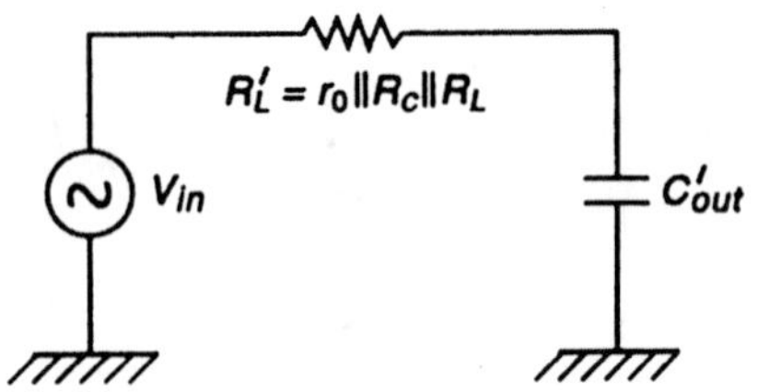

Fig 9.24. Output RC network for an RC coupled amplifier

$$fc = \frac{1}{2\pi R_L' C_{out}'} \quad \ldots(9.35)$$

At the ctitical frequency, the voltage gain is down by 3 dB. The gain will roll off further at – 20dB/decade.

The phase shift introduced will be

$$\phi = \tan^{-1}\left(\frac{R_L'}{C_{out}'}\right) \quad \ldots(9.35)$$

EXAMPLE 9.11 *Calculate the critical frequency for the amplifier in Example 9.9 (Fig. 9.22) corresponding to the output R-C network. [Assume C_{w_0} = 1 pF]*

SOLUTION The Miller output capacitance is

$$C_{out}\,(\text{Miller}) = C_{bc}\left(\frac{232{\cdot}76 + 1}{232{\cdot}76}\right)$$

$$\simeq 2 \text{ pF.}$$

$$C_{out}' = 1 + 2 = 3 \text{ pF.}$$

$$R_L' = R_C \,||\, R_L$$

$$= 2{\cdot}7\text{ k} \,||\, 2{\cdot}7\text{ k} = 1{\cdot}35\text{ k.}$$

The equivalent $R–C$ output network will be as shown in Fig. 9.25

The critical frequency is

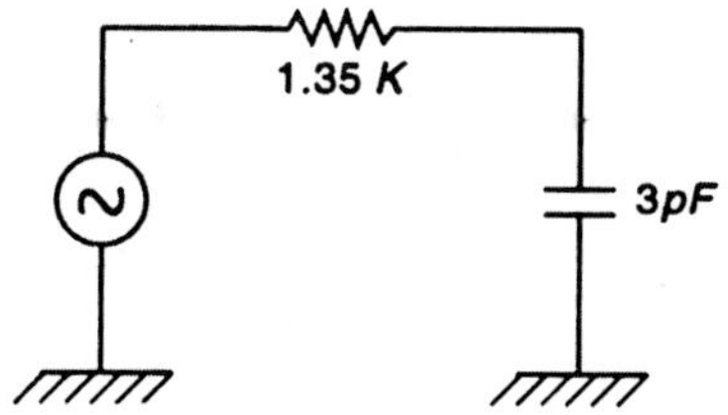

Fig 9.25. Reference Example 9.11

$$fc = \frac{1}{2\pi R_L' \cdot C_{out}'}$$

$$= \frac{1}{2\pi (1.35 \text{ k}) \times 3 \text{ pF}}$$

$$= 39.3 \text{ MHz.}$$

9.8 TOTAL HIGH FREQUENCY RESPONSE

Fig. 9.26 shows the high frequency response of an R–C coupled amplifier where both the effects of the input R–C network and output R–C network have been taken into consideration. The gain will be down by – 3 dB

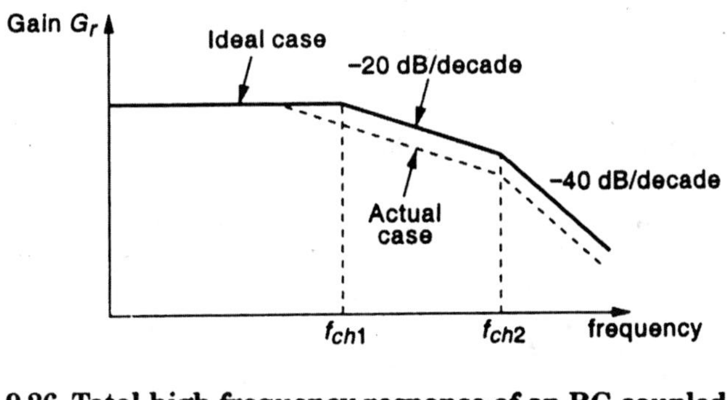

Fig 9.26. Total high frequency response of an RC coupled amplifier

at $f = f_{ch_1}$ and roll off at – 20 dB/decade till f_{ch_2}. After f_{ch_2} the gain will roll off at – 40 dB/decade. The frequency f_{ch_1} corresponding to network 1 is the dominant critical frequency and the input R-C network is the dominant one.

9.9 BAND WIDTH

The composite Bode plot for an ideal $R - C$ coupled amplifier is depicted in Fig. 9.27.

The generalized Bode Plot for a ideal amplifier of Fig. 9.27 depicts that the two dominant critical frequencies are f_{cl_1} and f_{ch_1}. **They are also termed as lower cut off (half power) frequency and upper cut off (half power) frequency respectively. The Bandwidth of the amplifier is the range of the frequencies where the gain is constant. It is normally taken as the difference between the two dominant cut off frequencies (or half power frequencies). Thus**

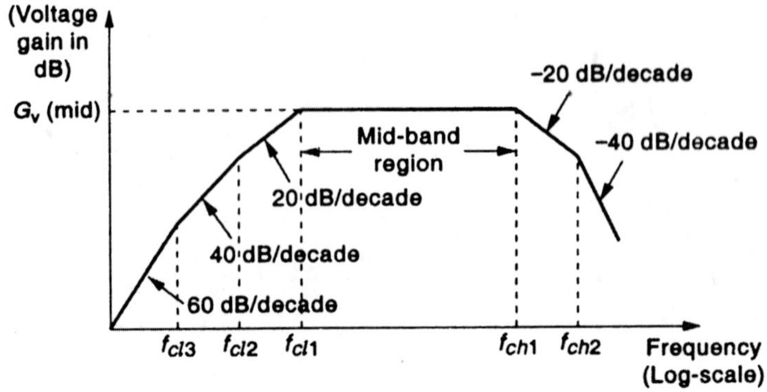

Fig 9.27. Generalized bode plot (Ideal case)

$$BW = \text{Bandwidth}$$

$$= fch_1 - fcl_1 \quad \ldots(9.36)$$

EXAMPLE 9.12 *What is the bandwidth of an amplifier if dominant critical frequencies are 20 Hz and 5 kHz.*

SOLUTION
$$BW = f_{ch_1} - f_{cl_1}$$
$$= 5000 \text{ Hz} - 20 \text{ Hz}$$
$$= 4980 \text{ Hz.}$$

Gain-Bandwidth Product

A very important characteristic of an amplifier is the gain bandwidth product. It is the product of the mid band gain and the bandwidth.

$$\text{Gain-}BW \text{ product} = (G_V)_{\text{mid}} \times BW$$

$$= (G_V)_{\text{mid}} \times (f_{ch_1} - f_{cl_1}) \quad \ldots(9.37)$$

If $f_{ch_1} >> f_{cl_1}$, then the gain-bandwidth product is

$$\text{Gain}-BW = (G_V)_{\text{mid}} \times f_{ch_1} \quad \ldots(9.38)$$

EXAMPLE 9.13 *A manufacturer specifies the f_T of a transistor as being equal to 150 MHz. If the midband voltage gain is 25, calculate the bandwidth that can be achieved.*

SOLUTION
$$f_T = G_{V(\text{mid})} \times BW$$

or, $$BW = \frac{f_T}{G_{V(\text{mid})}} = \frac{150 \text{ MHz}}{25} = 6 \text{ MHz.}$$

9.10. FREQUENCY RESPONSE OF FET AMPLIFIERS

Let us consider the *D-MOSFET* amplifier of Fig. 9.28.

With the capacitative coupling C_{C_1} of the input capacitor, the amplifier forms an input *RC* network. Another *RC* network is formed with capacitor C_{C_2}. With zero bias, if the capacitative reactance could be neglected, the midband voltage gain could be written as

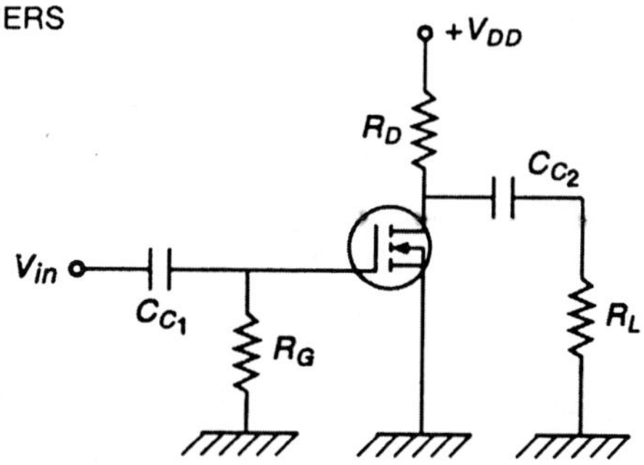

Fig 9.28. *D-MOSFET* amplifier (zero bias)

$$G_{V(\text{mid})} = g_m R_L' \qquad \text{...(9.39)}$$

where g_m is the transconductance and R_d is the a.c. drain resistance. For determining the low frequency response of the *FET* amplifier, the input *RC* and output *RC* networks could be considered and their effect determined.

Input *RC* network

Let us consider the input *RC* network of Fig. 9.29. With increasing frequency, the reactance of the coupling capacitor C_{C_1} decreases. When $R_{in} = X_{C_{C_1}}$, the gain is 3 dB less compared to the midband value.

Further the gain rolls off at 20 dB/decade. The critical frequency occurs when

$$R_{in} = X_{C_{C_1}} \qquad \text{...(9.40)}$$

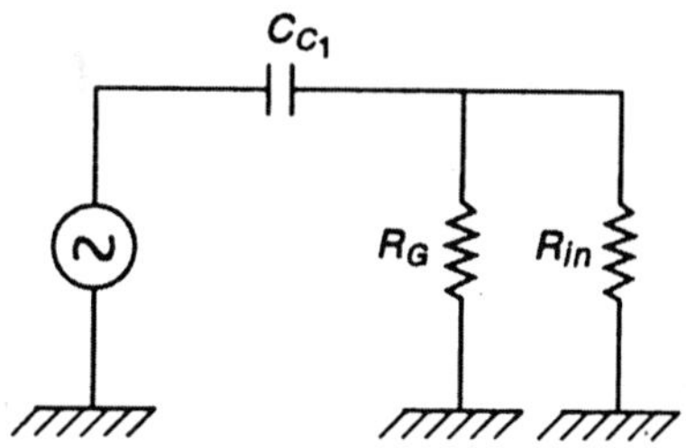

Fig 9.29. Input RC network for the *FET* amplifier of Fig. 9.28

and the critical frequency is

$$fc = \frac{1}{2\pi R_{in} C_{C_1}} \qquad \text{...(9.41)}$$

where R_{in} is the parallel combination of R_G and $R_{in\,(gate)}$ and

$$R_{in\,(gate)} = \left|\frac{V_{GS}}{I_{DSS}}\right| \qquad \text{...(9.42)}$$

Thus $$fc = \frac{(R_G + R_{in\,(gate)})}{2\pi R_G \,.\, R_{in\,(gate)} \cdot C_{C_1}} \qquad \text{...(9.43)}$$

and the phase shift $\phi = \tan^{-1}\left(\frac{X_{C_{C_1}}}{R_{in}}\right)$ ...(9.44)

Output *R-C* network

The output $R-C$ network is formed between C_{C_2}, the output coupling capacitor and the resistance looking in at the drain. This is depicted in Fig. 9.30.

For the output RC network, the critical frequency is

$$fc = \frac{1}{2\pi (R_D + R_L) C_{C_2}} \qquad \ldots(9.45)$$

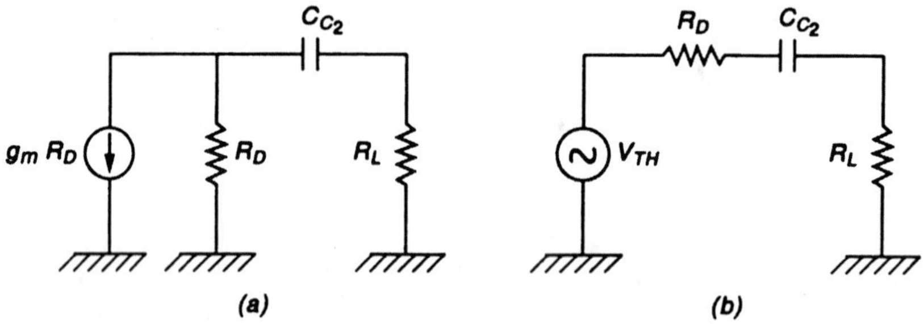

Fig 9.30. (a) Output RC network for the *FET* amplifier of Fig. 9.28 (b) Electrical equivalent circuit

and the phase shift ϕ is given by

$$\tan \phi = \frac{1}{wC_{C_2} (R_D + R_L)} \qquad \ldots(9.46)$$

At the critical frequency the value of ϕ is 45° and it increases towards 90° as the frequency decreases towards zero value.

EXAMPLE 9.14 *Determine the cut off frequency for the input RC network of Fig. 9.31.*

SOLUTION

$$R_{in} = R_G \,||\, R_{in\,(gate)}$$

$$R_{in\,(gate)} = \left|\frac{V_{GS}}{I_{GSS}}\right| = \left|\frac{12 \text{ V}}{40 \text{ nA}}\right|$$

$$= 300 \text{ M}\Omega$$

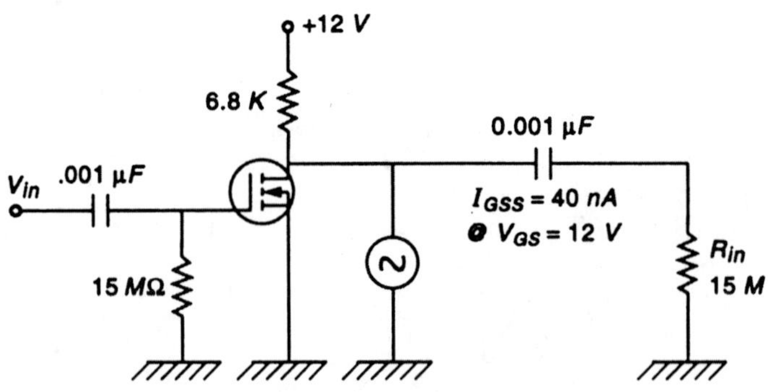

Fig 9.31. Reference Example 9.14

$$R_{in} = 15\ \text{M}\Omega \parallel 300\ \text{M}\Omega$$
$$= 14{\cdot}29\ \text{M}\Omega$$

and cut off frequency $= \dfrac{1}{2\pi R_{in} \cdot C_{C_1}}$

$$= \frac{1}{2\pi\,(14{\cdot}29\ \text{M}\Omega)\,(0{\cdot}001\ \mu\text{F})}$$
$$= 11{\cdot}13\ \text{Hz}$$

The critical frequency is very small due to the high input resistance.

EXAMPLE 9.15 *Find the critial frequency for the output RC network in Fig. 9.31.*

SOLUTION From equation (9.45)

$$f_c' = \frac{1}{2\pi\,(R_D + R_L)\,C_{C_2}}$$
$$= \frac{1}{2\pi\,(14{\cdot}29\ \text{M}\Omega + 6{\cdot}8\ \text{k}) \times {\cdot}001}$$
$$= \frac{1}{2\pi \times 14{\cdot}2968\ \text{M} \times {\cdot}001}$$
$$= 11{\cdot}13\ \text{Hz.}$$

9.11 HIGH FREQUENCY RESPONSE

The high frequency response of an *FET* amplifier is similar to that of *BJT* amplifier. The high frequency equivalent circuit is depicted in Fig. 9.32.

FET manufacturers specify in the data sheets the values of the three quantities C_{iss}, the input capacitance, C_{rss}, the reverse transfer capacitance and C_{oss}, the output capacitance. The value of C_{gs}, C_{ds} and C_{gd} may be obtained from the equations

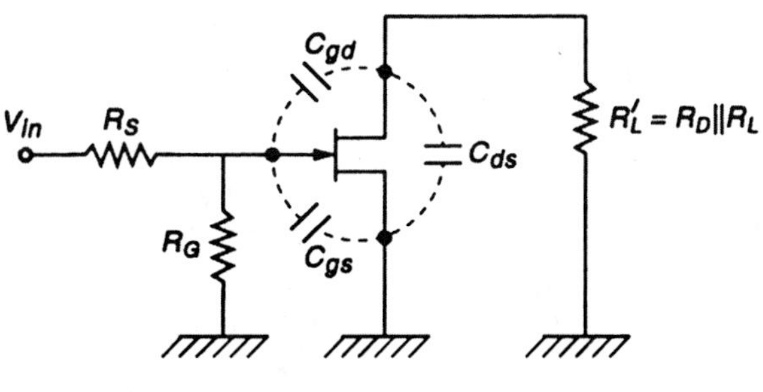

Fig 9.32. High frequency equivalent circuit

$$C_{gs} = C_{iss} - C_{rss} \quad \ldots(9.47)$$
$$C_{ds} = C_{oss} - C_{rss} \quad \ldots(9.48)$$
$$C_{gd} = C_{rss} \quad \ldots(9.49)$$

C_{oss} is also not often specified.

Fig. 9.33 shows the high frequency input *RC* network for the *FET* amplifier. It is basically a low pass filter. Since $R_{in} >> R_s$, the value of R_s determines in effect the input critical frequency.

$$f_c = \frac{1}{2\pi R_s C'} \qquad ...(9.50)$$

where $$C' = C_{gs} + C_{in\,(Miller)} + C_{w0} \qquad ...(9.51)$$

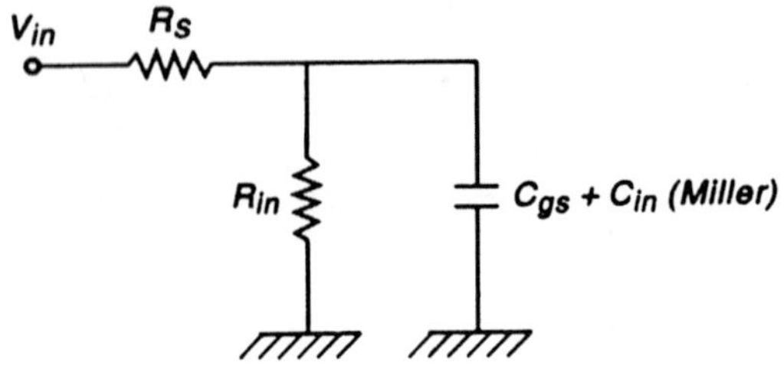

Fig 9.33. The high frequency input RC network

The input *RC* network reduces the midband voltage gain by 3 dB at f_c and further makes the gain to roll off at – 20 dB/decade beyond f_c.

EXAMPLE 9.16 *The data sheets for 2N3796 specifies C_{iss} = 5 pF, C_{rss} = 0·5 pF. Find the values of C_{gd}, C_{gs} and C_{ds}. [Assume C_{oss} = 2 pF].*

SOLUTION $C_{gd} = C_{rss} = 0{\cdot}5$ pF

$C_{gs} = C_{iss} - C_{rss} = (5 - 0{\cdot}5)$ pF $= 4{\cdot}5$ pF

$C_{ds} = C_{oss} - C_{rss} = (2 - 0{\cdot}5)$ pF $= 1{\cdot}5$ pF.

High Frequency Output *RC* network

The output *RC* network is developed by the output resistance looking in at the drain and the Miller output capacitance. The network is shown in Fig. 9.34.

For the network in Fig. 9.34, the critical frequency f_c is given as

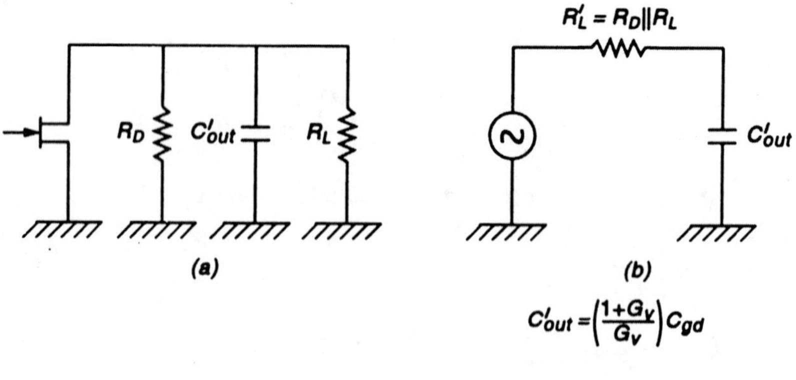

Fig 9.34. (a) High frequency output RC network (b) Electrical equivalent circuit

$$f_c = \frac{1}{2\pi R_L' C_{out}'} \quad ...(9.52)$$

where $$R_L' = R_D \| R_L = \frac{R_D \cdot R_L}{R_D + R_L} \quad ...(9.53)$$

and $$C_{out}' = \left(\frac{1 + G_V}{G_V}\right) C_{gd} = \text{Miller output capacitance} \quad ...(9.54)$$

and The phase angle $\phi = \tan^{-1} (R_L'/X_{C_{out}}')$ ...(9.55)

EXAMPLE 9.17 *Calculate the critical frequency of the input RC network for the amplifier of Fig. 9.35. C_{iss} = 6 pF, C_{rs} = 2·5 pF, g_m = 7500 μ Siemens and C_{wi} = 2 pF.*

SOLUTION $C_{gd} = C_{rss}$ = 2·5 pF

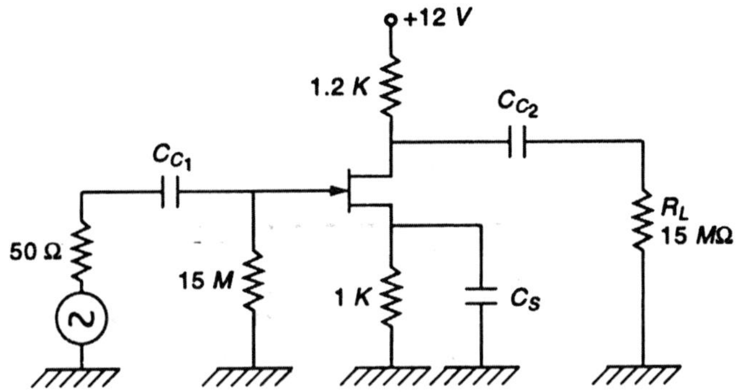

Fig 9.35. Reference Example 9.17

$C_{gs} = C_{iss} - C_{rss}$ = 6 − 2·5 = 3·5 pF

Now $R_L' = R_D \| R_L$ = (1·2 k ‖ 15 MΩ) = 1·2 kΩ

$G_V = g_m R_L'$ = (7500 μS) × 1·2 kΩ = 9·0

$C_{in\,(Miller)} = C_{gd} (1 + G_V)$ = 2·5 pF (1 + 9) = 25 pF

Total input capacitance

$C_{in}' = C_{gs} + C_{wi} + C_{in\,(Miller)}$
= 3·5 pF + 2 pF + 25 pF
= 30·5 pF.

The cut off frequency is thus

$$f_c = \frac{1}{2\pi R_s C_{in}'} = \frac{1}{2\pi (50\ \Omega)(30{\cdot}5\ \text{pF})}$$
= 104·36 MHz

EXMPLE 9.18 *Find the critical frequency in the Example of 9.17 if g_m has a value of 5000 μS instead of 7500 μS and C_{wi} is negligible.*

SOLUTION In this case $G_V = g_{in} R_L'$ = (5000 μS) × 1·2 kΩ = 6·0

$C_{in\,(Miller)} = C_{gd} (1 + G_V)$ = 2·5 × 6 = 15 pF

$$C_{in}' = C_{gs} + C_{in\,(Miller)}$$
$$= 3{\cdot}5 \text{ pF} + 15 \text{ pF}$$
$$= 18{\cdot}5 \text{ pF.}$$

The cut off frequency is

$$f_c = \frac{1}{2\pi\, R_s\, C_{in}'} = \frac{1}{2\pi\,(50\ \Omega)\,(18{\cdot}5 \text{ pF})}$$
$$= 172 \text{ MHz.}$$

EXAMPLE 9.19 *Calculate the cut off frequency for the output RC network in the amplifier of Fig. 9.35. [Assume $C_{w0} = 1{\cdot}0$ pF]*

SOLUTION For the network of Fig. 9.35,

$$R_L' = R_D \,\|\, R_L = 1{\cdot}2 \text{ k} \,\|\, 15 \text{ M}\Omega$$
$$\simeq 1{\cdot}2 \text{ k}\Omega$$

Equivalent output capacitance

$$C_{out\,(Miller)} = C_{gd}\left(\frac{1 + G_V}{G_V}\right) = 2{\cdot}5 \text{ pF}\left(\frac{1+9}{9}\right) = 2{\cdot}78 \text{ pF}$$
$$C_{out}' = C_{w0} + C_{out\,(Miller)} = 1 \text{ pF} + 2{\cdot}78 \text{ pF}$$
$$= 3{\cdot}78 \text{ pF}$$

The cut off or critical frequency is thus

$$f_c = \frac{1}{2\pi\, R_L'\, C_{out}'}$$
$$= \frac{1}{2\pi \times 1{\cdot}2 \text{ k}\Omega \times 3{\cdot}78 \text{ pF}}$$
$$= 35{\cdot}09 \text{ MHz.}$$

Review Questions

1. Why does the voltage gain of an *RC* coupled amplifier decreases at low frequencies ?
2. Why does the voltage gain of an *RC* coupled amplifier drops at high frequencies compared to mid band range of frequencies ?
3. Why does the voltage gain of a transformer coupled amplifier drops at low frequencies (compared to midband range voltage gain) ?
4. Why does the voltage gain of a transformer coupled amplifier drops at high frequencies (compared to midband voltage gain) ?
5. Discuss why the voltage gain of a direct coupled amplifier drops at high frequencies.
6. What do you mean by Miller effect input capacitance and Miller effect output capacitance ?
7. What is meant by decibel gain ?
8. What are cut off frequencies ?
9. Draw the frequency response curve of an *RC* coupled amplifier and label the critical points in it.
10. Derive equation 9.21.
11. Derive equation 9.22.

12. What is meant by "6 dB/octave" and "20 dB/decade". Have they the same meaning ? Explain.
13. What is meant by the dominant critical frequencies ?
14. What is a bandwidth ?
15. Draw the composite Bode plot of an *RC* coupled *BJT* amplifier and label the critical points. Show the bandwidth in the plot.
16. Derive equation 9.30.
17. Derive equation 9.35.
18. Draw the input *RC* network for the low frequency response of the *FET* amplifier.
19. Draw the output *RC* network for the high frequency response of an *FET* amplifier.
20. Draw the high frequency input *RC* network for the *FET* amplifier and derive the expression for cut off frequency.
21. Draw the high frequency output *RC* network for the *FET* amplifier and derive the expression for cut-off frequency.

Objective Type Questions

(Tick mark the correct statement)

1. The low frequency response of an *RC* coupled *BJT* or *FET* amplifier is determined by
(a) the transistor type (b) supply voltage
(c) coupling and bypass capacitors (d) interelectrode capacitances of the *BJT* or *FET*
2. The high frequency response of an *RC* coupled amplifier is determined by
(a) the coupling capacitors (b) the bypass capacitors
(c) the parasitic (inter electrode) capacitances and frequency dependence of β of the transistor
(d) supply voltage
3. The gain of a certain amplifer decreases by 12 dB when the frequency is decreased from 2·0 kHz to 20 Hz. The roll off is
(a) 6 dB/decade (b) 6 dB/octave
(c) 12 dB/decade (d) 12 dB/octave
4. The low frequency response of a transformer coupled *BJT* amplifier is determined by
(a) supply voltage
(b) shorting effect of magnetizing inductive reactance
(c) stray capacitances between the turns of primary and secondary
(d) variation of β with frequency for the transistor
5. The bandwidth of an amplifier is determined by
(a) the midband gain (b) the critical frequencies
(c) the roll of rate (d) coupling capacitors and bypass capacitors.
6. The high frequency response of the transformer coupled amplifier is determined by
(a) the shorting effect off the magnetizing inductive reactance
(b) parasitic capacitances of the transistor
(c) stray capacitance between the turns of the primary and the secondary.
(d) supply voltage
7. The Miller input capacitance of an amplifier is determined by
(a) midband gain and input capacitance (b) bypass capacitor
(c) coupling capacitors at the input (d) coupling capacitor at the output
8. The Miller output capacitance of the *BJT* amplifier is determined by
(a) bypass capacitors (b) coupling capacitor at the input
(c) coupling capacitor at the output (d) midband gain and output capacitance

9. The gain of an amplifier decreases by 20 dB when the frequency is increased from 100 kHz to 1 MHz. The roll off rate is
(a) 6 dB/octave
(b) 6 dB/decade
(c) 20 dB/octave
(d) none of the above

10. At the lower corner frequency
(a) the gain is 0 dB
(b) the gain 3 dB down from the midband gain (dB)
(c) the voltage gain is half the midband gain
(d) none of the above

11. At the upper critical frequency,
(a) the gain is 0 dB (mid band gain)
(b) the voltage gain is half the midband gain
(c) the gain is 3 dB down from the midband gain
(d) none of the above

12. An amplifier possesses the following critical frequencies 200Hz, 150 Hz, 20 kHz, 22 kHz. The bandwidth is
(a) 19850 Hz
(b) 21850 Hz
(c) 19800 Hz
(d) 21800 Hz

13. If the voltage gain of an amplifier is decreased, the bandwidth
(a) increases
(b) decreases
(c) is unaffected
(d) becomes poor

14. In the midband range of frequencies for an amplifier, the output voltage is 5 V rms. At the upper critical frequency, the r.m.s. output voltage is
(a) 2·5 V
(b) 3·54 V
(c) 7·07 V
(d) none of the above

15. At the lower corner (critical) frequency of an amplifier, the peak output voltage is 5V. The peak output voltage at the middle of the amplifier's bandwidth is
(a) 5V
(b) 7·07 V
(c) 3·18 V
(d) 3·535 V

16. For a transistor the f_T is 120 MHz. If the bandwidth 20 MHz, the voltage gain is
(a) 120
(b) 6
(c) 20
(d) 1

17. If $\frac{P_{out}}{P_{in}}$ for an amplifier is 100, the dB power gain is
(a) 20 dB
(b) 10 dB
(c) 6 dB
(d) 40 dB

18. The lower critical frequency of a direct coupled amplifier with no bypass capacitor is
(a) 0 Hz
(b) 20 Hz
(c) dependent on β
(d) none of the above

Exercise Problems

1. Calculate the following quantities in dB units
(a) $\frac{P_{out}}{P_{in}} = 150$
(b) $\frac{P_{out}}{P_{in}} = 50$
(c) $\frac{P_{out}}{P_{in}} = 20$
(d) $\frac{P_{out}}{P_{in}} = 15$
(e) $\frac{P_{out}}{P_{in}} = 5$
(f) $\frac{V_{out}}{V_{in}} = 150$
(g) $\frac{V_{out}}{V_{in}} = 100$
(h) $\frac{V_{out}}{V_{in}} = 50$
(i) $\frac{V_{out}}{V_{in}} = 5$
(j) $\frac{V_{out}}{V_{in}} = 1$
(k) $\frac{V_{out}}{V_{in}} = 0{\cdot}5$
(l) $\frac{V_{out}}{V_{in}} = 0{\cdot}4$
(m) $\frac{V_{out}}{V_{in}} = 0{\cdot}3$
(n) $\frac{V_{out}}{V_{in}} = 0{\cdot}02$
(o) $\frac{V_{out}}{V_{in}} = 0{\cdot}002$

[**Ans.** (a) 21·76 dB (b) 16·99 dB (c) 26 dB, (d) 11·76 dB (e) 6·99 dB (f) 43·52 dB (g) 40 dB (h) 34 dB (i) 14 dB (j) 0 dB (k) – 6 dB (l) – 7·96 dB (m) – 10·45 dB (n) – 34 dB (o) – 54 dB]

(a) the input Miller capacitance (b) output Miller capacitance if C_{bc} = 3 pF and C_{be} = 1 pF.

[**Ans.** (a) 111 pF (b) 1 pF]

3. Find the input and output Miller capacitance if the voltage gain is 100 instead of 36.

[**Ans.** (a) 303 pF (b) 1 pF]

4. Find the gain magnitude if the decibel power gain is 45 dB.

[**Ans.** 31622]

5. An amplifier has a wattage rating of 40 W. It is connected to a 16 Ω speaker.

(a) Find the input power required for full power output if the power gain is 12 dB.

(b) Find the input voltage for rated output if the amplifier voltage gain is 30 dB.

[**Ans.** (a) 2·52 W (b) 0·8 V]

6. Find the critical frequency for the amplifier circuit of Fig. E.9.1

[**Ans.** 338·64 Hz.]

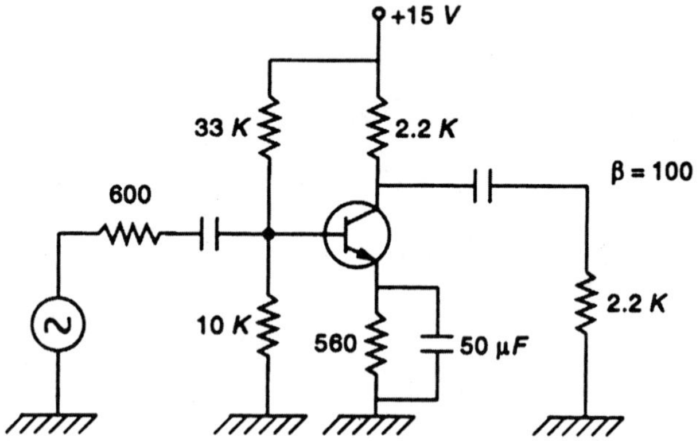

Fig E.9.1. Reference Exercise Problem 9.6

7. Find the total low frequency response for the amplifier in Fig. E9.2

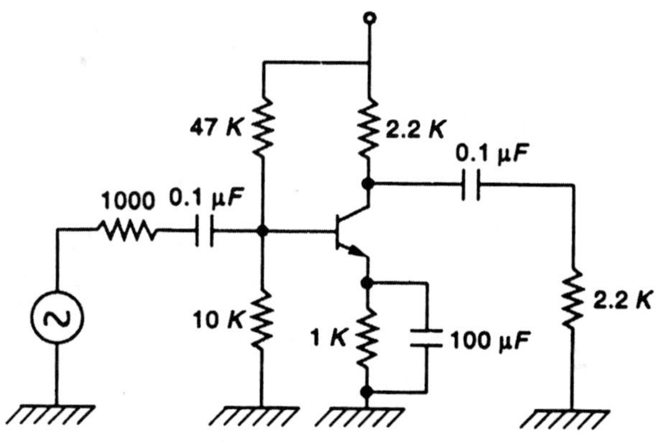

Fig E.9.2. Reference Exercise Problem 9.7

[**Hints :** (a) f_c input = 829·4 Hz (b) f_c (bypass) = 78·52 Hz (c) f_c (output) = 361·7 Hz.]

8. For the circuit in Fig. E9.3, develop the high frequency RC network and find the critical frequency. Assume $\beta = 100$ for the transistor, $C_{be} = 18$ pF, $C_{bc} = 3$ pF , $C_{wi} = 2$ pF.

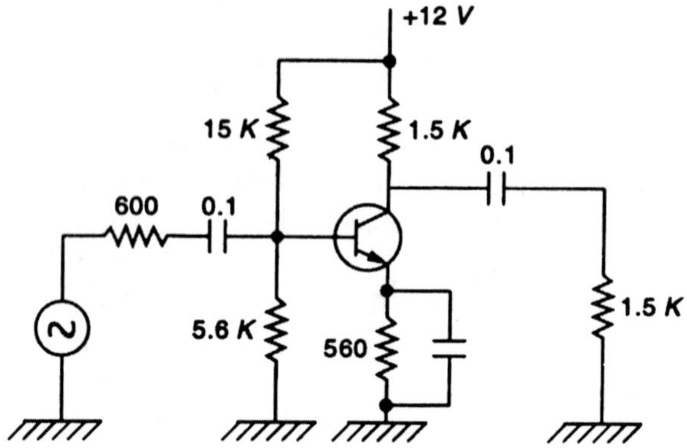

Fig E.9.3. Reference Exercise Problem 9.8

[**Ans.** Critical frequency = 1·465 MHz]

9. Refer Fig. E.9.3 and determine the critical frequency for the amplifier corresponding to the output RC network.

[Assume C_{w0} = wiring capacitance at the output = 1 pF].

[**Ans.** 53 MHz.]

10. What is the bandwidth of an amplifier if the dominant critical frequencies are 10 Hz and 10 kHz.

[**Ans.** 9990 Hz]

11. A manufacturer specifies the f_T of a transistor as being equal to 200 MHz. If the midband voltage gain is 20, find the bandwidth that can be achieved.

[**Ans.** 10 MHz]

12. Find the cut off frequency for the input RC network for the FET amplifer in Fig. E9.4.

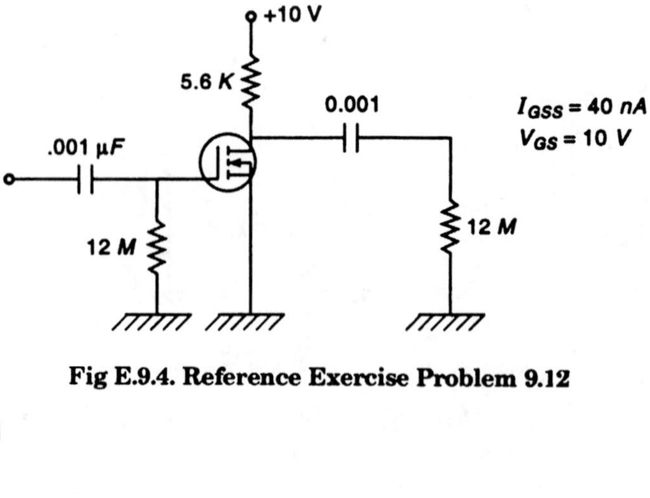

Fig E.9.4. Reference Exercise Problem 9.12

[**Ans.** 13·66 Hz]

13. Find the critical frequency for the output *RC* network of Fig. E9.4.

[**Ans.** 13·65 Hz]

14. The data sheets of a typical *FET* specifies C_{iss} = 6 pF, C_{rss} = 1 pF. Find C_{gs}, C_{gd} and C_{ds}. [Assume C_{oss} = 3 pF]

[**Ans.** c_{gs} = 5 pF, C_{gd} = 1 pF and C_{ds} = 2 pF]

15. Calculate the critical frequencies for the input *RC* network for the *FET* amplifier in Fig. E9.5. Assume C_{iss} = 6 pF, C_{rss} = 1 pF, C_{oss} = 3 pF, g_m = 7000 μS, C_{wi} = 2 pF.

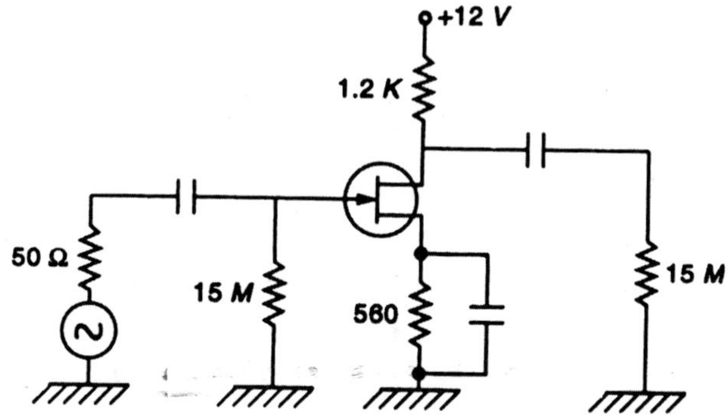

Fig E.9.5. Reference Exercise Problem 9.15

[**Ans.** 194 MHz]

16. Calculate the critical frequencies for the output *RC* network of the *FET* amplifier in Fig. E9.5. Assume C_{w0} = 1 pF.

[**Ans.** f_c = 62·5 MHz]

Chapter 10

FEEDBACK IN AMPLIFIERS

10.1 INTRODUCTION

By the term feeback we mean the transfer of energy from the output to the input of a system. If a fraction of the output signal in an amplifier is fed back to the input side and superimposed with the input signal, the amplfier is called a feedback amplifier. The performance of a feedback amplifier improves considerably over an amplifier without feedback.

Feedback introduced in an amplifier may be either **negative** feedback or **positive** feedback. **Negative feedback** is also called as **degenerative feedback or inverse feedback.** In negative feedback, the feedback is also called as **degenerative feedback or inverse feedback.** In negative fedback, the feedback signal decreases the magnitude of the input signal. There is another type of feedback which is called as positive feedback. In positive feedback, the feedback signal increases the magnitude of the input signal. **Positive feedback** is also termed as **regenerative feedback** or **direct feedback.**

Although, the composite gain or overall gain of an negative feedback amplifier is less, the negative feedback offers many advantages and is widely used in various applications. It improves stability of the gain. The positive feedback amplifier on the other hand provides instability of the gain and distortion of the signal. However, positive feedback is employed in oscillators, narrowband amplifiers and radio receivers.

10.2 PRINCIPLE OF FEEDBACK AMPLIFIERS

The voltage gain or amplification A for an amplifier is given by

$$A = \frac{V_{\text{out}}}{V_{\text{in}}} \quad \text{...(10.1)}$$

If we consider the feedback loop β for the amplifer and if V_0' is the output voltage with feedback, the voltage fed back is $\beta V_0'$. This voltage $\beta V_0'$ is applied to the input of the amplifier. So the total input voltage is $V_i + \beta V_0'$. Thus if A_{fb} is the voltage gain with feedback, then V_0' can be written as

$$V_0' = A\,(V_i + \beta V_0') \quad \text{...(10.2)}$$

or,

$$V_0'\,(1 - A\beta) = A\,.\,V_i \quad \text{...(10.3)}$$

The amplifier gain with feedback is thus given by

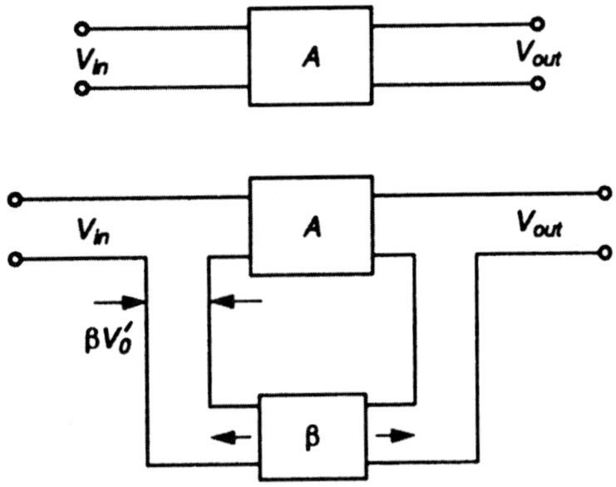

Fig 10.1. Schematic diagram illustrating the principle of feedback amplifiers

$$A_{fb} = \frac{V_0'}{V_i} = \frac{A}{1 - A\beta} \qquad ...(10.4)$$

Thus the voltage gain with feedback A_{fb} is greater than A by the factor $\frac{1}{(1 - A\beta)}$. The quantity β where

$$\beta = \frac{V_{fb}}{V_0'} \qquad (10.5)$$

is called the feedback ratio, feedback fraction, reverse transfer ratio or reverse transmission factor. β is a positive quantity for positive feedback while β is negative for negative feedback. The quantity $A\beta$ is called the loop gain, the loop transmission or return ratio. The gain with feedback can be written from equation (10.4) as

$$A_{fb} = \frac{A}{1 - A\beta} \text{ for positive or regenerative feedback} \qquad ...(10.6)$$

and

$$A_{fb} = \frac{A}{(1 + A\beta)} \text{ for negative or degenerative feedback} \qquad ...(10.7)$$

In the above equation both A and β can be either real (positive or negative) quantities or complex.

If the denominator $1 - A\beta = 0$, $A_{fb} \to \infty$, implying an infinite gain or that the amplifier giving an output signal even in the absence of an input signal. The amplifier then acts as an oscillator. If the feedback amplifier gain is expressed is decibels, then

$$F = 20 \log_{10} \frac{1}{|1 + A\beta|} \qquad ...(10.8)$$

The quantity F is positive for positive feedback and F is negative for degenerative type feedback.

Advantages of negative feedback

The advantages of a negative feedback amplifier are

(1) high stability in gain

(2) high fidelity

(3) higher band width

(4) Decrease in noise

(5) Reduced distortion (amplitude distortion, harmonic distortion, phase distortion and frequency distortion).

EXAMPLE 10.1 *In a feedback amplifier circuit, the input voltage is $V_i = 2\ mV$, $V_0' = 10\ V$ and $\beta V_0' = 200\ mV$. Find the values of A_{fb} and β.*

SOLUTION

$$A = \text{Open loop gain} = \frac{10\ \text{V}}{2\ \text{mV}} = 5000$$

$$A_{fb} = \text{closed loop gain} = \frac{10\ \text{V}}{202\ \text{mV}} \simeq 50$$

or, $$50 = \frac{A}{1 - A\beta}$$

yielding $$50 = \frac{5000}{1 - 5000\beta}$$

or, $$\beta = \frac{-99}{5000} \simeq -\frac{1}{50}$$

EXAMPLE 10.2 *For a negative feedback amplifier, $A = 300$, $\beta = 0{\cdot}01$ and input voltage = 25 mV. Calculate (i) the gain with feedback (ii) output voltage (iii) feedback factor (iv) feedback voltage.*

SOLUTION

(i) $A_{fb} = \dfrac{A}{1 + A\beta} = \dfrac{300}{1 + 300 \times 0{\cdot}01} = 75$

(ii) $V_0' = A_{fb} \cdot V_i = 75 \times 25\ \text{mV} = 1{\cdot}875\ \text{V}$

(iii) feedback factor $= A\beta = 0{\cdot}01 \times 300 = 3$

(iv) feedback voltage $= \beta V_0' = 0{\cdot}01 \times 1{\cdot}875 = 0{\cdot}0188\ \text{V}$

Gain Stability

The gain of amplifier with negative feedback is given by

$$A_{fb} = \frac{A}{1 + A\beta} \qquad ...(10.9)$$

Taking logarithms of both sides in equation (10.9), we get

$$ln\ A_{fb} = ln\ A - ln\ (1 + A\beta) \qquad ...(10.10)$$

Differentiating equation (10.10)

$$\frac{dA_{fb}}{A_{fb}} = \frac{dA}{A} - \frac{\beta \cdot dA}{1 + \beta A}$$

$$= \frac{1}{(1 + A\beta)} \frac{dA}{A} \qquad ...(10.11)$$

When $A\beta >> 1$, the above equation reduces to $\dfrac{dA_{fb}}{A_{fb}} = \dfrac{1}{A\beta}\dfrac{dA}{A}$ ...(10.12)

EXAMPLE 10.3 *A feedback amplifier has an open loop gain of 500 and a feedback factor of 0·1. If the open loop gain varies by 10%, find the percentage variation of closed loop gain.*

SOLUTION In this example $A = 500$

$$\beta = 0{\cdot}1$$

$$\frac{dA}{A} = 10\% = 0{\cdot}1$$

Thus $$\frac{dA_{fb}}{A_{fb}} = \frac{dA}{A} \times \frac{1}{A\beta} = 0{\cdot}1 \times \frac{1}{500 \times 0{\cdot}1}$$

$$= 0{\cdot}002 \text{ or } 0{\cdot}2\%$$

Thus, the closed loop gain is much more stable compared to the open loop gain.

Decrease in Distortion

Let us assume that the harmonic distortion voltage is v_D before negative feedback and the harmonic distortion voltage is v_D' after negative feedback is applied so that

$$v_D' = f\, v_D \qquad \text{...(10.13)}$$

Then the output distortion which is fed back to the input is

$$\beta \,.\, v_D' = \beta \,.\, f \,.\, v_D \qquad \text{...(10.14)}$$

Thus $$v_D' = v_D - \beta f \,.\, v_D \,.\, A \qquad \text{...(10.15)}$$

From (10.13) and (10.15), we get

$$f\, v_D = v_D - \beta f\, v_D \,.\, A \qquad \text{...(10.16)}$$

or, $$f = \frac{1}{1 + A\beta} \qquad \text{...(10.17)}$$

and $$v_D' = \frac{v_D}{1 + A\beta} \qquad \text{...(10.18)}$$

Equation (10.18) specifies that the negative feedback reduces the amplifier distortion by the factor of loop gain $(1 + A\beta)$. It should, however, be kept in mind that the distortion reduction is possible only when the distortion is generated in the amplifier itself and not when the distortion is originally present at the input.

EXAMPLE 10.4 *An amplifier with negative feedback has an open loop gain of 70. If the amplifier open loop gain increases by 5%, find the percentage change in closed loop gain. Given β = 10%.*

SOLUTION
$$A_{fb} = \frac{A}{1 + A\beta} = \frac{70}{1 + 0{\cdot}10 \times 70} = \frac{70}{8} = 8{\cdot}75$$

If the open loop gain increases by 5%

$$A_{fb}' = \frac{A'}{1 + A'\beta}$$

$$= \frac{73{\cdot}5}{1 + 0{\cdot}1 \times 73{\cdot}5} = \frac{73{\cdot}5}{1 + 7{\cdot}35} = 8{\cdot}80$$

Thus percentage change in closed loop gain

$$= \frac{A_{fb}' - A_{fb}}{A_{fb}} \times 100\% = \frac{8{\cdot}80 - 8{\cdot}75}{8{\cdot}75} \times 100\%$$

$$= \frac{{\cdot}05}{8{\cdot}75} \times 100\% = 0{\cdot}57\%$$

EXAMPLE 10.5 *An amplifier has a gain of 100 and distortion of 5%. Find (a) output signal voltage (b) distortion voltage (c) output voltage if the input signal voltage of 0·5 V is applied to the amplifier.*

SOLUTION (a) output signal voltage $= A \,.\, V_{in} = 100 \times 0{\cdot}5 = 50$ V.

(b) Distortion voltage $= D \,.\, V_{out} = 0{\cdot}05 \times 50 = 2{\cdot}5$ V.

(c) Amplifier output voltage $= V_{out} + D_{out}$

$= 50 + 2{\cdot}5 = 52{\cdot}5$ V.

10.3 FEEDBACK AMPLIFIER WITH MULTIPLE STAGES

Let us consider a two stage feedback amplifier each with a feedback factor of β and open loop gain of A. Then

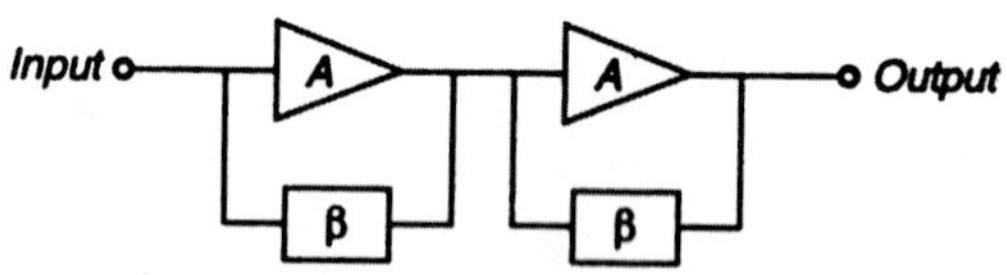

Fig 10.2. Feedback amplifier with multiple stages

The overall gain

$$= \left(\frac{A}{1 + A\beta}\right)\left(\frac{A}{1 + A\beta}\right) = \frac{A^2}{(1 + A\beta)^2} \quad \ldots(10.19)$$

For n stages the gain is $A_n = \dfrac{A^n}{(1 + A\beta)^n}$ ...(10.20)

EXAMPLE 10.6 *For a two stage amplifier, the second stage has 10% of the output voltage as negative feedback and has – 200 as forward gain. Find (a) the gain of the first stage (b) the second harmonic distortion, if the second stage introduces 2% second harmonic without feedback. Assume that the first stage does not introduce distortion.*

SOLUTION

(a) For second stage, $D_2' = \dfrac{D_2}{1 + \beta A_2} = \dfrac{0{\cdot}02}{1 + 200 \times 0{\cdot}1}$

$$= \frac{0{\cdot}02}{1 + 2} = 0{\cdot}006 = 0{\cdot}6\%$$

(b) For the second stage, gain with feedback is

$$A_2' = \frac{A_2}{1 + \beta A_2} = \frac{200}{1 + 200 \times 0{\cdot}1} = \frac{200}{20 + 1} = 9{\cdot}52$$

Thus $A_1 \times A_2' = 200$

or, $A_1 = \dfrac{200}{9{\cdot}52} = 21$

Higher Bandwidth

Using negative feedback, the gain of the amplifier is reduced. For an amplifier, the gain bandwidth product is constant. The bandwidth is the difference in 3 dB frequencies f_1 and f_2.

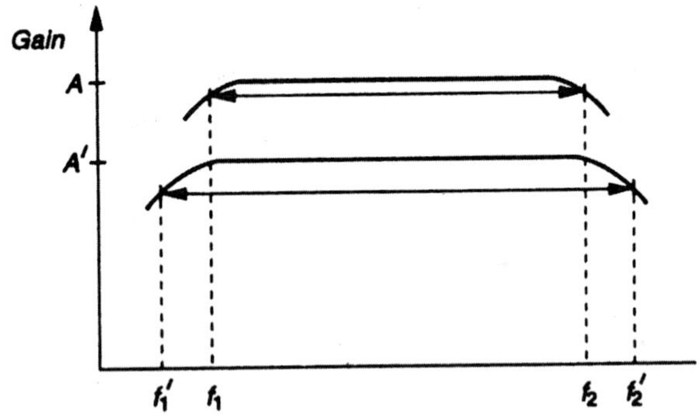

Fig 10.3. Figure illustrating that the gain-bandwidth product is constant

$$BW = f_2 - f_1$$

Also $\text{Gain} \times BW = f_T$

when the feedback gain is decreased, this amplifier has a higher bandwidth. The feedback amplifier has a higher **upper 3 dB** frequency and reduced **lower 3 dB** frequency.

EXAMPLE 10.7 *An RC coupled amplifer has a mid frequency gain of 250 and a frequency response from 100 Hz to 25 kHz. A negative feedback with $\beta = 0{\cdot}025$ is introduced into the circuit. Find the new bandwidth.*

SOLUTION

$$A_{fb} = \frac{A}{1 + A\beta} = \frac{250}{1 + 0{\cdot}025 \times 250} = 34{\cdot}48$$

$$f_1' = \frac{f_1}{1 + A\beta} = \frac{100}{1 + 0{\cdot}025 \times 25} = 13{\cdot}79 \text{ Hz}$$

$$f_2' = f_2 (1 + A\beta) = 25 \text{ kHz} (1 + 250 \times {\cdot}025)$$

$$= 181{\cdot}25 \text{ kHz}$$

Thus new $BW = f_2' - f_1' = 181{\cdot}25 \text{ kHz} - 13{\cdot}79 \text{ Hz}$

$$= 181{\cdot}25 \text{ kHz}$$

10.4 FEEDBACK AMPLIFIER CONFIGURATIONS

The feedback amplifier configurations depend on the basic ways of connecting the feedback signal. Both voltage and current can be connected to the input either in series or in parallel. There are four basic types namely

(A) Voltage series Feedback (Fig. 10.4)

(B) Voltage shunt Feedback (Fig. 10.5)

(C) Current series Feedback (Fig. 10.5) and

(D) Current shunt Feedback (Fig. 10.7)

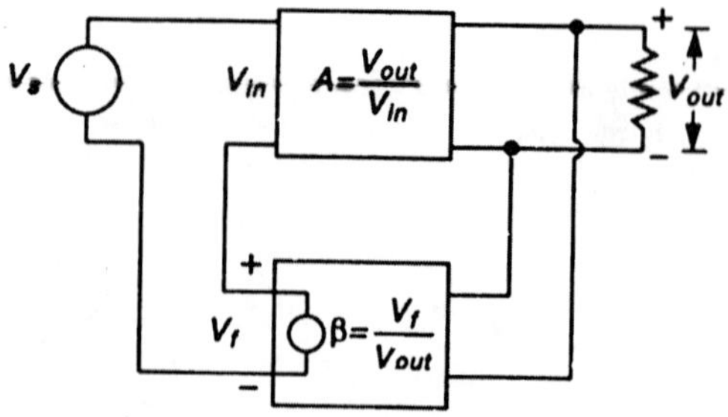

Fig 10.4. Voltage series feedback (closed loop gain = A_{fb} = V_{out}/V_s)

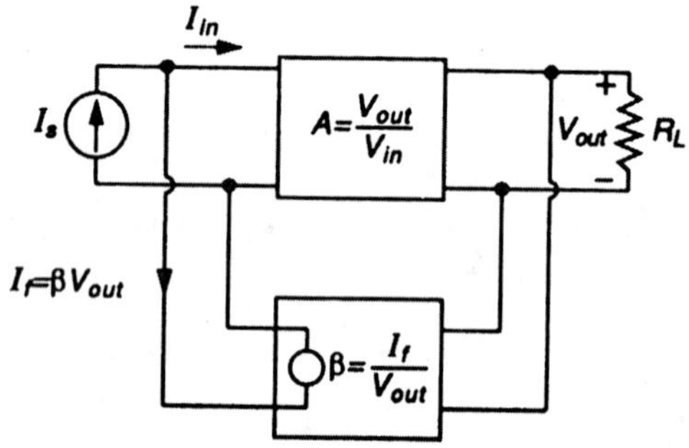

Fig 10.5. Voltage shunt feedback circuit (closed loop gain = A_{fb} = V_{out}/I_s)

In the above list, the term **voltage** refers to feedbacking a portion of the **output voltage** (across R_L) as input to the feedback amplifier network. The term **series** refers to connecting the feedback signal in **series** with the input signal voltage. Thus Fig. 10.4 results.

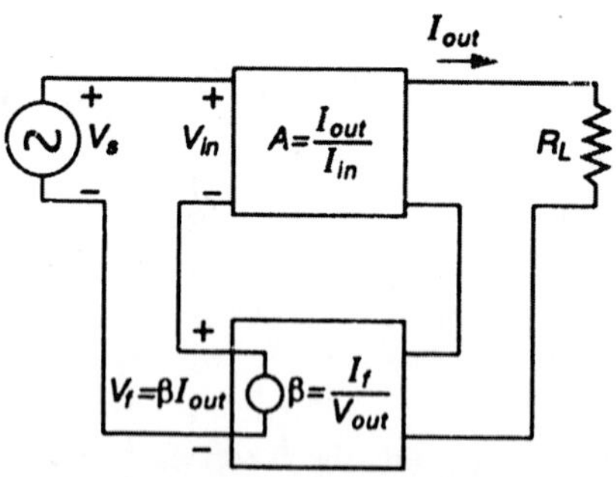

Fig 10.6. Current series feedback (closed loop gain = I_{out}/V_s)

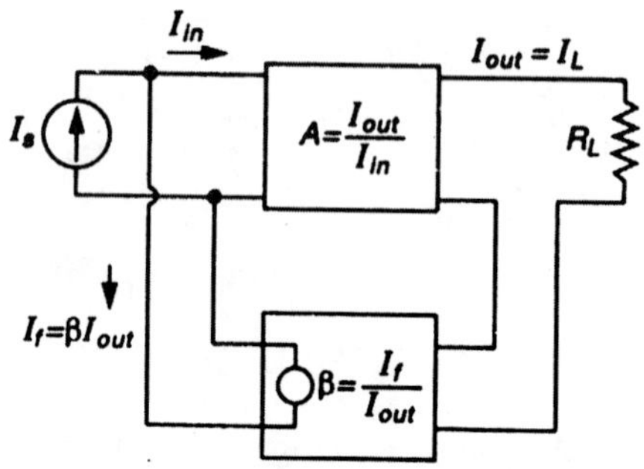

Fig 10.7. Current shunt feedback (closed loop gain = I_{out}/I_s)

Similarly, the term **shunt** is used to connecting the feedback signal in **shunt** (parallel) with the input current source. Thus we can construct Fig. 10.5. The use of the term **current** means taking a portion of the **output current** through the feedback network.

Therefore, the current series feedback network and current shunt feedback networks can be constructed as shown in Fig. 10.6 and 10.7 respectively. It will be apparent from Fig. 10.4 and 10.6 that, series feedback **increases the input resistance** and shunt feedback **decreases the output resistance** (Figs. 10.5 and 10.7).

It can also be noticed that using voltage feedback the output resistance is decreased while using current feedback, the output resistance is increased.

Voltage series feedback therefore, provides, high input and low output resistances and this feedback type is favoured in cascade amplifiers.

10.5 VOLTAGE SERIES FEEDBACK

Fig. 10.4 shows the circuit of voltage series feedback. For this circuit, we may write, the open loop gain

$$A = \frac{V_{\text{out}}}{V_s} = \frac{V_{\text{out}}}{V_{\text{in}}} \qquad ...(10.21)$$

If V_f is the feedback signal

$$V_{\text{in}} = V_s - V_f = V_s - \beta V_{\text{out}} \qquad ...(10.22)$$

Further, output voltage

$$V_{\text{out}} = A \cdot V_{\text{in}} \qquad ...(10.23)$$

$$= A\,(V_s - V_f)$$

$$= AV_s - A \cdot \beta V_{\text{out}} \qquad ...(10.24)$$

or, $$V_{\text{out}}\,(1 + A\beta) = AV_s \qquad ...(10.25)$$

or, $$A_{fb} = \frac{V_{\text{out}}}{V_s} = \frac{A}{1 + A\beta} \qquad ...(10.26)$$

Thus the gain with feedback is reduced by a factor $(1 + A\beta)$.

Input Impedance

From Fig. 10.8, we get

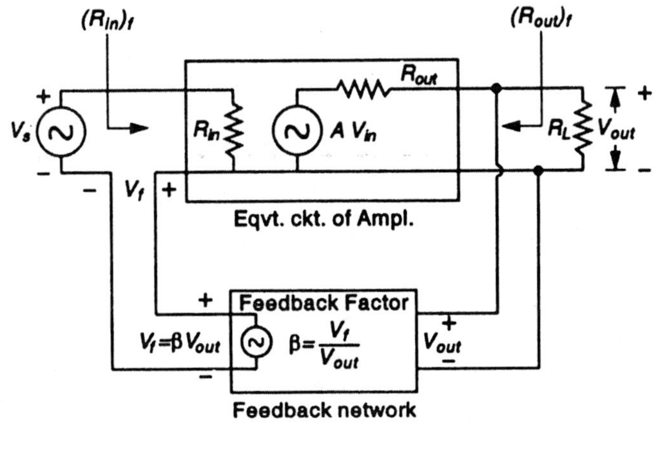

Fig 10.8. Circuit diagram of voltage series feedback connection for calculation of $(R_{in})_f$ and $(R_{out})_f$

$$I_{\text{in}} = \frac{V_{\text{in}}}{Z_{\text{in}}} \qquad \text{...(10.27)}$$

and $\quad V_{\text{in}} = V_s - V_f$

$$= V_s - \beta V_{\text{out}} \qquad \text{...(10.28)}$$

Thus

$$I_{\text{in}} Z_{\text{in}} = V_s - \beta V_{\text{out}}$$

$$= V_s - A\beta V_{\text{in}}$$

$$= V_s - A\beta Z_{\text{in}} I_{\text{in}} \qquad \text{...(10.29)}$$

or, $\quad V_s = I_{\text{in}} Z_{\text{in}} (1 + A\beta)$

Thus $\quad (Z_{\text{in}})_{fb} = \dfrac{V_s}{I_{\text{in}}} = Z_{\text{in}} (1 + A\beta) \qquad \text{...(10.30)}$

Thus the input impedance with feedback for a voltage series feedback connection is the product of the input impedance Z_{in} without feedback and the quantity $(1 + A\beta)$.

Calculation of Output Impedance with Feedback

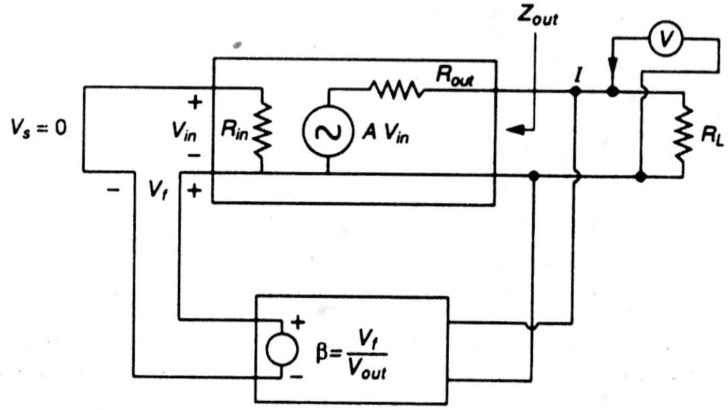

Fig 10.9. Calculation of output impedance with feedback.

Let us refer to Fig. 10.8 again. The calculation of output impedance can be carried out by applying a voltage V at the output while V_s is short circuited (Fig. 10.9).

The external voltage V gives rise to a current I.

From the output circuit,

$$V = A V_{\text{in}} + I Z_{\text{out}} \qquad \text{...(10.31)}$$

From the input side

$$V_s = 0 = V_{\text{in}} + V_f \qquad \text{...(10.32)}$$

or, $\quad V_{\text{in}} = - V_f \qquad \text{...(10.33)}$

Thus from equation (10.31)

$$V = A(-V_f) + I Z_{\text{out}} \qquad \text{...(10.34)}$$

$$= I Z_{\text{out}} - A\beta V_{\text{out}} \qquad \text{...(10.35)}$$

or, $\quad V + A\beta V = I Z_{\text{out}} \qquad \text{...(10.36)}$

or, $\quad V_{\text{out}} = I Z_{\text{out}} \dfrac{1}{(1 + A\beta)} \qquad \text{...(10.37)}$

or, $$(Z_{out})_f = \frac{V_{out}}{I} = \frac{Z_{out}}{(1 + A\beta)} \quad ...(10.38)$$

Thus the output impedance of a voltage series feedback is the same as the magnitude of the output impedance without feedback divided by the factor $(1 + A\beta)$.

10.6 VOLTAGE SHUNT FEEDBACK

For the voltage shunt feedback of Fig. 10.5, we may write the closed loop gain as

$$A_{fb} = \frac{V_{out}}{I_s} \quad ...(10.39)$$

with $$I_s = I_{in} + I_f \quad ...(10.40)$$

$$V_{out} = A \cdot I_{in} \quad ...(10.41)$$

Also $$I_f = \beta V_{out} \quad ...(10.42)$$

Thus, we get the closed loop gain as

$$A_{fb} = V_{out}/I_s$$

$$= (A \cdot I_{in})/(I_{in} + I_f)$$

$$= \frac{A \cdot I_{in}}{I_{in}(1 + A\beta)}$$

$$= \frac{A}{(1 + A\beta)} \quad ...(10.43)$$

A more detailed voltage shunt feedback circuit is depicted in Fig. 10.10.

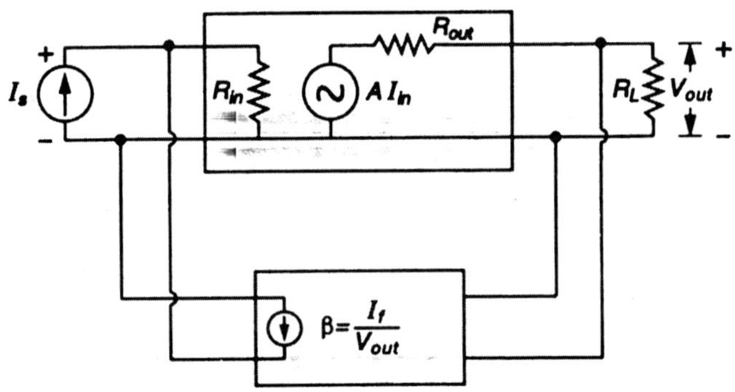

Fig 10.10. Circuit diagram of a voltage shunt feedback connection

The input impedance with feedback is

$$(Z_{in})_f = \frac{V_{in}}{I_s} \quad ...(10.44)$$

Also $$I_s = I_{in} + I_f \quad ...(10.45)$$

Further $$I_f = \beta V_{out} \quad ...(10.46)$$

Thus $$(Z_{in})_f = \frac{V_{in}/I_{in}}{\frac{I_{in}}{I_{in}} + \frac{\beta V_{out}}{I_{in}}} \qquad ...(10.47)$$

$$= \frac{Z_{in}}{1 + A\beta} \qquad ...(10.48)$$

The input impedance is thus reduced by a factor $(1 + A\beta)$.

For the voltage shunt feedback, the output impedance is the same as that of the voltage series feedback namely

$$(Z_{out})_{fb} = \frac{Z_{out}}{(1 + A\beta)} \qquad ...(10.49)$$

10.7 CURRENT SERIES FEEDBACK

For the circuit series feedback connection, we refer to Fig. 10.11.

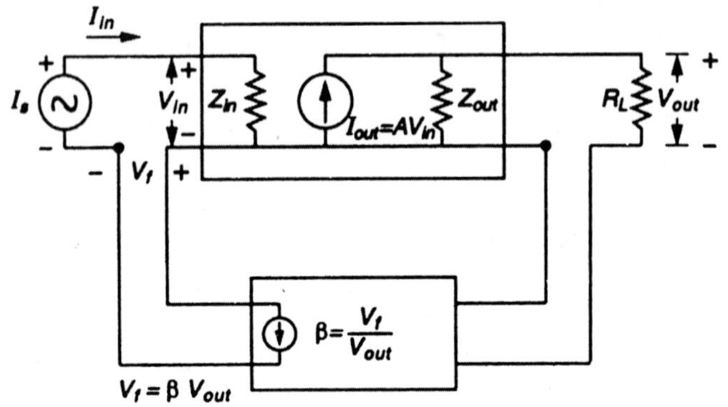

Fig 10.11. Current series feedback connection

The output impedance for the current series feedback can be calculated by shorting out V_s and applying at the output a voltage V.

Thus $$V_f = V_{in} \qquad ...(10.50)$$

and $$I = \frac{V}{Z_{out}} - A \,.\, V_{in} \qquad ...(10.51)$$

where $$V_{in} = \beta I \qquad ...(10.52)$$

or, $$I = \frac{V}{Z_{out}} - A\beta I \qquad ...(10.53)$$

or, $$Z_{out}\,(1 + \beta A)\, I = V \qquad ...(10.54)$$

or, $$(Z_{out})_f = \frac{V}{I} = Z_{out}\,(1 + \beta A) \qquad ...(10.55)$$

This yields, the output impedance with feedback

$$(Z_{out})_f = \frac{V}{I} = Z_{out}\,(1 + \beta A) \qquad ...(10.56)$$

Table 10.1 : The effect of feedback on various types of circuits is depicted

Circuit Type	*Gain without feedback*	*Feedback factor*	*Gain with feedback*	*Input Impedance with feedback*	*Output Impedance with feedback*
Voltage Series	$\frac{V_{out}}{V_{in}}$	$\frac{V_f}{V_{out}}$	$\frac{V_{out}}{V_s} = \frac{A}{(1 + A\beta)}$	$Z_{in}(1 + A\beta)$ INCREASES	$\frac{Z_{out}}{(1 + A\beta)}$ REDUCES
Voltage Shunt	$\frac{V_{out}}{I_{in}}$	$\frac{I_f}{V_{out}}$	$\frac{V_{out}}{I_s} = \frac{A}{(1 + A\beta)}$	$\frac{Z_{in}}{(1 + A\beta)}$ DECREASES	$\frac{Z_{out}}{(1 + A\beta)}$ REDUCES
Current Series	$\frac{I_{out}}{V_{in}}$	$\frac{V_f}{I_{out}}$	$\frac{I_{out}}{V_s} = \frac{A}{(1 + A\beta)}$	$Z_{in}(1 + A\beta)$ INCREASES	$Z_{out}(1 + A\beta)$ INCREASES
Current Shunt	$\frac{I_{out}}{I_{in}}$	$\frac{I_f}{I_{out}}$	$\frac{I_{out}}{I_s}$	$\frac{Z_{in}}{(1 + A\beta)}$ REDUCES	$Z_{out}(1 + A\beta)$ INCREASES

10.8 PRACTICAL FEEDBACK AMPLIFIER CIRCUITS

First of all, let us consider a voltage-series feedback circuit (emitter follower) of Fig. 10.12.

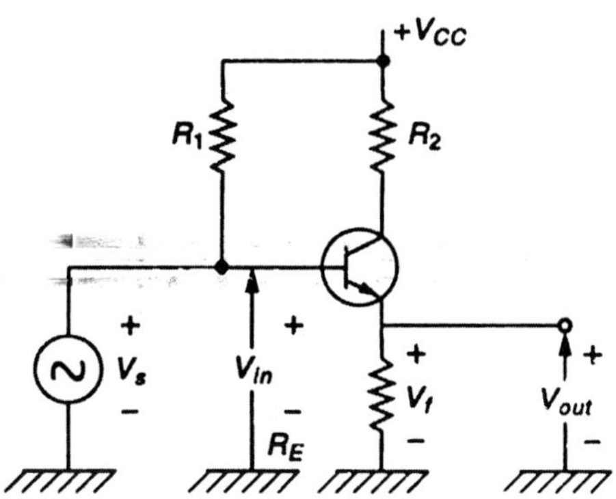

Fig 10.12. Emitter follower type voltage series feedback circuit

In the circuit, V_{in} is the same as V_s and V_{out} is the same as V_f.

Thus $\quad V_{out} = V_f$ ...(10.57)

and $\quad V_{in} = V_s$ ...(10.58)

Also $\quad V_{out} = h_{fe} \cdot I_b \cdot R_E$ ...(10.59)

Thus gain without feedback is

$$A = \frac{V_{out}}{V_s} = \frac{h_{fe} \cdot I_b \cdot R_E}{V_s} \qquad ...(10.60)$$

Also

$$I_b = \frac{V_s}{h_{ie}} \qquad ...(10.61)$$

and $$\beta = \frac{V_f}{V_0} = 1 \qquad ...(10.62)$$

From equation (10.60), we get

$$A = \frac{h_{fe} \cdot I_b \cdot R_E}{V_s}$$

$$= \frac{h_{fe} R_E (V_s/h_{ie})}{V_s}$$

$$= \frac{h_{fe} R_E}{h_{ie}} \qquad ...(10.63)$$

The amplifier gain with feedback can now be written as

$$A_{fb} = \frac{V_{out}}{V_s} = \frac{A}{1 + A\beta} \qquad ...(10.64)$$

From equations (10.63) and (10.64), we now get

$$A_{fb} = \frac{h_{fe} R_E/h_{ie}}{1 + \dfrac{h_{fe} R_E}{h_{ie}}}$$

$$= \frac{h_{fe} R_E}{h_{ie} + h_{fe} R_E} \qquad ...(10.65)$$

If $h_{ie} << h_{fe} R_E$, then

$$A_{fb} \simeq 1 \qquad ...(10.66)$$

Let us consider another voltage series feedback amplifier circuit of Fig. 10.13.

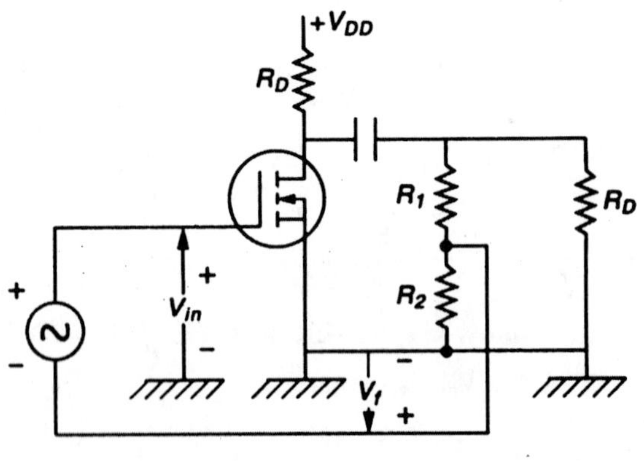

Fig 10.13. FET amplifier with voltage series feedback

For the amplifier of Fig. 10.13, the amplifier gain with feedback is

$$A_{fb} = \frac{A}{1 + A\beta} \qquad ...(10.67)$$

The value of A, the open loop gain is given by

$$A = \frac{V_{out}}{V_{in}} = -g_m R_L' \qquad ...(10.68)$$

...(10.69)

where $$R_L' = R_D' \,\|\, R_L \,\|\, (R_1 + R_2) \quad ...(10.70)$$

The feedback factor β is given by

$$\beta = \frac{V_f}{V_{out}} = \frac{-R_2}{R_1 + R_2} \quad ...(10.71)$$

Using the values A and β in equation (10.67), the gain with negative feedback is

$$A_f = \frac{A}{1 + A\beta} = \frac{-g_m R_L}{1 + g_m R_L \cdot \left(\frac{R_2}{R_1 + R_2}\right)} \quad ...(10.72)$$

If $$A\beta >> 1,$$

$$A_f \cong \frac{1}{\beta} = -\left(\frac{R_1 + R_2}{R_2}\right) \quad ...(10.73)$$

EXAMPLE 10.8 *Find the gain without and with feedback for the FET amplifier circuit of Fig. 10.13. The parameters are $R_L = 6{\cdot}8$ k, $R_D = 6{\cdot}8$ k, $R_s = 400$ k, $R_2 = 100$ k, $g_m = 5000$ μS.*

SOLUTION

$$R_L \simeq \frac{R_L R_D}{R_L + R_D} = \frac{6{\cdot}8 \text{ k} \times 6{\cdot}8 \text{ k}}{2 \times 6{\cdot}8 \text{ k}} = 3{\cdot}4 \text{ k}\Omega$$

where $R_1 + R_2 = 500$ kΩ has been neglected.

Now $A = -g_m R_L' = -(5000 \times 10^{-6}\text{ S}) \times 3{\cdot}4\text{k} = -17$

The feedback factor $\beta = \frac{-R_2}{(R_1 + R_2)} = \frac{-100 \text{ k}}{500 \text{ k}} = -{\cdot}02$

The closed loop gain is

$$A_{fb} = \frac{A}{1 + A\beta} = \frac{-17}{1 + (-0{\cdot}2)(-17)} = \frac{-17}{4{\cdot}4} = -3{\cdot}86$$

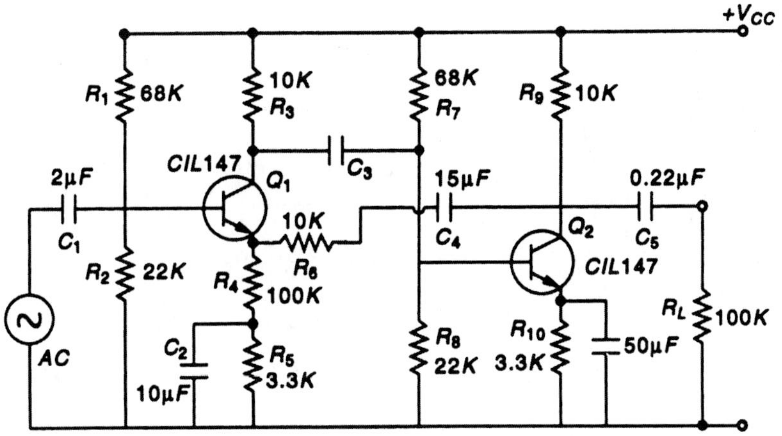

Fig 10.14. A practical two stage capacitor coupled voltage series feedback amplifier

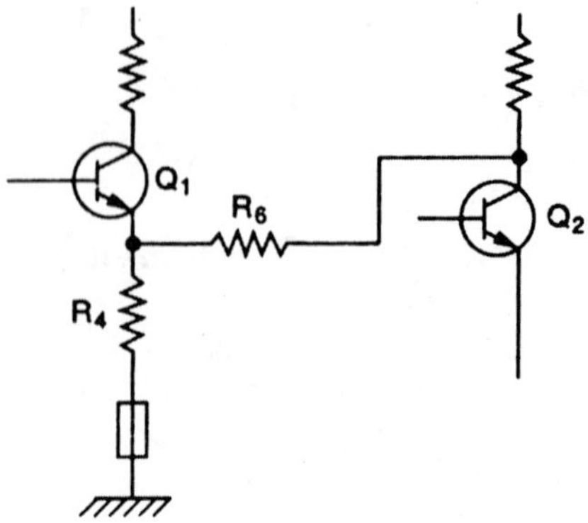

Fig 10.15. AC equivalent circuit for feedback network at low and medium range of frequencies ($X_{C_2} = 0$)

A practical two stage capacitor coupled voltage series feedback amplifier is depicted in Fig. 10.14 and its equivalent circuit in Fig. 10.15 at low and medium frequencies. For this circuit, the feedback factor is

$$\beta = \frac{R_4}{R_4 + R_6} \qquad \text{...(10.74)}$$

If $\qquad A\beta >> 1$

$$A_{fb} = \frac{A}{1 + A\beta} \simeq \frac{1}{\beta} \qquad \text{...(10.75)}$$

$$= \frac{R_4 + R_6}{R_4} = 1 + \frac{R_6}{R_4} \qquad \text{...(10.76)}$$

Typically $R_6 >> R_4$ and thus

$$A_{fb} \simeq \frac{R_6}{R_4} \qquad \text{...(10.77)}$$

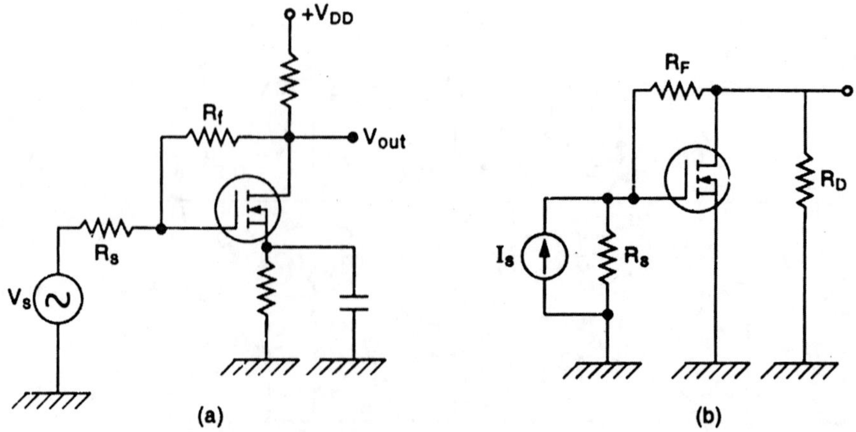

Fig 10.16. (a) Voltage shunt feedback amplifier circuit (b) Electrical equivalent circuit

10.9 VOLTAGE SHUNT FEEDBACK AMPLIFIER CIRCUIT

Let us consider the voltage shunt feedback amplifier circuit of Fig. 10.16. For the amplifier circuit with no feedback

$$A = \frac{V_{out}}{I_{in}} = -g_m R_D \cdot R_s \qquad ...(10.78)$$

The feedback factor is

$$\beta = \frac{I_f}{V_{out}} = -\frac{1}{R_F} \qquad ...(10.79)$$

With feedback, the gain of the circuit is

$$A_{fb} = \frac{V_{out}}{I_s} = \frac{A}{1 + A\beta}$$

$$= \frac{-g_m R_D R_s}{1 + (-1/R_F)(-g_m R_D R_s)} \qquad ...(10.80)$$

$$= \frac{-g_m R_F R_D R_s}{R_F + g_m R_D R_s} \qquad ...(10.81)$$

The closed loop voltage gain of this circuit is

$$(A_{fb})_V = \frac{V_{out}}{I_s} \cdot \frac{I_s}{V_{in}} = \frac{-g_m R_D \cdot R_F}{R_F + g_m R_D R_s} \qquad ...(10.82)$$

EXAMPLE 10.9 *Find the voltage gain with and without feedback for the circuit of Fig. 10.16. The values of the parameters are $R_D = 4{\cdot}7\ k\Omega$, $R_s = 1\ k\Omega$, $R_F = 10\ M\Omega$ and $g_m = 2\ mS$.*

SOLUTION Without feedback, the voltage gain is

$$A_V = -g_m R_D = -(2 \times 10^{-3})(4{\cdot}7 \times 10^3) = -9{\cdot}4$$

The gain with feedback is

$$(A_{fb})_{\text{Voltage}} = (-g_m R_D) \times \frac{R_F}{R_F + g_m R_s R_D}$$

$$= (-9{\cdot}4) \times \frac{10 \times 10^6}{10 \times 10^6 + (9{\cdot}4) \times 1 \times 10^3}$$

$$\simeq -9{\cdot}4.$$

EXAMPLE 10.10 *Repeat Example 10.9 when the gain $g_m = 5 \times 10^{-3}$ S and $R_F = 15\ k\Omega$ for an JFET amplifier.*

SOLUTION The voltage gain without feedback is

$$A_V = -g_m R_D = -5 \times 10^{-3} \times 4{\cdot}7 \times 10^3 = -23{\cdot}5$$

The voltage gain with feedback

$$(A_V)_{fb} = -g_m R_D \frac{R_F}{R_F + g_m R_D R_s}$$

$$= \frac{-23{\cdot}5 \times 15 \times 10^3}{15 \times 10^3 + 23{\cdot}5 \times 1 \times 10^3}$$

$$= (-23{\cdot}5) \times \frac{15}{38{\cdot}5}$$

$$= -9{\cdot}16$$

10.10 CURRENT SERIES FEEDBACK

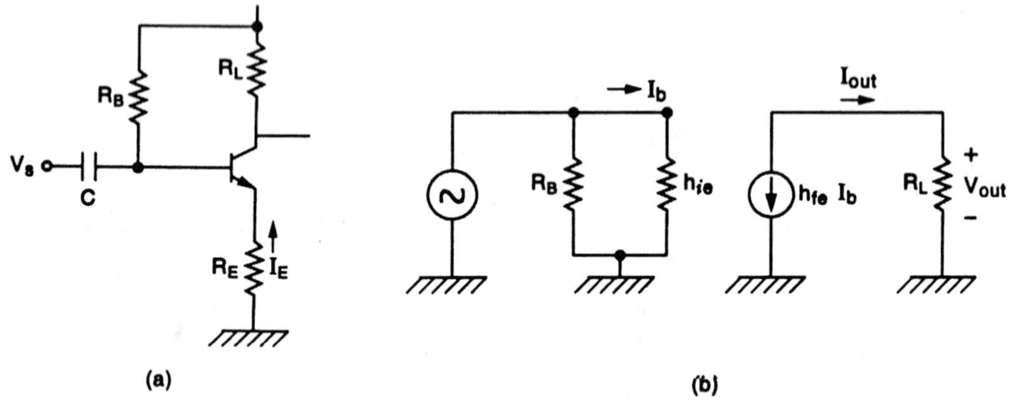

Fig 10.17. (a) Circuit diagram of current series feedback (b) Electrical equivalent circuit

The circuit diagram of a current series feedback type network is depicted in Fig. 10.17. The voltage drop across R_E is $R_E\, I_E = R_E\, I_{out}$. And this voltage drop reduces the signal voltage V_s. Thus the output is reduced. The current series feedback can be removed by removing R_E or bypassing it through a capacitance.

The gain without feedback $= A = \dfrac{I_{out}}{V_{in}} = \dfrac{-I_b\, h_{fe}}{I_b\,(h_{ie} + R_E)}$

$$= \frac{-h_{fe}}{h_{ie} + R_E} \quad \ldots(10.83)$$

The feedback factor $\beta = \dfrac{V_f}{I_{out}} = \dfrac{-I_E\, R_E}{I_{out}} = -R_E.$ $\quad\ldots(10.84)$

The input impedance $= Z_{in} = R_B \,||\, (h_{ie} + R_E)$

$$\approx h_{ie} + R_E \quad \ldots(10.85)$$

The output impedance is $Z_{out} = R_L$ $\quad\ldots(10.86)$

The closed loop gain is

$$A_{fb} = \frac{I_{out}}{V_s} = \frac{A}{1 + A\beta} = \frac{-h_{fe}/h_{ie}}{1 + (-R_E)\left(\dfrac{-h_{fe}}{h_{ie} + R_E}\right)}$$

$$\simeq \frac{-h_{fe}}{h_{fe}\, R_E + h_{ie}} \quad \ldots(10.87)$$

The voltage gain with feedback is

$$(A_V)_{fb} = \frac{V_{out}}{V_s} = \frac{I_{out}}{V_s} \cdot R_L$$

$$\simeq \frac{h_{fe} R_L}{h_{ie} + h_{fe} R_E} \qquad ...(10.88)$$

EXAMPLE 10.11 *Find the voltage gain for the circuit of Fig. 10.18 (consider $h_{fe} = 150$ and $h_{ie} = 1000\ \Omega$ for the transistor).*

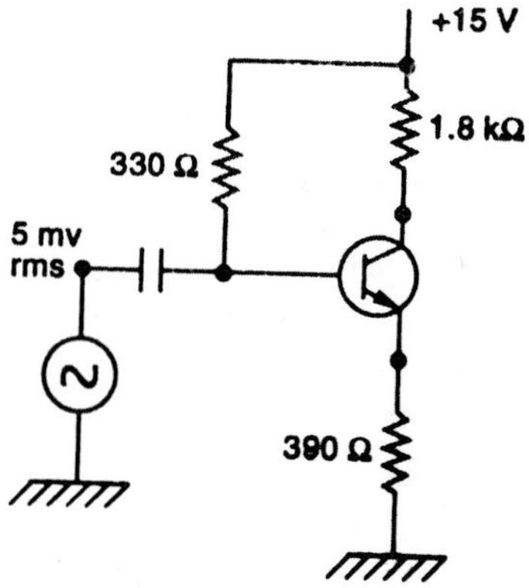

Fig 10.18. Reference Example 10.10

SOLUTION

$$\text{Gain } A = \frac{I_{out}}{V_{in}} = \frac{-h_{fe}}{h_{ie} + R_E} = \frac{-150}{1000 + 390} = -0{\cdot}108$$

$$\text{Feedback factor } \beta = \frac{V_f}{I_{out}} = -R_E = -390$$

$$1 + A\beta = 1 + (-0{\cdot}108)(-390) = 43{\cdot}09$$

The closed loop gain is

$$A_{fb} = \frac{I_{out}}{V_s} = \frac{A}{1 + A\beta}$$

$$= \frac{-0{\cdot}108}{43{\cdot}09} = -2{\cdot}51 \times 10^{-3}$$

The closed loop voltage gain is

$$(A_V)_{fb} = \frac{V_{out}}{V_s} = A_{fb} \cdot R_L = (-2{\cdot}51 \times 10^{-3})(1{\cdot}8 \times 10^3)$$

$$= -4{\cdot}51$$

EXAMPLE 10.12 *The midband gain of an amplifier is 200. For the feedback transfer ratio of $\beta = 0{\cdot}5$, find the midband gain.*

SOLUTION $$A_{fb} = \frac{A_0}{1 + \beta A_0} = \frac{200}{1 + 0{\cdot}05 \times 200} = 18{\cdot}18$$

EXAMPLE 10.13 *For the problem in Example 10.11, find the lower cut off frequency with negative feedback implemented, if the lower cut off frequency without feedback is $f_L = 25$ Hz.*

SOLUTION $f_{Lf} = \dfrac{f_L}{1 + A_0\beta} = \dfrac{25}{1 + 0{\cdot}05 \times 200} = 2{\cdot}28$ Hz.

EXAMPLE 10.14 *If the upper cut off frequency is 100 kHz for the amplifier in Example 10.11, find the upper cut off frequency with negative feedback.*

SOLUTION $(f_H)_f = f_H(1 + A_0\beta) = (100 \times 10^3) \times [1 + 0{\cdot}05 \times 200]$
$= 1{\cdot}1$ MHz.

Review Questions

1. What do you understand by a feedback amplifier ?
2. Distinguish between positive and negative feedback amplifiers.
3. Derive the expression for gain of a feedback amplifier with feedback factor β.
4. What are the advantages of negative feedback ?
5. What do you understand by "gain stability"?
6. Derive an expression for distortion in a feedback amplifier.
7. What do you understand by gain-bandwidth product ? Find its value in negative feedback amplifier.
8. Distinguish between voltage series feedback and voltage shunt feedback in a feedback amplifier. Explain them with suitable circuit diagrams.
9. Distinguish bewteen current series feedback and current shunt feedback in an amplifier. Illustrate with the help of suitable circuit diagrams.
10. Derive expressions for the input impedance of a voltage series feedback and voltage shunt feedback.
11. Derive expressions for the input impedance of a current series feedback and current shunt feedback.
12. Draw the circuit diagram of a *FET* amplifier with voltage series feedback.
13. Derive equation (10.82).
14. Derive equation (10.87).

Objective Type Questions

Choose the correct or most appropriate answer.

1. Using a feedback amplifier it is possible to
   (a) increase the gain (b) control the output
   (c) decrease its input (d) stabilize the gain
2. The following is not an advantage of a negative feedback amplifier
   (a) reduction in noise (b) less phase distortion
   (c) less harmonic distortion (d) lower *BW*
3. The following is a feature of a negative feedback amplifier
   (a) high distortion (b) high noise
   (c) high bandwidth (d) high fidelity
4. The gain of an amplifier with negative feedback is given by
   (a) $\dfrac{A}{1 - A\beta}$ (b) $\dfrac{A}{1 + A\beta}$

(c) $\frac{1}{A\beta}$ (d) $A\beta$

5. The distortion with negative feedback in an amplifier (if D is the distortion without feedback) is given by
(a) $\frac{D}{1-A\beta}$ (b) $D(1+A\beta)$
(c) $D(1-A\beta)$ (d) $\frac{D}{1+A\beta}$

6. A negative feedback amplifier has $A = 100$, $\beta = 0{\cdot}03$ and $V_{in} = 100$ mV. The feedback voltage is
(a) 0·03 V (b) 0·06 V
(c) 0·075 V (d) 0·15 V

7. A negative feedback amplifier has a gain of 100. If the amplifier gain without feedback is 200, the feedback factor is
(a) 0·001 (b) 0·002
(c) 0·004 (d) 0·005

8. An amplifier with 10% negative feedback has an open loop gain of 50. If the open loop gain increases by 10%, the % change in closed loop gain is
(a) 1·56% (b) 0·78%
(c) 0·01% (d) 0·02%

9. An amplfier possesses an open loop gain of 400 and feedback factor of 0·1. If the open loop gain changes by 10%, the percentage change in closed loop gain is
(a) 0·01% (b) 0·05%
(c) 0·5% (d) 1·0%

10. In a voltage series negative feedback amplifier, the input resistance R_i and output resistance R_0 (Compared to amplifier without feedback) respectively
(a) increases, increases (b) increases, decreases
(c) decreases, increases (d) decreases, decreases

11. In a current series negative feedback amplifier, the input resistance R_i and output resistance R_0 (Compared to amplifier without feedback) respectively
(a) increases, increases (b) increases, decreases
(c) decreases, increases (d) decreases, decreases

12. In a voltage shunt negative feedback amplifier, the input resistance R_i and output resistance R_0, (Compared to amplifier without feedback) respectively
(a) increases, increases (b) increases, decreases
(c) decreases, increases (d) decreases, decreases

13. In a current shunt negative feedback amplifier, the input resistance R_i and output resistance R_0 (Compared to amplifier without feedback) respectively
(a) increases, increases (b) increases, decreases
(c) decreases, increases (d) decreases, decreases

14. In a voltage series negative feedback amplifier, the voltage gain and bandwidth (Compared to amplifier without feedback) respectively
(a) increases, increases (b) increases, decreases
(c) decreases, increases (d) decreases, decreases

Exercise Problems

1. A negative feedback amplifier has $A = 200$, $\beta = 0{\cdot}01$ and $V_{in} = 5$ mV. Find (a) the gain with feedback (b) output voltage (c) feedback factor (d) feedback voltage.

2. In a feedback amplifier, the input voltage is $V_{in} = 2$ mV, output voltage with feedback $= 10\text{ V} = V_0'$ $\beta V_0' = 300$ mV. Find the values of A_{fb} and β.

3. For a negative feedback amplifier $A = 200$, $\beta = 0{\cdot}015$, output voltage with feedback $V_0' = 1V$. Find (a) the gain with feedback. (b) feedback factor (c) input voltage (d) feedback voltage.
4. A feedback amplifier has an open loop gain of 400 and feedback factor = 0·1. If the open loop gain changes by 2%, determine the percentage change in closed loop gain.
5. An amplifier has a gain of 200 and distortion of 6%. Find (a) output signal voltage (b) distortion voltage (c) output voltage if the input signal of 0·25 V is applied to the amplifier.
6. For a two stage amplifier, the gain is 200, the second stage having 10% of the output voltage as negative feedback and has – 200 as forward gain. Find (a) the gain of the first stage (b) the second harmonic distortion if the second stage increases 2% second harmonic without feedback. Assume that the first stage does not introduce distortion.
7. An *RC* coupled amplifier has a mid frequency gain of 200 and a frequency response from 100 Hz to 30 kHz. A negative feedback with $\beta = 0{\cdot}03$ is introduced into the circuit. Find the new bandwidth.
8. The midband gain of an amplifier is 275. For the feedback transfer ratio of $\beta = 0{\cdot}04$, find the midband gain.
9. For the problem of 8, find the lower cut off frequency with negative feedback amplifier if the lower cut of frequency without feedback is 33 Hz.
10. If the upper cut off frequency is 100 kHz with the amplifier of Problem 8, find the upper cut off frequency with negative feedback.
11. An amplifier has a closed loop gain of 500,000 and input resistance to the base of the first transistor of $Z_b = 1$ kΩ. Negative feeback to be used to increase Z_b to 2 MΩ. Find β and the closed loop gain of the circuit with negative feedback.

Chapter 11

OSCILLATORS AND MULTIVIBRATORS

11.1 INTRODUCTION

An oscillator produces a sinusoidal or other repetitive waveforms at the output even though no external signal is applied at the input. A typical oscillator circuit may be constructed using a solid state device (like *BJT*, *FET*, *OP AMP*) and passive elements. An oscillator, in fact, converts or transforms d.c. applied power of the power supply source to a.c. power. Oscillators have various applications in radio, TV transmitters and receivers and in various measurement systems.

Oscillators can be classified as

(1) Sinusoidal oscillators and

(2) Relaxation oscillators

In a sinusoidal oscillator, the output waveform is sinusoidal or nearly sinusoidal and the oscillations have a constant frequency. If the oscillations of waveforms are nonsinusoidal in nature we get a type of oscillator termed as relaxation oscillator.

A basic oscillator circuit concept is depicted in Fig. 11.1.

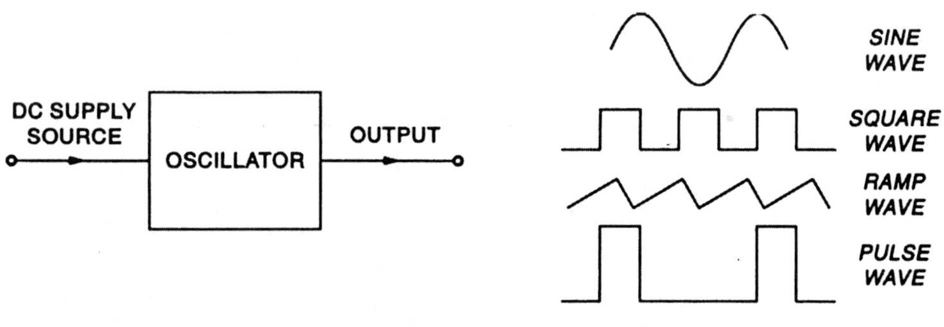

Fig 11.1. Schematic of oscillator circuit with different possible output waveforms

An oscillator is called a feedback oscillator, if a positive feedback amplifier is used with an overall amplification of infinity. The output voltage of a feedback oscillator can be either sinusoidal or non sinusoidal. Another oscillator type is possible which is known as negative resistance oscillator. In **resistance oscillator** the negative slope of the current voltage characteristics of cert utilized. Some such solid state devices are (a) tunnel diodes (b) unijunction transistors etc. Negative resistance oscillators can also provide sinusoidal or nonsinusoidal output waveforms. Feedback type sinusoidal oscillators can be classified as *LC* (inductor-capacitor) and *RC* (resistor-capacitor) oscillators. Oscillators can also be classified on the basis of the frequency range of the generated signals. Audio frequency oscillators generate signals in the audio frequency range. Radio frequency oscillators generate signals in the radio frequency range namely (20 kHz to 30 MHz). VHF (Very high frequency) oscillators generate signals in the VHF range namely 30 MHz to 300 MHz. UHF or Ultra high frequency oscillators generate signals in the 300 MHz to 3 GHz range. The frequency range 3 GHz to 30 GHz is called microwave frequency range. Oscillators that generatė waveform in this frequency range are called mirowave oscillators.

RC oscillators are normally employed for generation of audio frequency signals. A number of different types of *LC* oscillators are available (like tuned collector, Harteley, Colpitts etc.). *LC* oscillators depend on the practical values of *L* and *C* and their use mainly in the generation of radio frequency signals. Sometimes crystal oscillators (which depend for its operation on the piezoelectric properties of certain crystals like Quartz, tourmaline etc.) are employed. The main advantage of crystal oscillators is their high frequency stability properties.

11.2 BASIC OSCILLATOR PRINCIPLES

In a feedback type oscillator, the **overall gain** is made **infinity** by **positive feedback**. In the negative resistance oscillator, the desired oscillations are provided by the **negative resistance** property of the solid state devices. In the oscillator circuit, the amplitude of voltage or current oscillation decays with time due to the dissipation of energy in the resistance contained in the circuit. The negative resistance compensates for the positive resistance of the circuit.

In an oscillator, **no external signal is applied.** With the switching ON of the circuit, **a small noise voltage is produced.** The frequency spectrum of noise voltages are very wide. The noise spectrum contains the desired signal component at the voltage required by the oscillator. The **initial voltage component** at the input is thus obtained from the **noise voltage**. The output energy is provided by the d.c. voltage supply. This is shown pictorially in Fig. 11.2.

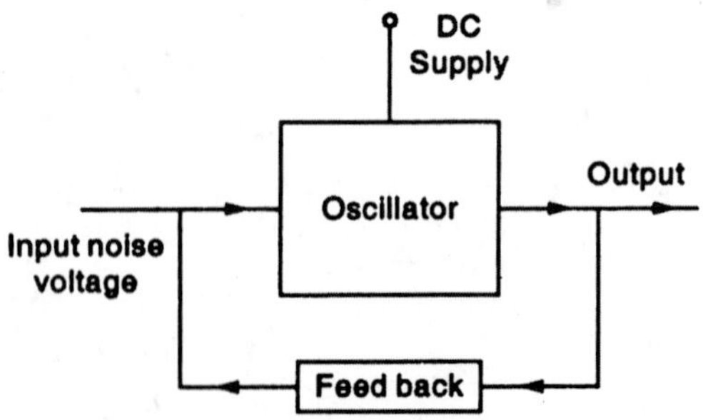

Fig 11.2. Pictorial representation showing help extended by noise signals in starting the oscillations

For a positive feedback amplifier, the overall voltage gain is written as

$$A_{fb} = \frac{A}{1 - A\beta} \qquad \ldots(11.1)$$

where $A\beta$ is the loop gain. In equation (11.1) when $A\beta = 1$,

$$A_{fb} \to \infty \qquad ...(11.2)$$

This implies that the amplifier provides an output signal even if there is no applied signal. The amplifer is also called an oscillator. To understand the properties clearly, let us assume that a transient disturbance provides an output signal V_0 although no external signal is applied. A fraction of V_0 (namely βV_0) is fed back to the input. This voltage βV_0 is then amplified by the gain A and the output signal becomes $A\beta V_0$. If this quantity $A\beta V_0$ becomes equal to V_0

$$A\beta V_0 = V_0 \qquad ...(11.3)$$

the output voltage regenerates itself. We say that the amplifier oscillates. The condition that the loop gain

$$A\beta = 1 \qquad ...(11.4)$$

is also called the **Barkhausen criterion**. This condition says that

$$|A\beta| = 1 \qquad ...(11.5)$$

and the phase angle of $A\beta$ is zero or integral multiple of 360°. The two basic condition for oscillation of such a feedback amplifier are

(1) Feedback regenerative (2) $A\beta = 1$

$A\beta$ is normally a complex quantity. For positive feedback, $|1 - A\beta| < 1$ and the amplifier may break into oscillations. For an amplifier to be stable, the response to a transient disturbance should decay with time. However, if the response to a transient stimulus persists or grows as time elapses, then the system is unstable. Ultimately the response is limited by nonlinearities in the circuit. The Nyquist condition provides the condition for instability of the amplifier.

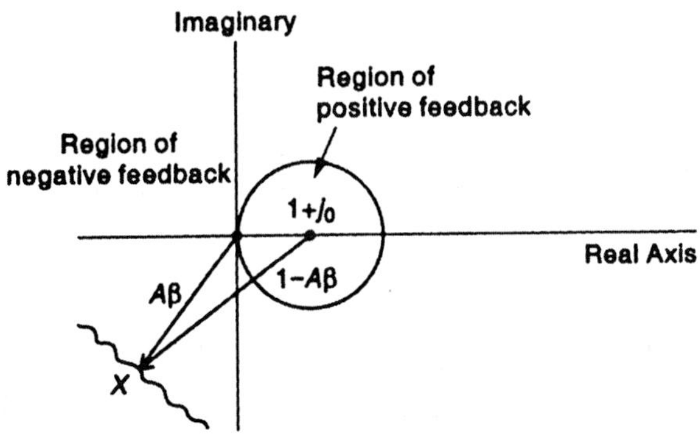

Fig 11.3. Graph illustrating the regions of positive and negative feedback in the complex plane

Let us consider Fig. 11.3 which depicts the graph illustrating the regions of positive and negative feedback in the complex phase. Let us consider a point X on the locus of $A\beta$ while frequency is varied. If X lies outside the circle of unit radius about the point $(1 + j0)$, $|1 - A\beta| > 1$ and the feedback is negative. If X lies inside the circle, we find $| -A\beta| < 1$ and this implies positive feedback. If however X is located at $(1 + j0)$, the gain with feedback will be infinite and oscillation will take place. The locus of the phasor $A\beta$ will trace a closed curve in the complex plane as the frequency changes from $-\infty$ to $+\infty$. Nyquist criterion specifies that if this closed curve encloses the point $(1 + j0)$, the amplifier will have instability and will oscillate. On the other hand if the curve does not pass through or enclose the point $(1 + j0)$, the amplifier will be stable. This means that a positive feedback amplifier will not oscillate unless Nyquist's criterion is satisfied.

When the oscillations grow in a unstable amplifier, the output signal amplitude grows with time. However, as time passes, the nonlinearity of the devices comes into play and the oscillation amplitudes are contained and a steady state is built up. In the steady state, the loop gain is unity and the oscillations are sustained. The frequency is determined by the *LC* and *RC* circuits of the oscillator. The *LC* and *RC* circuits are also called as frequency determining networks. The requirements of a feedback oscillator are

(1) **an amplifer**

(2) **positive feedback**

(3) **a frequency determining *LC* or *RC* network**

(4) ***dc* power supply as energy source.**

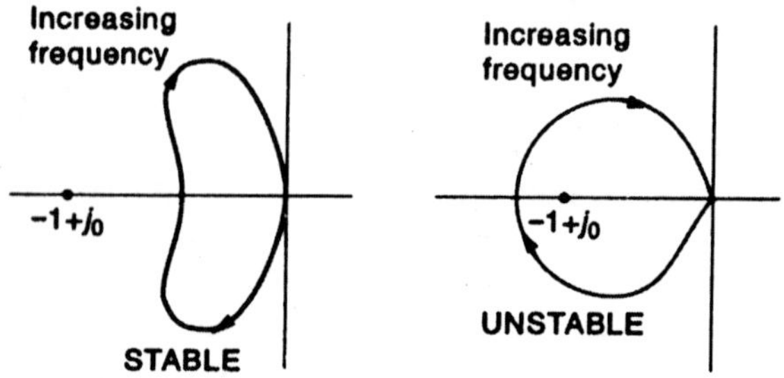

Fig 11.4. Nyquist plots for (a) stable (b) unstable system

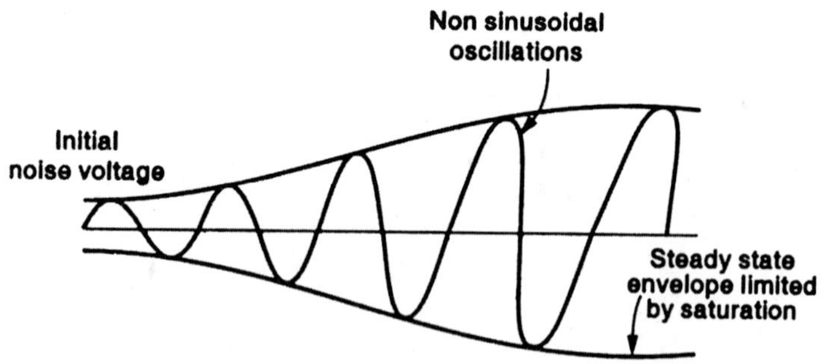

Fig 11.5. Building up of oscillations from the initial noise voltage

The circuit diagram of a tuned collector oscillator is depicted in Fig. 11.6. The circuit is a common emitter configuration. The resistor R_f provides the feedback path. The frequency determining network is the *LC* tuned circuit formed by L_p and C_p.

The resistors R_1, R_2 and R_3 provide the d.c. biasing circuit for the transistor. The capacitor C_1 provides a shunt path for the alternating component of current. The resistor R_B has a high value. Besides providing regenerative feedback, it reduces the loading of the collector circuit by the low resistance input circuit of the transistor. It also reduces the input nonlinear distortion. The secondary L_s of the transformer carries a small current such that the voltage drop in the transformer primary is very small.

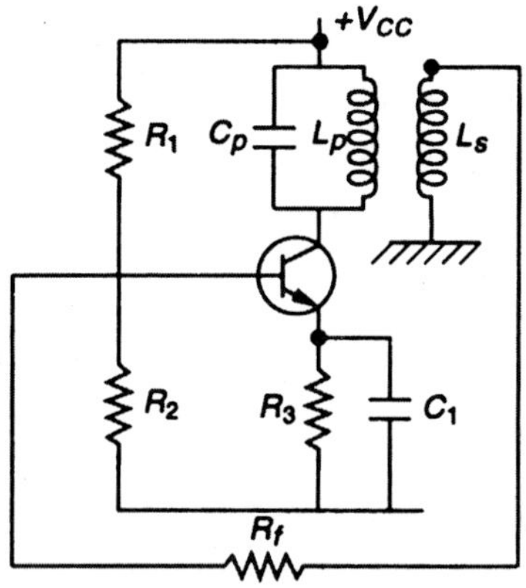

Fig 11.6. Circuit diagram of a tuned collector oscillator

The *CE* transistor amplifer provides a phase shift of 180° between the input and output voltages. The transformer provides another 180° needed for oscillation. The frequency of oscillation is given by

$$f = \frac{1}{2\pi \sqrt{L_p C_p}} \quad ...(11.6)$$

The frequency of oscillation can thus be varied by varying either L_p or C_p. With the available values of L_p and C_p, the tuned collector oscillators generate oscillations in the *RF* range. The tank circuit comprising L_p and C_p has a high Q since the resistance of the transformer primary is very small. At resonance, the resistance offered by the tank circuit is

$$R_t = \frac{1}{CR} \quad ...(11.7)$$

Fig. 11.7 and 11.8 depict two variations of the Harteley oscillator employing a *BJT* and an *FET* respectively.

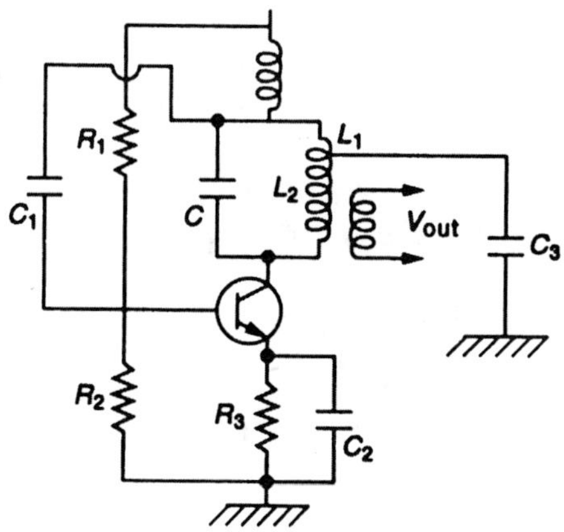

Fig 11.7. Transistorized Harteley oscillator

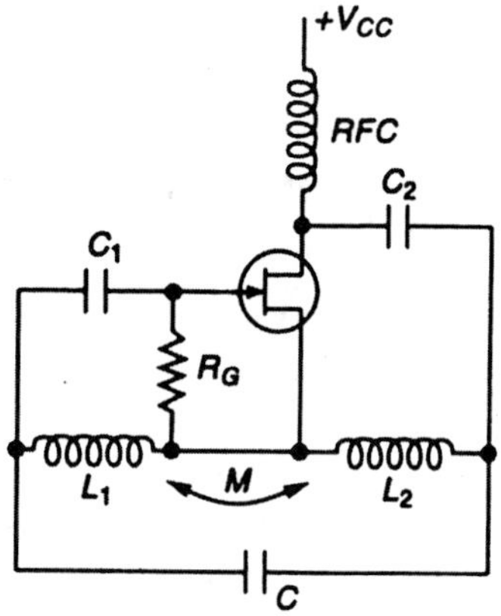

Fig 11.8. *FET* type Harteley oscillator

In the Harteley oscillator, the circuit frequency for oscillation is given by

$$f_0 = \frac{1}{2\pi \sqrt{L_e C}} \qquad \text{...(11.8)}$$

where $L_e = L_1 + L_2 + 2M$ is the equivalent inductance.

The inductor L_1 and L_2 determine the attenuation of the feedback network namely

$$\beta = \frac{L_2}{L_1} \qquad \text{...(11.9)}$$

For starting of oscillations

$$A_V > (L_1/L_2) \qquad \text{...(11.10)}$$

If the tank circuit is loaded, the value of Q is decreased and f_0 decreases.

The Harteley oscillators normally generate oscillation in the r.f. frequency range. The frequency of oscillations can be varied by changing L or C or both. Smooth variation of C is possible but smooth variation in the value of L_1 or L_2 is difficult. Since variable capacitors possess capacitances in the range 50 pF to 500 pF, (a ratio of 10 : 1), the frequency of oscillation can be varied in the ratio of 3 : 1.

11.3 THE COLPITTS OSCILLATORS

The Colpitts oscillator employ an *LC* feedback loop to provide the necessary phase shift. The approximate frequency of oscillation is

$$f_0 = \frac{1}{2\pi \sqrt{LC_e}} \qquad \text{...(11.11)}$$

where C_e is the equivalent series capacitance

$$C_e = \left(\frac{C_1 C_2}{C_1 + C_2}\right) \qquad \text{...(11.12)}$$

The attennuation factor β for the Colpitts oscillator is given by the expression

$$\beta = \frac{V_f}{V_{out}} \simeq \frac{1/\omega C_1}{1/\omega C_2} = \frac{C_2}{C_1} \qquad \text{...(11.13)}$$

Since, according to Barkhausen criterion $A_v \beta > 1$

$$A_V \geq \frac{C_1}{C_2} \qquad \text{...(11.14)}$$

for the starting of oscillations.

The input impedance of the transistor loads the resonant feedback circuit and reduces the Q value of the circuit. If Q is not large, the resonant frequency for the parallel resonant circuit will be

$$f_0 = \frac{1}{2\pi \sqrt{LC_e}} = \sqrt{\frac{Q^2}{Q^2 + 1}} \qquad \text{...(11.15)}$$

Fig. 11.9 depicts the circuit diagram for the transistorized Colpitts oscillator. Using a *FET* instead of the *BJT*, the loading effect can be reduced appreciably. The oscillator loading effect is shown in Fig. 11.11. Besides Fig. 11.12 also shows the transformer coupling circuit for reduced loading effect.

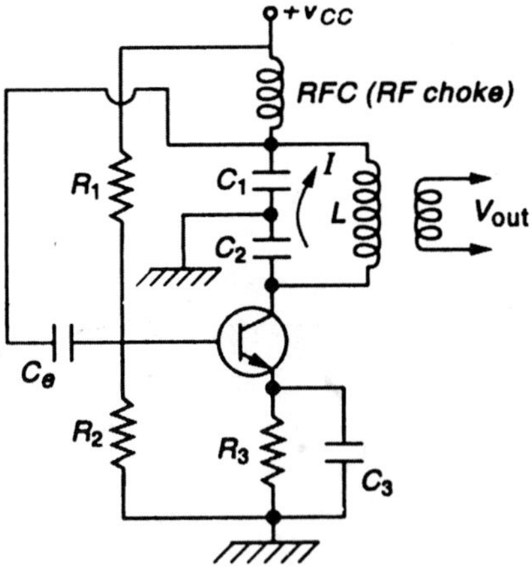

Fig 11.9. Transistorized Colpitts oscillator

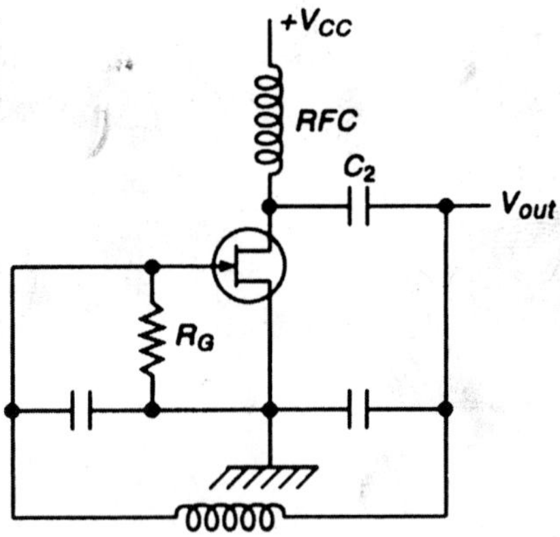

Fig 11.10. Colpitts oscillator employing *FET*

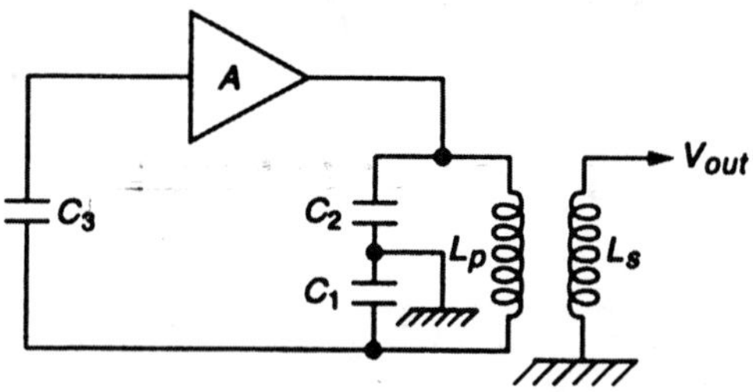

Fig 11.11. Schematic diagram depicting oscillator loading effect

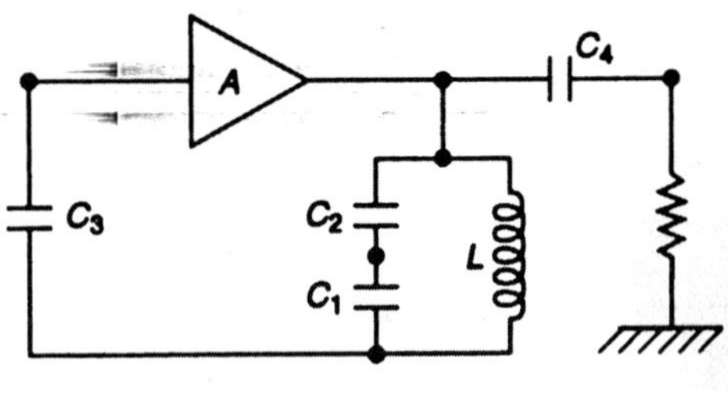

Fig 11.12. Transformer coupling for reduced loading effect

EXAMPLE 11.1 *Find the frequency of oscillation for the Colpitts oscillator shown in Fig. E11.1.*

SOLUTION C_e = Equivalent capacitance

$$= \frac{(0{\cdot}02\ \mu\text{F})\ (0{\cdot}2\ \mu\text{F})}{0{\cdot}22\ \mu\text{F}} = 0{\cdot}0182\ \mu\text{F}$$

$$f_0 = \frac{1}{2\pi\sqrt{LC_e}} = \frac{1}{2\pi\sqrt{(20\ \text{mH})\ (0{\cdot}0182)}}$$

$$= \frac{1}{2\pi\sqrt{2 \times {\cdot}0182 \times 10^{-9}}} = 8{\cdot}34\ \text{kHz}$$

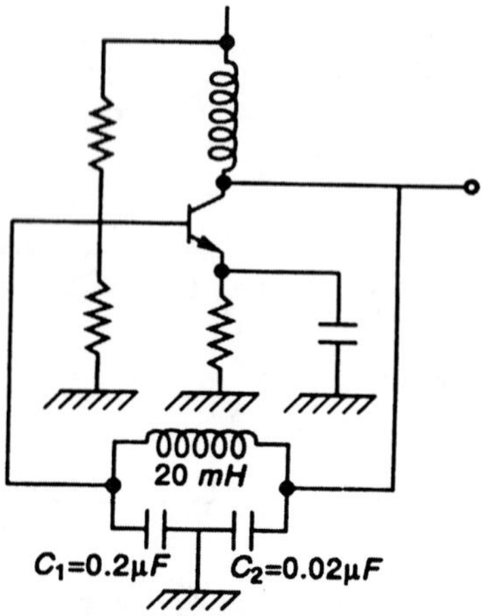

Fig E11.1 Reference Example 11.1

EXAMPLE 11.2 *Find the frequency of oscillation for the circuit in Fig. E11.1 if Q has a value (a) 10 (b) 5.*

SOLUTION

$$\text{(a) } f_0 = \frac{\sqrt{\dfrac{Q^2}{Q^2+1}}}{2\pi\sqrt{LC_e}} = \frac{1}{2\pi\sqrt{20\text{ mH}\times 0{\cdot}0182\ \mu\text{F}}}\times\sqrt{\frac{100}{101}} = 8300\text{ Hz}$$

$$\text{(b) } f_0 = \frac{1}{2\pi\sqrt{LC}}\sqrt{\frac{25}{25+1}} = \frac{1}{2\pi\sqrt{(20\text{ mH}\times 0{\cdot}0182\ \mu\text{F})}} = \sqrt{\frac{25}{26}} = 8179\text{ Hz.}$$

11.4 CRYSTAL OSCILLATOR

In a crystal oscillator, a piezoelectric crystal is employed in the feedback loop to control the frequency. The crystal oscillator provides a very stable and accurate type of oscillator.

The Quartz crystal exhibits the piezoelectric effect. When a mechanical stress is applied across the faces of the crystal, a difference of potential develops across the opposite faces of the crystal. Alternatively, a voltage applied across one set of faces of the crystal causes mechanical distortion in crystal shape. The electrical equivalent circuit of the crystal provides a resonant circuit. The crystal equivalent circuit has both a series resonant frequency (f_1) and an antiresonant frequency (f_2).

The Q of the crystal is typically 20,000 although Q values as high 5×10^5 can be achieved. The series resonant frequency takes place when the reactances of the inductor and capacitor in the series *RLC* leg is equal and opposite. ($X_L = X_C$). The impedance of the series *RLC* leg is *R* at resonance. The antiresonant frequency (f_2) occurs when the reactance of the series resonant leg equals the reactance of the capacitor C_p. At series resonance,

$$\omega_1 L - \frac{1}{\omega_1 C} = 0 \qquad \text{...(11.16)}$$

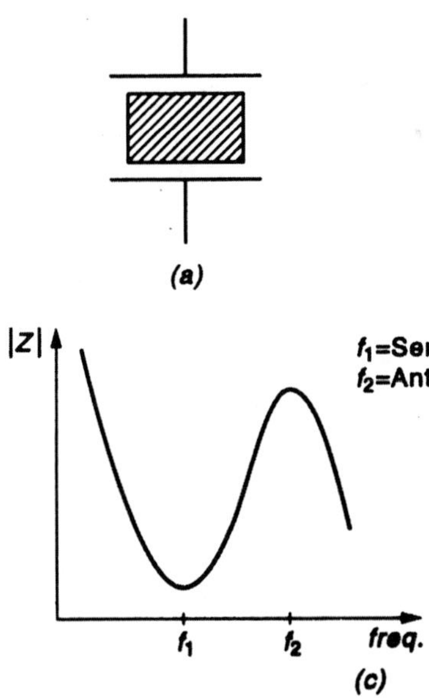

Fig 11.13. Variation of crystal impedance with frequency (a) crystal oscillator (e) electrical equivalent circuit (c) crystal impedance vs. frequency curve

$$f_1 = \frac{\omega_1}{2\pi} = \frac{1}{2\pi\sqrt{LC}} \qquad \ldots(11.17)$$

Similarly, the parallel resonance frequency, f_2 is defined by

$$\omega_2 L - \frac{1}{\omega_2 C} = \frac{1}{\omega_2 C_p} \qquad \ldots(11.18)$$

or,
$$f_2 = \frac{\omega_2}{2\pi} = \frac{1}{2\pi}\sqrt{\frac{C + C_p}{LCC_p}} \qquad \ldots(11.19)$$

The fundamental frequency of a crystal is the smallest frequency at which it is naturally resonant. The fundamental frequency depends upon the mechanical dimensions of the crystal, type of cut, but is inversely proportional to the thickness of the crystal slab. Since a crystal slab can not be very small (it will be fractured then), there is an upper limit to the fundamental freuqency. The upper limit is normally limited to 20 MHz. For high frequencies, the crystal should be operated in the overtone mode (integer multiples of the fundamental frequency). Crystal oscillators typically range in frequency from 10 kHz to 20 MHz (fundamental frequencies). The extremely high values of the Q factor (20,000 to 500,000) and stability of the Quartz crystal with respect to time and temperature is the main reason for the extremely high frequency stability of the crystal oscillator in the range from few kHz to several MHz. Two typical crystal oscillator circuits are depicted in Fig. 11.14.

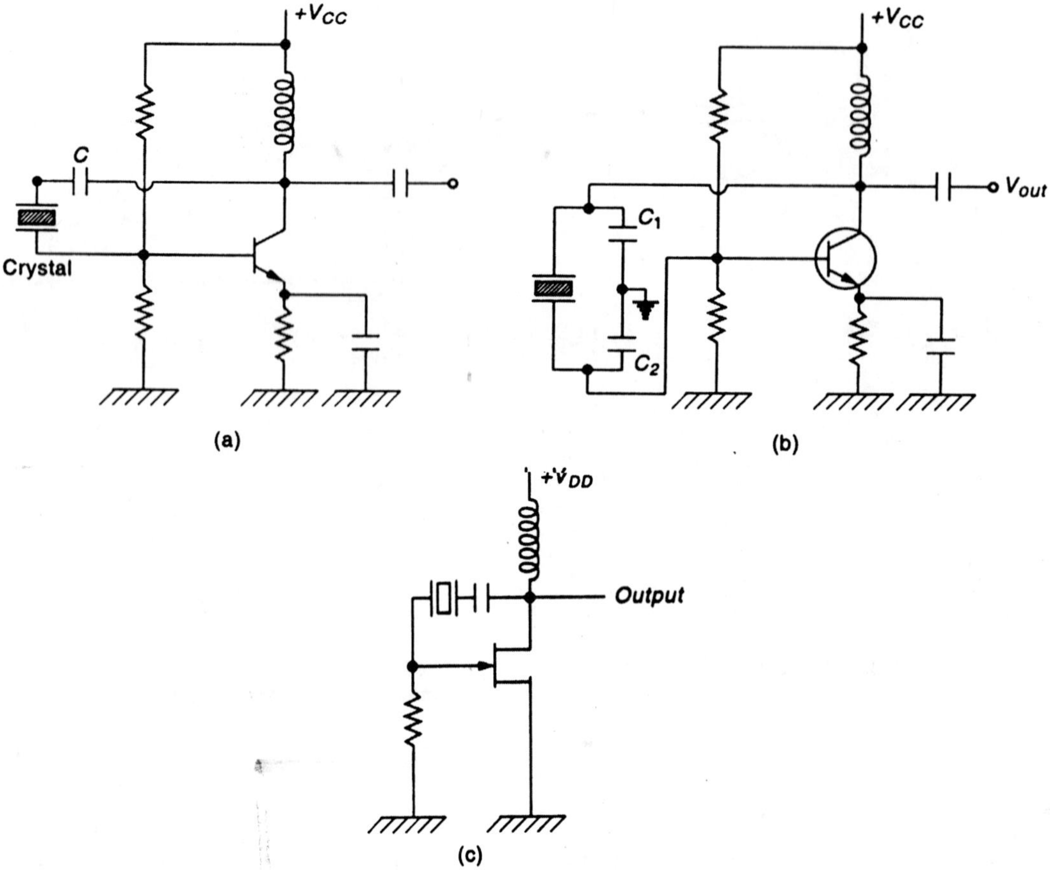

Fig 11.14. Typical crystal oscillator circuits

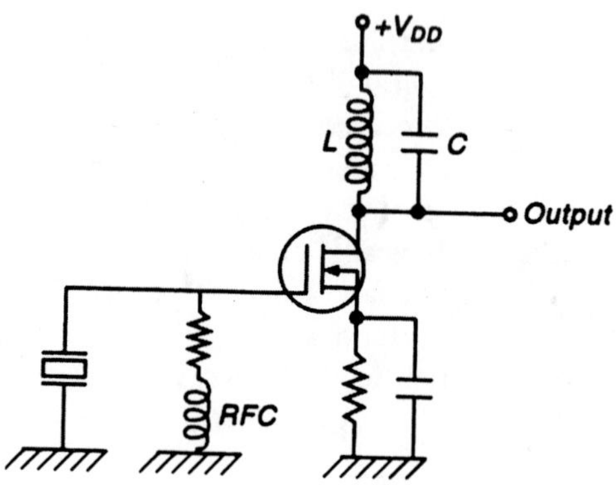

Fig 11.15. *MOSFET* crystal oscillator circuit

EXAMPLE 11.3 *In a piezoelectric crystal oscillator, the oscillation frequency is 3·8 MHz. For this frequency the equivalent inductance and series resistance are 0·2 H and 6000 ohm respectively. Find the value of Q of the crystal.*

SOLUTION The quality factor is

$$Q = \frac{\omega L}{R} = \frac{2\pi f L}{R} = \frac{2\pi \times 3{\cdot}8 \times 10^6 \times 0{\cdot}2}{6000} = 796$$

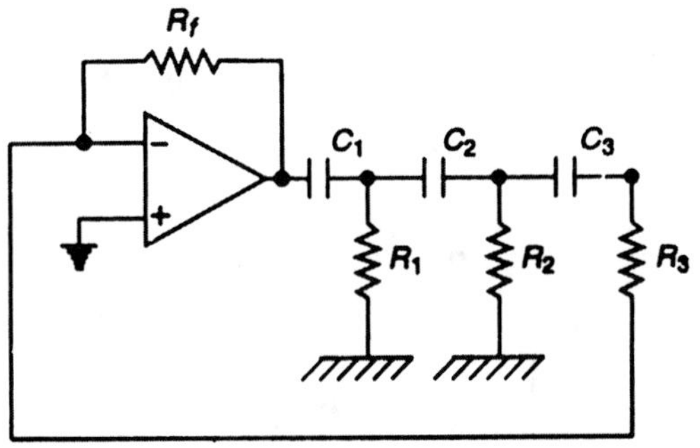

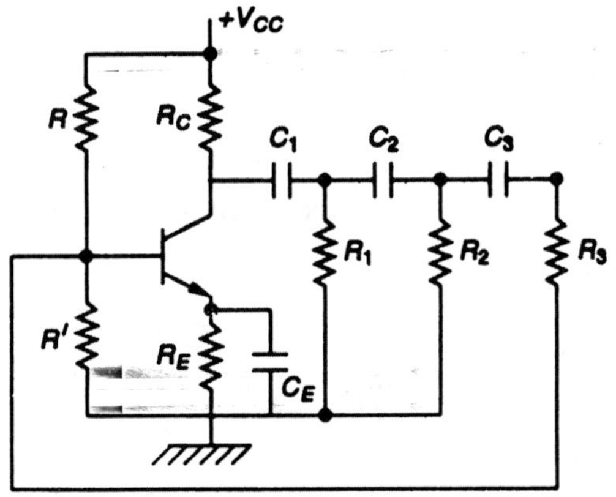

Fig 11.16. Phase shift oscillator circuits (a) circuit using an operational amplifier (b) circuit employing a transistor

11.5 PHASE SHIFT OSCILLATOR

In a phase shift oscillator, an external resistor capacitor network is employed to feedback a portion of the ac amplifier output to the input. Since the single stage amplifier has an internal phase shift of 180°, another 180° phase shift is provided by the feedback network such that the amplifier can work as an oscillator. The *CR* network comprising three equal sections of *RC* networks is connected between the amplifier output and input. The phase shift network depicted works as a phase lead network. The total phase shift of the network is – 180° + 180° = 0°. A phase lag network that provides a phase shift of 180° would work equally well since – 180° – 180° = – 360° = 0.

The frequency of operation of the circuit depends on the values of the resistors and capacitors in the phase shift network. Assuming $R_1 = R_2 = R_3 = R$ and $C_1 = C_2 = C_3 = C$, the frequency of operation can be derived as follows.

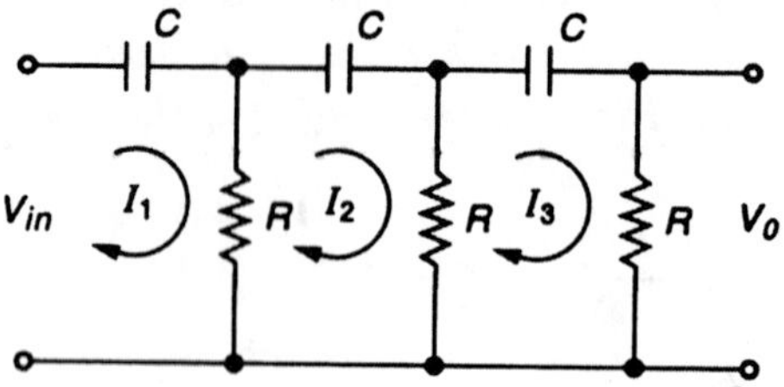

Fig 11.16A. Electrical equivalent of the phase lag network in Fig. 11.16

Using Kirchhoff's law, we may write

$$(R - j/\omega C)\, I_1 - RI_2 + 0I_3 = V_{in} \quad \text{...(11.20)}$$

$$-RI_1 + \left(2R - \frac{j}{\omega C}\right) I_2 - RI_3 = 0 \quad \text{...(11.21)}$$

$$0I_1 - RI_2 + \left(2R - \frac{j}{\omega C}\right) I_3 = 0 \quad \text{...(11.22)}$$

Solving for I_3 using the determinant method, we find

$$I_3 = \frac{\begin{vmatrix} (R - j/\omega C) & -R & V_{in} \\ -R & (2R - j/\omega C) & 0 \\ 0 & -R & 0 \end{vmatrix}}{\begin{vmatrix} (R - j/\omega C) & -R & 0 \\ -R & (R - j/\omega C) & -R \\ 0 & -R & (2R - j/\omega C) \end{vmatrix}} \quad \text{...(11.23)}$$

which when simplified yields

$$I_3 = \frac{R^2 V_{in}}{\left(R - \frac{j}{\omega C}\right)\left(2R - \frac{j}{\omega C}\right)^2 - R^2\left(2R - \frac{j}{\omega C}\right) - \left(R - \frac{1}{\omega C}\right) R^2} \quad \text{...(11.24)}$$

Putting this value in the expression for the amplifier gain one gets

$$\frac{V_0}{V_{in}} = \frac{1}{\left(1 - \frac{5}{4\pi^2 f^2 R^2 C^2}\right)^2 - j\left(\frac{6}{\omega RC} - \frac{1}{\omega^3 R^3 C^3}\right)} \quad \text{...(11.25)}$$

In the phase shift amplifier, the phase shift of the *RC* network must be equal to 180°. This condition will be satisfied when the imaginary term becomes zero at the oscillation frequency.Thus

$$\frac{6}{\omega_0 RC} - \frac{1}{\omega_0 3R^3 C^3} = 0 \quad \text{...(11.26)}$$

or,

$$\omega_0 = \frac{1}{\sqrt{6}\, RC} \quad \text{...(11.27)}$$

which gives the frequency of oscillation as

$$f_0 = \frac{1}{2\pi\sqrt{6}\, RC} = \frac{0{\cdot}065}{RC} \quad \text{...(11.28)}$$

We can also find from expression (11.25), that at the oscillation frequency

$$\frac{V_0}{V_{in}} = \frac{1}{1 - \frac{5}{\omega_0^2 R^2 C^2}} = \frac{1}{1 - \frac{5}{\frac{1}{6}}} = \frac{-1}{29} \qquad \ldots(11.29)$$

The magnitude of attenuation of the feedback network is

$$\beta = \frac{1}{29} \qquad \ldots(11.30)$$

The negative sign in equation (11.29) occurs due to phase inversion.

EXAMPLE 11.4 *Find the frequency of oscillation for the circuit in Fig. 11.16 (b) if $R_1 = R_2 = R_3 = 4{\cdot}7\ k\Omega$ and $C_1 = C_2 = C_3 = 2{\cdot}2\ \mu F$.*

SOLUTION

$$f_0 = \frac{1}{2\pi\sqrt{6}\,RC} = \frac{0{\cdot}065}{(4{\cdot}7 \times 10^3)(2{\cdot}2 \times 10^{-9})}$$

$$= 6{\cdot}286 \text{ kHz}$$

EXAMPLE 11.5 *What should be oscillation frequency if $C_1 = C_2 = C_3 = 4{\cdot}7$ mμF instead of 2·2 mμF. Find also the value of R_f in Fig. 11.16 (a) if $\frac{R_f}{R_3}$ = Gain of the OP AMP.*

SOLUTION

$$f_0 = \frac{1}{2\pi\sqrt{6}\,RC} = \frac{0{\cdot}065}{(4{\cdot}7 \times 10^3)(4{\cdot}7 \times 10^{-9})}$$

$$= 2{\cdot}94 \text{ kHz.}$$

$$R_f = 29\ R_3 = 29 \times 4{\cdot}7\ k\Omega = 136{\cdot}3\ k\Omega.$$

11.6 WIEN BRIDGE OSCILLATOR

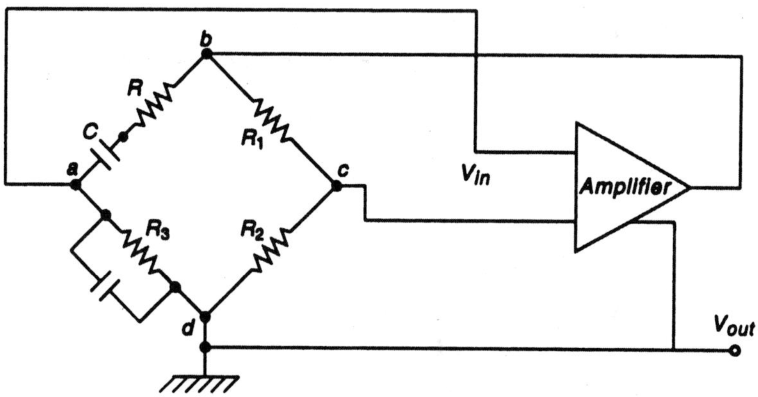

Fig 11.17. The Wien bridge oscillator circuit

Fig. 11.17 depicts the circuit diagram of a Wien bridge oscillator. The output voltage of the bridge network works as the input voltage of the amplifier. The magnitude of $V_i = 0$ when the bridge in balanced. However, for sustained oscillations to be maintained in the circuit, V_i should be non vanishing. The bridge is thus slightly unbalanced by adjustment of the ratio R_1 and R_2. Since the phase shift introduced by the bridge is zero, no phase shift should be introduced by the amplifier. A two stage common emitter transistor amplifier should thus fulfil this condition.

For balancing of the bridge, the condition is

$$\frac{R_1}{R_2} = \frac{R + \dfrac{1}{j\omega C}}{\dfrac{R\,\dfrac{1}{j\omega C}}{\left(R + \dfrac{1}{j\omega C}\right)}} \qquad \text{...(11.31)}$$

Simplification of equation (11.31) yields,

$$\frac{R_1}{R_2} = 2 + j\,\frac{R^2 - \dfrac{1}{\omega^2 C^2}}{(R/\omega C)} \qquad \text{...(11.32)}$$

Equating the real and imaginary terms on either side we get,

$$\frac{R_1}{R_2} = 2 \qquad \text{...(11.33)}$$

and $$R^2 - \frac{1}{\omega^2 C^2} = 0 \qquad \text{...(11.34)}$$

The frequency of oscillation is thus

$$\omega = \frac{1}{RC} \qquad \text{...(11.35)}$$

and $$f = \frac{1}{2\pi RC} \qquad \text{...(11.36)}$$

When the bridge is balanced,

$$V_{ad} = V_{dc} = \frac{1}{3} V_{out} \qquad \text{...(11.37)}$$

So, at balance

$$\frac{V_{ad}}{V_{out}} = \frac{V_{dc}}{V_{out}} = \frac{1}{3} \qquad \text{...(11.38)}$$

Since the input voltage $V_{in} \neq 0$ for oscillations to take place,

$$\frac{R_2}{R_1 + R_2} = \frac{1}{3} - \frac{1}{\delta} \qquad \text{...(11.39)}$$

a value less than $\frac{1}{3}$, where $\delta > 3$

Thus $$\frac{V_{dc}}{V_0} = \frac{1}{3} - \frac{1}{\delta} \qquad \text{...(11.40)}$$

The feedback voltage is given as

$$V_i = V_{ad} - V_{dc} = \frac{V_0}{\delta} \qquad \text{...(11.41)}$$

Since V_i and V_{out} are in phase,

$$\beta = \frac{V_i}{V_{out}} = \frac{1}{\delta} \quad ...(11.42)$$

Since $$A\beta = 1 \quad ...(11.43)$$

The value of gain A is $\delta > 3$ ...(11.44)

The Wien bridge oscillators can be employed in the audio frequency range. The frequency variation can be achieved by variation of the capacitors C.

The Wien bridge oscillator provides the following advantages.

(1) The circuit has good frequency stability and the sinusoidal wave shape is distortion free.

(2) The oscillation frequency is variable.

(3) A high gain can be achieved as the amplifer is a two stage amplifier (total phase shift = 360°).

(4) The amplitude stability can be improved by using a thermistor in place of R_2.

11.7 MULTIVIBRATORS

Multivibrators generate non sinusoidal waveform. They may also be called a relaxation oscillators. The multivibrator is a two stage amplifier or oscillator operating in two modes of operation. The active element of one multivibrator stage is driven to saturation while the other is cut off. Multivibrators can be classified to three different types.

1. **Astable or Free running**
2. **Bistable (also called ECCLES JORDAN CIRCUIT or FLIP FLOP)**
3. **Monostable (also called as SINGLE SHOT or SINGLE SWING or ONE SHOT).**

An astable multivibrator alternates automatically between two states.

In a bistable multivibrator an external trigger is needed for transition from one state to another. In the monostable multivibrator one of the states is the normal or quiscent state while the other is the transient state. When a trigger pulse is applied, the circuit goes to the transient state from the quiscent state and returns to the initial quiescent state after a definite time interval determined by the circuit components.

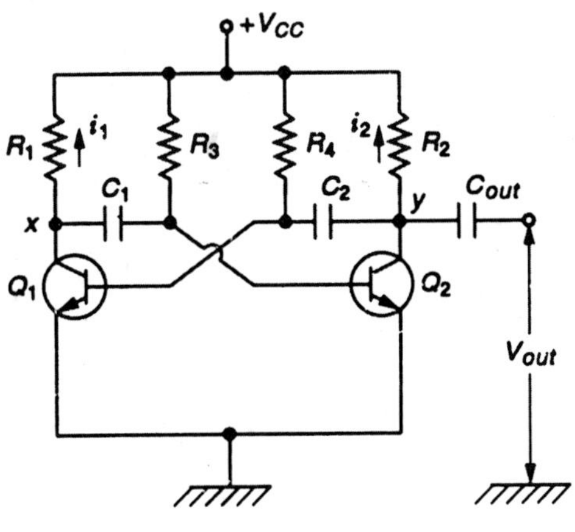

Fig 11.18. Circuit diagram of an astable multivibrator

11.7.1 Astable Multivibrator

Let us first consider the astable multivibrator of Fig. 11.18. In the circuit two identical *NPN* transistors Q_1 and Q_2 have been employed. In the circuit $R_1 = R_2$, $R_3 = R_4$ and $C_1 = C_2$. Although the transistors are identical and the components are chosen to have nearly identical values for R_1 an R_2, R_3 and R_4 and C_1 and C_2) it is only practical that there will be some slight unbalance in the two transistor circuits as a result of which one of the transistors will be drawing more current compared to the other when the switch is first turned on. Let us assume that after the switch is first turned on, at a certain instant the transistor Q_1 is drawing more current compared to the other transistor Q_2 *i.e.* $i_1 > i_2$. This means that the voltage drop across R_1 will be more compared to the voltage drop across R_2. Thus the voltage at the collector of transistor Q_1, $(V_C)_{Q_1} = V_x$ will be lower compared to the voltage at the collector of transistor Q_2, $(V_C)_{Q_2} = V_y$. That is $V_x < V_y$. This lower value of V_x will mean a lower potential at the base of transistor Q_2 compared to that at the base of transistor Q_1 ($V_{B_2} < V_{B_1}$). This means a lowering of the collector current of transistor Q_2. The fall or decrease in i_2, means a lower voltage drop across R_2. That mean V_y increases. A rise in the value of V_y implies higher potential at the base of Q_1. As a result the collector current of Q_1 increases. Thus the intial higher value of i_1 leads to a further increase of i_1 and decrease in i_2. The circuit, has thus a cumulative effect in that i_1 goes on increasing while i_2 falls further and further. The series of actions continues, driving quickly Q_1 to saturation while Q_2 is driven to cut off. The collector voltage waveforms for the astable multivibrator are depicted in Fig. 11.19.

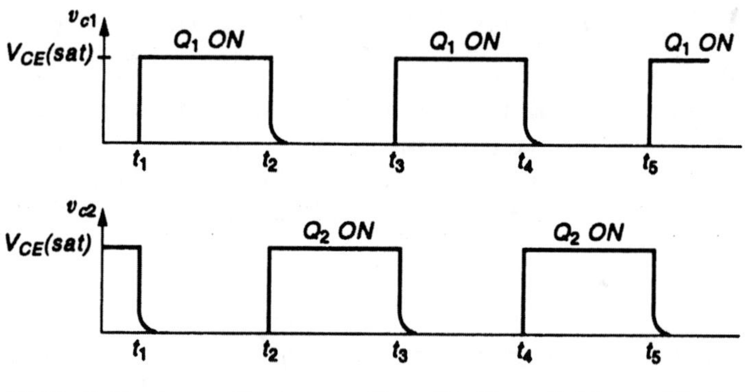

Fig 11.19. Collector voltage waveform for the astable multivibrator

When Q_1 is in saturation, $V_{C_1} = V_{CE}$ (sat) whereas when Q_2 is cut off, $V_{C_1} = V_{CC}$. It will be apparant from the Figure, that at any instant of time, either (i) Q_1 is ON, Q_2 is OFF and (2) Q_1 is OFF and Q_2 is ON.

The frequency of operation is

$$f = \frac{1}{T} = \frac{1}{t_3 - t_1} = \frac{1}{t_4 - t_2} \qquad ...(11.45)$$

The frequency can be altered by altering R_3, R_4, C_1 and C_2. The collector voltage waveform can be obtained through capacitor C_{out}.

11.7.2 Bistable Multivibrator

The circuit diagram of a bistable multivibrator is depicted in Fig. 11.20.

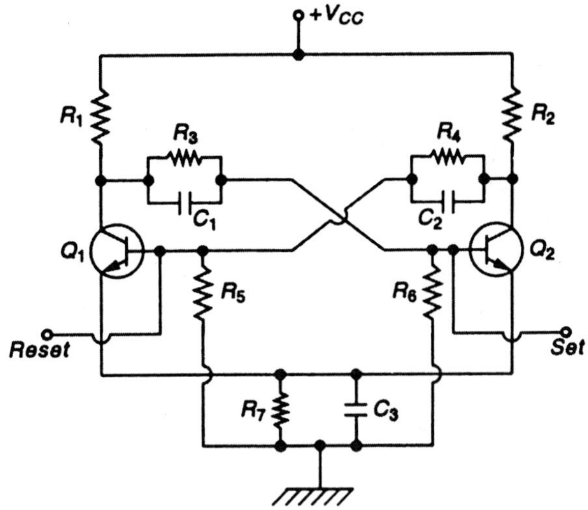

Fig 11.20. Circuit diagram of a bistable multivibrator

The bistable multivibrator may be constructed by using the two transistors Q_1 and Q_2. It will be seen from the circuit of Fig. 11.20 that the base emiter bias of transistor Q_1 is determined by the voltage source V_{CC} along with the resistors R_2, R_4, R_5 and R_7. Similarly the base emitter bias of transistor Q_2 is determined by the resistors R_1, R_3, R_6 and R_7 along with V_{CC}. C_3 is a bypass capacitor connected across R_7. In the circuit of Fig. 11.20, the coupling between Q_1 and Q_2 introduces positive feedback. Because the characteristics of two transistors can never become identical, one of the transistors say Q_1 will start conduction a bit earlier compared to the other transistor. This means the collector current of Q_1 will be greater than that of the collector curent of Q_2. A higher value of collector current i_1 for transistor Q_1 means that the voltage drop i_1R_1 acorss R_1 will be more and the collector voltage V_{c1} of transistor Q_1 will be low. The collector voltage of Q_2 will be higher as the collector current i_2 of transistor Q_2 is lower. The higher collector voltage of Q_2 is coupled to the base of Q_1. Thus the voltage at the base of Q_1 will be higher thus increasing the collector current of Q_1. Thus our initial assumption of a higher value of i_1, the collector current of Q_1 lead to the conclusion that the value of i_1 will increase further. This is a cumulative process. And Q_1 will go into saturation. And Q_2 will go to cut off. This is a stable state where Q_1 is in SATURATION and Q_2 is CUT OFF. If a positive trigger pulse is applied to the input, Q_2 conducts and goes into saturation Q_1 is driven to cut off. Alternatively if a negative trigger pulse is applied to the input R, Q_1 will go to cut off and Q_2 will enter saturation. The positive and negative trigger pulses are shown in Fig. 11.21.

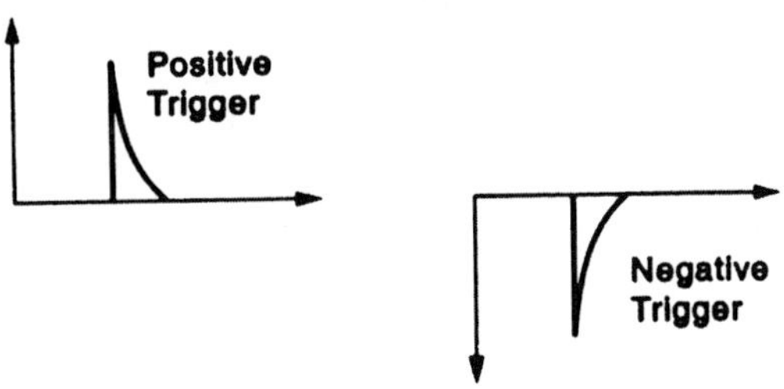

Fig 11.21. Positive and negative trigger pulses

The two stable states of Fig. 11.20 are thus

(a) Q_1 ON (SATURATION), Q_2 OFF

(2) Q_1 OFF, Q_2 ON (SATURATION).

The capacitors C_1 and C_2 are speed up capacitors. They assist in rapid transitions from one stable state to another when a suitable trigger pulse is applied.

11.7.3 Monostable Multivibrator

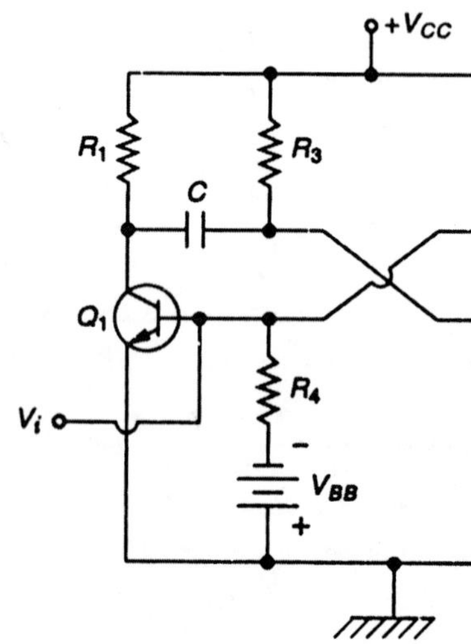

Fig 11.22. Circuit diagram of a monostable multivibrator

For 11.22 shows the circuit diagram of a monostable multivibrator. Without the application of an external trigger pulse, Q_1 is cut off since it is reverse biosed by V_{BB} and R_4. For transistor Q_2, the forward biasing is achieved by V_{CC} and R_3. The transistor Q_2 is thus driven into saturation. This is the Quiscent stage for the monostable multivibrator.

Let a positive trigger pulse be applied to the base of Q_1. Q_1 conducts and the voltage at the collector of Q_1 drops due to increase in collector current of Q_1. Since the base of Q_2 is connected to the collectes of Q_2 through R_5, the voltage at the base of Q_2 also decreases. This lowers the collector current of Q_2 and as a result the voltage at the collector of Q_2 increases. An increased collector voltage of Q_2 increases the drive of Q_1 and the collector current of Q_1 increases. This process goes on again and again and ultimately Q_1 is driven to saturation and Q_2 to cut off. This is the transient state of the circuit. When Q_1 is saturated, C_1 charges towards V_{CC} through R_3. The voltage at the ... transistor Q_2 therefore increases with a time constant of CR_3.

The trigger voltage pulses and the output voltage V_{out} are plotted in Fig. 11.23 as functions of time.

When Q_2 emitter base voltage is higher than the cut in value, Q_2 conducts. The resulting voltage drop across R_2 increases the voltage at the collector of Q_2 and thus the base of Q_1. The collector current of Q_1

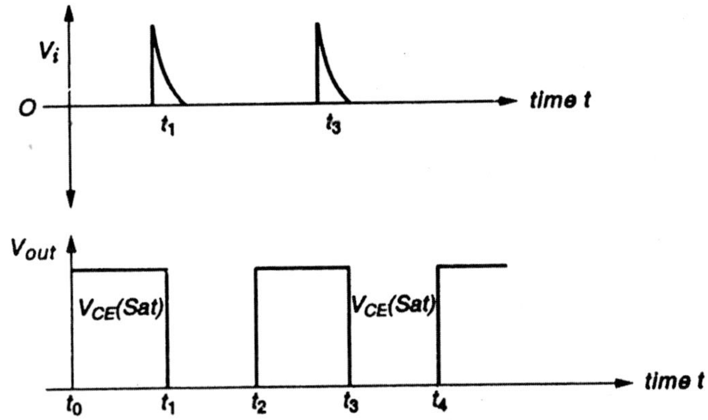

Fig 11.23 Trigger Pulses and output voltage waveform for a monostable multivibrator

thus rises. The rapid cummulative process drives Q_1 to cut off and Q_2 to saturation. The circuit returns back to the Quiscent state after a time interval determined by the time constant CR_3.

EXAMPLE 11.6 *A Wien bridge oscillator operates between 40 Hz to 40 kHz. The variable capacitance can be varied over the range 40 pF to 400 pF. Find the resistance values needed. If the gain of the amplifier is 10, what should be the value of resistance in the other arms of the bridge if $R_2 = 7$ kΩ.*

SOLUTION The frequency of oscillation of a Wien bridge oscillator is

$$f = \frac{1}{2\pi RC}$$

$f = 40$ Hz when $C = 400$ pF. Thus

$$40 = \frac{1}{2\pi R \times 400 \times 10^{-12}}$$

or, $$R = 9{\cdot}947 \times 10^6$$
$$\approx 9{\cdot}95 \text{ M}\Omega$$

The capacitance can change in the ratio 400 pF : 40 pF = 10 : 1. Thus the oscillator frequency span will be 40 Hz to 400 Hz using $R = 9{\cdot}95$ MΩ. For the range 4 kHz to 40 kHz, the resistance value required is $R = 99{\cdot}5$ kΩ. The three resistors needed to cover the entire span of frequencies is $R = 9{\cdot}95$ MΩ (40 – 400 Hz range), 995 kΩ (400 Hz – 4 kHz range) and 99·5 kΩ (4 kHz – 40 kHz range).

From equation (11.39) we get

$$\frac{R_2}{R_1 + R_2} = \frac{1}{3} - \frac{1}{\delta}$$

Also from equations (11.42) and (11.43)

$$A\beta = 1$$

$$\beta = \frac{1}{A} = \frac{1}{10} = \frac{1}{\delta}$$

$$\therefore \qquad \delta = 10$$

Also $$1 + \frac{R_1}{R_2} = \frac{1}{\frac{1}{3} - \frac{1}{10}} = \frac{30}{7}$$

or, $$\frac{R_1}{R_2} = \frac{23}{7}$$

The value of R_1 is

$$R_1 = \frac{23}{7} \times 7 \text{ k}\Omega = 23 \text{ k}\Omega$$

EXAMPLE 11.7 *In the astable multivibrator circuit of Fig. 11.18, $R_1 = R_2 = 1{\cdot}5$ k, $R_3 = R_4 = 12$ kΩ, $C_1 = C_2 = 0{\cdot}068$ μF. Find the frequency of oscillation of the circuit.*

SOLUTION $$T_1 = T_2 = 0{\cdot}693\,(R_3C_1) = 0{\cdot}693 \times 12 \text{ k} \times 0{\cdot}068\ \mu\text{F}$$

$$= 0{\cdot}693 \times 12 \times 10^3 \times 0{\cdot}068 \times 10^{-6}$$

$$= 0{\cdot}565 \times 10^3$$

$$\therefore \quad T = T_1 + T_2 = 1{\cdot}13 \times 10^{-3}$$

$$\therefore \quad f = \frac{1}{T} = \frac{1}{1{\cdot}13 \times 10^{-3}} = 884 \text{ Hz.}$$

Review Questions

1. What is an "oscillator" ?
2. What is a relaxation oscillator ? Distinguish between sinusoidal and relaxation oscillators.
3. What do you understand by a negative resistance oscillator ?
4. Discuss the basic principle of operation of an oscillator.
5. What is Barkhausen criterion ?
6. What is Nyquist condition ? Draw the Nyquist plots for stable and unstable systems.
7. What is a tuned collector oscillator ? Draw its circuit diagram and and explain the operating principles.
8. Explain, with suitable diagram the principle of operation of a Harteley oscillator.
9. Explain, with suitable diagrams, the principle of operation of a Colpitt's oscillator.
10. What is a crystal oscillator ? Explain its principles of operation.
11. Draw the electrical equivalent circuit of a quartz crystal. Mention some important features of a crystal oscillator.
12. What is a phase shift oscillator ? Draw the circuit diagram of a phase shift oscillator and explain its principle of operation.
13. Derive the expression for the oscillation frequency of a phase shift oscillator. Also derive the value of β needed in the phase shift oscillator.
14. What is a Wien bridge oscillator ? Explain, with circuit diagrams, its principles of operation.
15. Derive the expression for oscillations of the Wien bridge oscillator.
16. Mention some important advantages of the Wien bridge oscillator.
17. What do you understand by the term "multivibrator"?

18. Distinguish between astable, bistable and monostable multivibrators.

Objective Type Questions

Choose the correct (or the most appropriate) answer.

1. Barkhausen criterion specifies that
(a) $A\beta = 1$ (b) $A\beta = -1$
(c) $A\beta = 0$ (d) $\frac{A}{1 - A\beta} = 1$

2. The locus of the phasor $A\beta$ in a feedback will trace a closed curve in the complex plane if the frequency is changed from $-\infty$ to $+\infty$. Nyquist criterion specifies that the amplifier will oscillate if the curve encloses the point
(a) $-1 + j0$ (b) $1 + j0$
(c) $-1 - j1$ (d) $1 + j1$

3. An oscillator operating in the frequency range in the 30 MHz to 3000 MHz is called an
(a) *R* F oscillator (b) *VHF* oscillator
(c) microwave oscillator (d) *UHF* oscillator

5. The resistance offered by the tank cicuit in a tuned collector oscillator is
(a) $\frac{C}{LR}$ (b) $\frac{RC}{L}$
(c) $\frac{R}{LC}$ (d) $\frac{L}{CR}$

6. A Harteley oscillator employs a
(a) tapped inductor (b) tapped capacitor
(c) negative resistance property (d) *RC* network

7. A Colpitts oscillator employs a
(a) tapped inductor (b) tapped capacitor
(c) negative resistance property (d) *RC* network

8. For a Harteley oscillator, the frequency of oscillation is
(a) $f_0 = \frac{1}{2\pi\sqrt{(L_1 + L_2)\,C}}$ (b) $f_0 = \frac{1}{2\pi\sqrt{(L_1 + L_2 + 2M)\,C}}$
(c) $f_0 = \frac{1}{2\pi\sqrt{(L_1 + L_2 + M)\,C}}$ (d) $f_0 = \frac{1}{2\pi\sqrt{(L_1 + L_2 - M)\,C}}$

9. For a Colpitt's oscillator, the frequency of oscillation is of the form
(a) $f_0 = \frac{1}{2\pi\sqrt{L\,(C_1 + C_2)}}$ (b) $f_0 = \frac{1}{2\pi\sqrt{(L_1 + L_2)\,C}}$
(c) $f_0 = \frac{1}{2\pi\sqrt{L \cdot \frac{C_1C_2}{C_1 + C_2}}}$ (d) $f_0 = \frac{1}{2\pi\sqrt{(L_1 + L_2)\frac{C_1C_2}{C_1 + C_2}}}$

10. The fundamental frequency of a crystal in a crystal oscillator
(a) can be increased by increasing the thickness of the crystal
(b) can be increased by decreasing the thickness of the crystal
(c) is independent of the thickness of the crystal.

11. Typical *Q* values of a Quartz crystal are
(a) 10 (b) 50
(c) 100 (d) 20,000

12. In a phase shift oscillator, a phase lag network is needed with
(a) 1 stage (b) 2 stages
(c) 3 stages of *RC* network

13. For a phase shift oscillator, the angular frequency of oscillation is

(a) $\omega_0 = \frac{1}{RC}$ (b) $\omega_0 = \frac{1}{\sqrt{3}\,RC}$

(c) $\omega_0 = \frac{1}{2RC}$ (d) $\omega_0 = \frac{1}{\sqrt{6}\,RC}$

14. The condition for oscillation in an RC network reduces to

(a) $\beta = 1$ (b) $\beta = \frac{1}{6}$

(c) $\beta = \frac{1}{15}$ (d) $\beta = \frac{1}{29}$

15. In Wien bridge oscillator, the angular frequency of oscillation is given by

(a) $\omega_0 = \frac{1}{2RC}$ (b) $\omega_0 = \frac{1}{\sqrt{6}\,RC}$

(c) $\omega_0 = \frac{1}{\sqrt{3}\,RC}$ (d) $\omega_0 = \frac{1}{RC}$

Exercise Problems

1. Find the frequency of oscillation for the Colpitts oscillator shown in Fig. 11.13. Assume $C_1 = 0{\cdot}22\ \mu F$, $C_2 = 0{\cdot}03\ \mu F$, $L = 25$ mH.
2. Find the frequency of oscillation for the circuit of Fig. 11.13 if Q has a value of (a) 5 (b) 7·5 (c) 15
3. In a piezoelectric crystal oscillator, the frequency of oscillation is 4 MHz. For this frequency, the equivalent inductance and series resistance are 0·1 H and 5000 ohms respectively. Find the value of Q of the crystal..
4. Design a phase shift oscillator to have an oscillation frequency of 2 kHz. Use a transistor circuit with $V_{CC} = +\ 15$ V.
5. Design a Colpitts oscillator to give $f = 50$ kHz. Use a 200 mH inductor.
6. Design a Wien bridge oscillator to have a an oscillation frequency of 15 kHz.
7. Find the frequency of oscillation for the circuit of Fig. 11.16 (b) if $R_1 = R_2 = R_3 = 5\ k\Omega$ and $C_1 = C_2 = C_3 = 2$ mμF.
8. What should be the frequency of oscillation if $C_1 = C_2 = C_3 = 5{\cdot}6$ mμF instead of 2 mμF. Find also the value of R_f in Fig. 11.16 (b) if $\frac{R_f}{R_3}$ = gain of the operational amplifier.
9. A Wien bridge oscillator operates between 50 Hz and 50 kHz. The variable capacitance can be varied over the range of 50 pF to 500 pF. Find the resistance values needed. If the gain of the amplifier is 10, what should be the value of the resistance in the other arms of the bridge if $R_2 = 5\ k\Omega$.
10. In the astable multivibrator circuit of Fig. 11.18, $R_1 = R_2 = 2{\cdot}5\ k\Omega$, $R_3 = R_4 = 10\ k\Omega$, $C_1 = C_2 = 0{\cdot}056\ \mu F$, find the frequency of oscillation of the circuit.

Chapter 12

MODULATION AND DEMODULATION

12.1 INTRODUCTION

The intelligence or message signal (or base band) signal is a signal near dc (zero)frequency. Many information sources generate analog output signals. These are signals which are continuous in time and amplitude. Two common examples are speech signals and video signals.

Analog signals can be converted into digital signals by sampling and quantizing the signal amplitude. This digital signal can then be encoded and transmitted through the communication channel. At the receiving end the digital signal is received and converted back to the analog form. Alternatively, the digital signal can be stored on magnetic tape or optical media for storage and future use. By the term modulation, we mean impressing the analog signal on the amplitude, the phase or the frequency of a sinusoidal carrier. The message signal is the modulation signal. A signal whose frequency range is near the value $f = 0$(dc) is called the baseband signal.

The modulation of a carrier $c(t)$ by the message signal $m(t)$ is carried out to achieve one or more of the following desired objectives.

(1) **To translate a low frequency signal to the passband of the communication channel.**

(2) **To make possible simultaneous transmission of signals from a number of different message sources.**

(3) **To expand the bandwidth of the transmitted signal and to increase its noise immunity in transmission over a noisy channel. The objecives (1) and (2) are met by all of the modulation methods while the objective (3) is met by angle modulation.**

The modulation method followed depends on number of factors. The major factors are

(1) **Nature of the message signal (its bandwidth)**

(2) **Characteristics of the communication channel (like the medium of propagation, wire line, free space, fiber optic communication). It will also depend on the type of channel disturbance (noise, interference from other signals etc.).**

(3) **Received signal fidelity**

(4) **Cost effectiveness**

12.2 MODULATION

The message signal is also called the **modulating signal or modulating wave** while the high frequency wave, whose characteristics are altered is called the **carrier signal** or **carrier wave.** The combined signal or wave is called **modulated** wave.

Normally, the carrier signal is sinusoidal in nature so that it can be represented by

$$c(t) = A_C \cos \theta(t) \quad \ldots(12.1)$$

where $c(t)$ is the instanteneous value, A_c is the amplitude or peak value and $\theta(t)$ is the instantaneous angular displacement of the carrier wave.

The instantaneous angular velocity is

$$\omega(t) = \frac{d\theta}{dt} \quad \ldots(12.2)$$

Integrating equation (12.2), it is easy to find that

$$\theta(t) = \int \omega(t)\, dt + \phi \quad \ldots(12.3)$$

where ϕ is the constant of integration. From equation (12.3),

$$\theta(t) = \omega_c + \phi \quad \ldots(12.4)$$

Thus, carrier signal can be represented by

$$c(t) = A_c \cos(\omega_c t + \phi) \quad \ldots(12.5)$$

If $\phi = 0$ at $t = 0$, the above equation is represented as

$$c(t) = A_c \cos \omega_c t \quad \ldots(12.6)$$

Modulation of a carrier wave can be performed in two different ways (1) by changing the amplitude A_C of the carrier and (2) by changing the total phase angle $\theta(t)$. The first process is called as amplitude modulation (*AM*).

12.3 FREQUENCY SPECTRUM

Radio frequencies and optical frequencies for different communication services extend from very low frequencies to extra high frequencies to light shown in Table 12.1.

Table 12.1

Frequency Range	*Disignation*	*Wavelength range*	*Uses*
3–30 kHz	Very Low Frequency (VLF)	100km–10km	Long distance telephony, broadcasting
30–300 kHz	Low Frequency (LF)	10 km – 1km	Aeronautical, navigation, Long distance point to point service, other navigational aids, sound broadcasting, line carrier system
300 kHz–3MHz	Medium Frequency (MF)	1 km – 100 m	AM broadcasting, Ship-shore communication
3 MHz–30MHz	High Frequency (HF)	100 m – 10 m	Business, Amateur radio, International radio, citizen's band

30–300 MHz	Very High Frequency (VHF)	10 m – 1 m	Mobile, aeronatical, VHF TV, FM Broadcating Mobile radio
300 MHz–3 GHz	Ultra high Frequency (UHF)	1 m – 10 cm	UHF TV, Radio
3 GHz–30 GHz	Super High Frequency (SHF)	10 cm – 1 cm	Experimental, Navigation, Satellite to satellite, Microwave relay, Earth satellite, Radar
30 GHz–300 GHz	Extra High Frequency (EHF)	1 cm – 1 mm	Microwave communiction, Radar, Experimental
10^{14}–10^{15} Hz	Light wave	10^{-6} m	Fiber optic communation, Local Area Network (LAN) Fiber Distributed Data Interface (FDDI) Asynchronous Transfer Mode (ATM) network, Syncronous optical NET work (SONETs)

The bandwidth required for common analog systems are shown as follows :

Message Signal	*Bandwidth*	*Use*
Voice	3·5 kHz	Single Telephone channel
Music	10 kHz	AM Radio Broadcasting
Music	200 kHz	FM
Television	6 MHz	Television Broadcasting station

Telephone links normally require a bandwidth of 3500 Hz because most of the energy in normal speech is contained in the frequencies below this value. Messages are normally intelligible and individual voices recognizable with this bandwidth. The range of frequencies upto 3·5 kHz is the baseband of voice signals. The AM broadcast band ranges from 500 kHz to 1500 kHz. AM broadcasting stations normally transmit mesages from 100 to 5000 Hz. AM stations thus requires a bandwidth of 10 kHz and their carrier frequencies are separated by 10 kHz. For high quality music reproduction a bandwidth upto 15 kHz is necessary. However, highly sensitive ear can detect upto 20,000 Hz. The frequency modulation system requires a bandwidth of 200 kHz to transmit the signal.

Video signals, however, require high bandwidths. Commercial TV channels have a bandwidth of 6 MHz. The highest video frequency actually transmitted is below 4·2 MHz.

When the analog signals are transmitted digitally, the bit rate will depend on the rate at which the analog signal is sampled and the coding method. The **sampling theorem states that an analog signal can be transmited accurately if the sampling rate is at least twice the highest frequency content of the signal.** Thus a telephone channel containing upto 4000 Hz is sampled at 8000/second. In the coding procedure 8 bits are necessary to describe the amplitude of each sample so that 64000 bits per second (bps) or 64 kbps is transmitted for a single telephone message.

For commercial TV broadcast the data rate assuming a 6 MHz bandwidth and sampling at twice the highest frequency content and encoding at 8 bits per sample is $2 \times 6 \times 8 = 96$ Mbps. If several signals are multiplexed, the data rate would be several hundred Megabits per second. Because the video information is contained in a BW less than 6 MHz, the 96 Mbps rate can be reduced. If we consider a bandwidth of 4·5

MHz, sample at twice the rate and encode by using 9 bits/sample, the data rate would be 81 Mbps. An accompanying audio signal of 15 kHz, sampled at 30 kbps and encoded using 8 bits/sample would require a rate of 240 kbps. The total signal rate for this system is 81·24 Mbps. It can be sent over a standard DS-3C telephone line, which operates at 91·053 Mbps data rate.

The digital transmission rate of the Telephone system in the USA ranges between 64 kbps for a single voice channel, to about 565 Mbps for 8064 vioce channels.

The data rate can be reduced for digitized video image by **compressed digital video (CDV).** Fundamentally, the process eliminates transmission of redundent data. As an example, the background of a scene may remain constant over a number of frames. We may omit transmitting the background of the frame with each frame, thereby reducing the required data rate . Using compression methods, it is possible to reduce the data rate from 100 Mbps to 5 Mbps. Digitized high definition television signals normally require 1 Gbps (Giga bits per second) for transmission. Compression reduces the necessary data rate to 20 Mbps.

For high speed transmission of data, the fiber distributed data interface (FDDI) operates at a data rate of 100 Mbps. Another type of LAN, the high performance parallel interface operates at 800 Mbps. In fiber optic transmissions data rate upto several Gbps is possible. However for data rates from 10 Mbps to several Gbps, improved circuits, light emitting diodes and light detectors must be used. A strong optical signal is necessary at the receiver for high signal to noise ratios and low bit error rates.

12.4 AMPLITUDE MODULATION

In amplitude modulation, the amplitude of the carrier is altered by the modulating signal. The change in amplitude from the unmodulated signal is directly proportional to the instantaneous value of the modulating signal. The change is independent of its frequency.

Considering a sinuosidal modulating signal, the modulating voltage can be represented as

$$e_m = A_m \cos \omega_m t \qquad \text{...(12.7)}$$

where A_m is the amplitude and ω_m is the angular frequency of the modulating signal. ω_m is much lower compared to ω_c.

The modulated carrier can be represented as

$$e(t) = A_c + K_a A_m \cos \omega_m t \qquad \text{...(12.8)}$$

where K_a is the constant of proportionality. The amplitude modulated carrier can be represented as

$$(e)_{am} = e(t) \cos \omega_c t \qquad \text{...(12.9)}$$

$$= [A_c + K_a A_m \cos \omega_m t] \cos \omega_c t$$

$$= A_c (1 + m \cos \omega_m t) \cos \omega_c t \qquad \text{...(12.10)}$$

The quantity m is called the **modulation index or modulation factor or the depth of modulation.** It can be written as

$$m = \frac{E_{max} - E_{min}}{E_{max} + E_{min}} = \frac{E_{max} - E_{min}}{2A_c} \qquad \text{...(12.11)}$$

where E_{max} is the maximum amplitude of the modulated carrier and E_{min}, the minimum amplitude of the modulated carrier. The quantity A_c is the amplitude of the carrier.

The percent modulation is expressed as

$$m\,(\%) = m \times 100\% \qquad \text{...(12.12)}$$

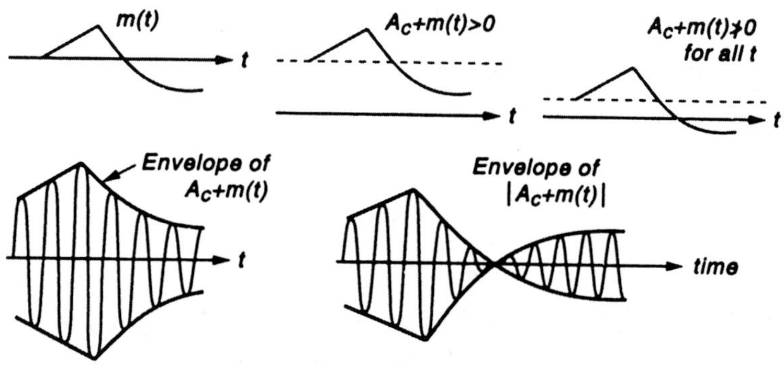

Fig 12.1. Amplitude modulation by a modulation signal *m*(*t*)

Figs 12.1 and 12.2 depicts the waveform of the modulated carrier wave for amplitude modulation. It will be seen that the envelope of the modulated carrier is the same as that of the modulating signal. It is called the modulation envelope. The value of the modulation index can thus be obtained from the modulated carrier. The parameter m expresses the degree to which the signal is modulated. m can lie between 0 and 1. The higher the value of m_a the stronger is the degree with which the intelligence has been superimposed on the carrier. The amplitude of the modulated carrier is twice the unmodulated carrier amplitude on positive and zero on negative excursions of the modulation envelope for 100 per cent modulation. Overmodulation takes place when m is greater than 1. If over modulation takes place, the signal is distorted.

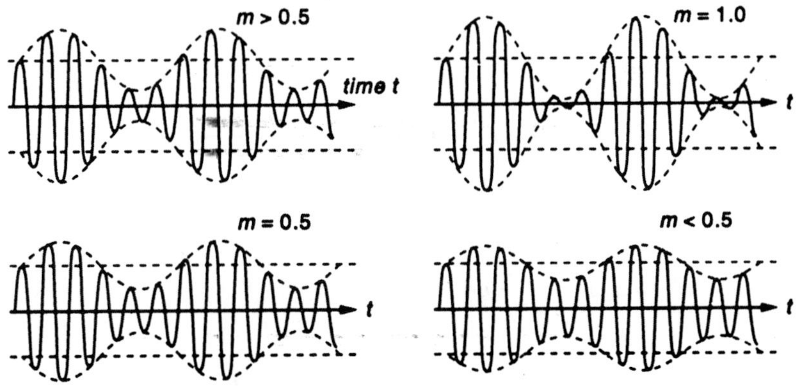

Fig 12.2. Amplitude modulated signal using sinusoidal modulation signal (tone modulation)

$$(e)_{AM} = A_c \cos\omega_c t + \frac{1}{2} m A_c \cos(\omega_c + \omega_m) t + \frac{1}{2} m A_c \cos(\omega_c - \omega_m) t \quad \ldots(12.13)$$

$$= A_c \cos 2\pi f_c t + \frac{1}{2} A_c \cos 2\pi (f_c + f_m) t + \frac{1}{2} m A_c \cos 2\pi (f_c - f_m) t \quad \ldots(12.14)$$

Equation (12.13) shows that amplitude modulated signal consist of the carrier frequency and two other frequencies which is equal to the carrier frequency ± modulation frequency. The sum frequency term $(f_c + f_m)$ is called the upper side frequency while the difference frequency term is called the lower side frequency. The relative amplitudes of the three frequencies is shown in Fig. 12.3. This plot of relative amplitude against frequency is also called the frequency spectrum. The message signal or intelligence is contained in the side frequency components and not in the carrier frequency component.

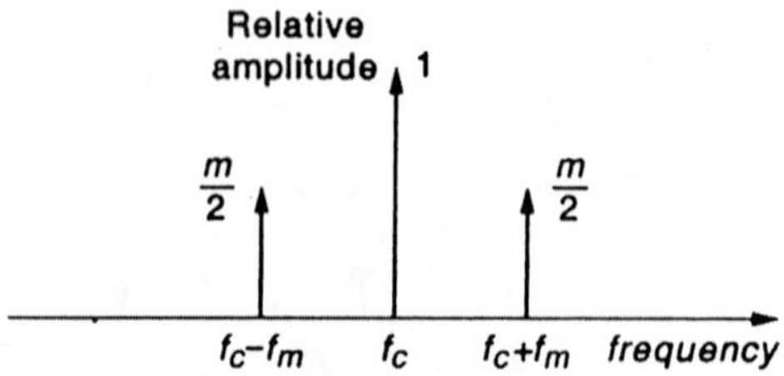

Fig 12.3. Frequency spectrum of the amplitude modulated carrier for sinusoidal modulation signal of Fig 12.2

12.4.1 Frequency Band

The message or intelligence signal does not normally comprise a single frequency sinusoid but rather a band of frequencies. For example for radio broadcast the highest frequency component is in 5 kHz. Thus the frequency spectrum will comprise two sidebands. The band of frequencies above the carrier frequency is called the upper sideband while the band of frequencies below the carrier frequency is called the lower sideband. Obviously, the band width or channel width is twice the highest frequency of the modulating signal. For signal frequency modulating signal, the bandwidth is twice the modulation frequency. If the 670 kHz carrier is amplitude modulated by an audio having the highest carrier frequency of 5 kHz, the lower sideband will range from 665 Hz, while the upper sideband will range from 670 to 675 kHz. The bandwidth for AM will be 675 − 665 = 10 kHz.

12.4.2 Power in Amplitude Modulated Wave

If A_c is the carrier amplitude, the r.m.s value is $\frac{A_c}{V_2}$ and the power dissipated in a resistor R is

$$P_0 = \frac{(A_c/\sqrt{2})^2}{R} = \frac{A_c^2}{2R} \quad \text{...(12.15)}$$

The amplitude of the side frequency component is $\frac{mA_c}{2}$. Thus, the power in the lower side frequency component is

$$P_l = \left(\frac{mA_c}{2} \times \frac{1}{\sqrt{2}}\right)^2 \cdot R = \frac{m^2A_c^2}{8R} \quad \text{...(12.16)}$$

Similarly, the power contained in the upper side frequency component is

$$P_u = \frac{m^2A_c^2}{8R} \quad \text{...(12.17)}$$

The total power in the side frequency components is

$$P_s = P_l + P_u = \frac{m^2A_c^2}{8R} + \frac{m^2A_c^2}{8R} = \frac{m^2A_c^2}{4R} \quad \text{...(12.18)}$$

The power in the side frequency component is thus proportional to the square of the modulation index (*i.e.* m^2). For 50% modulation

$$P_s = \frac{A_c^2}{16R} \quad \text{...(12.19)}$$

whereas for m = 100%, the value of power in the side frequency components is

$$P_s = \frac{A_c^2}{4R} \quad \text{...(12.20)}$$

It can be checked from equation (12.15), that for 100% modulation, the power in the side frequency component is half the power in the carrier frequency components.

The total power in the amplitude modulated signal is

$$P = P_0 + P_s = \frac{A_c^2}{2R} + \frac{m^2 A_c^2}{4R} = P_c\left(1 + \frac{m^2}{2}\right) \quad \text{...(12.21)}$$

The power in the amplitude modulated signal is greater than in the unmodulated carrier. When modulation is 100%, the total power is 50% more than in the unmodulated carrier.

12.4.3 AM wave generation

The block diagram for generating an AM wave is illustrated in Fig. 12.4.

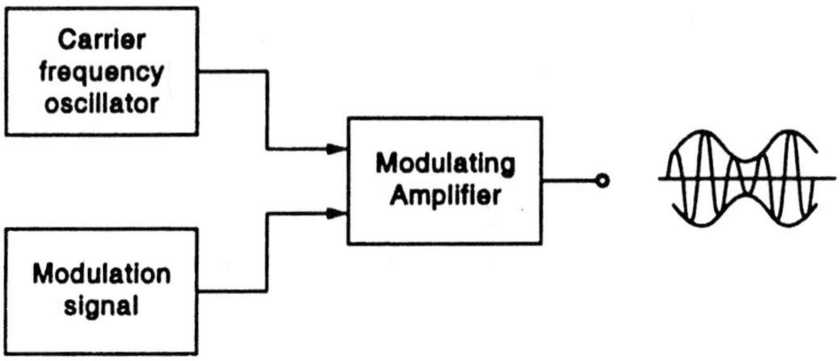

Fig 12.4. Amplitude modulation technique

In the Fig. 12.4, the carrier frequency and the message signal are fed to the amplifier. The modulation occurs in the active device used in the amplifier.

Fig. 12.5 depicts the diagram of a typical modulator circuit using a common emitter amplifier. The message (or modulation) signal is connected to the emitter of the transistor while the carrier signal is connected to the base of the amplifier. The capacitor C_1 has a very low impedance for the lower modulation frequency. The capacitor C_2 and C_6 tunes the primary circuits of transformer T_1 and T_2 at the resonant frequency. The resistors R_1, R_2, R_3, R_E and V_{CC} establish the Q point for the circuit. The capacitors C_3 bypass the resistor R_2 and C_4 bypass the resistor R_E at the carrier frequency.

The output voltage at the primary of the transformer consists of the carrier frequency f_c and the terms $f_c + f_m$ and $f_c - f_m$. The output of the modulator is thus an amplitude modulated signal.

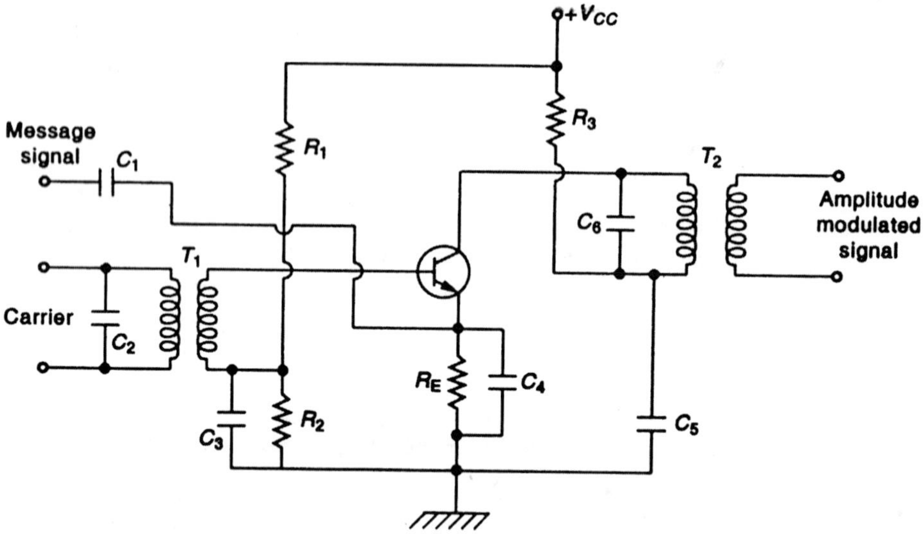

Fig 12.5. An amplitude modulation circuit

EXAMPLE 12.1 *The peak-to-peak voltage of an amplitude modulated signal is 10 V maximum and 3 V minimum. Find the percent modulation. What is the amplitude of the unmodulated carrier ?*

SOLUTION

$$m = \frac{E_{max} - E_{min}}{E_{max} + E_{min}} = \frac{(10-3)/2}{(10+3)/2} = 0{\cdot}534$$

or percent modulation = 53·4%

If A_c is the amplitude of the unmodulated carrier,

$$m = \frac{E_{max} - E_{min}}{2A_c}$$

or, $$A_c = \frac{E_{max} - E_{min}}{2m} = \frac{10-3}{2 \times \frac{7}{13}} = 6{\cdot}5 \text{ Volts.}$$

EXAMPLE 12.2 *A 1000 kHz carrier wave is amplitude modulated by audio signals between 450 and 1650 Hz. Find (i) frequency span of each side band (ii) maximum upper side frequency (iii) minimum lower side frequency (iv) channel width.*

SOLUTION (i) Frequency span of each sideband

= 1650 – 450 = 1200 Hz

(ii) Maximum upper side frequency = 1000 kHz + 1650 Hz

= 1001·65 kHz

(iii) Minimum lower side frequency = 1000 kHz – 1650 Hz

= 998·35 kHz

(iv) Channel width = (1001·65 − 998·35) = 3·3 kHz.

EXAMPLE 12.3 *Find the power developed by an AM wave in a load of 180 Ω when the peak voltage of the carrier is 90 V and the modulation index is 50%.*

SOLUTION

Unmodulated carrier power $= P_c = \dfrac{V_c^2}{2R}$

$$= \frac{(90)^2}{2 \times 180} = 22 \cdot 5 \text{ W}$$

Total power developed by the AM wave is

$$P_t = P_c\left(1 + \frac{m^2}{2}\right) = 22 \cdot 5\left(1 + \frac{(0 \cdot 5)^2}{2}\right) = 25 \cdot 3125 \text{ W.}$$

Power in the sideband $= P_c \cdot \dfrac{m^2}{2} = 22 \cdot 5 \times \dfrac{(0 \cdot 5)^2}{2}$

$$= 2 \cdot 8125 \text{ W.}$$

EXAMPLE 12.4 *A modulating signal consists of a triangular wave having zero d.c. component and peak-to-peak voltage of 12V. It is used to amplitude modulate a carrier of peak voltage 9V. Find the depth of modulation and the ratio of the side lengths* $\dfrac{L_1}{L_2}$ *of the corresponding trapezoidal pattern.*

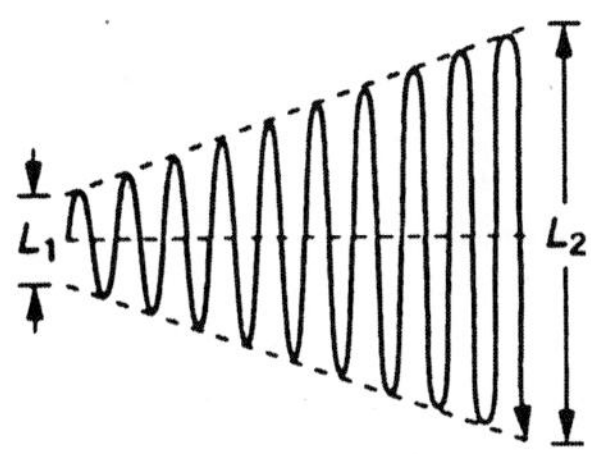

Fig E. 12.4. Reference example 12.4

SOLUTION

$$E_{max} = 9 + \frac{12}{2} = 15 \text{ V}$$

$$E_{min} = 9 - \frac{12}{2} = 3 \text{ V}$$

$$m = \left(\frac{15 - 3}{15 + 3}\right) \times 100\% = 66 \cdot 66\%$$

From Fig. E 12.4

L_1 is proportional to 15 V and L_2 to 3V.

Thus $\dfrac{L_1}{L_2} = \dfrac{3}{15} = \dfrac{1}{5}$

EXAMPLE 12.5 *An amplitude modulated signal is produced by a message signal shown in Fig. E 12.5. Sketch the AM pattern and find the modulation index if* $A_c + KA_m \cos \omega_m t \ngtr 0$

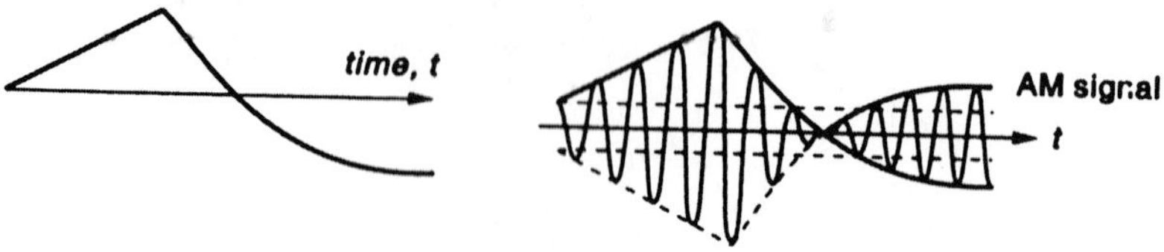

Fig E.12.5. Reference example 12.5

SOLUTION The plot is shown in Fig. E 12.5

$A_c + A_m \cos \omega_m t \not> 0$

or, $A_c < kA_m \cos \omega_m t$

$$\text{Modulation index} = \frac{KA_m}{A_c} > 1$$

% Modulation index > 100%. **(over modulation)**

EXAMPLE 12.6 *A carrier wave of frequency 9 MHz and peak value 9 V is amplitude modulated by 10 kHz sine wave o amplitude 4·5 V. Find the modulation index and sketch the frequency spectrum.*

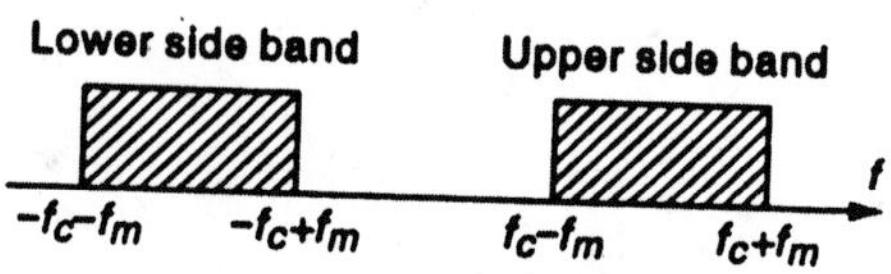

Fig E.12.6. Reference example 12.6

SOLUTION $m = \frac{4 \cdot 5}{9} = 50\%$

The side frequencies are $9 \pm 0{\cdot}01 = 9{\cdot}01$and 8·99 MHz.

The amplitude of each side frequency is $0{\cdot}5 \times \frac{9}{2} = 2{\cdot}25$.

The frequency spectrum is shown in Fig. E 12.6.

EXAMPLE 12.7 *The r.m.s. antenna current of a radio transmitter is 12 Amps when unmodulated and 14 Amps when sinusoidally modulated. Find the depth of modulation.*

SOLUTION The effective or r.m.s. voltage E of the modulated wave is

$$\frac{A_c^2}{R} = P_c$$

$$\frac{E^2}{R} = P_c\left(1 + \frac{m^2}{2}\right) = \frac{E_c^2}{2}\left(1 + \frac{m^2}{2}\right)$$

or, $$E = E_c \sqrt{1 + \frac{m^2}{2}}$$

$$I = I_c \sqrt{1 + \frac{m^2}{2}}$$

where E = r.m.s. voltage of the modulated wave

E_c = r.m.s. voltage of the unmodulated carrier

I = r.m.s. current of the modulated wave

I_c = r.m.s. current of the unmodulated carrier

$$m = \sqrt{2\left(\frac{E}{E_c}\right)^2 - 1}$$

Also $$m = \sqrt{2\left(\frac{I}{I_c}\right)^2 - 1}$$

Thus putting $I = 14$ Amps and $I_c = 12$ Amps, we get

$$m = \sqrt{2\left(\frac{14}{12}\right)^2 - 1} = 0{\cdot}8498 \text{ or } 84{\cdot}98\%$$

EXAMPLE 12.8 *Find the audio power necessary for AM of a 10 kW carrier modulated to a depth of 75% if the efficiency of the modulator is 65%.*

SOLUTION Total sideband power $= \frac{1}{2} m^2 P_c = \frac{1}{2}(0{\cdot}75)^2 \times 10$ kW

$= 2{\cdot}812$ kW

Since the efficiency of the modulator system is 65%, the required audio power is $\frac{2812}{0{\cdot}65} = 4{\cdot}326$ kW.

EXAMPLE 12.9 *In a certain transmitter, the unmodulated carrier power is 12 kW and it can be modulated to a maximum of 75% by a pure sinusoidal signal before overloading takes place. Find the value to which the carrier power can be raised if the modulation limit is 45%.*

SOLUTION $$P_t = P_c\left(1 + \frac{m^2}{2}\right) = 12\left(1 + \frac{(0{\cdot}75)^2}{2}\right)$$

$= 15{\cdot}375$ kW

When $m = 0{\cdot}45$, P_t is still $15{\cdot}375$ kW. Hence the new value of carrier power is

$$15{\cdot}375 = P_c\left(1 + \frac{m^2}{2}\right) = P_c\left(1 + \frac{(0{\cdot}45)^2}{2}\right)$$

or, $$P_c = \frac{15{\cdot}375}{1{\cdot}10125} = 13{\cdot}96 \text{ kW}.$$

12.5 VARIOUS AMPLITUDE MODULATION METHODS

In AM generation, the signal consists of one signal carrier and two sidebands. We however know that the information is actually carried by the sideband and it is not necessary to transmit all of the above to retrieve the message signal. Considerable amount of bandwidth and transmitter power can be saved if other form of transmission are accepted. We consider below the different systems.

12.5.1 Double-Sideband-Suppressed Carrier (DSB-SC) system

The spectrum of DSB-SC signal is depicted in Fig. 12.6. The DSB-SC signal can be generated by two AM modulators arranged in a balanced configuration as depicted in the circuit diagram of Fig. 12.7.

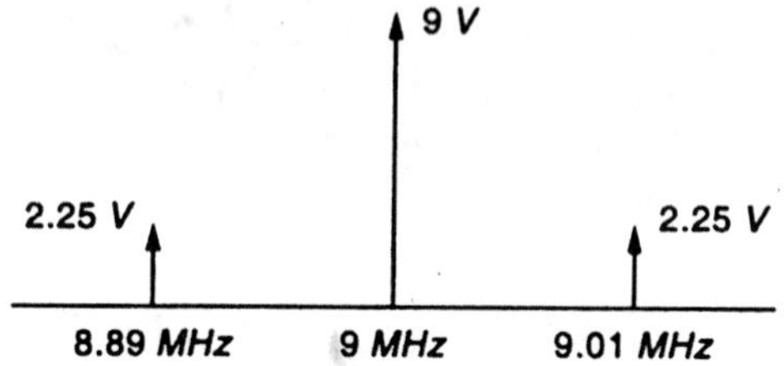

Fig 12.6. Spectrum of double-sideband suppressed carrier modulation

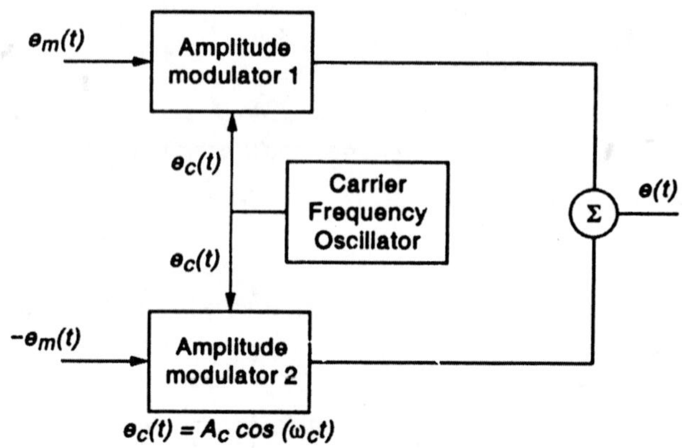

Fig 12.7. Circuit diagram for generating DSB-SC signals

The output of the two modulators can be expressed as

$$e_1(t) = A_c(1 + ke_m(t)) \cos(\omega_c t) \quad \text{...(12.22)}$$

and

$$e_2(t) = A_c(1 - ke_m(t)) \cos(\omega_c t) \quad \text{...(12.23)}$$

Thus, the output of the balanced modulator is

$$e(t) = e_1(t) - e_2(t) = 2A_c\, k\, e_m(t) \cos(\omega_c t) \quad \text{...(12.24)}$$

Thus the output of the balanced modulator is the product of the modulating wave and the carrier which represents the DSB-SC signal.

Fig. 12.8 (a) depicts the circuit diagram of a switching modulator of DSB-SC. It is a shunt bridge modulator where the diodes D_1, D_3 and D_3, D_4 are matched pairs. When the signal cos $\omega_c t$ has polarity such that terminal b is positive with respect to terminal d, all four diodes conduct if we assume $A_c >> e_m(t)$. Since D_1 and D_2 are matched the potential at points a and c are the same and the input to the bandpass filter is shorted during this half cycle. In the next half cycle, the terminal b is negative with respect to terminal d and all four diodes are open. Thus the diode bridge connects $e_m(t)$ to the input of the bandpass filter. This switching on and off of $e_m(t)$ repeats for each cycle of the carrier. The bandpass filter input is $e_m(t)\, k(t)$

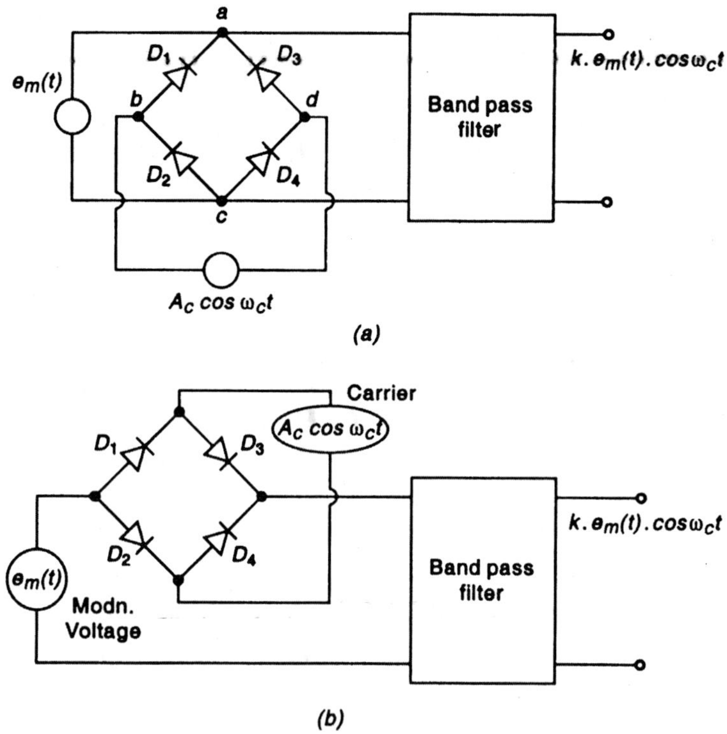

Fig 12.8. (a) Circuit diagram of a switching modulator for DSB-SC (b) Circuit of series bridge diode modulator.

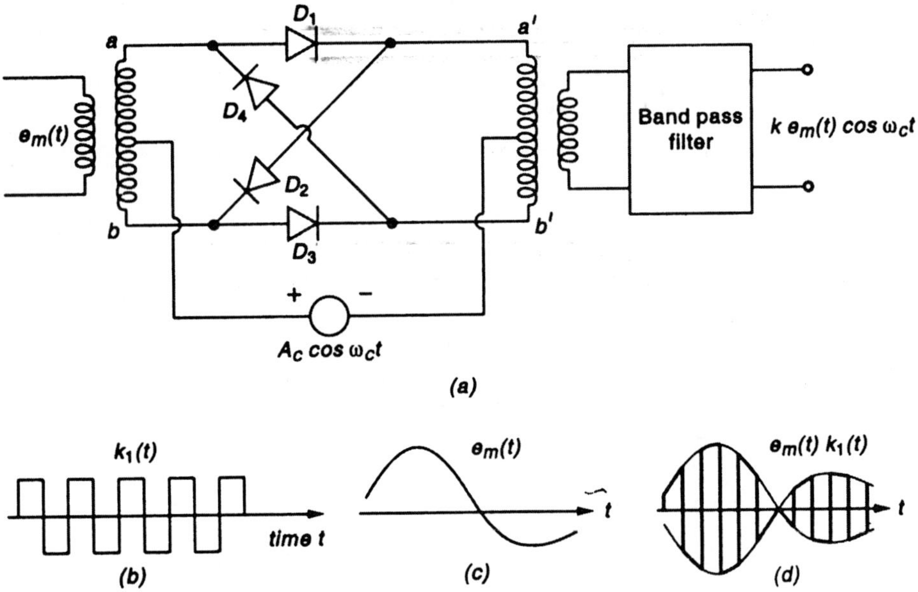

Fig 12.9 (a) Ring modulator circuit (b) Square wave pulse train (c) modulation signal (d) output AM wave

and the output is the desired modulating signal $e_m(t) \cos \omega_c t$. Fig. 12.8 (b) shows the circuit of a series bridge diode modulator. This bridge also achieves the same purpose by peroidically interrupting the input to the bandpass filter.

Figure 12.9 shows the circuit diagram of a ring modulator. During the positive half cycle, diodes D_1 and D_3 conduct while D_2 and D_4 are open and the output is proportional to $e_m(t)$. In the negative half cycle, diodes D_1 and D_3 are open while D_2 and D_4 conduct. Thus terminal a is connected to $b'$and b to a'. During this negative half cycle, the output is proportional to $-e_m(t)$. Thus $e_m(t)$ is as if multiplied by $k_1(t)$ as shown in Fig. 12.9 (b). The modulation signal and the modulated output is depicted in Fig. 12.9 (c) and (d) respectively.

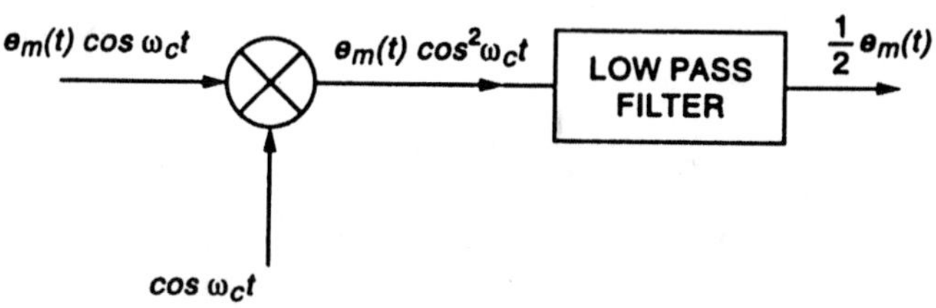

Fig 12.10. DSB-SC detector scheme

The DSB-SC modulation shifts or translates the frequency spectrum by ω_c. To recover the original message signal, demodulation or detection is carried out by a circuit as shown in Fig. 12.10. The modulated carrier spectrum is shifted again by $\pm \omega_c$. The output of the mixing circuit yields an additional term of $\pm 2\omega_c$. This term can be removed easily by a low pass filter. The method of retrieving the baseband signal in Fig. 12.10 is known as synchronous or coherent detection where a carrier of exactly the same frequency or phase is employed as was used for modulation.

12.5.2 AM Signal Generation

An amplitude modulated signal can be generated by a DSB-SC generator if the modulating signal $e_m(t)$ is replaced by $e_c + e_m$. An amplitude modulated signal generator is shown in Fig. 12.11. The input is $A_c \cos \omega_c t + e_m(t)$. With $A_c >> e_m(t)$, the switching action is controlled by $A_c \cos \omega_c t$. The diode opens and

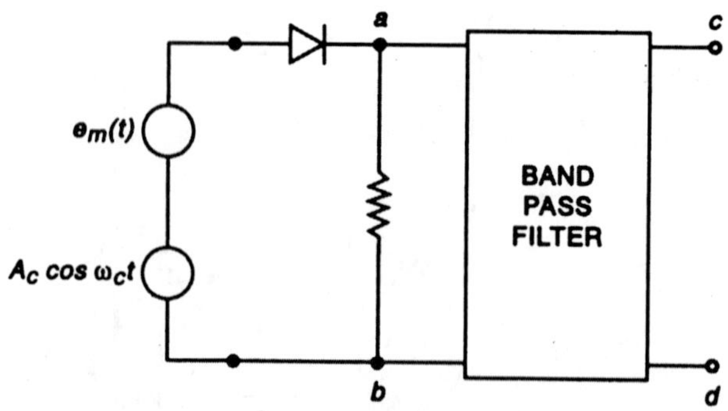

Fig 12.11. AM generator

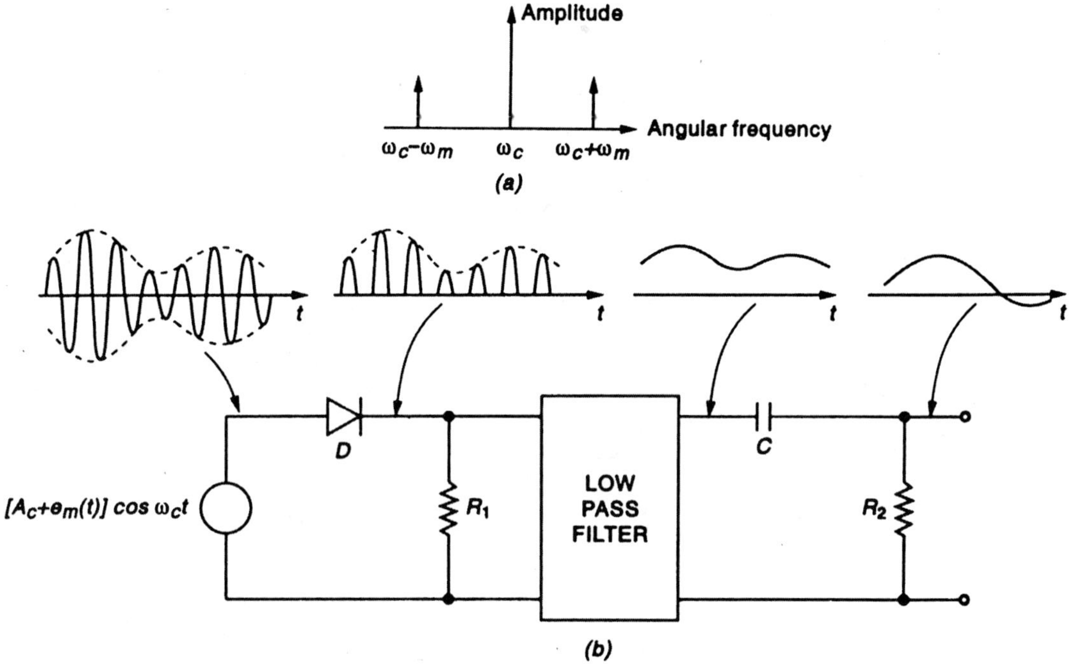

Fig 12.12. (a) Frequency spectrum (b) Circuit schematic of a rectifier detector.

shorts periodically with $\cos \omega_c t$. This is the same action as that of multiplying the input signal $[A_c \cos \omega_c t + e_m(t)]$ by $k(t)$. Thus the voltage between a and b is

$$e_{ab} = [e_m(t) + A_c \cos \omega_c t] \cdot k(t)$$

$$= [e_m(t) + A_c \cos \omega_c t]\left[\frac{1}{2} + \frac{2}{\pi}\left(\cos \omega_c t - \frac{1}{3}\cos 3\omega_c t + \frac{1}{5}\cos 5\omega_c t - \ldots\right)\right]$$

$$= \frac{A_c}{2}\cos \omega_c t + \frac{2}{\pi} e_m(t) \cos \omega_c t + X \qquad \ldots(12.25)$$

where X contains terms which may be eliminated by the bandpass filter.

Fig. 12.12 (b) shows the circuit schematic of a rectifier type detector for AM. In the circuit of Fig. 12.12 (b), the diode blocks the negative part of the AM wave. The output across R_1 is the rectified AM voltage. When the voltage across R_1 is applied to a low pass filter of cut off frequency B, the output is $\left[\frac{A_c + e_m(t)}{\pi}\right]$. The capacitor C blocks the d.c. term and the output is $\frac{e_m(t)}{\pi}$ where $e_m(t)$ is the modulation voltage.

The AM wave can also be demodulated by the envelope detector circuit of Fig. 12.13.

In the circuit the output follows the envelope of the input due to the presence of the RC network. An SSB signal generator is depicted in Fig. 12.14. The blocks labelled $-\frac{\pi}{2}$ are $\frac{\pi}{2}$ phase shifters. A synchronous detection circuit can be used to demodulate an SSB-SC signal.

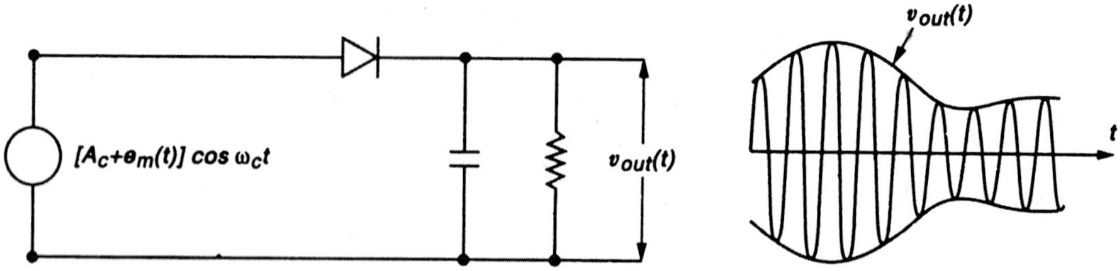

Fig 12.13. Envelope detector circuit

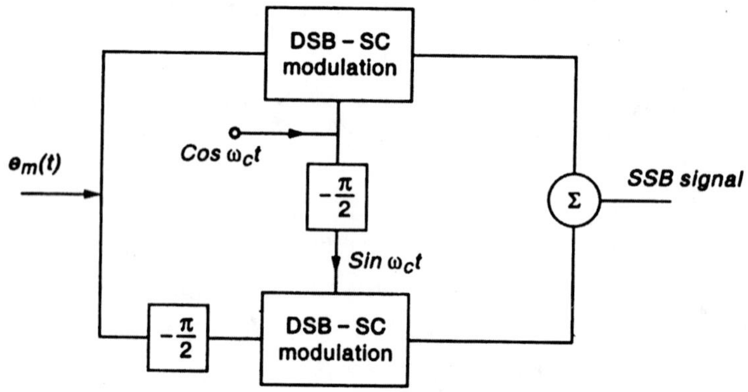

Fig 12.14. Envelope detector circuit

12.5.3 Quadrature Amplitude Modulation

DSB signals require a bandwidth which is twice that of SSB signals. This disadvantage can be eliminated by transmitting two DSB signals using carriers of the same frequency but with a phase difference of 90°. Both modulated signals occupy the same band. The two baseband signals can be then seperated at the receiver by synchronous detection using two local carriers in phase quadrature (90° phase difference).

The transmission rate in quadrature amplitude multiplexing can be enhanced by using M-ary Q AM. This is also known as amplitude phase shift keying. For $M = 16$, the following 16 pulses (16 symbols) are employed. A properly shaped baseband pulse $e(t)$ is sent as

$$= a_i\, e(t) \cos(2\pi f_c t) + b_i\, e(t) \sin(2\pi f_c t)$$

$$e_i(t) = r_i\, e(t) \cos(\omega_c t + \theta_i) \qquad \ldots(12.26)$$

where $\quad i = 1, 2, 3, \ldots 16$

In Fig. 12.15 (b), there are 16 possible combinations of (r_i, θ_i). There are 16 possible combinations of four binary digits. Thus, every possible four bits sequence is sent by a particular (r_i, θ_i). Therefore, one single pulse $r_i\, e(t) \cos(2\pi f_c t + \theta_i)$ transmits four bits. A normal transmission line which can send 2400 bits/second now transmit 9600 bits.

The modulaton and demodulation process is carried out by the system in Fig. 12.15 (a).

The inputs are $e_{mi}(t) = a_i\, e(t)$ $\qquad \ldots(12.27)$

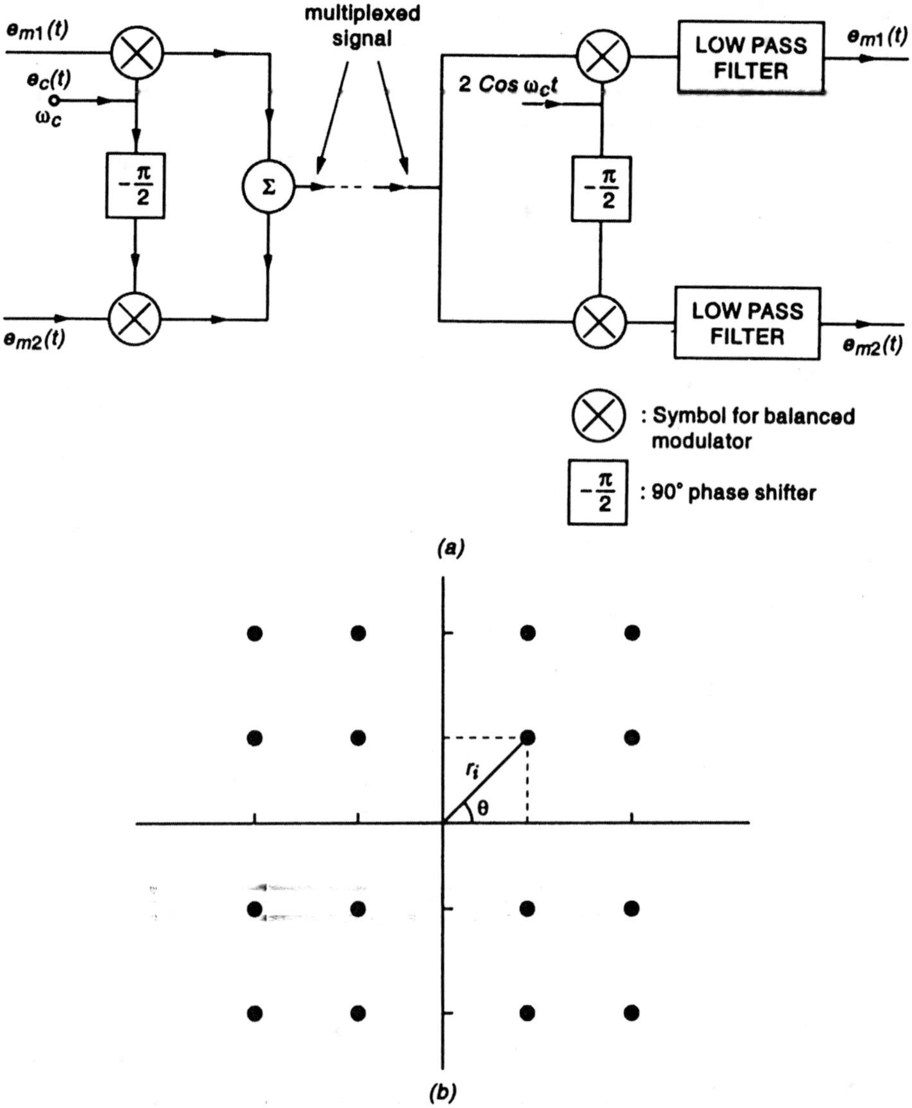

Fig 12.15. (a) Quadrature multiplexing (b) 16 point quadrature amplitude multiplexing (QAM)

and $$e_{m_2}(t) = b_i\, e(t) \qquad \ldots(12.28)$$

The two outputs of the modulator will be

$$\left.\begin{aligned} e_{01} &= a_i\, e(t) \\ \text{and} \quad e_{02} &= b_i\, e(t) \end{aligned}\right\} \qquad \ldots(12.29)$$

If a_i and b_i are known, the four transmitted bits can be found.

In a telephone line, at each end a modulator and a demodulator are required to transmit and receive data. A 'modem' is one unit that is obtained by packaging a modulator and a demodulator.

12.5.4 Single Sideband Suppressed Carrier (SSB-SC) Modulation

In the case of amplitude modulation, if the carrier and one of the sidebands are suppressed, no intelligence or information (message signal) is lost. The reason behind such suppression in SSB-SC system is that the transmission bandwidth now required is equal to just the message bandwidth. This is the case of single sideband suppressed carrier modulation. Fig. 12.16 depicts the spectrum of such a signal.

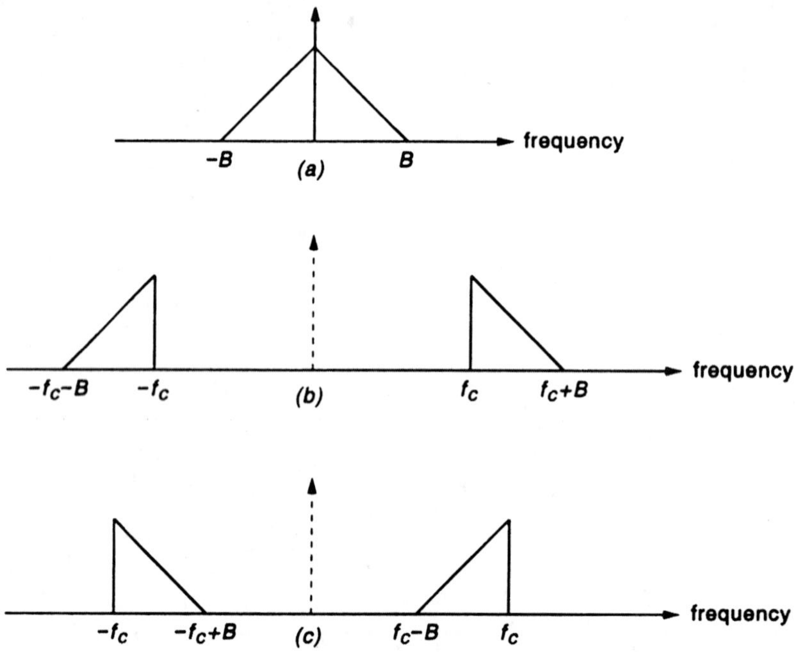

Fig 12.16. Spectrum of (a) baseband signal, (b) SSB-SC with upper sideband (c) SSB-SC with lower sideband

An SSB-SC signal can be generated by the filter method of generation of SSB-SC signal as depicted in Fig. 12.17. The bandpass filter passband should be the same as that of the desired sideband. Further the transition band of the filter (the distance between the transmission band and the stop band) should not exceed twice the main frequency component present in the baseband. This may be realised by using high Q(Q values between 1000 and 2000) filters.

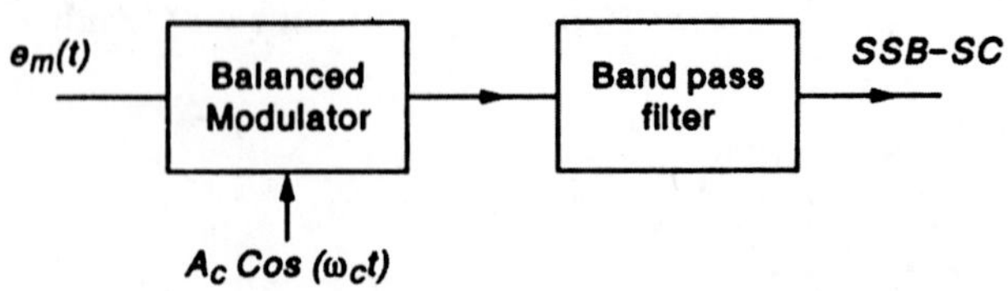

Fig 12.17. Block schematic for SSB-SC generation

A synchronous detection circuit may be employed for demodulation of the SSB-SC signal as depicted Fig. 12.18.

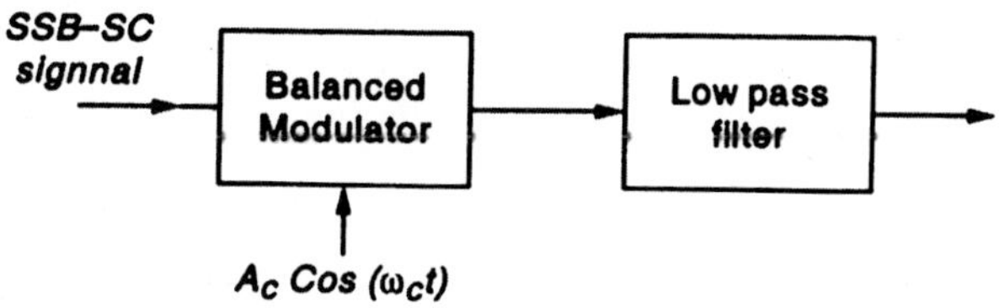

Fig 12.18. Block schematic for SSB-SC demodulation

12.5.5 Vestigial Sideband (VSB) Systems

In vestigial sideband system, instead of suppressing or rejecting one sideband completely, a gradual cut off of one sideband is allowed. The gradual cut off is compensated by a vestige (or portion of the other sideband) as depicted in Fig. 12.19. When signals contain component of frequency at extremely low frequencies (such

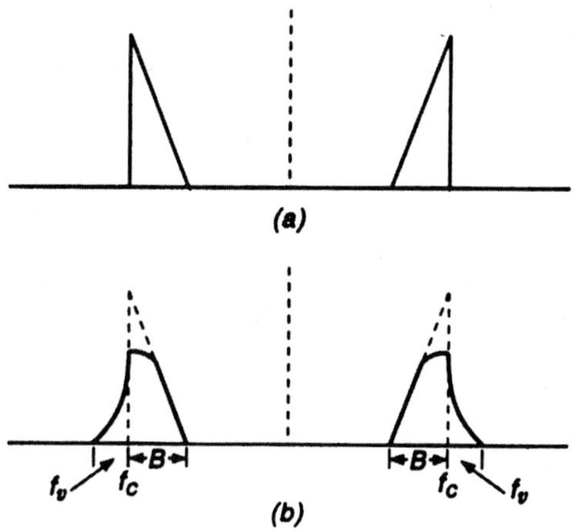

Fig 12.19. Spectrum of (a) SSB-SC signal (b) VSB signal

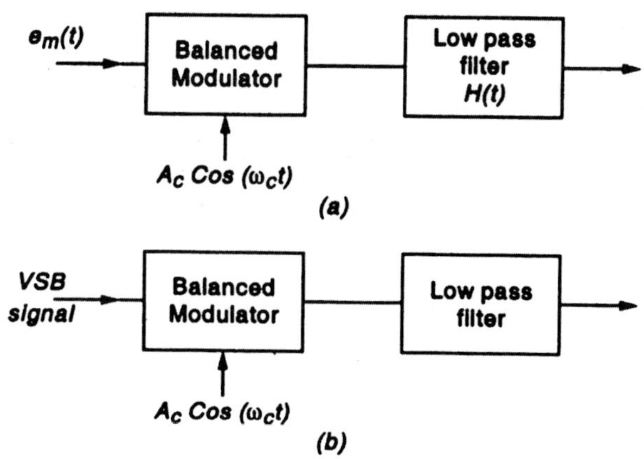

Fig 12.20. VSB signal generation and detection

as in television or telegraph signals), the upper and lower sideband of the translated signal meet at the carrier frequency. In such cases, it becomes very difficult to isolate one band from the other as in SSB-SC. The VSB scheme becomes suitable in such cases. The VSB-SC signal generation and detection block schematic is depicted in Fig. 12.20.

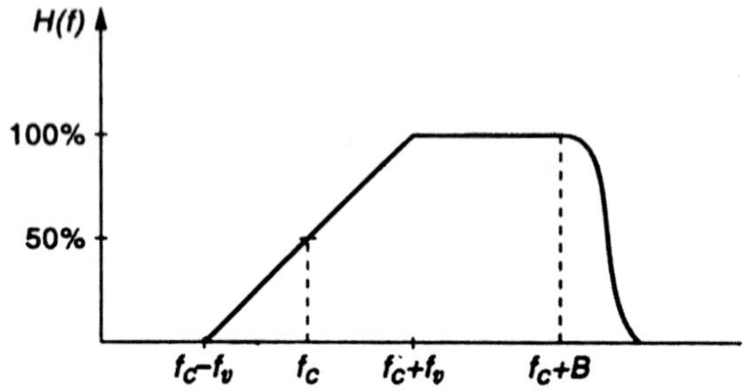

Fig 12.21. Amplitude response of VSB filter

12.6 COMPARISON OF VARIOUS AM SYSTEMS

1. The amplitude modulation (AM) system are advantageous over AM-SC systems in that the detectors required (retifier or envelope detectors) are **simpler** and **cheaper** compared to those needed for suppressed carrier systems. Further, AM signals can be easily generated at high power levels. In the case of suppressed carrier systems, the balanced modulators needed are difficult to design.

2. Suppressed carrier systems, however, need lesser power to transmit the same information. Since the carrier power is high in AM, the transmitter power in high and the transmitter is more expensive. For suppressed carrier systems however, the receiver is more complex. The receiver required in AM is simpler and cheaper. Thus, for public broadcast system, AM is the preferred modulation.

3. The effect of **selective fading** occurs due to different attenuation and phase shifts of different frequencies. **Selective fading will badly distort an AM signal.** The effect of selective fading is more pronounced at higher frequencies. The AM system is thus employed for medium frequency transmission.

If we compare now a SSB_SC system with a DSB_SC system we can note that

(a) **The bandwidth required in SSB-SC is half that of DSB-SC.**

(b) **Frequency and phase errors in the local carrier at the detection stage have more pronounced effects in DSB-SC than in SSB-SC systems.**

(c) **Selective fading affects SSB-SC system less than the DSB-SC system.**

It should, however, be noted that the generation of high level SSB is more difficult than high level DSB signals. This difficulty can be removed in VSB.

12.7 FREQUENCY MODULATION

In frequency modulation, the information signal is utilized to vary the carrier frequency. The important advantages of frequency modulation (compared to AM) are

(1) **Increase in signal to noise ratio for the same transmitter power.**

(2) **Elimination of certain types of interference at the receiver.**

(3) **Modulation power needed is small as the modulation takes place at a low level power stage of the transmitter.**

In FM, the information or message signal $e_m(t)$ varies with the carrier frequency. The voltage $e_m(t)$ may be employed to control the frequency of an oscillator. In a properly designed oscillator, the change in carrier frequency is proportional to the modulating voltage. Thus the instantaneous frequency can be written as

$$f_i(t) = f_c + k_f e_m(t) \qquad \ldots(12.30)$$

where f_c is the carrier frequency, k_f is a constant called the **frequency deviation constant.** The unit of k_f is Hertz/volt or Hz/V. Fig. 12.22 depicts the frequency modulation for (a) sinusoidal modulation (b) square wave modulation.

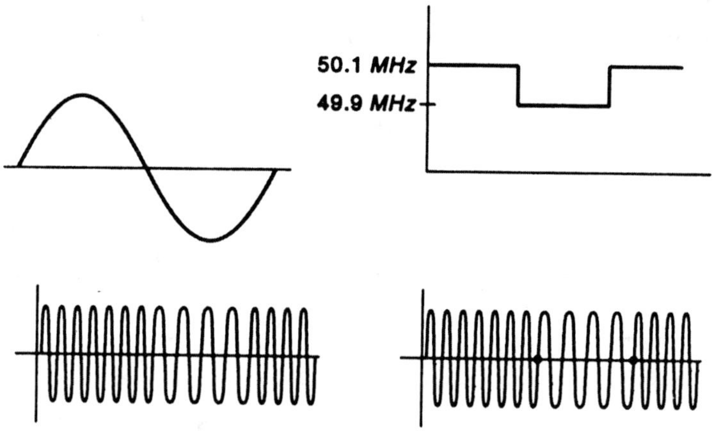

Fig 12.22. Frequency modulated wave for (a) sinusoidal modulation (b) Square wave modulation

If we consider sinusoidal modulation, the instantaneous frequency will be

$$\begin{aligned} f_i(t) &= f_c + k_f e_m(t) \\ &= f_c + k_f A_m \cos 2\pi f_m t \\ &= f_c + \Delta f \cos 2\pi f_m t \end{aligned} \qquad \ldots(12.31)$$

where the maximum frequency deviation will be

$$\Delta f = k_f A_m \qquad \ldots(12.32)$$

The expression for sinusoidally modulated FM will be

$$\begin{aligned} e(t) &= A_c \cos\left\{2\pi f_c t + 2\pi \Delta f \int_0^t \cos(2\pi f_m t)\, dt\right\} \\ &= A_c \cos\left\{2\pi f_c t + \frac{\Delta f}{f_m} \sin 2\pi f_m t\right\} \end{aligned} \qquad \ldots(12.33)$$

The modulation index for FM, usually denoted by β is obtainable as

$$\beta = \frac{\Delta f}{f_m} \qquad \ldots(12.34)$$

The expression for the sinusoidally modulated FM is therefore,

$$e(t) = A_c \cos(2\pi f_c t + \beta \sin 2\pi f_m t) \qquad \ldots(12.35)$$

The quantity β is also known as the **deviation ratio** of the FM. β can also be expressed as

$$\beta = m_f = \frac{(\Delta f)_{\max}}{(f_m)_{\max}} = \frac{k_f A_m}{2\pi\,(f_m)_{\max}} = \frac{k_f A_m}{(\omega_m)_{\max}} \qquad \ldots(12.36)$$

It should be kept in mind that the maximum frequency deviation $(\Delta f)_{\max}$ is directly proportional to A_m, the amplitude of the modulating wave but it is independent of frequency. Hence the frequency of FM wave alters with time while the amplitude remains constant.

Equation (12.35) may be analyzed by Fourier methods so as to obtain the spectrum. The trigonometric series contains a carrier frequency term $J_0(\beta) A_c \cos \omega_c t$, a pair of first frequencies $J_1(\beta) A_c \cos(\omega_c \pm \omega_m) t$, a pair of side frequencies $J_2(\beta) A_c \cos(\omega_c \pm 2\omega_m) t$, a third pair of side frequencies $J_3(\beta) A_c \cos(\omega_c \pm 3\omega_m) t$ and so on. The amplitudes $J_n(\beta)$ are called the Bessel function of the first kind of order n. The values of $J_n(\beta)$ are available in tabular and graphical form as also in computer software (like Mathcad). Some of the values are depicted in Table 12.2 for $A_m = 1$ volts.

Table 12.2 : Bessel Functions for sinusoidally frequency modulated signal.

Modulation Index β Carrier		*Side Frequency*									
		1st	*2nd*	*3nd*	*4th*	*5th*	*6th*	*7th*	*8th*	*9th*	*10th*
	J_0	J_1	J_2	J_3	J_4	J_5	J_6	J_7	J_8	J_9	J_{10}
0·25	0·98	0·12	0·01								
0·5	0·94	0·24	0·03								
1·0	0·77	0·44	0·11	0·02							
2·0	0·22	0·58	0·35	0·13	0·03	0·01					
3·0	– 0·26	0·34	0·49	0·31	0·13	0·04	0·01				
4·0	– 0·4	– 0·07	0·36	0·43	0·28	0·13	0·05	0·02			
5·5	0	– ·34	– ·12	·26	·40	·32	0·19	·09	0·03	0·01	

If can be seen from the above table, that for $\beta = 0{\cdot}25$, the spectral components are

Carrier (f_c) $J_0(0{\cdot}25) = 0{\cdot}98$

First order frequencies $(f_c \pm f_m)$ $J_1(0{\cdot}25) = 0{\cdot}12$

Second order frequencies $(f_c \pm 2f_m)$ $J_2(0{\cdot}25) = 0{\cdot}01$

Similarly for $\beta = 1{\cdot}0$, the spectral components are

Carrier (f_c) $J_0(1{\cdot}0) = 0{\cdot}77$

First order frequencies $(f_c \pm f_m)$ $J_1(1{\cdot}0) = 0{\cdot}44$

Second order frequencies $(f_c \pm 2f_m)$ $J_2(1{\cdot}0) = 0{\cdot}11$

Third order frequencies $(f_c \pm 3f_m)$ $J_3(1{\cdot}0) = 0{\cdot}02$

The spectra for sinusoidal FM with (a) $\beta = 0{\cdot}25$ (b) $\beta = 1{\cdot}0$ and (c) $\beta = 5{\cdot}5$ are depicted in Fig. 12.23 below.

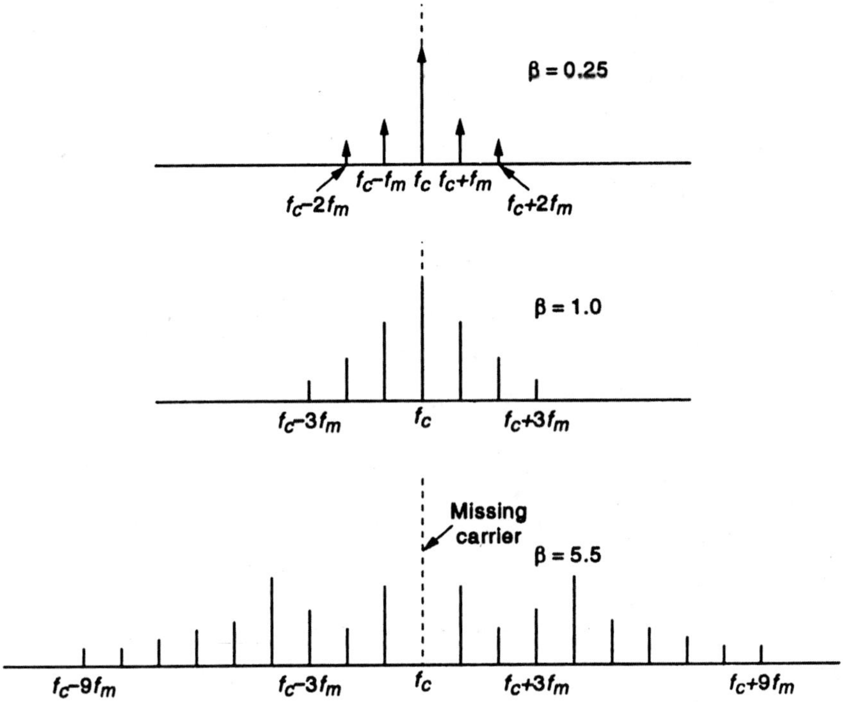

Fig 12.23. Frequency spectra for (a) β = 0·25 (b) β = 1·0 and (c) β = 5·5

EXAMPLE 12.10 *Find the modulation index when an FM transmitter sends a 150 MHz carrier wave frequency modulated by a 10 kHz sinusoidal waveform. Find also the significant pairs of side frequencies, channel width required for these three side frequency pairs.*

SOLUTION

(i) $\beta = \dfrac{\Delta f}{fm} = \dfrac{25 \text{ kHz}}{10 \text{ kHz}} = 2{\cdot}5$

(ii) The three significant side frequency pairs are

150 MHz ± 10 kHz, 150 MHz ± 20 kHz and 150 MHz ± 30 kHz.

The spectra of various values of β shown in Fig. 12.23 also illustrate that since the spectral lines are spaced by f_m, the bandwidth occupied by the spectrum is

$$B_{FM} = 2\,(\Delta f + n\,f_m) \qquad \text{...(12.37)}$$

where n is the highest order of frequencies for which the amplitude is significant.

The peak voltage of the spectral components can be denoted by

$$E_n = J_n\,(\beta)\,.\,A_c \qquad \text{...(12.38)}$$

The r.m.s. values can be expressed as

$$(E_n)_{rm.s} = J_n\,(\beta)\,(A_c)_{rms} \qquad \text{...(12.39)}$$

If R is the load resistance, the average power in a spectral component is

$$P_n = \frac{(E_n)_{rms}}{R} \qquad \ldots(12.40)$$

In terms of r.m.s. values, the total average power is

$$P_T = \frac{E_0^2}{R} + \frac{2}{R}(E_1^2 + E_2^2 + \ldots\ldots) \qquad \ldots(12.41)$$

In terms of the unmodulated carrier and Bessel function co-efficients,

$$P_T = \frac{A_c^2 J_0^2(\beta)}{R} + \frac{2A_c^2}{R}[J_1^2(\beta) + J_2^2(\beta) + \ldots\ldots]$$

$$= P_c[J_0^2(\beta) + 2J_1^2(\beta) + 2J_2^2(\beta) + \ldots\ldots] \qquad \ldots(12.42)$$

The unmodulated power is $P_c = \dfrac{E_c^2}{R}$ ...(12.43)

EXAMPLE 12.11 *Find the bandwidth required in frequency modulated wave transmission, if $\Delta f = 75$ kHz and $f_m = 20$ kHz.*

SOLUTION Since there are 7 pairs of side frequencies (refer Table 12.2), we get the bandwidth as

$$B = 2(\Delta f + f_m)$$

$$= 2(75 + 20) \text{ kHz} = 190 \text{ kHz}.$$

EXAMPLE 12.12 *For a frequency modulation signal, the peak frequency deviation is 6 kHz and the frequency of the modulating signal is 1·5 kHz. Find the average power output by summing the powers of all side frequency components. The unmodulated carrier power is 25 Watts.*

SOLUTION Modulation index $\beta = \dfrac{6 \text{ kHz}}{1{\cdot}5 \text{ kHz}} = 4{\cdot}0$

From Table 12.2, we find that there are seven pairs of side frequencies and that

$J_0(4) = -0{\cdot}4$, $J_1(4) = -0{\cdot}07$, $J_2(4) = 0{\cdot}36$,

$J_3(4) = 0{\cdot}43$, $J_4(4) = 0{\cdot}28$, $J_5(4) = 0{\cdot}13$, $J_6(4) = 0{\cdot}05$ and $J_7(4) = 0{\cdot}02$

The total carrier power is

$$P_T = 25[(-0{\cdot}4)^2 + 2\{(-0{\cdot}07)^2 + (0{\cdot}26)^2 + (0{\cdot}43)^2$$
$$+ (0{\cdot}238)^2 + (0{\cdot}13)^2 + (0{\cdot}05)^2 + (0{\cdot}02)^2\}]$$
$$= 25[0{\cdot}16 + 2\{0{\cdot}0049 + 0{\cdot}1296 + 0{\cdot}1849 + 0{\cdot}0784$$
$$+ 0{\cdot}0169 + 0{\cdot}0025 + 0{\cdot}0004\}]$$
$$= 25 \times {\cdot}9952 = 25 \text{ W}$$

Fig. 12.24 depicts a method for generation of FM wave by the Clapp oscillator method. Here the modulating voltage is applied to a voltage dependent capacitor (**a varactor diode or varicap**) which in turn controls the frequency of an oscillator. The bias for the varactor is adjusted by the variable potentiometer. The RFC does not allow the radio frequency carrier to enter into the modulation circuit. Detection of FM signal may be carried out by employing a circuit that develops a voltage which varies linearly with the frequency of the input signal. FM signal can be detected by a slope detector or a Foster Seely discriminator.

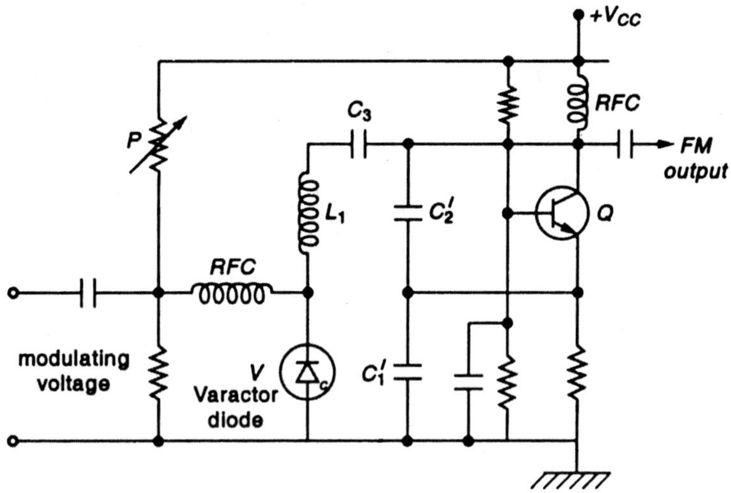

Fig 12.24. FM signal generation by Clapp oscillator method

12.8 PHASE MODULATION

Let us consider a modulation voltage

$$e_m(t) = A_m \cos \omega_m t \quad \text{...(12.44)}$$

and carrier signal

$$e_c(t) = A_c \cos(\omega_c t + \phi_c) \quad \text{...(12.45)}$$

In phase modulation, the information signal or modulating voltage is used to vary the angle of phase lead or lag. Mathematically, the phase modulation can be expressed as

$$\phi(t) = \phi_c + k_p e_m(t) \quad \text{...(12.46)}$$

where k_p is called the phase deviation constant and it has units of radians/volt.

The equation for the phase modulation wave is

$$e_p(t) = A_c \cos(\omega_c t + k_p e_m(t)) \quad \text{...(12.47)}$$

From equation (12.46), the instantaneous frequency can be written as

$$f_i(t) = \frac{1}{2\pi}\frac{d\phi}{dt} \quad \text{...(12.48)}$$

$$= f_c + \frac{k_p}{2\pi}\frac{de_m(t)}{dt} \quad \text{...(12.49)}$$

The quantity $\phi(t)$ is equation (12.46) can be written as

$$\phi(t) = \phi_c + k_p A_m \cos \omega_m t \quad \text{...(12.50)}$$

$$= \phi_c + \Delta\phi \cos \omega_m t$$

The maximum value of phase deviation $\Delta\phi$ is proportional to the peak modulating signal and is

$$\Delta\phi = k_p A_m \quad \text{...(12.51)}$$

The equation for the sinusoidal PM is

$$e(t) = A_c \cos(\omega_c t + \Delta\phi \sin \omega_m t) \quad \text{...(12.52)}$$

The maximum deviation in frequency from the modulated frequency is

$$\Delta f = [f_i\ (t) - f_c]_{max} \qquad ...(12.53)$$

$$= m_p f_m$$

$$= k_p A_m f_m \qquad ...(12.54)$$

Thus the maximum value of frequency deviation for PM wave is proportional to both the amplitude and frequency of the modulating signal. Compare this situation with that of a FM wave where Δf is proportional to only amplitude of the information or message signal. It should be noted that there is a phase shift of 90° of the modulating voltage in the FM and PM waves. The modulation indices are also different in the two cases. The frequency spectra of PM signal consists of an infinite number of side frequency pairs seperated from the carrier frequency by f_m, $2f_m$, $3f_m$ and so on. The amplitudes are given in terms of Bessel function of the first kind J_n ($\Delta\phi$). In the case of PM, argument is the maximum phase deviation $\Delta\phi$ instead of the frequency modulation index β for FM case. For analog message or information signals, phase modulation is employed mainly as a stage for generation of frequency modulation as shown in Fig. 12.25.

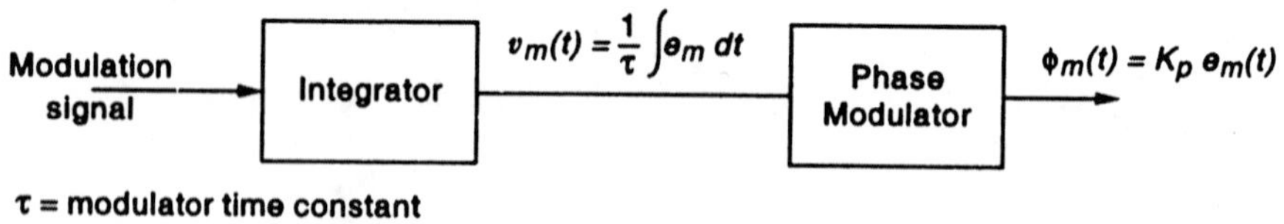

Fig 12.25. Method of obtaining FM from PM

12.9 PULSE MODULATION

Pulse modulation can be of different types like pulse amplitude modulation (PAM), pulse code modulation (PCM), pulse width modulation (PWM) and pulse position modulation (PPM).

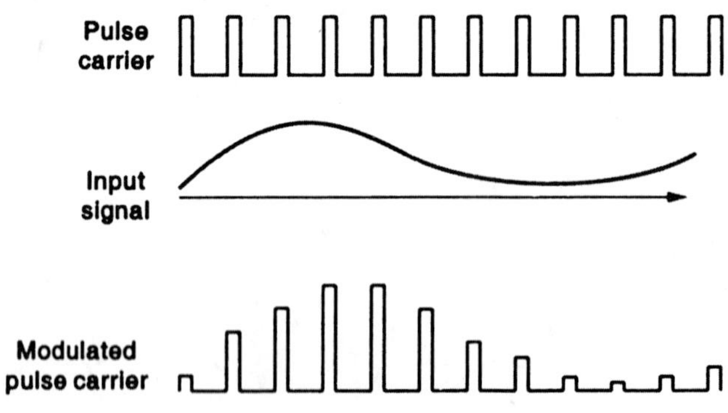

Fig 12.26. Pulse amplitude modulation

In pulse amplitude modulation, the information signal is used to control the amplitude of a pulse carrier (Fig. 12.26). Appropriate filtering is necessary to reconstruct the signal at the receiver end. **The pulse repetition rate must be at least twice the signal frequency according to sampling theorem.** For telephone voice signals which has a bandwidth of 4000 Hz, a pulse carrier of about 16 kHz which is four timer higher than the highest frequency of the input signal may be employed.

Fig. 12.27 shows the principle of pulse code modulation. As in the case of pulse amplitude modulation, the input is sampled. The magnitude of each sample is then converted to a binary code. If one uses four

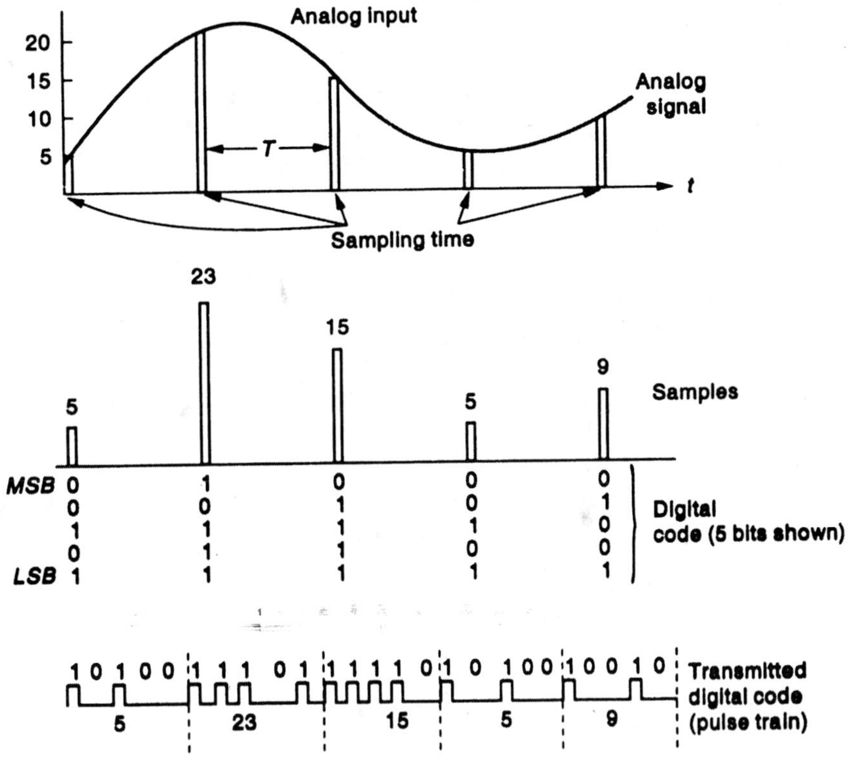

Fig 12.27. Pulse code modulation

samples for each input cycle, and each sample is converted to a 8 bit code, the number of bits for each input cycle is $4 \times 8 = 32$. For a 4 kHz analog signal, one needs $4000 \times 32 = 128$ kp/sec. PCM is normally employed in long distance telephone transmission. Besides PAM and PCM, pulse position modulation (PPM) and pulse width modulation (PWM) are also used. In PPM, the position of the pulse with respect to a reference time pulse denotes the magnitude of the analog samples., In pulse width modulation (PWM), the pulse width represents the amplitude of the sample.

Review Questions

1. What is meant by modulation ? What do you understand by demodulation ?
2. What are the advantages of modulating a low frequency or message signal ?
3. What do you understand by medium frequency (MF), high frequency (HF) and very high frequency (VHF) band ? Mention typical communication applications of each of the band.
4. What is the bandwidth required for common voice, music and television signals ?
5. What is sampling theorem ?
6. What do you understand by the term depth of modulation in AM system ?
7. Draw the modulated carrier waveform when a sinusoidal modulation signal is employed in AM systems for $m = 0{\cdot}5$, $m = 1$ and $m > 1$.

8. Draw the soectrum of an amplitude modulated carrier.

9. Show that for 100% modulation, the ratio of power in side frequency components to the carrier power is 50%.

10. Draw the circuit diagram of a typical modulator using a common emitter amplifier.

11. What is double side band-suppressed carrier system ? What are its advantages over AM system.

12. Draw a schematic circuit diagram for generation of DSB-SC signals ? Discuss how DSB-SC signal can be detected by a ring modulator.

13. Draw the schematic circuit diagram of a rectifier detector and explain how an AM signal can be demodulated.

14. What is an SSB signal ? Draw the circuit schematic of an SSB signal generator.

15. What do you understand by Quadrature ampltitude modulation ? What is a 'modem' ?

16. What do you understand by SSB-SC (single sideband suppressed carrier) modulation ? Draw the block schematic for SSB-SC signal generation and demodulation.

17. What is vestigial sideband system ? Discuss how a VSB signal can be generated. Draw the amplitude response characteristics of a VSB filter.

18. What is selective fading ? What is its effect in AM transmission ?

19. Write a short essay comparing the various AM systems.

20. What is meant by frequency modulation ? Draw the waveform for a frequency modulated wave for (a) sinusoidal (b) square wave modulation.

21. Derive equation (12.33).

22. Show that for a frequency modulated wave, the maximum frequency deviation is directly proportional to the amplitude of the modulating wave but is independent of frequency.

23. Draw the frequency spectra for a frequency modulated wave if (i) $\beta = 0{\cdot}5$ (ii) $\beta = 2{\cdot}0$ and (iii) $\beta = 3{\cdot}0$.

24. What is phase modultion ?

25. Derive equation (12.52).

26. Derive equation (12.54).

27. What do you understand by pulse amplitude modulation ? Draw a typical waveform for a pulse amplitude modulated wave when a carrier pulse train is modulated by a modulation signal of the form $E_{dc} + A_m \sin \omega_m t$.

28. What do you understand by pulse code modulation ? Draw the waveform of the transmitted digital code when the analog input is as depicted in Fig. R 12.28.

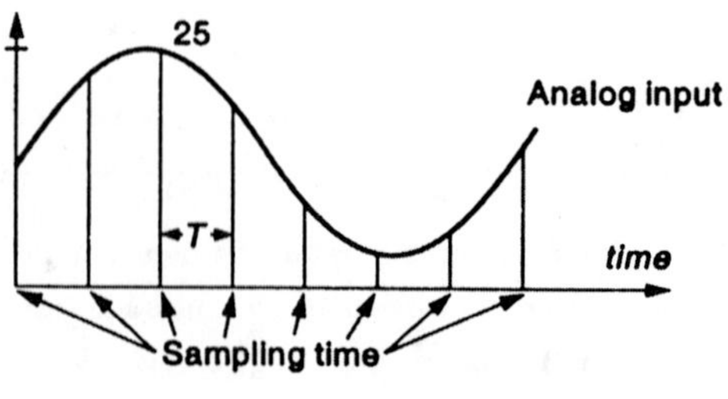

Fig 12.28. Reference review question 12.28

Objective Type Questions

(Tick (√) mark the correct or most appropriate statement)

1. By the term modulation is meant
   (a) impressing a carrier signal on the amplitude, phase or frequency of the message
   (b) impressing the message signal on the amplitude, the phase or the frequency of the carrier
   (c) mixing the message signal with the carrier signal
   (d) none of the above
2. By the term baseband we mean
   (a) a signal where value is near zero frequency (dc) (b) any information signal
   (c) modulated signal (d) signal which has the maximum carrier power.
3. Amplitude modulation means
   (a) that the amplitude of the modulated carrier is variable with time
   (b) that the amplitude of the modulation signal is variable with time
   (c) that a high frequency signal is translated to the passband of a communication channel.
   (d) none of the above
4. In DSB-SC transmission, the frequency which is not transmitted is
   (a) the upper side frequency (b) the lower side frequency
   (c) audio frequency (d) carrier frequency
5. The AM broadcast band ranges from
   (a) 3-30 kHz (b) 30-300 kHz
   (c) 300 kHz to 3 MHz (d) 500-1500 kHz
6. In an DSB-SC system with a depth of modulation of 100%, the percentage power saving will be
   (a) 25% (b) 33·3%
   (c) 66·6% (d) 100%
7. The bandwidth required for voice transmission in single telephone channel is
   (a) 3·5 kHz (b) 5 kHz
   (c) 10 kHz (d) 20 kHz
8. The bandwidth required for music in AM radio broadcasting is
   (a) 5 kHz (b) 10 kHz
   (c) 20 kHz (d) 200 kHz
9. The bandwidth required for FM radio broadcast is
   (a) 10 kHz (b) 20 kHz
   (c) 75 kHz (d) 200 kHz
10. For TV broadcasting, the bandwidth required is
    (a) 4·5 MHz (b) 6 MHz
    (c) 5·75 MHz (d) none of the above
11. In a certain transmitter, the unmodulated carrier power is 1000 W and it is modulated with a simusoidal signal with a depth of modulation of 50%. The total power in the modulated wave will be
    (a) 1000 W (b) 1125 W
    (c) 1200 W (d) 1500 W
12 An audio signal 5 sin (3000 πt) amplitude modulates a signal 15 sin ($2\pi \times 10^6 t$). The depth of modulation is
    (a) 20% (b) 33·3%
    (c) 50% (d) 100%
13. In the above question the two side frequencies are
    (a) 997 kHz, 1003 kHz (b) 995·5 kHz, 1004·5 kHz
    (c) 998·5 kHz, 1001·5 kHz (d) 1·5 kHz, 1000 kHz
14. For the question above, the total power is the two side frequencies compared to that at the carrier frequency is
    (a) ·055 (b) 0·16

(c) 0·11 (d) 0·333

15. The total bandwidth required to transmit the AM wave in objective question 12 will be
(a) 1·5 kHz (b) 4·5 kHz
(c) 9 kHz (d) 3 kHz

16. For the objective question 12, the total bandwidth required to transmit by SSB-SC method will be
(a) 1·5 kHz (b) 2·25 kHz
(c) 4·5 kHz (d) 0·75 kHz

17. A modulation signal 12 sin $(2\pi \times 4{\cdot}5 \times 10^6 t)$ modulates a carrier 2·4 sin $(2\pi \times 94 \times 10^6 t)$. For VSB transmission the bandwidth required would be
(a) 9 MHz (b) 4·5 MHz
(c) 11·25 MHz (d) 5·65 MHz.

18. An AM wave has a modulation index of 100%. If the transmission is done by DSB-SC method, the percentage power saving will be
(a) 50% (b) 66·6%
(c) 75% (d) 95%

19. For the amplitude modulated wave depicted in Fig. 12·19, the modulation index will be
(a) zero (b) less than 100%
(c) equal to 100% (d) greater than 100%.

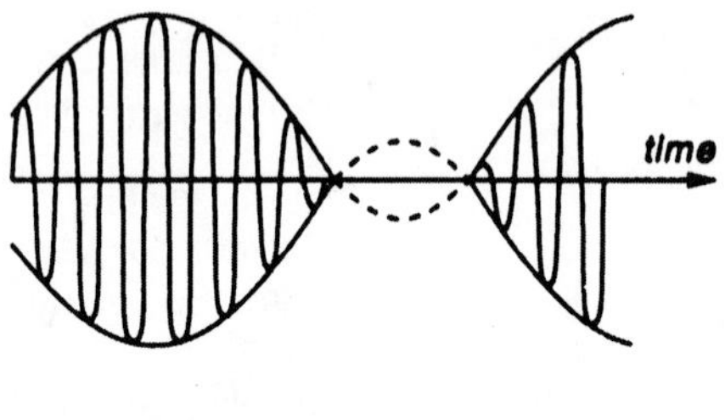

Fig 12.29 Reference objective question 19

20. An AM signal wave can be detected by
(a) a rectifier detector (b) an envelope detector
(c) a synchronous detector (d) both (a) and (b)

21. Two modulated signals having carriers of the same frequency but a phase difference of 90° is transmitted. The modulation system is called
(a) DSB-SC modulation (b) SSB-SC modulation
(c) Amplitude modulation (d) Quadrature amplitude modulation

22. QAM signals can be detected by
(a) envelope detector circuit
(b) rectifier detector circuit
(c) synchronous detection using two local carriers in phase quadrature
(d) none of the above.

23. In VSB transmission, the bandwidth required is the same as
(a) AM system (b) DSB-SC system
(c) SSB-SC system (d) 25% more than the SSB system

24. The effect of selective fading
(a) is more pronounced at lower frequencies (b) is more pronounced at higher frequencies
(c) is more pronounced at higher frequencies (e) none of the above

25. In a modulation system, if the modulation frequency is trebled the index of modulation is also trebled. The modulation system is
(a) AM (b) FM
(c) PM (d) Both PM & FM

26. The disadvantage of FM over AM is

(a) high modulating power is necessary
(b) high transmitter power is needed
(c) large bandwidth is needed
(d) noise is high for high frequency signals

27. FM broadcast lies in
(a) HF band
(b) VHF band
(c) UHF band
(d) SHF band

28. An FM wave is represented by

$e = 10 \sin \{5{\cdot}56 \times 10^8 t + 2 \sin (5655\, t)\}$

Its carrier frequency is
(a) 56·5 MHz
(b) 90 MHz
(c) 180 MHz
(d) 113 MHz

Exercise Problems

1. The peak-to-peak voltage of an amplitude modulated signal is 7 V maximum and 1 V minimum. Find the percent modulation. What is the amplitude of the unmodulated carrier ?

[**Ans.** m = 75%, 4 V]

2. A 850 kHz carrier wave is amplitude modulated by audio signals between 500 and 2000 Hz. Find (i) frequency span of each sideband (ii) maximum upper side frequency (iii) minimum lower side frequency (iii) channel width.

[**Ans.** 1500 Hz, 1002 kHz, 998 kHz, 4 kHz]

3. Find the power developed by an AM wave in a load of 100 Ω when the peak voltage of the carrier is 60 V and the modulation index is 60%.

[**Ans.** Unmodulated carrier power = 18 W, Total power in AM wave = 21·24 W, Power in sideband = 3·24 W]

4. A modulating signal consists of a triangular wave having zero d.c. component and a peak-to-peak voltage of 10 V. It is used to amplitude modulate a carrier of peak voltage 10 V. Find the depth of modulation.

[**Ans.** m = 50%]

5. An amplitude modulated signal is produced by a message signal shown in Fig. E 12.5. Find the AM pattern.

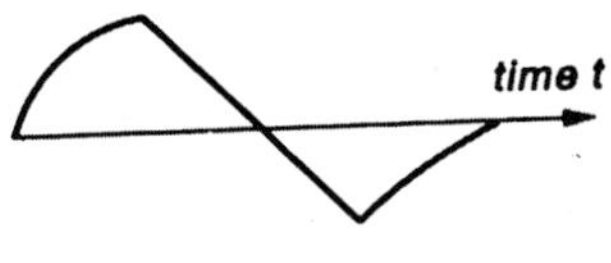

Fig E.12.5. Reference exercise problem 5

6. A carrier wave of frequency 10 MHz and peak value 10 V is amplitude modulated by a sine wave information signal of amplitude 5V. If the modulation frequency is 10 kHz, find the modulation index. Also sketch the frequency spectrum.

[**Ans.** m = 50%]

7. The r.m.s. antenna current of radio transmitter is 10 Amps when unmodulated. Find the r.m.s antenna current when modulated sinusoidally with a modulation index of 60%.

[**Ans.** 10·86 A]

8. Find the audio power necessary for AM for a 5 kW carrier modulated to a depth of 70%, if the efficiency of the modulator is 60%.

[**Ans.** 2·042 kW]

9. An FM transmitter sends a 120 MHz carrier wave frequency modulated by a 15 kHz sinewave waveform. The maximum frequency deviation is 30 kHz. Find (i) the modulation index.

(ii) three significant pairs of side frequencies

(iii) channel width required for these three side frequency pairs.

[**Ans.** (i) 2·0 (ii) 120 MHz ± 15 kHz, 120 MHz ± 30 kHz, 120 MHz ± 45 kHz (iii) 90 kHz]

10. A diode detector has a load resistance of 0·3 MΩ and capacitance of 100 pF. If the amplitude modulated input voltage has a depth of modulation of 0·6, what is the highest modulation frequency that can be detected with tolerable distortion ?

$$\left[\text{Ans. } f_{mh} = \frac{1}{2\pi\, mRC} = 8840 \text{ Hz}\right]$$

11. Find the bandwidth required in FM transmission if $\Delta f = 75$ kHz and $f_m = 15$ kHz.

[**Ans.** 180 kHz]

12. For a frequency modulated signal, the peak frequency deviation is 6 kHz and the frequency of the modulating signal is 2 kHz. Find the average power output by summing the powers of all side frequency components. The unmodulated carrier power is 20 Watts. What is the carrier power in the modulated wave ?

[**Ans.** 20 Watts, 1·352 Watts]

Chapter 13

INTEGRATED CIRCUITS

13.1 INTRODUCTION

During the last four decades, tremendous progress has been made in the development of of semiconductor device fabrication and technology. The first integrated circuit was fabricated by J.S. Kilby in 1958. The IC fabrication technology has undergone various phases of development through SSI (Small scale integration), M.S.I. (Medium scale integration) LSI (large scale integration), VLSI (Very large scale integration) (VLSI). The present day trend is to have high yield rate (number of good elements in a batch), higher level of automation and higher density levels. The process steps required for fabrication of ICs have undergone tremendous development and change over the years.

The first necessity in the fabrication of ICs is to have a high purity semiconductor material. The most important semiconductor material is silicon. The compound semiconductor gallium arsenide (Ga As) is also suitable for fabrication of special types of integrated circuits. The material silicon have almost been anonymous with ICs. For fabrication of ICs, semiconductors must have high purity. An impurity level of smaller than 1 part in 1 billion (1 in 1,000,000,000 or 1 in 10^9) is required.

The raw material source (semiconductor compound, SiO_2 for example) undergoes a series of chemical reactions and a zone refining process for preparation of a polycrystalline crystal of the highest purity level. For fabrication of Integrated circuits (or for that matter diodes and transistors), single crystal materials are needed. The difference between a polycrystalline crystal and a single crystal will be apparent from Fig. 13.1. In a polycrystalline material, the crystal atoms are oriented haphazardly with no regular geometrical patterns as one moves from one grain to another. The different crystalline grains are separated by grain boundaries. In a single crystal, there is a regular geometric lattice structure.

In actual practice silicon is obtained by reduction in a graphite boat at a temperature of approximately 1800°C when SiO_2 is combined with carbon :

$$SiO_2 + 2C \rightarrow Si + 2CO \qquad ...(13.1)$$

Silicon of about 98% purity is obtained by this process. The silicon so produced is commonly known as metallurgical silicon (MG-Si). It is used in steel and aluminium industry. The total production of silicon worldwide is about 500,000 tonnes. The production process for metallurgical silicon is shown in Fig. 13.2.

The metallurgical grade silicon is chemically reduced to form trichlorosilane and hydrogen by a exothermic reaction.

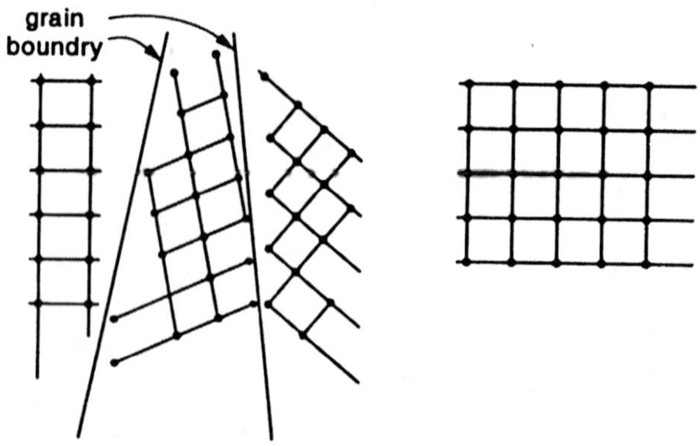

Fig 13.1. (a) Polycrystalline (b) single crystal semiconductor

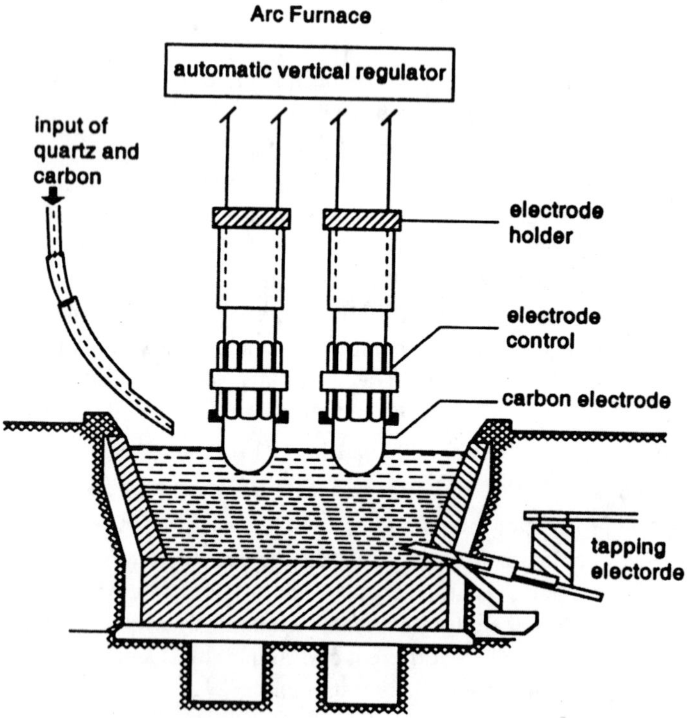

Fig 13.2. Production method of metallurgical silicon

$$Si + 3HCL \rightarrow SiHCl_3 + H_2 \qquad \ldots(13.2)$$

The trichlorosilane is liquid at a temperature of 30°C and can be easily separated.

Polycrystalline silicon can then be obtained by chemical vapour deposition (CVD process). A silicon bar (in the form of an U shape) is electrically heated to a temperature of 1350°C. When hydrogen (of high purity) is mixed with the trichlorosilane in the reactor, the trichrolosilane is reduced to silicon. The chemical reaction process is as follows.

$$4\ SiHCl_3 + 2H_2 \rightarrow 3Si + SiCl_4 + 8HCl \quad ...(13.3)$$

By this process high purity polycrystalline silicon is produced. The production of polycrystalline silicon by the process is depicted in Fig. 13.3.

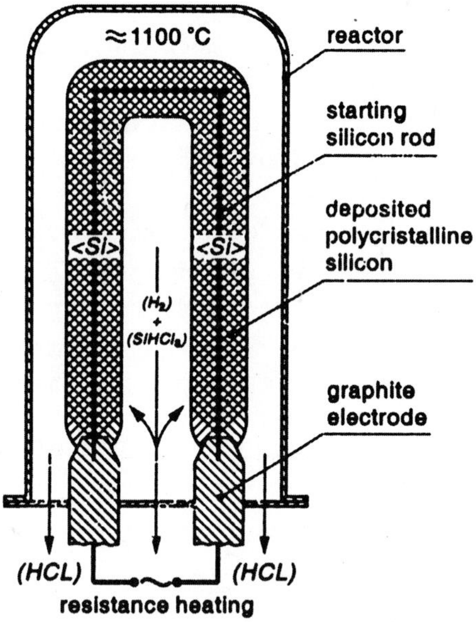

Fig 13.3. Production of polycrystalline silicon

The formation of a single crystal of silicon using the Czochralski method is followed next. In the Czochralski method, the polycrystaline silicon is heated RF heating coils. When a seed crystal (single crystal silicon) is immersed in the molten silicon and rotated and gradually taken out, single crystal silicon is grown

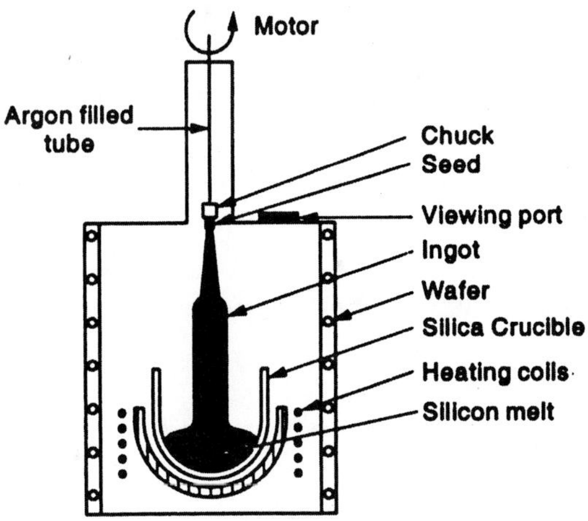

Fig 13.4. Czochralski Oven

on the seed. This is depicted in Fig. 13.4. The resulting single crystal is 6 to 30 inches in length and 1 to 10 inches in diameter.

13.2 ADVANTAGES OF INTEGRATED CIRCUITS

The integrated circuits possess the following important advantages over standard circuits with discrete components.

1. **Very small physical size (microminiature size).** Only a very powerful microscope is able to distinguish the various components and their interconnections.

2. **High reliability :** The very high reliability of ICs occur due to the absence of soldered connections. in the number of interconnections. A typical IC logic gate is 100 times more reliable compared to a transistor logic gate.

3. **Lower cost :** The cost reduction takes place due to simultaneous processing of a large number of ICs.

4. **Smaller weight :** As the packing density is very high, the weight is reduced substantantially. This is a very important advantage when military and space applications are considered.

5. **Suitability for small signal operation :** Since the physical distance between the different components of an integrated circuit is small, the chance of stray electrial pick up is almost nil. This makes them ideal for small signal operation.

6. **Low power consumption :** Due to their physically small size, the ICs are more suitable for low power operation than bulky discrete circuits.

7. **Ease of replacement :** It is very much economical to replace an IC than repair it.

13.2.1 Disadvantages of Integrated Circuits

1. Coils and inductors can not be fabricated.
2. They can handle only limited amount of power.
3. ICs function at typically small voltages.
4. They are delicate and may be damaged by rough handling or high temperature.
5. Isolation between components is poor for high density monolithic integrated circuit.
6. New set of masks are required for changes in circuit design.

13.3 INTEGRATION SCALES

Depending on the complexity of the circuit, ICs are classified according to the circuit or component density. For digital circuits a large number of similar logic gates may be necessary.

(a) **Small Scale Integration (or SSI)**

This refers to ICs where the **number of circuits** contained in one package is less than **30** (or **number of components** less than **50**)

(b) **Medium Scale Integation (or MSI)**

This refers to ICs with **number of circuits between 30 and 100** (or number of components between 50 and 500)

(c) **Large Scale Integration (or LSI)**

This refers to ICs with number of circuits between 100 and 100,000 (or number of components between 500 to 300, 000).

(d) **Very large Scale Integration (VLSI)**

This refers to ICs with circuit density in excess of 100,000.

13.4 IC CLASSIFICATION BY STRUCTURES

(1) Monolithic Integrated Circuit

The word monolithic refers to 'single'stone' or 'single solid structure'. In this integrated circuit, all circuit components (both active and passive) are produced inseperably within a single continuous piece of silicon crystal called **wafer** (or substrate). All the components are arranged in the same chip at different locations. The diodes and transistors are fabricated by epitaxial/diffusion techniques at appropriate locations in the wafer. The interconnection between different components is made on the surface of the silicon wafer and external connection wires are taken out to the terminals.

The monolithic ICs are mass produced, economical and are in widespread use.

(2) Thin film and thick film ICs

Compared to monolithic integrated circuits, these ICs are not fabricated within a single slilcon wafer but on the surface of an insulating subtrate such as glass or ceramic. The difference between thin film and thick film ICs is **not on the relative** thickness of the films but rather on the method of deposition of the film. In the film ICs, only the passive components are formed on the insulating glass or ceramic surface. The active elements (diodes, transistors) are added externally as discrete components to complete the circuits. It is possible that the discrete active elements are formed using monolithic techniques. We now consider separately the thin film and thick film ICs.

(a) Thin film ICs

In such ICs, a conducting material film is deposited on the (glass or arsanic) substrate through a suitable mask. Resistors and conductors are formed by depositing materials of different resistivity while varying the width and thickness of the film. Capacitors are formed by sandwitching a thin insulating oxide film in between two conducting layers. Inductors of low values may be formed by deposition of a spiral form of film. The active elements such as diodes and transistors are added externally and interconnected by wire bond.

Thin films may be produced by either of (a) **vacuum eveporation** or (b) **cathode sputtering techniques.**

In the vacuum deposition technique, a vapourised material is deposited on the substrate by using suitable masks. In the cathode sputtering method, atoms from the cathode of a desired film material is deposited on the substrate which is located in between the cathode and the anode.

(b) Thick film ICs

These circuits are also referred to as **printed thin film circuits**. The technique called the silk screen printing technique is employed to create the desired circuit on the surface of the substrate. The screens are typically of fine stainless steel wire mesh and the inks are pastes (of pulverized glass and aluminium) which possess conductive, resistive or dielectric properties. After the printing is over, the circuits are fixed at high temperature in a furnace to fuse the film on the substrate. The active elements are added externally as discrete components.

(3) Hybrid ICs

The hybrid ICs are formed by interconnecting a number of individual chips or a combination of film and monolithic IC techniques. In hybrid or multichip ICs, the active elements are first formed on the silicon wafer using monolithic techniques. The wafer is subsequently covered by an insulating layer such as SiO_2. Film techniques are then used to fabricate the passive components on the SiO_2 surface. Metallic connections between the active and passive components take place through windows in SiO_2 layer.

Advantages and Disadvantages of Film Circuits

1. **Better isolation**

Film circuits provide between isolation between components because the components are formed on the surface of an insulator.

2. **Flexibility in circuit design**

Since active elements are added externally, greater flexibility in circuit design is possible.

3. **Broader range of values for passive components**

In film ICs, a wider range of component values and better tolerances is possible as compared to monolithic integrated circuits.

4. **Better high frequency performance**

A better high frequency performance is possible with film ICs.

Disadvantages

The disadvantages of film ICs are

1. Active components can not be fabricated.
2. provide lesser reliability.
3. higher cost of longer physical size and weight.

The various integrated circuits can be classified as depicted in Fig. 13.5.

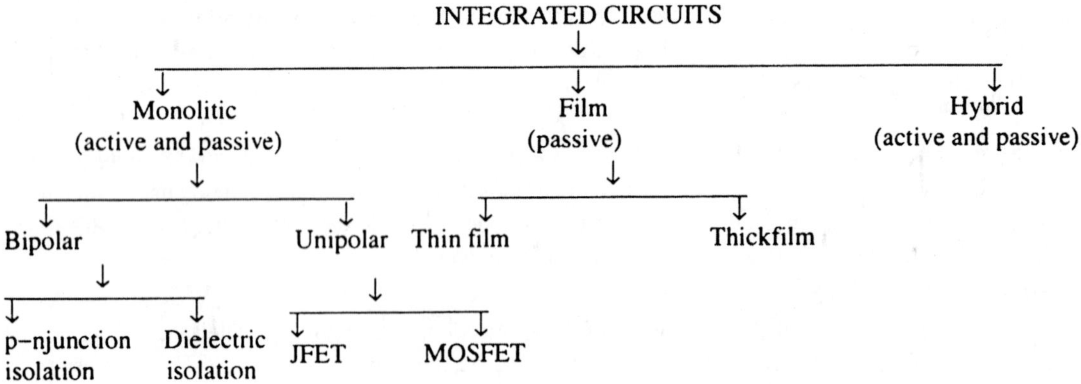

Fig. 13.5. Classification of ICs depending on fabrication method

The integrated circuits can also be classified depending on general functions. They may be divided into two important categories namely (1) analog (2) digital.

INTEGRATED CIRCUITS

↓

ANALOG IC	**DIGITAL IC**
Example	**Example**
1. operational amplifiers (μA741, CA3078)	1. logic gates
2. voltage regulators (μA723)	2. counters
3. voltage comparators (CA 139A)	3. flip flops
4. small signal amplifiers	4. clock
5. power amplifiers (CA3020)	5. memory chip
6. broadband amplifiers (CA3002)	6. calculator chip
7. multipliers	7. microprocessor
8. microwave amplifiers.	(8085, 80486, 80586, Pentium)
9. TV receiver circuits (CA 3065, CA 3068)	

Fig. 13.6. Classification of ICs based on general functions

13.5 MANUFACTURER'S DESIGNATION OF INTEGRATED CIRCUITS

Each manufacturer specifies a specific code and a type number for the ICs produced by it. Linear integrated circuits are also termed as analog integrated circuits because their input and output values may assume a continuous range of values and the output voltage is typically proportional to the input voltage. The operational amplifier (OP AMP) is by far the most widely used and versatile type linear integrated circuit. It will be discussed separately in the next chapter.

The 741 operational amplifier is very common and National semiconductor uses the code LM741 while Motorola specifies it as MC 1741, Fairchild specifies it μA741. They are all equivalents and all these operational amplifier have the same specificiations.

Many IC manufacturers specify the absolute maximum rating system, which is defined by the Industry Standard (JEDEC) statement.

Absolute maximum ratings are the limiting values for operational and environmental conditions applicable to any electron device of a specified type as defined by published data and should not be exceeded under the most possible conditions.

13.6 BATCH PRODUCTION OF INTEGRATED CIRCUITS

ICs are fabricated in a batch. A typical 10 cm diameter can be easily visualized as being comprised by 10,000 rectangular chips of sides approximately 0·9 mm. Each of the IC chips may contain some terms to several thousand components. If 10 such IC wafers are processed in a batch, then, 100,000 ICs can be fabricated simultaneously. Some of the IC chips may be faulty due to imperfections of the manufacturing process. Even if the yield rate (i.e. percentage of defect free chips/total number of chips in the wafer) is 75%, can be seen that a huge 75,000 ICs are produced in one single batch.

The fabrication of discrete devices such as diode, transistor or IC can be done by the same technology. The various fabrication processes usually take place through a single plane and the technology is called the

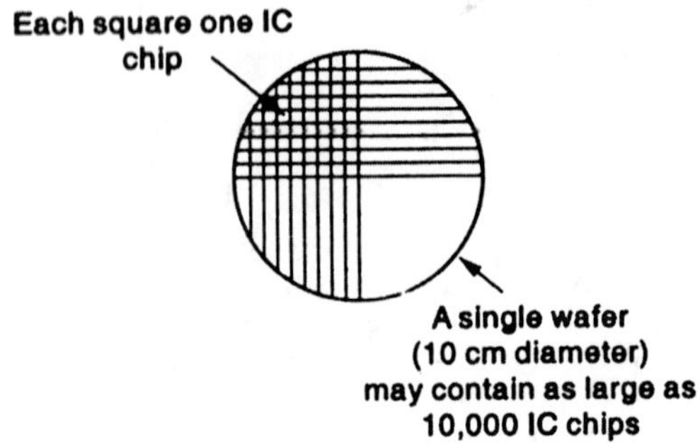

Fig 13.7. Batch production of ICs

lanar technology. The simple circuit of Fig. 13·8 can be fabricated by silicon planar technology whose cross ctional view is depicted in Fig. 13.9.

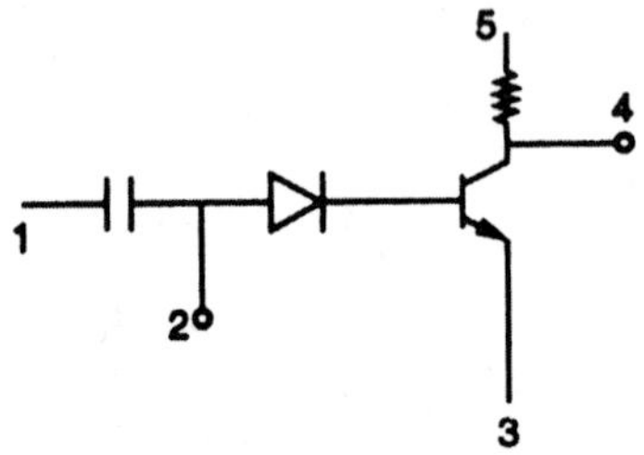

Fig 13.8. A representative circuit

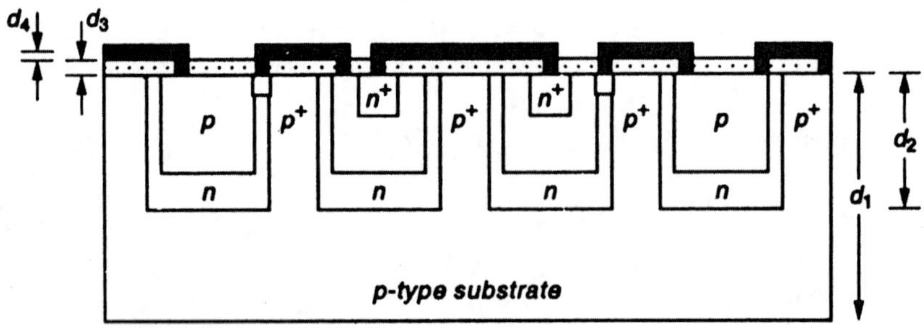

$d_1 \simeq 250$ μm, $d_2 = 5$ *to* 30 μm, $d_3 = 0.02$ *to* 2 μm, $d_4 = 1$ μm

Fig 13.9. Cross-sectional view of a typical integrated circuit

In a general IC, the substrate of silicon is *p*-type with a thickness of the order of 250 μm. A diffused *n*-type region is then formed in the surface with a thickness of the order of 5-30 μm. All passive and active components are fabricated within this layer using selective diffusion of impurities by using suitable masks. A silicon dioxide layer of thickness approximately 0·02-2 μm is formed over the surface of the substrate by selective oxidation. A further metallisation layer is needed ($d_4 \approx 1$ μm) for interconnection patterns.

IC terminology

Some very common processes used in IC are

1. Wafer preparation
2. Epitaxial growth
3. Oxidation
4. Photolithography
5. Selective Diffusion of impurities
6. Ion implantation
7. Isolation methods
8. Aluminium (Al) metallisation
9. Assembly porocessing, testing
10. Packing.

1. **Wafer preparation**

The starting material is very high purity silicon in the form of a cylinder 10 to 12 cm in diameter and 1 metre in length of **steel gray colour**. The single crystal rod is sawed to produce wafers 10-12 cm in diameter and 250 μm thick (1 μm ≡ 10^{-6} m). The surface is then polished to mirror finish. The intrinsic electrical and mechanical properties of the wafer depend upon the crystal orientation the doping density etc. The doping is strictly controlled during crystal growth.

2. **Oxidation :** By the term oxidation, we refer to silicon reacting with oxygen to form silicon dioxide. The oxidation process is carried out in furnaces at a temperature of 1000 to 1200°C. To prevent contaminants to pollute the process, the oxidation is carried out in ultraclean environments. Oxygen is introduced as a high purity gas (**dry oxide**) or as water vapour (**wet oxide process**). The oxide formed has dielectric constant of 3·5. it can be used in capacitors. Besides, the oxide can be used as a good mask against many impurities. The oxide thus protects the silicon surface from contaminants. The oxide is a thin transparent film over silicon. It can be used as good mask for fabrication of integrated circuits.

3. **Diffusion :** By the term diffusion, we mean movement of atom through the crystal lattice. In IC fabrication, the term diffusion is referred to introduction of impurity atoms (**acceptors and donors)** into silicon and change the doping density. The diffusion process is carried out in furnaces at a temperature of about 1000 to 1200°C. The impurity diffusion takes place to a depth which depends on both temperature and time allowed. In Silicon, the two most common type of impurities are **Boron** (which acts as **acceptor)** and phosphorous (which acts as **donor**).

Both of them are masked by SiO_2.

4. **Ion implantation :** For precise and controlled introduction of impurities at room temperature, the ion implantation process is employed. In the ion implantation process, ions of desired material are produced and accelarated by a high voltage. The ions then penetrate into the silicon material and produces accurate and reproducible impurity profiles. The depth of penetration of ions depends on the beam current. The ion implantation process is more expensive (than diffusion process) and is used if accurate and precise control of dopant is necessary for device fabrication.

5. **Chemical Vapour Deposition (CVD) :** In the CVD process, gases or vapours are chemically reacted, leading to formation of a solid on a substrate. The advantage of CVD is that the growth is faster and occurs at lower temperatures compared to other processes.

If CVD is employed for growth of silicon **over single crystal material, it is referred to as epitaxy.** The grown epitaxial layer is thin but has the same crystal oxientation as the underlying wafer. The grown layer is also single crystal and various semiconductor devices can be fabricated on it. If the wafer is polycrystalline, the grown layer is also polycrystalline.

6. **Metallization :** The process of metallisation involves the deposition of a metal (aluminium). In integrated circuit techniques, the metallisation is needed to interconnect the various components of the IC (transistors, diodes, capacitors, resistors) to form the desired circuit function.

7. **Photolithography :** The process involves selective etching of SiO_2 layer with the help of a photographic mask, photoresists and etchants. The process is used to define transistor regions and also to isolate one transistor from another.

8. **Circuit Testing :** Each IC is tested on the wafer by placing suitable probes on the bonding pad. Faulty chips are marked (with a dot of ink) and discarded after the wafer has been scribed and broken down into individual ICs.

9. **Mounting and packing :** The individual ICs are very small and brittle. They are cemented or soldered to a gold plated header through which lead have already been connected.

The various types of IC packages are depicted in Fig. 13.10.

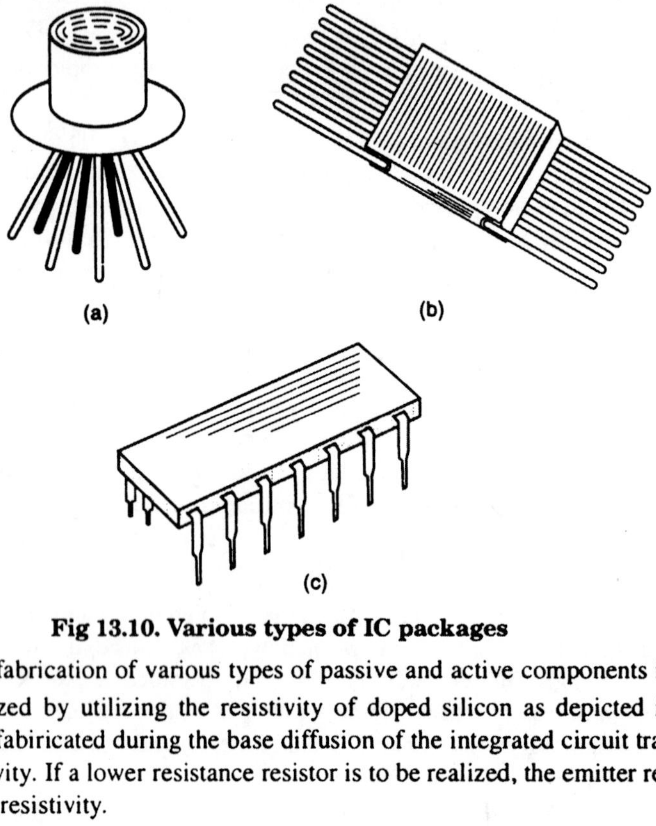

Fig 13.10. Various types of IC packages

Fig. 13.11 depicts the fabrication of various types of passive and active components in IC.

Resistors can be realized by utilizing the resistivity of doped silicon as depicted in Fig. 13.11 (a). Typically, the resistors are fabiricated during the base diffusion of the integrated circuit transistor because it provides the highest resistivity. If a lower resistance resistor is to be realized, the emitter region is employed because it has much lower resistivity.

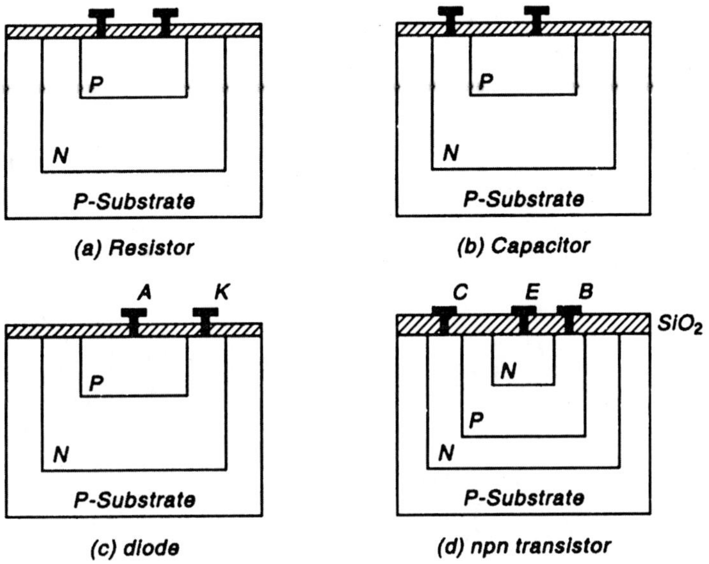

Fig 13.11. Fabrication of various passive and active components in IC

Let us consider a square semiconducting material of size $l \times l$ of resistivity ρ thickness t and cross section $S = l \times t$ as depicted in Fig. 13.12.

The sheet resistance of the material is

$$R_s = \frac{\rho l}{l \,.\, t} = \frac{\rho}{t} \text{ (ohm/square)} \qquad \text{...(13.4)}$$

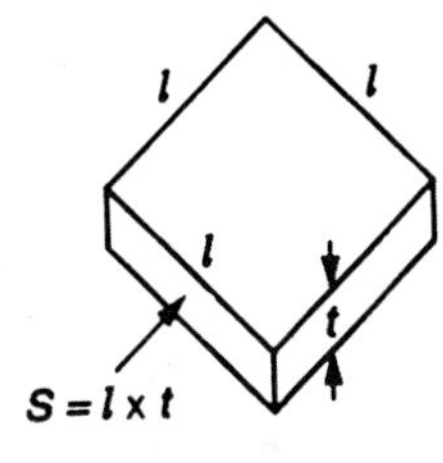

Fig 13.12. Square sheet of semiconducting material

The sheet reistance R_s depends on the diffusion characteristics of the material and is independent of the size of the semiconducting material. The resistance of the diffused resistor can be written in terms of the surface dimensions namely L and W

$$R = \rho \frac{L}{Wt} = R_s \cdot \frac{L}{W} \qquad \text{...(13.5)}$$

The quantity $\frac{L}{W}$ is called the aspect ratio of the surface geometry. It indicates the number of squares contained in the resistor. A base resistor having values between 20 Ω and 400 kΩ can be fabricated by medium resistivity base region (250 Ω/square). A lower resistor value can be realized by using low resistivity (high doped) n-type emitter diffusion. Typical resistivity values are 10 Ω/square. Emitter resistors in the range of 1 Ω to 1 kΩ can be fabricated by this method.

EXAMPLE 13.1 *Design (a) 0·5 kΩ (b) 10 kΩ diffused resistors.*

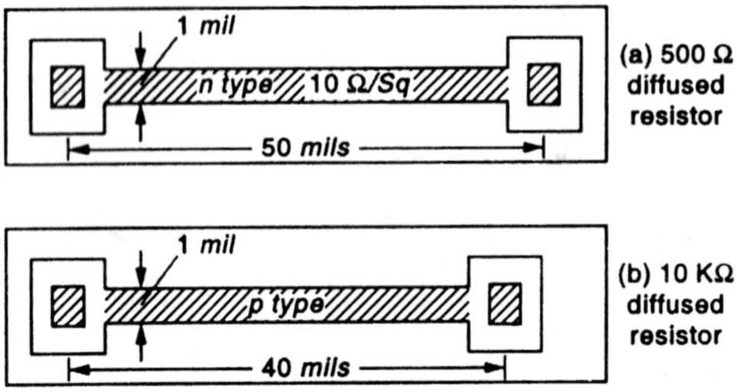

Fig 13.13. Reference Example 13.1

SOLUTION (a) The sheet resistance of n-type emitter diffusion is 10 Ω/square.

Then $$\frac{L}{W} = \frac{R}{R_s} = \frac{500}{10} = \frac{50}{1}$$

Thus a 500 Ω diffused resistor can be fabricated by using a pattern of 50 mils long by 1 mil wide.

(b) The sheet resistance of p-type base diffusion is 250 Ω/square.

Then $$\frac{L}{W} = \frac{R}{R_s} = \frac{10\text{ k}\Omega}{250\ \Omega} = \frac{40}{1}$$

So a 10 kΩ resistor can be obtained by forming a pattern of 40 mils long by 1 mil wide.

Capacitors : Capacitors in integrated circuits can be obtained by using the transition capacitance of a reverse biased p–n junction. The capacitance is dependent on the reverse voltage across the p–n junction. Typically its value is limited to 100 pF.

A capacitor can also be formed by utilizing the surface SiO_2 layer as a dielectric between two metallization layers.

Integrated Inductors : Inductors, transformers or chokes are difficult to fabricate in the planar 'monolithic' integrated circuit. Circuit designers often replace an R–L network with an R–C networks so that it can be integrated with monolithic integrated circuits. In applications of *RF* and *IF* circuits, where inductors are needed, external inductors are added to the IC package. In thin film hybrid microwave integrated circuits, thin film inductors spirals of value upto 250 nH are formed.

Diodes : The collector-base junction of the transistor can be used as diode. In p-type substrates, the collector acts as the cathode and the p-type base region acts as the anode. For high speed switching, the emitter-base transistor junction is used as the diode.

Transistors : The p-type substrate acts as a supporting structure for the transistor. The collector is formed by planar diffusion process by using suitable masks and photolithographic technique. The base regions and the emitter regions are subsequently formed using mask 2 and mask 3 respectively. After the emitter diffusion is over, a protective SiO_2 layer is formed over the surface of the monolithic IC. Contacts are then made to the different transistor regions.

13.7 LIMITATIONS OF THE INTEGRATED CIRCUITS

1. Large capacitances (greater than 200 pF) can not be fabricated.
2. Resistors outside the range 1 Ω - 1 MΩ cannot be formed.
3. *p–n–p* transistor performance is poor.
4. IC components are voltage sensitive and possess large temperature co-efficients.
5. Power dissipation capacity is low.
6. The performance of ICs at high frequencies is poor.

Review Questions

1. What do you understand by the terms LSI, MSI, VLSI ?
2. Which basic material is employed in the fabrication of monolithic ICs. ?
3. What is the difference between a polycrystalline and a single crystal semiconductor material ?
4. Discuss the production method of metallurgical silicon.
5. Discuss the method of obtaining polycrystalline Silicon from trichlorosilane.
6. What do you understand by monolithic ICs ?
7. What do you understand by thin film ICs ?
8. What do you understand by thick flim ICs ?
9. Mention the advantages of film circuits ?
10. What are the disadvantages of film circuits.
11. Mention some examples of analog ICs.
12. Mention some examples of digital ICs.
13. What is meant by planar technology ?
14. What is meant by oxidation ?
15. What is meant by diffusion ?
16. What do you understand by the chemical vapour deposition (CVD) process ?
17. What do you understand by ion implantation ?
18. What is photolithography ?
19. What is sheet resistance of semiconducting material ?
20. Mention some limitations of ICs.

Objective Type Questions

Tick mark (√) the correct answer or the best alternative.

1. The first integrated circuit was developed by
(a) Shockely, Bardeen and Brattain (b) Sir J.J. Thompson
(c) W. von Klitzing (d) J.S. Kilby
2. A monotlithic IC is developed in
(a) Single crystal of silicon (b) in a polycrstalline silicon
(c) in a amorphous silicon (d) Si_3N_4

3. The most commonly used semiconductor material for fabrication of ICs is
(a) Si (b) Ge
(c) Ga As (d) CdS

4. Polycrystalline Silicon can be fabricated by
(a) reducing trichlorosilane with H_2 at a temperature of 1000°C
(b) reducing trichlorosilane with H_2 at a temperature of 1350°C
(c) reducing trichlorosilane withO_2 at a temperature of 1000°C
(d) reducing trichlorosilane with N_2 at a temperature of 1000°C.

5. Single crystal silicon is obtained
(a) directly by reducing trichlorosilane in a heated furnace.
(b) by heating polycrystalline silicon and immersing a high purity polycrystalline silicon rod in the melt
(c) by heating polycrystalline silicon and immersing a single crystal silicon rod in the melt
(d) by reducing SiO_2 in a graphite boat at a temperature of 1800°C

6. In monolithic ICs, all the components are fabricated by
(a) sputtering (b) hydrogen reduction
(c) oxidation (d) diffusion process

7. The most important advantage of monolithic ICs is
(a) low cost (b) very high density
(c) very high reliability (d) low power consumption

8. Thin film and thick film ICs are fabricated within a
(a) silicon layer (b) inside a ceramic/glass base
(c) on the surface of a ceramic or glass substrate (d) within SiO_2

9. Thin film ICs
(a) are thinner compared to thick film ICs. (b) are prepared by ion implantation technique
(c) silk screen printing technique (d) are prepared by epitaxial methods

10. A typical operational amplifier type number is
(a) CA 3065 (b) CA 3020
(c) μA 741 (d) μA 723

11. A typical voltage regulator is
(a) μA 723 (b) 555
(c) CA 3065 (d) NE 565

12. In silicon, the acceptor and donor impurities introduced for doping are respectively
(a) boron and phosphorous (b) phosphorous and arsenic
(c) phosphorous and boron (d) asrsenic and boron

13. In epitaxy,
(a) a new layer is grown on the surface of the silicon wafer
(b) the surface layer is converted to *n* or *p* type
(c) the surface layer doping density is altered
(d) the surface layer doping concentration and impurity type both are altered.

14. The sheet resistance of a silicon bar depend on
(a) diffusion temperature of the material (b) dimensions of the semiconductor material
(c) diffusion temperature and diffusion time (d) resistivitly of silicon and the dimensions of the bar.

15. The following materials are not fabricated in monolithic ICs
(a) diffused resistor (b) capacitor
(c) inductors (d) transistors

16. A thin layer of conducting material on the surface of monolithic ICs is obtained by
(a) epitaxy (b) diffusion
(c) ion implantation (d) sputtering or evaporation

17. A depth upto about 0·1 μm can be precisely controlled by
(a) epitaxy (b) diffusion
(c) ion implantation (d) chemical vapour deposition

Chapter 14

OPERATIONAL AMPLIFIERS

14.1 INTRODUCTION

The term operational amplifier stands for a special purpose amplifier which is intended to be used with external feedback components for proper operations. The operational amplifier (abbreviated as OP AMP) now enjoys a place in electronics as ubiquitous as the transistor. The first successful operational amplifier was developed by Robert J. Widlar, μA709.

The principles of the operational amplifier can be properly understood if the principles of differential amplifier is understood. Fig. 14.1 represents the circuit diagram of a bipolar transistor differential amplifier.

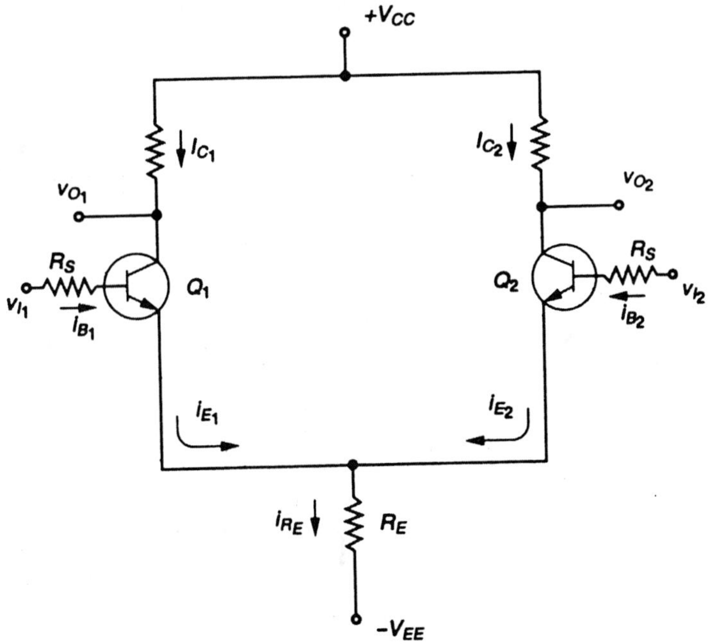

Fig. 14.1 Bipolar transistor differential amplifier.

In the circuit, there are two transistors Q_1 and Q_2. In the circuit diagram, the voltages common to both inputs cause the two collector voltages to move in one direction, whereas the voltages differing on the two inputs cause the two collector voltages to move in opposition. The voltage common to both the inputs is called the common mode input voltage $(v_I)_{CM}$ whereas the voltage common to both outputs is called the common mode output voltage $(v_O)_{CM}$.

If v_{I_1} is the voltage at the base of transistor Q_1 and v_{I_2} the voltage at the base of transistor Q_2, then

$$(v_I)_{CM} = \text{Common mode input voltage}$$

$$= \frac{v_{I_1} + v_{I_2}}{2} \qquad ...(14.1)$$

If $\qquad v_{I_1} = v_{I_2}$ then

$$(v_I)_{CM} = v_{I_1} = v_{I_2} \qquad ...(14.2)$$

The voltage $(v_O)_{CM}$ is similarly given by

$$(v_O)_{CM} = \frac{v_{O1} + v_{O2}}{2} \qquad ...(14.3)$$

where v_{O1} is the voltage at the collector of transistor Q_1 and v_{O2} is the voltage at the collector of transistor Q_2. When the two inputs are tied together, $v_{O1} = v_{O2}$ if circuit components are matched in nature and the two transistors are identical.

EXAMPLE 14.1 *A differential amplifier has a common mode output voltage of + 5·4 V when the input common mode voltage is zero and have an output common mode voltage of + 5·0 V when the input common mode voltage is + 1.05 V (i) Find the value of v_{01} if v_{02} is + 5·4 V.*

SOLUTION

$$v_{ICM} = \frac{0+0}{2} = 0 \text{ V}$$

$$v_{OCM} = +5{\cdot}4 \text{ V}$$

$$+5{\cdot}4 = \frac{+5{\cdot}4 + v_{01}}{2}$$

or, $$v_{01} = +5{\cdot}4 \text{ V}$$

EXAMPLE 14.2 *In the above problem, if $v_{I_1} = +1{\cdot}15$ V, $v_{I_2} = +0{\cdot}95$ V, $v_{01} = +2{\cdot}0$ V and $v_{02} = +8{\cdot}0$ V find v_{ICM} and v_{OCM}.*

SOLUTION

$$v_{ICM} = \frac{+1{\cdot}15 + 0{\cdot}95}{2} = +1{\cdot}05$$

$$v_{OCM} = \frac{2{\cdot}0 + 8{\cdot}0}{2} = +5{\cdot}0 \text{ V}.$$

EXAMPLE 14.3 *In a differential amplifier, v_{I_1} is grounded while v_{I_2} is connected to an a.c. voltage of 5 sin $(2\pi \times 0{\cdot}2\, t)$ mV. The output common mode voltage is (a) + 5·0 V when $v_{ICM} = 0$,*

(b) + 5·01 V when $v_{ICM} = -2$ mV and

(c) + 4·99 V when $v_{ICM} = +2$ mV. Find the the input and output common mode voltages.

SOLUTION $\Delta v_{ICM} = \dfrac{0 + 5 \sin (0{\cdot}4\, \pi t)\, mV}{2} = 2{\cdot}5 \sin (0{\cdot}4\, \pi t)\, mV.$

This is a sinusoidal wave with peak instantaneous voltage of $\pm 2{\cdot}5\, mV$. The collector voltage decreases as the base voltage increases. This results in a phase reversal on the output. Thus $\Delta v_{OCM} = + 5\text{ V} - 10 \sin (0{\cdot}4\, \pi t)\, mV$.

In a differential amplifier, the common mode voltage normally have two input voltage varying at different rates. This produces a changing d.c. or an a.c. common mode voltage. In such a case, between two different time instants in equation (14.1).

$$\Delta v_{ICM} = \frac{\Delta v_{I_1} + \Delta v_{I_2}}{2} \qquad \text{...(14.4)}$$

and

$$\Delta v_{OCM} = \frac{\Delta v_{01} + \Delta v_{02}}{2} \qquad \text{...(14.5)}$$

EXAMPLE 14.4 *Find the common mode gain in a differential amplifier if the following input and output voltages are established at different instants of time.*

Instant	v_{I_1}	v_{I_2}	$v_{out\ 1}$	$v_{out\ 2}$
$\tau 1$	− 3·3	− 3·7	+ 4·3	+ 4·5
$\tau 2$	+ 1·9	+ 1·1	+ 2·7	+ 3·1

SOLUTION In this case

$$(\Delta v_I)_{CM} = \frac{\{1{\cdot}9 - (-3{\cdot}3)\} + \{1{\cdot}1 - (-3{\cdot}7)\}}{2} = \frac{5{\cdot}2 + 4{\cdot}8}{2}$$

$$= 5 \text{ V}$$

$$(\Delta v_{out})_{CM} = \frac{\{2{\cdot}7 - 4{\cdot}3\} + (3{\cdot}1 - 4{\cdot}5)}{2} = \frac{-1{\cdot}6 - 1{\cdot}4}{2}$$

$$= -1{\cdot}5 \text{ V.}$$

Thus, if the input common mode voltage increases by + 5 V, the output common mode voltage decreases by 1·5 V.

or

$$A_{CM} = \frac{-1{\cdot}5 \text{ V}}{+5{\cdot}0 \text{ V}} = -0{\cdot}30$$

The term 'operational amplifier' came from the fact that the 'op amp' in addition to its circuit applications as an amplifier was used in analog computers to perform different types of mathematical operations like addition, subtraction, differentiation, integration etc. Fig. 14.2 depicts the circuit diagram of a very popular 741 operational amplifier. The seven terminals of the operational amplifier are :

(a) INVERTING INPUT

(b) NON-INVERTING INPUT

(c) OUTPUT

(d) V^+

(e) V^-

(f) OFFSET NULLS

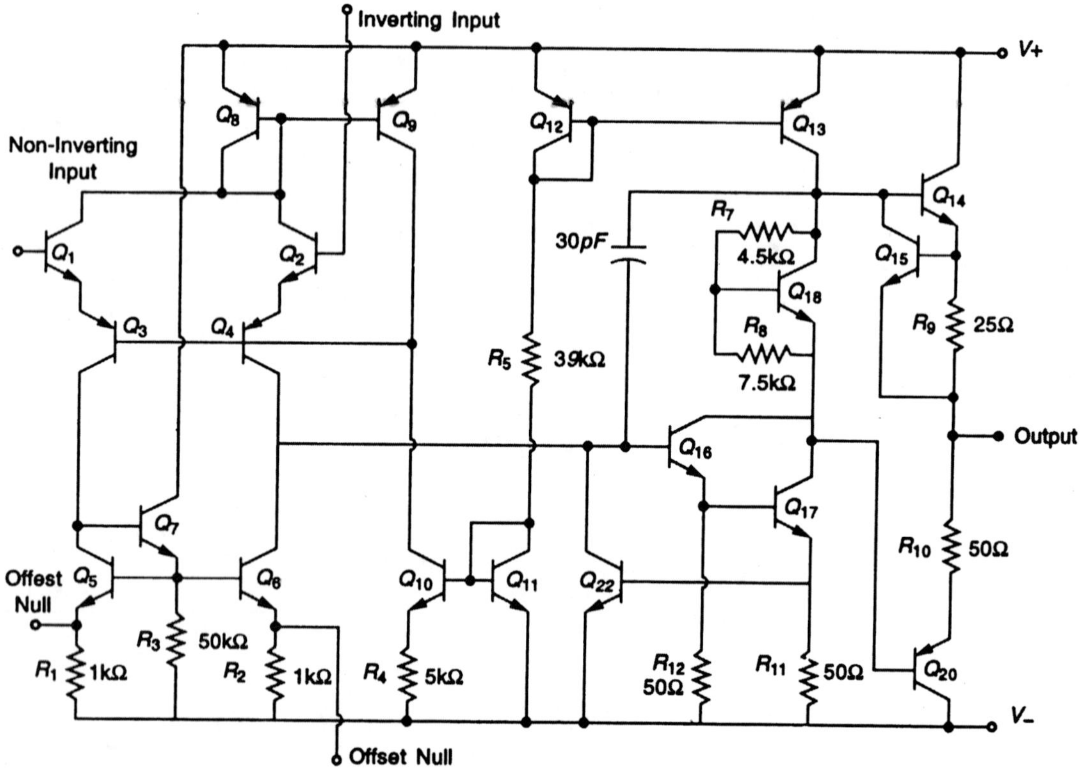

Fig. 14.2 Circuit diagram of a IC μA741 operational amplifier

For a user of the operational amplifier, the terminal properties are important rather than the analysis or the internal circuit performance.

The circuit designers often need to know the black box approach of the elementary circuit analysis. Thus the detailed knowledge regarding the terminal characteristics will be the prime requirement so far as the applications are concerned. The integrated circuit chip (IC) chip operational amplifier contains a single semiconductor chip are OP AMP (single), two OP AMPs and four OP AMPs (quad) package. Typically the chips contain 8, 10 or 14 pins. An μA741 OP AMP comes in 8 pin can, 8 pin dual-in-line (DIP) or a 10 pin flat package or 14 pin DIP. The μA747 is a dual 741 is available in 10 pin or 14 pin DIP. Four common types of packages are depicted in Figure 14.3.

The terminal marked as 'inverting input terminal' is labelled as '–' while the terminal marked as 'non inverting terminal' is labelled as '+'. The significance of the '–' sign is that a voltage applied between this '–' terminal and ground appears in amplified form at the output terminal but with the its polarity reversed. Similarly a voltage applied between '+' terminal and ground appears at the output in amplified form without polarity reversal. If a voltage v_1 is applied at the inverting input and a voltage v_2 is applied at the non inverting input, then the voltage at the output terminal is

$$v_{out} = A_V (v_2 - v_1) \qquad ...(14.6)$$

where A_V is called the difference mode voltage gain or amplification (or open loop voltage gain).

The terminal marked V^+ is the terminal to which the positive of the battery V_{dc1} is connected while the

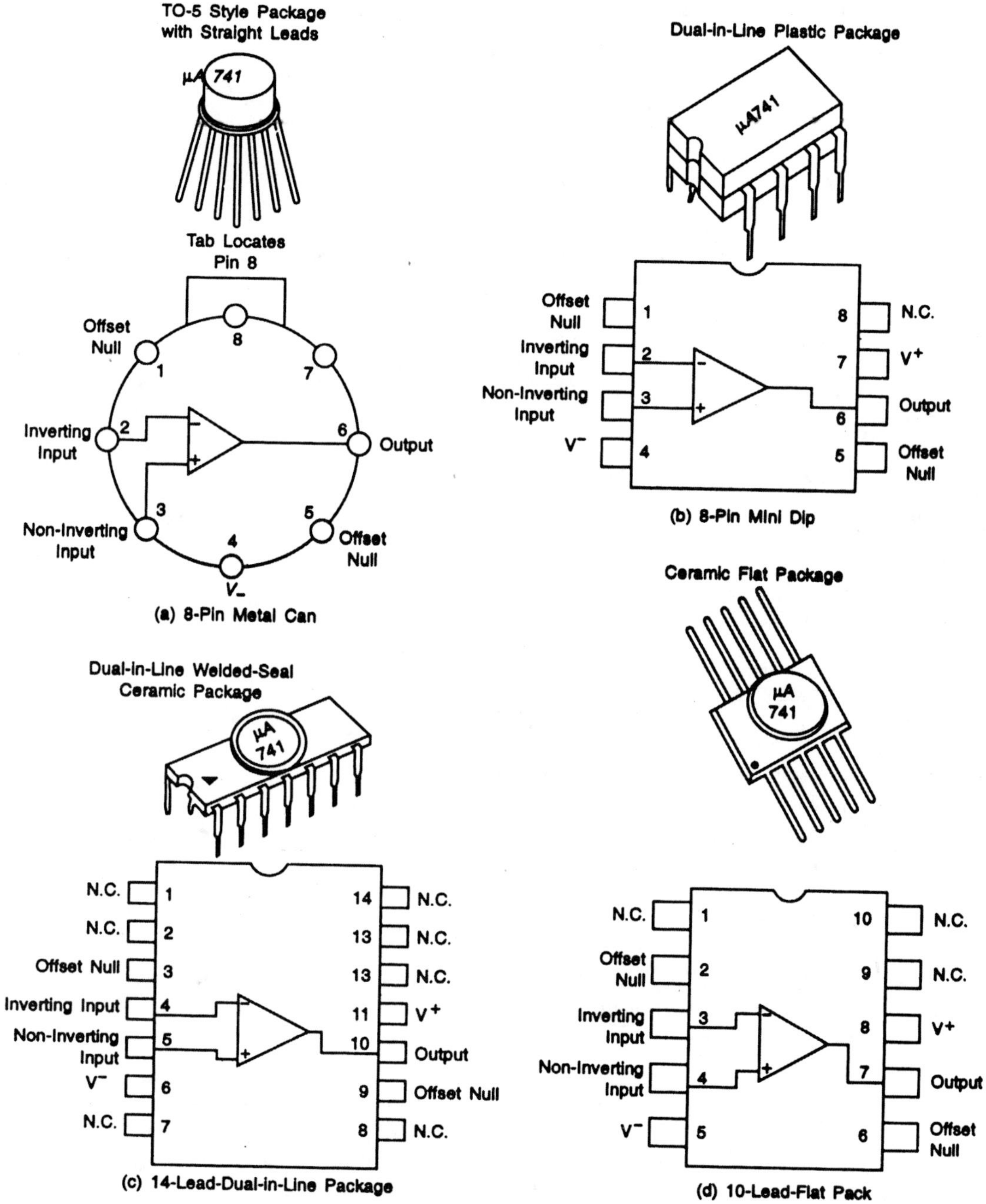

Fig. 14.3 Various IC packages along with connection diagram(a) 8pin metal can (b) 8-pin mini dual in line package (DIP) (c) 14 pin DIP (d) 10 lead flat packpack

negative is connected to the ground. Similarly the terminal V^- is the terminal to which the negative of the battery V_{dc2} is connected while the positive is connected to ground. This can be understood from Fig. 14.4.

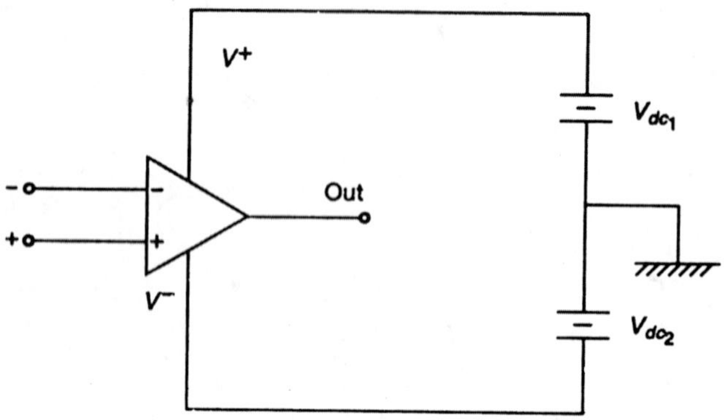

Fig. 14.4 OP AMP schematic diagram

Characteristics of an Operational Amplifier

An ideal operational amplifier can be thought of as possessing the following properties :

1. Voltage gain which is infinite ($A_V \rightarrow \infty$)
2. Input impedance $Z_{in} \rightarrow \infty$
3. Output impedance $Z_{out} \rightarrow 0$
4. Bandwidth $BW \rightarrow \infty$
5. Characteristics not dependent on temperature
6. Perfect balance condition or

output voltage = 0 when $v_{i_1} = v_{i_2}$

A practical operational amplifier has characteristics which are somewhat different from the above values for ideal operational amplifier. An ideal operational amplifier is a theoretical concept and cannot be achieved in practice. The popular 741C operational amplifier has an open loop voltage gain = 2,00,000 an input impedance Z_{in} = 2 MΩ and output impedance Z_{out} = 75 Ω.

14.2 OP AMP BASICS

Common Mode Operation : When the same input signals are applied to both the inputs; common mode operation takes place. Ideally, the two inputs are equally amplified and because they give rise to opposite polarity signals, these signals cancel and the output voltage is zero. In the practical case, however, a small output signal results.

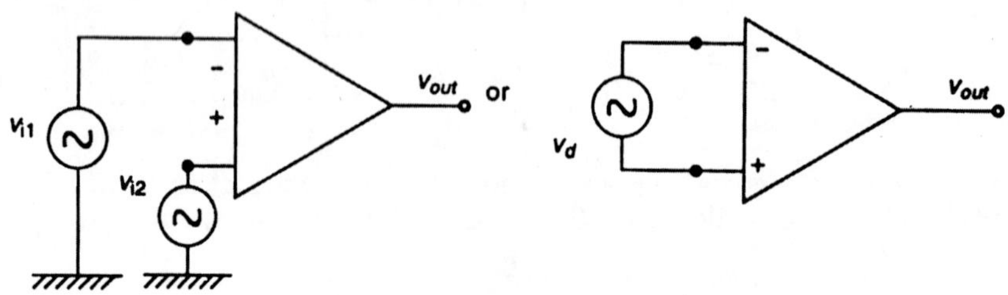

Fig. 14.5 Differential mode connection for OP AMP

Common Mode Rejection : One significant property of differential mode connection (Fig. 14.5), is that the voltage signals which are opposite at the inputs are highly amplified while the voltage signals which are common to the two inputs are only very feebly amplified. The operation can be summed up by saying that the difference signal is amplified while the common mode signal is rejected. Normally noise (an unwanted voltage signal) is common to both the inputs. Thus the differential mode of connection provides attenuation (or rejection) to the unwanted inputs while providing an amplified output to the difference signal. This property of the operational amplifier is known as common mode rejection.

Differential and Common Mode Operation : In the differential mode of connection, the circuit exhibits its capability to greatly amplify signals which are opposite in the two inputs while feebly amplifying signals that are common to both inputs. In fact, the output voltage of an operational amplifier consists of two components (i) a differential mode component and (ii) a common mode component. The differential mode input voltage is

$$v_d = v_{i2} - v_{i1} \qquad \text{...(14.7)}$$

while the common mode input voltage is

$$v_c = \frac{1}{2}(v_{i1} + v_{i2}) \qquad \text{...(14.8)}$$

Since any voltage signal can have both inphase and out-of-phase components, the output voltage can be expressed as

$$v_{out} = A_d\, v_d + A_c\, v_c \qquad \text{...(14.9)}$$

where $A_d = (A_v)_d$ is the differential voltage gain of the operational amplifier and $A_c = (A_v)_c$ is the common mode voltage gain of the operational amplifier.

In the ideal case $A_d \to \infty$ and $A_c \to 0$. In a practical operational amplifier, a figure of merit called the Common Mode Rejection Ratio (CMRR) is defined as follows

$$\text{CMRR} = \frac{|A_d|}{|A_c|} \qquad \text{...(14.10)}$$

Since A_d should be large and A_c very small, the operational amplifier is so designed that CMRR has a value much larger than unity. The value of CMRR can also be expressed in decibels as

$$(\text{CMRR})_{db} = 20 \log_{10} \frac{A_d}{A_c}$$

EXAMPLE 14.5 *In an operational amplifier $A_d = 15000$ and $A_c = 15$. Find CMRR is decibels.*

SOLUTION

$$\text{CMRR} = \frac{A_d}{A_c} = \frac{15000}{15} = 1000$$

$$(\text{CMRR})_{db} = 20 \log_{10} 1000 = 60 \text{ dB}$$

From equations (14.7) and (14.8), one may write,

$$v_{i1} = v_1 = v_c + \frac{1}{2} v_d \qquad \text{...(14.11)}$$

and

$$v_{i2} = v_2 = v_c - \frac{1}{2} v_d \qquad \text{...(14.12)}$$

Offset Error Voltage and Currents

An ideal operational amplifier is perfectly balanced *i.e.* $v_{out} = 0$ when $v_{i1} = v_{i2}$. A practical operational amplifier, however depicts a property in which there is a mismatch due to imbalance of the built in transistors following the inverting and non inverting input terminals. Such a mismatch gives rise to unequal bias currents at the inputs of the transistors. This requires application of an input offset voltage so as to balance the output.

The input bias current is the average of the currents flowing to the two input terminals of a balanced amplifier. The input bias current is given by

$$i_B = \frac{i_1 + i_2}{2} \text{ when } v_{out} = 0 \quad ...(14.13)$$

The input offset current is difference between the currents i_1 and i_2 flowing into the input terminals of a balanced amplifier. This gives

$$i_{i0} = i_1 - i_2 \text{ when } v_{out} = 0 \quad ...(14.14)$$

The input offset voltage $(v_i)_{off}$ is the voltage that should be applied between the two input terminals that produces a zero output voltage [Fig. 14.6(a)]. The output offset $(v_0)_{off}$ is the voltage at the output terminal when the two input terminals are grounded [Fig. 14.6(b)].

Practical OP AMPs have arrangements to balance the offset voltage by offset null arrangement.

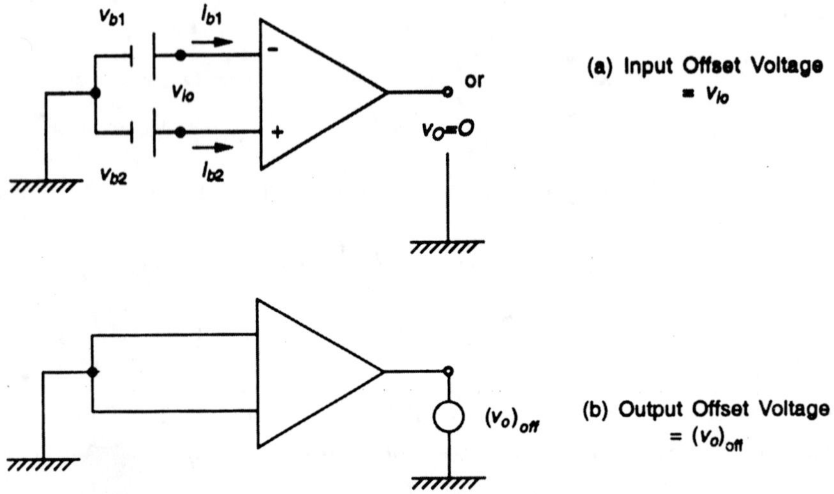

Fig. 14.6 Measurement of input offset voltage and output offset voltage

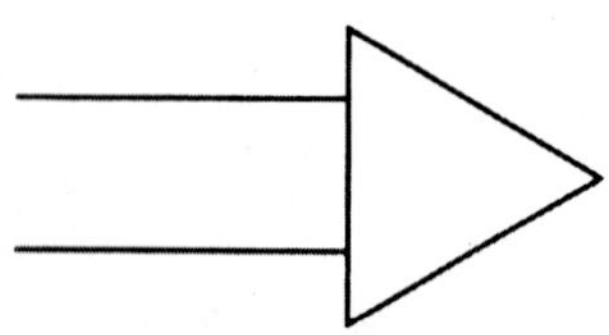

Fig. 14.7 Offset null arrangement

Slew Rate

The maximum rate of change of the output voltage when a step input voltage is applied to the input of an OP AMP is called the slew rate.

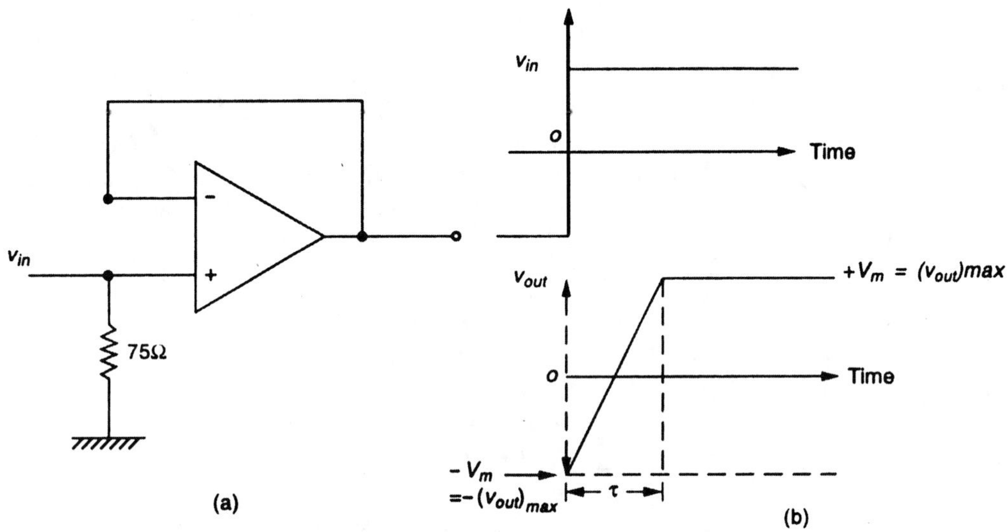

Fig. 14.8 Determination of slew rate (a) measurement circuit (b) Output voltage versus input voltage relationship

An input voltage pulse is applied to the input as depicted in Fig. 14.8(b) and the corresponding output voltage curve is depicted in Fig. 14.8(b). The slew rate is given by

$$\text{Slew rate} = \frac{\Delta v_{out}}{\tau} \qquad \text{...(14.15)}$$

where $\Delta v_{out} = V_m - (-V_m) = 2\ V_m.$

The slew rate is expressed in volts per microsecond.

EXAMPLE 14.6 *The output voltage of an operational amplifier appears as shown in Fig. 14.9. Find the slew rate.*

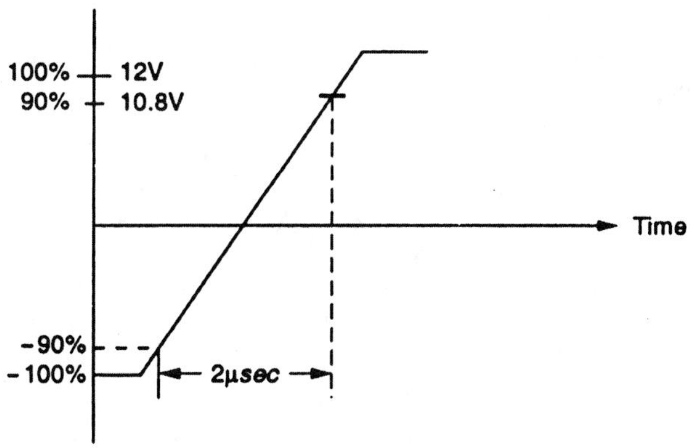

Fig. 14.9 Variation of output.

SOLUTION The output voltage proceeds from – 12 V to + 12 V. Considering the 10% and 90% points, the output voltage changes from – 10·8 V to + 10·8 V in 2 μsec. So the slew rate is

$$\frac{\Delta V}{\Delta t} = \frac{+10{\cdot}8 - (-10{\cdot}8)}{2\ \mu\text{sec}} = \frac{21{\cdot}6}{2} = 10{\cdot}8\ \text{V}/\mu\text{sec}\cdot$$

Frequency response

The internal amplifier stages that constitute an OP AMP have voltage gains limited by junction capacitances. The OP AMP has no internal coupling capacitors. However, the low frequency response ranges to the lower limit or zero frequency or d.c. A typical OP AMP frequency response characteristics is depicted in Fig. 14.10 below :

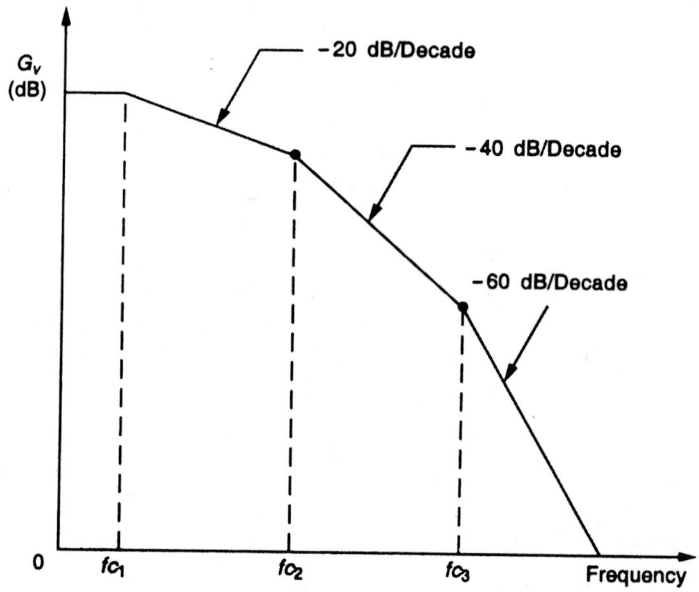

Fig. 14.10 OP AMP open loop frequency response characteristics

Comparison of OP AMP parameters

A comparison of values of the parameters of some common OP AMPs is depicted in Table 14.1. The values correspond to a temperature of 25°C.

Table 14.1

Parameter	OP AMP Chip			
	741C	LM218	HA2539	CLC300A
Open loop gain	2,00,000	2,00,000	1,50,000	
Input offset voltage	1 mV	2 mV	3 mV	10 mV
Input bias current	80 nA	120 nA	5 μA	10 μA
Input offset current	20 nA	6 nA	1 μA	
Input impedance	$2 \times 10^6\ \Omega$	$3 \times 10^6\ \Omega$	10 *k*Ω	100 kΩ/3 pf
Output impedance	75 Ω	–	–	–
Slew rate	1*V*/*μs*	70*V*/*μs*	600*V*/*μs*	3000V/μs
CMRR	90 dB	100 dB	60 dB	64 dB

Virtual Ground

The output voltage of an operational amplifier is limited by the supply voltage of typically, a few volts. The voltage gains are very high. If $v_{out} = -10$ V and $A_V = 2{,}00{,}000$, the input voltage can be calculated to have a magnitude

$$V_i = \frac{-v_{out}}{A_V} = \frac{10\text{ V}}{200000} = 0{\cdot}05\text{ mV} \qquad \text{...(14.16)}$$

Compared to other input and output voltages, the value of v_i is small and may be considered as 0V. It should be noted here that although $v_i \approx 0$V, it is not exactly 0V. The fact that $v_i \approx 0$V leads to the concept that at the amplifier input (points *a* and *b*) are virtual short circuit or virtual ground exists.

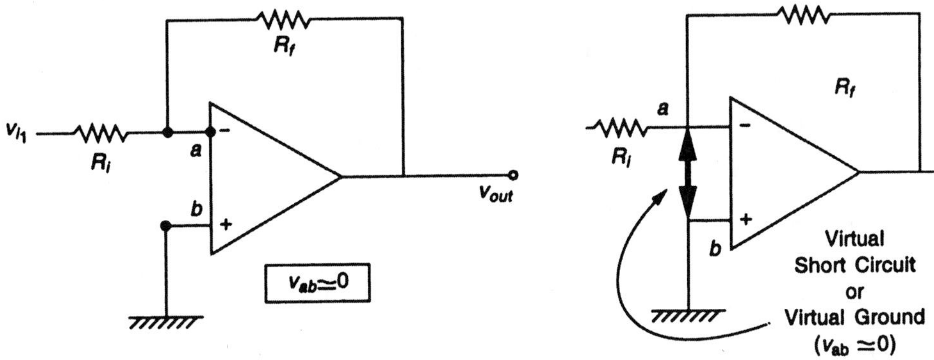

Fig. 14.11 Concept of virtual ground.

14.3 INVERTING AMPLIFIER

The circuit diagram of Fig. 14.12 represents an inverting amplifier.

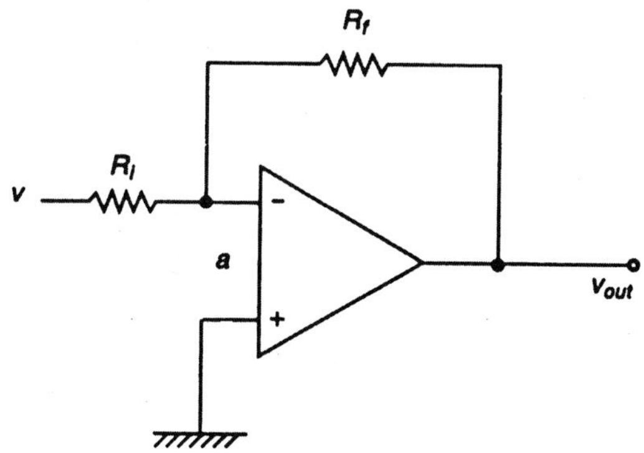

Fig. 14.12 Circuit diagram of an inverting amplifier

The current i through the resistance R_i can be written as

$$i = \frac{v_i - v_a}{R_i} \quad ...(14.17)$$

The current that flows through R_i is the same as the current that flows through R_f because the input impedance is very high ($Z_i \to \infty$) Kirchhoff's current law at point a yields

$$\frac{v_i - v_a}{R_i} = \frac{v_a - v_0}{R_f} \quad ...(14.18)$$

As the point a is at virtual ground, $v_a \approx 0$.

Thus $$\frac{v_i}{R_i} = -\frac{v_0}{R_f} \quad ...(14.19)$$

The ratio $\frac{v_0}{v_i}$ represents the closed loop gain of the amplifier. In this case then

$$(A_v)_d = \text{closed loop gain} = \frac{v_0}{v_i} = -\frac{R_f}{R_i} \quad ...(14.20)$$

The inverting amplifier can act as a **scale changer** by using precision resistor.

The negative sign implies that the output voltage is inverted with respect to the input voltage. The input resistance of the amplifier is

$$R_{in} = \frac{v_i}{i} = \frac{v_i}{(v_i - v_a)/R_i} \approx R_i \quad ...(14.21)$$

since $v_a \approx 0$ (VIRTUAL GROUND).

It should be noted that the **closed loop gain is independent of the op amp's internal open loop gain. The negative feedback stabilises the gain.**

EXAMPLE 14.7 *Find the value of R_f necessary in Fig. E14.7 so as to have a closed loop voltage gain of 50.*

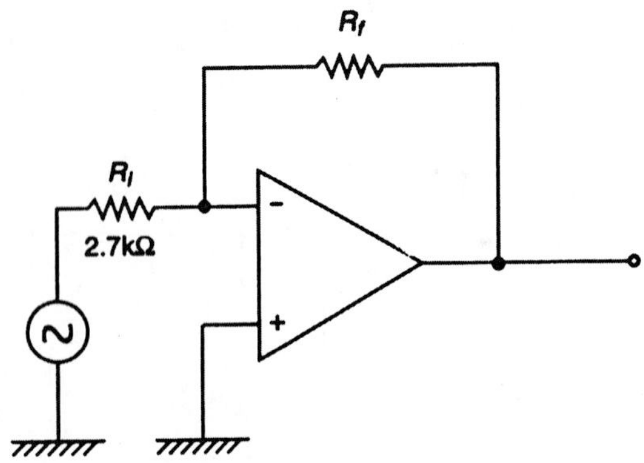

Fig. E14.7 Reference example

SOLUTION $$(A_v)_{cl} = -\frac{R_f}{R_i}$$

$$R_f = (A_v)_{cl} \cdot R_i$$
$$= (50) \cdot 2{\cdot}7 \text{ k}\Omega$$
$$= 135 \text{ k}\Omega$$

EXAMPLE 14.8 *Find the input and output impedances for the operational amplifier in the above example if the open loop gain is 200,000. Z_{in} = 5MΩ and Z_{out} = 50 Ω.*

SOLUTION
$$Z_{in} = R_i = 2{\cdot}7 \text{ k}\Omega$$
$$Z_{out} = (Z_{out})_{Miller} \,||\, Z_{out} \quad ...(14.22)$$
$$\text{Now } (Z_{out})_{Miller} = \left[\frac{(A_v)_{Ol}}{(A_v)_{O1} + 1}\right] R_f$$
$$\simeq R_f$$
$$\text{Thus } Z_{out} = R_f \,||\, Z_{out}$$
$$= 500 \,\Omega \,||\, 135 \text{ k}\Omega$$
$$\simeq 50 \,\Omega \quad ...(14.23)$$

Non inverting amplifier

The circuit diagram of a non inverting amplifier is given in Fig. 14.13. It will be seen that the input voltage is applied to the non inverting terminal. Since the OP AMP voltage gain is very high, the potential

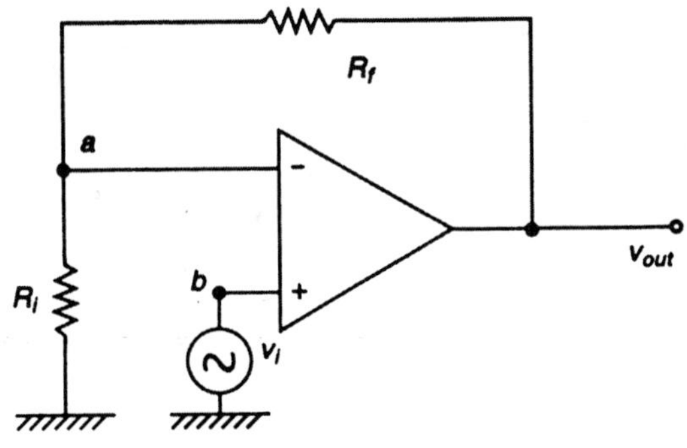

Fig. 14.13 Non inverting amplifier circuit.

at point a is the same as the potential at point b or $v_a = v_b = v_i$. Since the input impedance is very high, the current flowing to the OP AMP is very very small so as to be negligible. Thus applying Kirchhoff's current law to point a we get

$$\frac{v_0 - v_i}{R_f} = \frac{v_i}{R_i} \quad ...(14.24)$$

or
$$\frac{v_0}{v_i} = 1 + \frac{R_f}{R_i} \quad ...(14.25)$$

This is the closed loop gain of the non inverting amplifier. Since the gain is positive, there is no phase reversal between the input voltage and the output voltage.

In order to calculate the input impedance of the non-inverting amplifier let us consider Fig. 14.14.

If the OP AMP input impedance is not assumed infinite, then

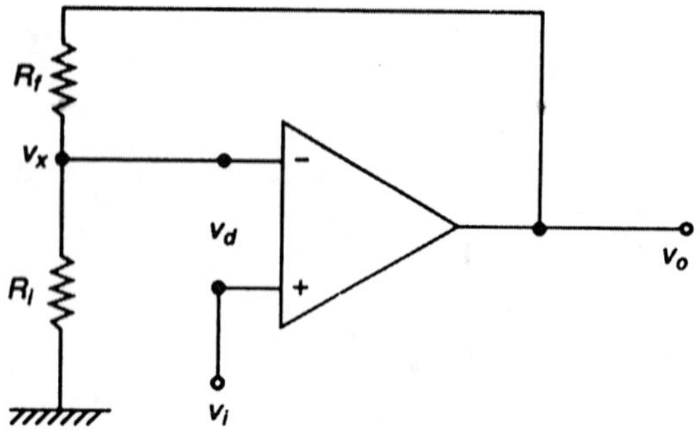

Fig. 14.14 Non inverting amplifier assuming a small differential voltage between the inputs.

$$v_i = v_d + v_x \quad \text{...(14.26)}$$

where $v_d \neq 0$ is a small differential voltage.

Now $$v_x = \frac{R_i}{R_i + R_f} \cdot v_0 \quad \text{...(14.27)}$$

$$= X\, v_0 \quad \text{...(14.28)}$$

where $$X = \frac{R_i}{R_i + R_f} \quad \text{...(14.29)}$$

Now, $$v_i = v_d + X\, v_0 \quad \text{...(14.30)}$$

$$= v_d + (A_v)_{0l}\, X\, v_d \quad \text{...(14.31)}$$

since $$v_{out} = v_0 = (A_v)_{0l}\, v_d$$

where $(A_v)_{0l}$ is the open loop voltage gain

Thus $$v_i = v_d\,[1 + (A_v)_{01}\, X] \quad \text{...(14.32)}$$

Now open loop input impedance is

$$\frac{v_i}{I_i} = \frac{v_d\,[1 + (A_v)_{0l} \cdot X]}{I_i} = \frac{I_i\, Z_i\,\{1 + (A_v)_{0l} \cdot X\}}{I_i}$$

$$\frac{V_i}{I_i} = \{1 + (A_v)_{0l} \cdot X\}\, Z_{in} \quad \text{...(14.33)}$$

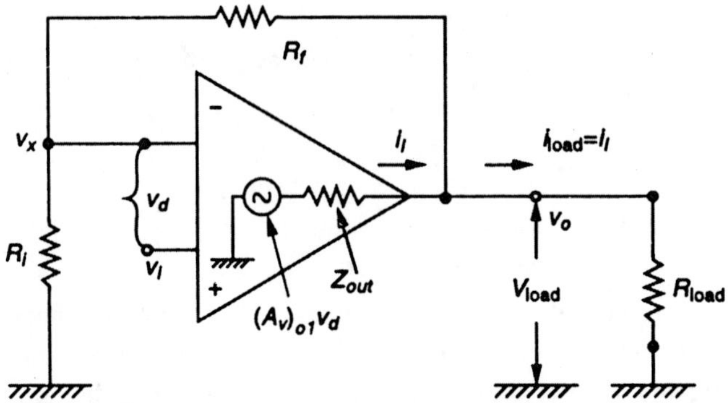

Fig. 14.15 Non inverting amplifier schematic for calculation of output impedence

Equation (14.33) thus reveals that the input impedance of the amplifier configuration with negative feedback is much higher compared to the internal input impedance of the OP AMP itself (without feedback).

The output impedance of the non-inverting amplifier is developed with the help of the Fig. 14.15.

Application of Kirchhoff's law to the circuit yields.

$$v_0 = (A_v)_{0l} \cdot v_d - Z_{out} \cdot i_l \qquad ...(14.34)$$

If $v_{load} << (A_v)_{0l} \cdot v_d$, then

$$v_0 \simeq (A_v)_{0l} \cdot (v_i - v_f) \qquad ...(14.35)$$

Substituting $X\, v_0$ for v_f, we get

$$v_0 \cong (A_v)_{0l}\, (v_i - X\, v_0) \qquad ...(14.36)$$

This can also be written as

$$v_0 = (A_v)_{0l} \cdot v_i - (A_v)_{0l} \cdot X\, v_0 \qquad ...(14.37)$$

Using the symbol A instead of $(A_v)_{0l}$ we may write

$$v_0 = A\, v_i - A\, X\, v_0 \qquad ...(14.38)$$

or

$$A\, v_i = (1 + XA)\, v_0$$
$$= (1 + XA)\, Z_{out} \cdot i_{load}$$
$$= (1 + XA)\, Z_{on} \cdot i_l \qquad ...(14.39)$$

where Z_{on} is the output impedance of the non inverting amplifier.

The internal output impedance of the operational amplifier is thus

$$Z_{out} = (1 + AX)\, Z_{on}$$

or

$$(Z_{on})_n = \frac{Z_{out}}{1 + AX} \qquad ...(14.40)$$

The equation (14.40), thus shows that the internal output impedance of the non inverting OP AMP configuration is much smaller compared to the internal output impedance of the OP AMP itself (without feedback).

EXAMPLE 14.9 *Find the input and output impedances for the amplifier depicted in Fig. 14.16 below. The OP AMP data sheet gives Z_{in} = 1·5 MΩ, Z_{out} = 70 Ω, $(A_v)_{0l}$ = 1,75,000. Also determine the closed loop voltage gain.*

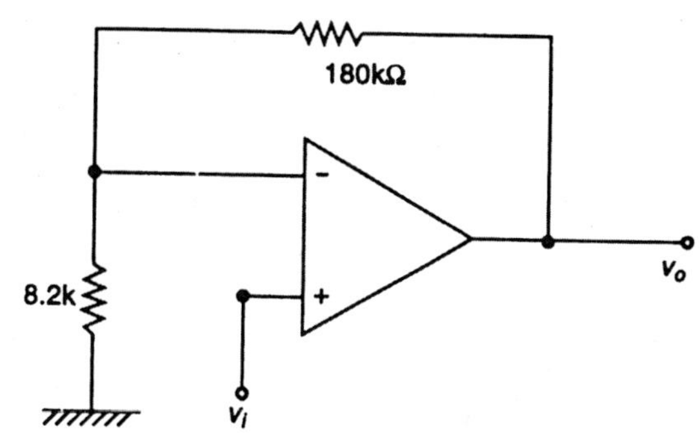

Fig. E14.9 Reference example 14.9.

SOLUTION For this circuit $X = \frac{R_i}{R_i + R_f} = \frac{8 \cdot 2 \text{ k}\Omega}{8 \cdot 2 \text{ k} + 180 \text{ k}} = \frac{8 \cdot 2}{188 \cdot 2} = 0 \cdot 0436$

$$(Z_{in})_n = [1 + A \cdot X] \, Z_{in} = (1 + 175000 \times 0 \cdot 0436) \, [1 \cdot 5 \text{ M } \Omega]$$

$$= (1 + 7630) \, 1 \cdot 5 \text{ M } \Omega = 11446 \text{ M } \Omega$$

$$(Z_{out})_n = \frac{Z_{out}}{1 + A \cdot X} = \frac{70}{1 + 75000 \times \cdot 0436} = 0 \cdot 0092 \, \Omega$$

The closed loop voltage gain $(A_v)_{cl} = \frac{1}{0 \cdot 0436} \simeq 23$

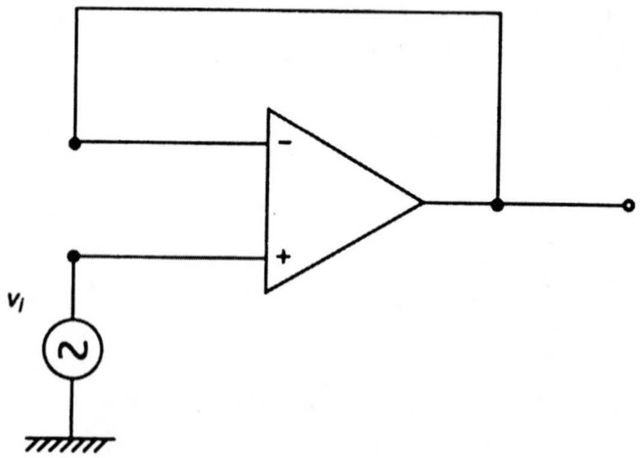

Fig. 14.16 Voltage follower circuit using OP AMP

Voltage Follower

Figure 14.16 depicts the circuit diagram of a voltage follower. The voltage follower is a special case of the non inverting amplifier configuration. For this case the closed loop gain is

$$(A_v)_{cl} = \frac{1}{X} = 1 \qquad \ldots(14.41)$$

The input impedance and output impedances of the voltage follower can be calculated from the equations for the non inverting amplifier configuration with $X = 1$.

$$(Z_v)_{VF} = [1 + (A_v)_{Ol}] \, Z_{in} \qquad \ldots(14.42)$$

$$(Z_0)_{VF} = \frac{Z_0}{1 + (A_v)_{Ol}} \qquad \ldots(14.43)$$

EXAMPLE 14.10 *An OP AMP has $(A_v)_{Ol}$ =175,000 Z_{out} = 70 Ω and Z_{in} = 1·5 M Ω is used as a voltage follower. Find the input and output impedances.*

SOLUTION

$$(Z_i)_{VF} = [1 + (A_v)_{Ol}] \, Z_{in} = [1 + 175000] \times 1 \cdot 5 \text{ M } \Omega \cong 262500 \text{ M } \Omega$$

$$(Z_v)_{VF} = \frac{Z_{out}}{1 + A} = \frac{70}{1 + 175000} \cong 0 \cdot 0004 \, \Omega$$

Phase Shifter

The phase shifter circuit depicted in Fig. 14.17, obtained by replacing R_i by an impedance Z_i and R_f by an impedance Z_f in Fig. 14.11. For this circuit

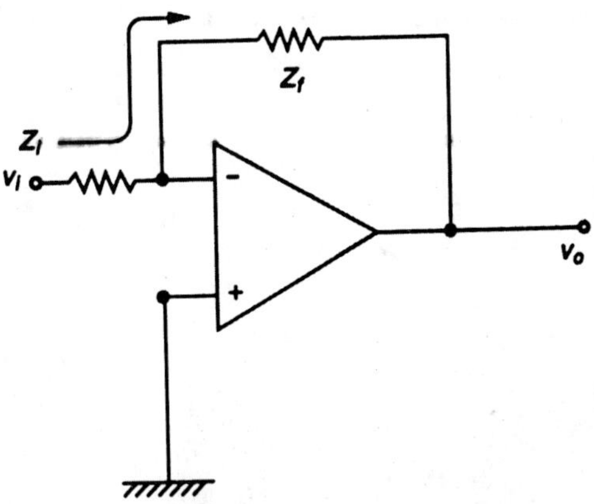

Fig.14.17 Phase shifter circuit

$$\frac{v_0}{v_i} = -\frac{Z_f}{Z_i} = -\frac{|Z_f| \exp(j\,\phi_f)}{|Z_i| \exp(j\,\phi_i)}$$

$$= \left|\frac{Z_f}{Z_i}\right| \exp\{j\,[\pi + \phi_f - \phi_i]\,\} \qquad ...(14.44)$$

If $|Z_f| = |Z_i|$ then the equation (14.44) reduces to

$$\frac{v_0}{v_i} = \exp[j\,\{\pi + \phi_f - \phi_i\}] \qquad ...(14.45)$$

$$= \exp[j\,\phi] \qquad ...(14.46)$$

Thus, this circuit can introduce a phase shift $\phi = (\pi + \phi_f - \phi_i)$ between the input and output voltages. The value of ϕ ranges between 0° and 360°.

14.4 ADDER OR SUMMING AMPLIFIER

The adder (also called summing amplifier) is depicted in Fig. 14.18. From virtual ground concept, the

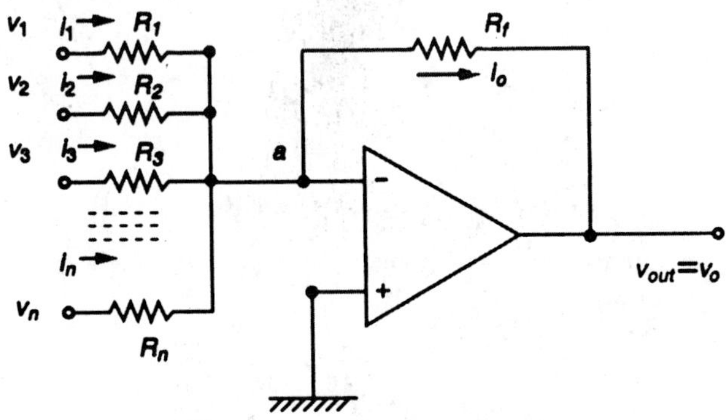

Fig. 14.18 Adder circuit

potential at point a is zero (GROUND). If the currents delivered by the input voltages $v_1, v_2, v_3, \ldots v_n$ are $i_1, i_2, i_3, \ldots i_n$ respectively and i_0 the current in the feedback path, then

$$i_0 = i_1 + i_2 + i_3 + \ldots + i_{in} \quad \ldots(14.47)$$

or,

$$-\frac{v_0}{R_f} = \frac{v_1}{R_1} + \frac{v_2}{R_2} + \frac{v_3}{R_3} + \ldots + \frac{v_n}{R_n} \quad \ldots(14.48)$$

or

$$v_0 = -\left\{\frac{R_f}{R_1} v_1 + \frac{R_f}{R_2} v_2 + \ldots + \frac{R_f}{R_n} v_n\right\} \quad \ldots(14.49)$$

If

$$R_1 = R_2 = R_3 = \ldots = R_n = R,$$

$$v = -\frac{R_f}{R}(v_1 + v_2 + v_3 + \ldots + v_n) \quad \ldots(14.50)$$

with $R_f = R$, equation (14.50) reduces to

$$v_0 = -(v_1 + v_2 + v_3 + \ldots + v_n) \quad \ldots(14.51)$$

This circuit is termed as an adder or summing amplifier because the output voltage is numerically equal to the sum of the input voltages.

If the algebraic sum of the input voltage is very small then R_f is kept larger compared to R. The sum voltage can then be obtained by dividing v_0 by R_f/R.

14.5 DIFFERENTIAL AMPLIFIER

A differential amplifier circuit is depicted in Fig. 14.19. It amplifies the difference between the two input voltages. The voltage v_1 is applied to the inverting input terminal while the voltage v_2 is applied to the non

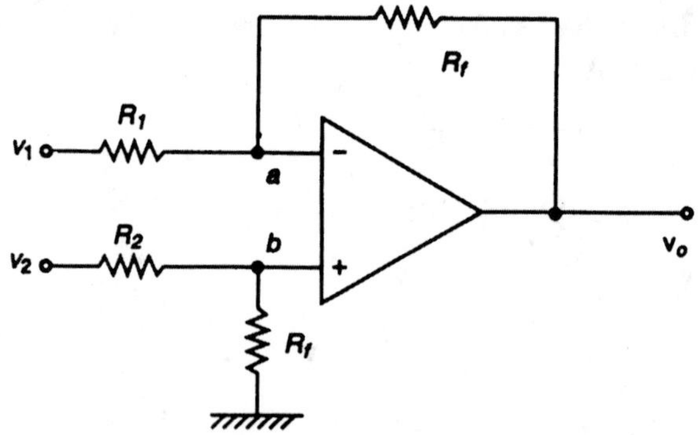

Fig. 14.19 Differential amplifier circuit

inverting input terminal. The output voltage is v_0. The voltage gain of the OP AMP being infinite, the points a and b will be at the same potential

$$v_a = v_b \quad \ldots(14.52)$$

Applying Kirchhoff's current law at a and b, we obtain this

$$\frac{v_1 - v_a}{R_1} = \frac{v_a - v_0}{R_f} \quad \ldots(14.53)$$

and $$\frac{v_2 - v_a}{R_1} = \frac{v_a}{R_f} \qquad ...(14.54)$$

where the input impedance of the OP AMP is infinite.

Thus $$v_0 = \frac{R_f}{R_1}(v_2 - v_1) \qquad ...(14.55)$$

If $$R_f = R_1$$

$$v_0 = v_2 - v_1 \qquad ...(14.56)$$

If, however, $R_f > R_1$, the amplifier gain is $\frac{R_f}{R_1}$.

14.6 OSCILLATOR

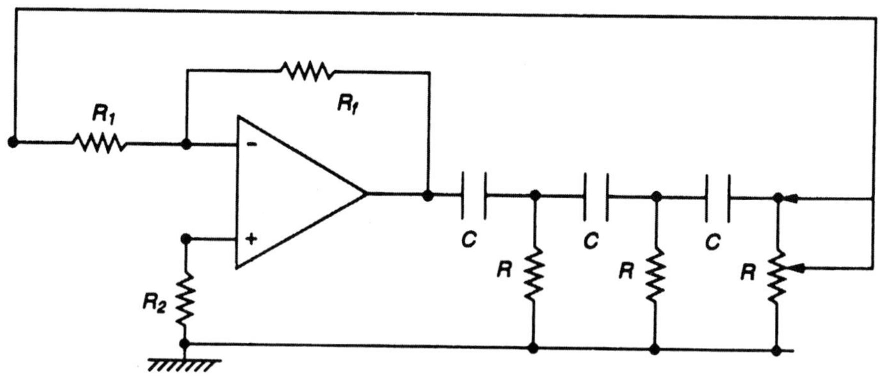

Fig. 14.20 Phase shift oscillator

Due to its high amplification and high bandwidth, the OP AMP may be used in oscillator circuits as well. Fig. 14.20 depicts a phase shift oscillator. The 3-section RC network is connected the output of the OP AMP. Each of the three sections of the three RC networks in feedback loop may provide a phase shift θ where the maximum value of θ can approach 90°. The phase shift oscillator provides oscillation at a frequency at which the phase shift provided by the RC network is 180°. The balance 180° is provided by the OP AMP. For the circuit, the frequency of oscillation is

$$f_r = \frac{2}{2\pi\sqrt{6}\,RC} \qquad ...(14.57)$$

and the attenuation β of the three section RC network is

$$\beta = \frac{1}{29} \qquad ...(14.58)$$

EXAMPLE 14.11 *Find the frequency of oscillation for the circuit of the phase shift oscillator shown in Fig. 14.21. Find also the value of R_f for oscillations to take place.*

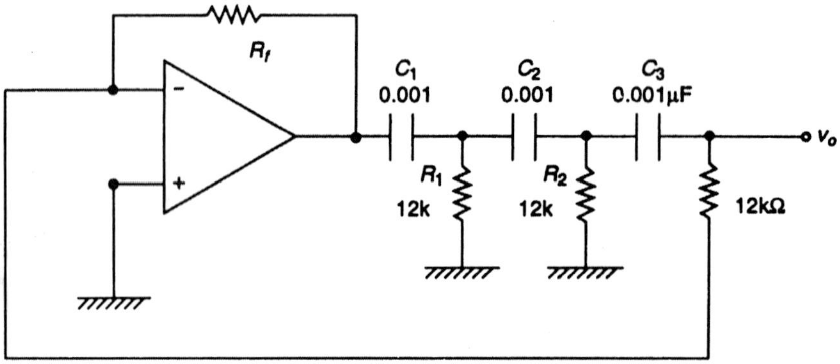

Fig. 14.21 Phase shift oscillator (Example 14.11)

SOLUTION
$$f_r = \frac{1}{2\pi\sqrt{6}\,RC} = \frac{1}{2\pi\sqrt{6}\,(12\ k\Omega)\,(0{\cdot}001\ \mu F)} \cong 5{\cdot}42\ \text{kHz}$$

$$\frac{R_f}{R_3} = 29$$

or
$$R_f = 29 \times 12\ k\Omega = 348\ k\Omega$$

For the Wien bridge oscillator, depicted in Fig. 14.22, the frequency of oscillation is

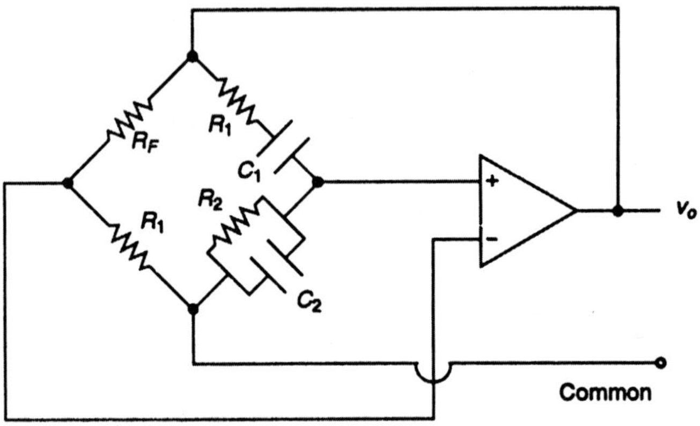

Fig. 14.22 Wien bridge oscillator

$$f_0 = \frac{1}{2\pi\sqrt{R_1R_2C_1C_2}} \qquad ...(14.59)$$

If $R_1 = R_2 = R$ and $C_1 = C_2 = C$, the frequency of oscillation is

$$f_0 = \frac{1}{2\pi RC} \qquad ...(14.60)$$

Also the non inverting gain must be 3 for oscillation.

14.7 DIFFERENTIATOR

The basic differentiator circuit is depicted in Fig. 14.23 in which an input voltage v_i is applied. The voltage

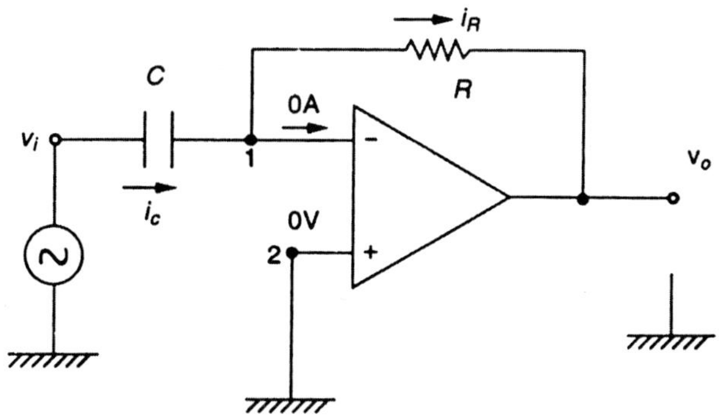

Fig. 14.23 Differentiator circuit

across the capacitor v_C will also be v_i due to VIRTUAL GROUND across points 1 and 2. Also the current flowing to the inverting input is negligible. Thus we can write

$$v_i = v_C \quad ...(14.61)$$

Differentiating with respect to time we get

$$\frac{dv_i}{dt} = \frac{1}{C}\frac{dQ}{dt} = \frac{i_C}{C} \quad ...(14.62)$$

Also $$i_C = i_R = -\frac{v_0}{R} \quad ...(14.63)$$

Thus $$v_0 = -CR\frac{dv_i}{dt} \quad ...(14.63)$$

The input output relationship for different types of inputs is shown in Fig. 14.24.

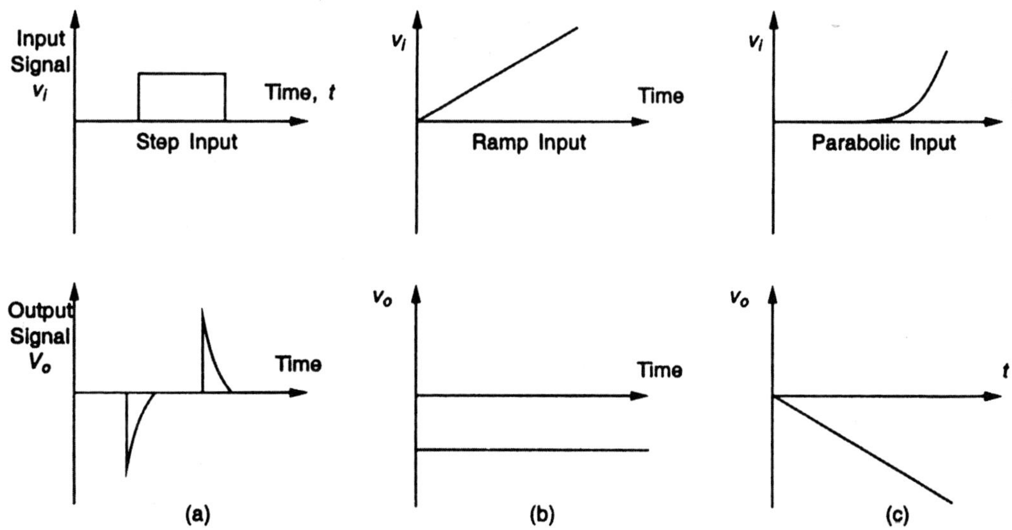

Fig. 14.24 Input-output relationships for the differentiator (a) step input (b) ramp input (c) parabolic input

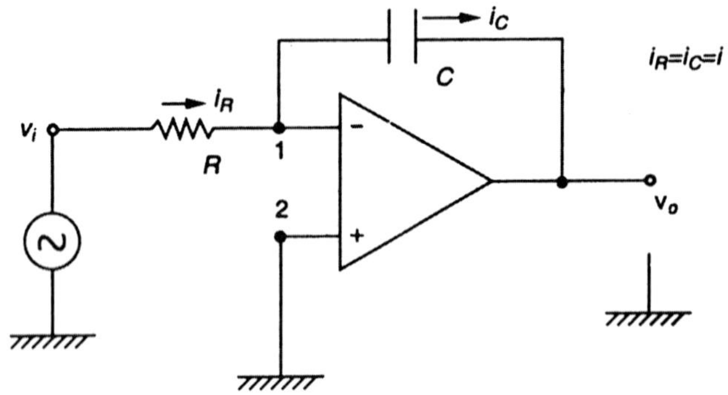

Fig. 14.25 OP AMP integrator circuit.

14.8 INTEGRATOR

An operational amplifier integrator circuit is depicted in Fig. 14.25. In order to analyse the circuit, let us keep in mind that due to the presence of VIRTUAL GROUND at the inverting input 1, the voltage $V_{12} = 0$. Also due to the very high input impedance ($Z_i \to \infty$), the current entering into the inverting input is zero. Thus

$$V_{12} = 0 \quad ...(14.64)$$

$$i_R = i_C = i \quad ...(14.65)$$

Also $$v_i = v_R = i \, . \, R \quad ...(14.66)$$

Therefore $$v_0 = -\frac{1}{C}\int i \, dt = -\frac{1}{CR}\int_0^t v_i \, dt \quad ...(14.67)$$

The output voltage is proportional to time integral of the input voltage with a proportionality constant ($-1/CR$). The input-output relationship for the integrator is shown in Fig. 14.26.

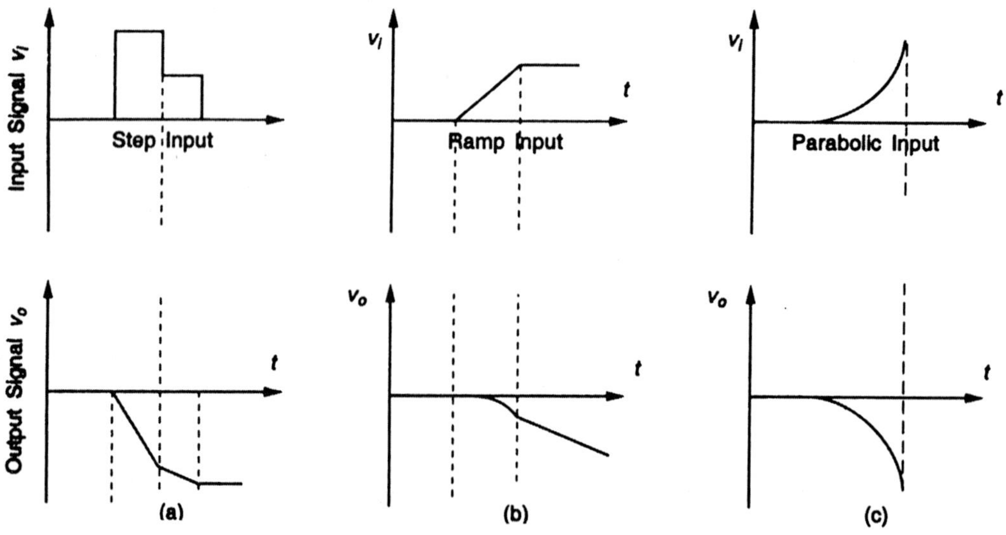

Fig. 14.26 Input- output relationships for the integrator (a) step input (b) ramp input (c) parabolic input

14.9 CURRENT TO VOLTAGE CONVERTER

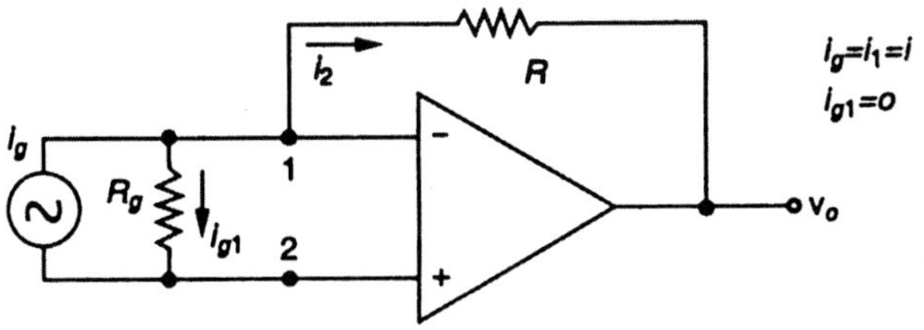

Fig. 14.27 Current to voltage converter

The current voltage converter is depicted in Fig. 14.27. In the circuit, due to the presence of the VIRTUAL GROUND, the voltage V_{12} is zero. Further, the current entering into the inverting input is zero due to the high input impedance. Also the current in R_g is zero since $V_{12} = 0$. Therefore, we can write,

$$V_{12} = 0 \quad ...(14.68)$$

$$i_{g_1} = \text{Current in } R_g = 0. \quad ..(14.69)$$

and $$i_g = i_2 = i \quad ...(14.70)$$

Further $$v_0 = -\, i_2 R = -\, i R = -\, i_g R \quad ...(14.71)$$

Thus, the output voltage is proportional to the input current. Hence the name current-to-voltage converter.

14.10 VOLTAGE TO CURRENT CONVERTER

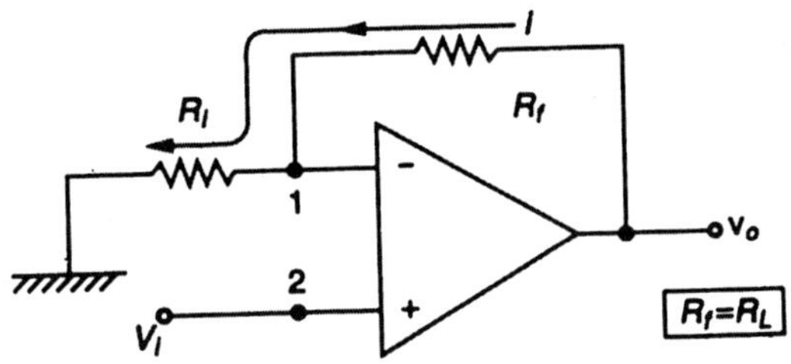

Fig. 14.28 voltage to current converter

The voltage to current converter is shown in Fig. 14.28. In this case, the current i flowing in the circuit can be written as

$$i = \frac{v_i}{R_1} \quad ..(14.72)$$

since points 1 and 2 are at the same potential (due to very high OP AMP Gain $G \to \infty$). Equation (14.72) thus states that the current i flowing through R_f (which in this case is the load resistance) depends on v_i and not on R_f (or load resistance). Some applicaions of voltage to current converters include analog-to-digital converters (ADCs) and drivers of the deflection coils of cathode ray tubes in television.

14.11 LOGARITHMIC CONVERTERS

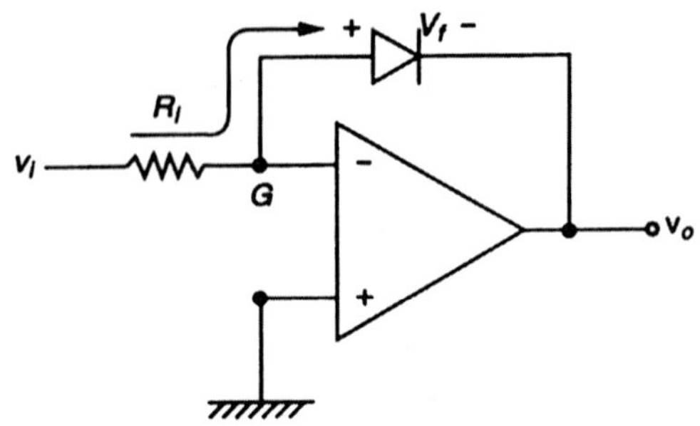

Fig. 14.29 A logarithmic amplifier

A logarithmic converter circuit is depicted in Fig. 14.29. For a diode the relationship between current and voltage is given by

$$i = I_0 \left[\exp\left(\frac{e V_f}{\eta K_B T} \right) - 1 \right] \quad ...(14.73)$$

where I_0 is the reverse saturation current, V_f is the forward voltage drop. If $\left(\frac{e V_f}{\eta K_B T} \right) >> 1$

or $i >> I_0$, we have

$$i = I_0 \exp\left(\frac{e V_f}{\eta K_B T} \right) \quad ...(14.74)$$

Taking logarithm of both sides

$$\ln\left(\frac{i}{I_0} \right) = \left(\frac{e V_f}{\eta K_B T} \right) \quad ...(14.75)$$

or

$$V_f = \frac{\eta K_B T}{e} \ln\left(\frac{i}{I_0} \right) \quad ...(14.76)$$

Since 1 is the virtual ground point, current i is given by

$$i = \frac{v_i}{R_i} \quad ...(14.77)$$

and

$$v_0 = -V_f = -\left(\frac{\eta K_B T}{e} \right) \ln\left(\frac{v_i}{I_0 R_i} \right) \quad ...(14.78)$$

The output voltage is thus proportional to the logarithm of $\left(\frac{v_i}{I_0 R_i} \right)$ and hence the circuit is called a logarithmic converter.

EXAMPLE 14.12 *Design an adder / subtractor circuit that will provide gains of – 5, – 2, + 5 and $+\frac{1}{3}$ for the voltage v_1, v_2, v_3, v_4 respectively.*

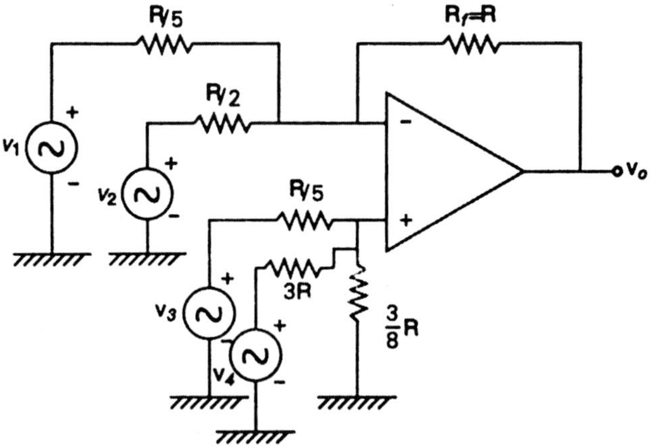

Fig. 14.30 Reference Example 14.12

SOLUTION Assuming that the OP AMP gain G is very high ($G \to \infty$), input impedance Z_i very high ($Z_i \to \infty$) output impedance very low ($Z_0 \to 0$), the circuit is shown in Fig. 14.30.

The output voltage is given by

$$v_0 = -5v_1 - 2v_2 + 5v_3 + \frac{1}{3}v_4 \quad ...(14.79)$$

For a generalized functional circuit we may write (vide Fig. 14.31)

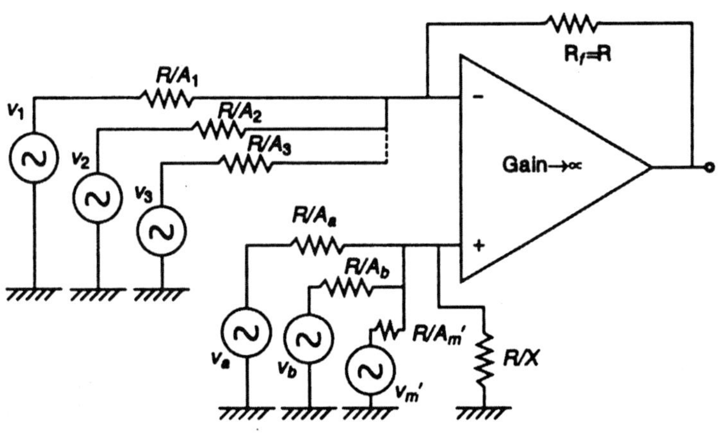

Fig. 14.31 A generalized adder/subtractor circuit

where voltage sources v_1 to v_n are connected to the inverting input terminal and voltage sources $v_a, v_b, \ldots v_m$ connected to the non-inverting input terminal. Assuming $A_1, A_2, \ldots A_n$ and $A_a, A_b, A_c, \ldots A_{m'}$ the gain magnitudes, output voltage of the circuit is

$$v_0 = -\sum_{i=1}^{n} A_i v_i + \sum_{i=a}^{m'} A_j \ v_j \qquad ...(14.80)$$

and
$$X = X_n - X_p \qquad ...(14.81)$$

where
$$X_n = 1 + \sum_{i=1}^{n} A_i = 1 + \text{Sum of negative gains} \qquad ...(14.82)$$

and
$$X_p = \sum_{j=a}^{m'} A_j = \text{ Sum of positive gains} \qquad ...(14.83)$$

EXAMPLE 14.13 *(a) What is the normal output voltage for the circuit in Fig. 14.31 ?*

(b) What is the output voltage if R_3 opens ?

(c) What happens if R_5 opens ?

SOLUTION (a) $V_0 = -(2V + 1V + 0.5V + 0.2V) = -3.7V$

(b) $V_0 = -(2V + 1V + 0.2V) = -3.2V$

(c) If the resistor R_5 opens, the circuit becomes that of a comparator and the output becomes equal to $-V_{max}$.

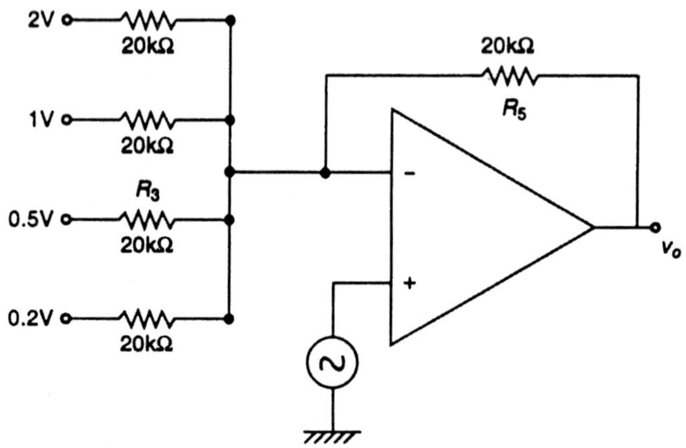

Fig. 14.32 Reference example 14.14

EXAMPLE 14.14 *Find out the output voltage for the circuit in Fig. 14.33. Also find out the output voltage if R_3 opens.*

SOLUTION

(a) $v_0 = -\dfrac{10k\Omega}{50k\Omega}(2V + 1V + 5.5V + 2.2V + 1.1V)$

$= -\dfrac{1}{5}(11.8V)$

$= -2.36V$

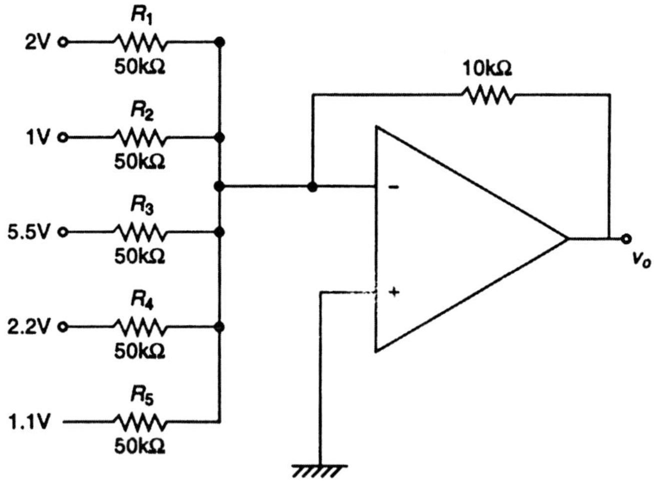

Fig. 14.33

(b) $v_0 = -\frac{10k\Omega}{50k\Omega}(2V + 1V + 2{\cdot}2V + 1{\cdot}1V)$

$= -\frac{1}{5}(6{\cdot}3V)$

$= -1{\cdot}26V.$

Review Questions

1. What do you understand by an 'operational amplifier' ?
2. What is a differential amplifier ?
3. What is common mode input voltage and common mode output voltage in a differential amplifier ?
4. What are the different terminals of an operational amplifier ?
5. What is meant by 'offset null' ?
6. What should be the features of an 'Ideal Operational Amplifier' ?
7. What do you understand by common mode gain and differential mode gain ?
8. What do you understand by common mode rejection ?
9. Discuss how the input offset voltage and output offset voltage of an operational amplifier can be measured?
10. What is slew rate ? Discuss how slew rate can be measured in an operational amplifier.
11. Draw the open loop frequency response characteristics of an operational amplifier.
12. What is the value of open loop gain, input impedance and output impedance of an operational amplifier 741?
13. What do you understand by 'Virtual Ground' in an operational amplifier ?
14. What is a non inverting amplifier ? Derive an expression for its voltage gain.
15. Derive the value of the input impedance and output impedance of a non inverting amplifier circuit.
16. What is a voltage follower circuit ? What is its closed loop voltage gain ? What is the output impedance ?

17. What is a phase shifter circuit ?
18. Draw the circuit diagram of an adder.
19. What is a differential amplifier ? Derive an expression for its voltage gain.
20. Draw the circuit diagram of a phase shift oscillator using an operational amplifier.
21. What is a differentiator ? Derive the expression for the output voltage of a differentiator. Draw the output voltage wave shapes if the following types of waveforms are input to a differentiator. (Step Input, RAMP, Parabolic input).
22. What is an integrator ? Derive the expression for the output voltage of an integrator. Draw the output voltage waveform if the following types of waveshapes are input to an integrator (step input, RAMP, Parabolic input).
23. Draw the schematic diagram of a voltage to current converter and explain the principles of operation.
24. What is a logarithmic converter ? Draw the circuit diagram of a logarithmic converter using operational amplifier.
25. What is a voltage-to-current converter ? Draw the circuit diagram of a voltage-to-current converter using an operational amplifier.

Objective Type Questions

Tick mark (√) the correct or the most appropriate answer.

1. An OP AMP can be classified as
   (a) linear IC (b) digital IC
   (c) RC coupled circuit (d) positive feedback circuit
2. An ideal OP AMP has
   (a) infinite voltage gain (b) infinite input resistance
   (c) zero output resistance (d) all of the above
3. The popularity of OP AMPs stems from the fact that
   (a) they are of small size
   (b) they possess a very high voltage gain
   (c) they have high input impedance
   (d) their external properties in a circuit can be changed to suit an application.
4. The voltage gain of an OP AMP in the open loop condition is of the order of
   (a) 10 (b) 100
   (c) 1000 (d) 1000,000
5. In the common mode application of an OP AMP
   (a) both inputs are connected to Earth potential (b) an identical signal appears at both the inputs
   (c) the outputs are tied together (d) the outputs signals are in phase
6. Common mode gain in an operational amplifier is
   (a) very low (b) very high
   (c) always unity (d) moderate
7. If the differential voltage gain is 4700 and the common mode gain is 0·47, the CMRR is
   (a) 40 dB (b) 10,000
   (c) 80 dB (d) both *b* and *c*
8. In an OP AMP, the bias currents are 46 μA and 45·1 μA. The input offset current is
   (a) 900 nA (b) 91·1 μA
   (c) 45·55 μA (d) none of the above
9. The output of an OP AMP increases by 5V in 15 μ seconds. The slew rate is
   (a) $75 \frac{V}{\mu s}$ (b) $3 \frac{V}{\mu s}$

(c) $0{\cdot}33 \frac{\text{V}}{\mu\text{s}}$ (d) none of the above

10. If negative feedback is employed in an OP AMP circuit, then
(a) the OP AMP oscillates (b) the gain is increased
(c) the voltage gain is decreased (d) none of the above

11. For a non inverting amplifier circuit the input resistor R_i has a value of 10 kΩ and R_f = 480 kΩ. The closed loop gain is
(a) 47 (b) 48
(c) 200,000 (d) 1000

12. In a non inverting OP AMP circuit, the resistor R_f is open. The voltage gain will
(a) decrease (b) become zero
(c) increase (d) will remain constant

13. The output impedance of 741C OP AMP is of the order of
(a) 1 Ω (b) 75 Ω
(c) 1000 Ω (d) infinity

14. The slew rate of 741*C* OP AMP is of the order (in units of *V*/μs) of
(a) 1 (b) 10
(c) 100 (d) 1000

15. If *G* be the open loop voltage gain of an OP AMP and Z_0 the output impedance of the OP AMP, the output impedance of a voltage follower using the OP AMP can be written as
(a) $\frac{Z_0}{G}$ (b) $Z_0 \cdot G$
(c) $\frac{Z_0}{1+G}$ (d) $Z_0(1+G)$

16. The frequency of oscillation of an Wien bridge oscillator using OP AMP is given by a formula of the nature
(a) $f_0 = \frac{1}{RC}$ (b) $f_0 = \frac{1}{2\pi RC}$
(c) $f_0 = RC$ (d) $f_0 = 2\pi RC$

17. The Bandwidth of an ideal OP AMP is of the order of
(a) 10 kHz (b) near zero
(c) infinity (d) 100 MHz.

18. The common mode rejection ratio (CMRR) of an OP AMP is
(a) $\frac{A_d}{A_c}$ (b) $\frac{A_c}{A_d}$
(c) $\frac{A_d}{A_c}$ (d) $\frac{A_0}{A_c}$
where A_0, A_d, A_c represent respectively the open loop, differential and common mode voltage gains.

19. For an OP AMP with negative feedback, the output is
(a) equal to input (b) fed back to non inverting input
(c) fed back to inverting input (d) decreased

Exercise Problems

1. A differential amplifier has a common mode output voltage of + 4·7V when the input common mode voltage is zero and has an output common mode voltage of + 4·3V when the input common mode voltage is + 1V (i) Calculate the value of v_{01} if v_{02} is + 4·7V (ii) Calculate v_{ICM} and v_{OCM} when v_{I_1} = + 1V, v_{I_2} = + 0·9V and v_{01} = + 2V and v_{02} = + 7V (vide ref. Fig. 14.1).

2. In the above problem 1 if, v_{I_1} = + 1·1V, v_{I_2} = + 0·95V, v_{01} = + 2V and v_{02} = + 8·5V, find v_{ICM} and v_{OCM}.

3. In a differential amplifier, v_{I_1} is grounded while v_{I_2} is connected to an a.c. voltage of $2{\cdot}1 \sin(2\pi \times 0{\cdot}21t)$ mV. The output common mode voltage is + 5·5V when $v_{ICM} = 0$; + 5·68V when $v_{ICM} = -2{\cdot}1$ mV and + 5·66V when $v_{ICM} = +2{\cdot}1$ mV. Find the input and output common mode voltages.

4. Find the common mode gain in a differential amplifier if the following input and output voltages are established at different points of time

Instant	v_{I_1}	v_{I_2}	$v_{out\,1}$	$v_{out\,2}$
t_1	− 3·36V	− 3·76V	+ 4·76V	+ 4·96V
t_2	+ 1·90V	+ 1·18V	+ 2·5V	+ 3·2V

5. In an operational amplifier, the differential mode gain A_d = 19000 and the common mode gain is A_c = 21. Find the CMRR in decibles.
(**Ans.** 59·13 dB)

6. The output voltage of an operational amplifier appears as depicted in Fig. P14.6. Find the slew rate.

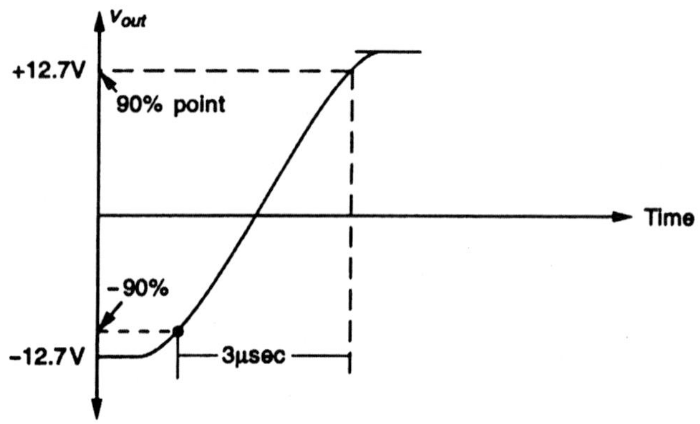

Fig. P14.6 Reference Problem 6

(**Ans.** 8·46V / μsec)

7. Find the value of R_f needed in Fig. P14.7 so as to have a closed loop voltage gain of 21.

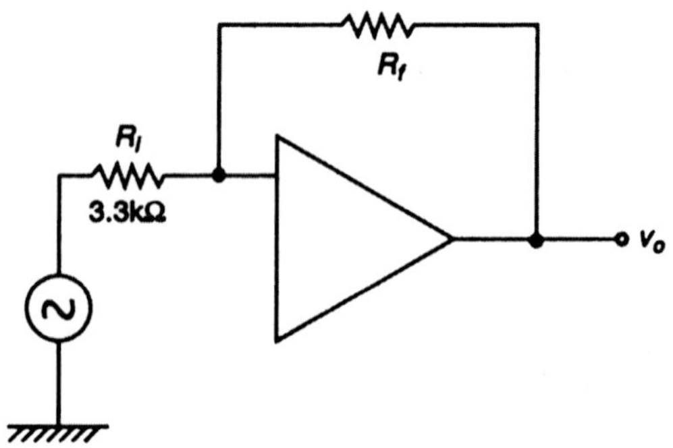

Fig. P14.7 Reference Problem 7

(**Ans.** 69·3 kΩ)

8. Find the input and output impedances for the operational amplifier in the above problem if the open loop gain is 250,000, Z_{in} = 5·7 MΩ and Z_{out} = 39 Ω.

(**Ans.** Z_{in} = 3·3 kΩ, Z_{out} = 39 Ω)

9. ·Calculate the input and output impedances for the amplifier shown in Fig. P14.19 below.
The OP AMP data sheet shows Z_{in} = 1·9 MΩ, Z_{out} = 57 Ω, $(A_v)_{Ol}$ = 190000. Find also the closed loop gain.

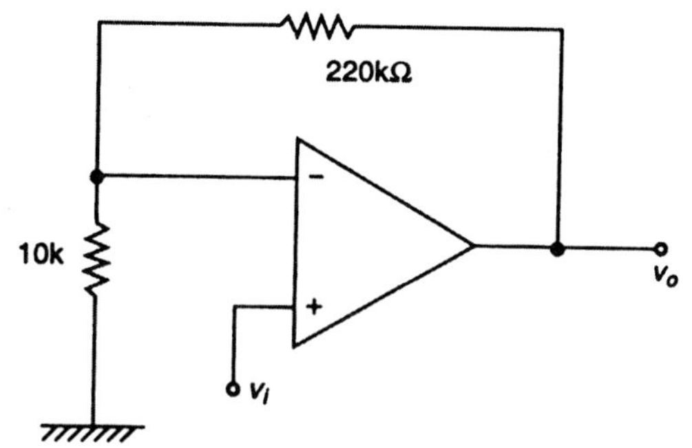

Fig. P14.9 Reference Problem 9

(**Ans.** Z_{in} = 15697 MΩ, Z_{out} = 0·0069 Ω, $(A_v)_{cl}$ = 23)

10. An OP AMP having values : $(A_v)_{Ol}$ = 190,000, Z_{out} = 57 Ω and Z_{in} = 1·0 MΩ is used as a voltage follower. Calculate the input impedance and output impedance.
(**Ans.** 190,000 MΩ, 0·0003 Ω)

11. Find the frequency of oscillation for the circuit of a phase shift oscillator in Fig. P14.11.
Also calculate the value of R_f for oscillations to take place.

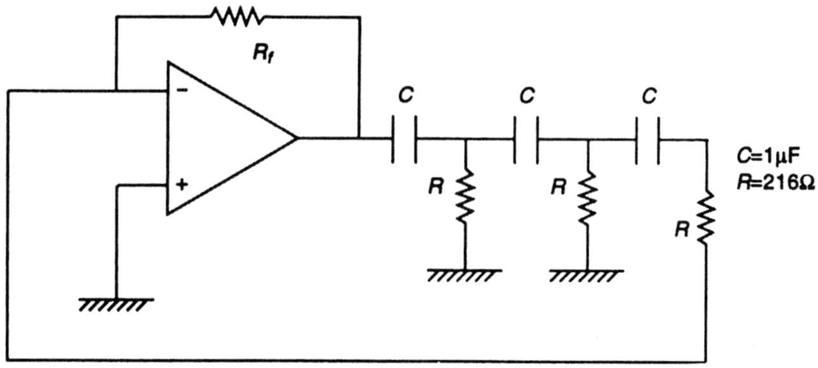

Fig. P14.11

(**Ans.** 300 Hz, 6264 Ω)

12. Calculate the value of the output voltage for the circuit of Fig. P14.12.

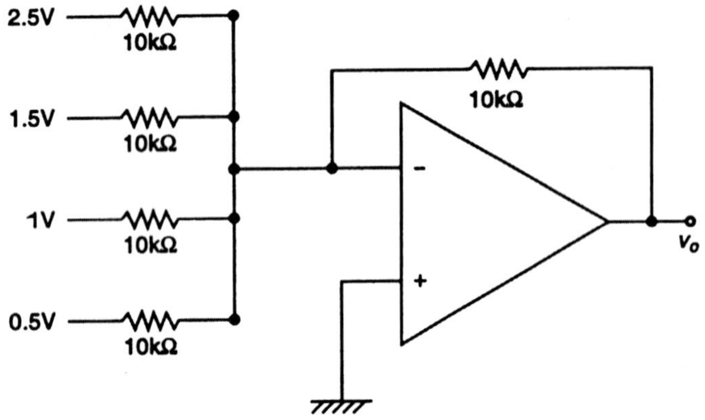

Fig. P14.12

(**Ans.** – 5·5V)

Chapter 15

TELEVISION

15.1 INTRODUCTION

The term television stands for vision from a distance. In television, visual images are transmitted as electrical signals. The image may be considered as being subdivided into a large number of small squares, called as picture elements or pixels. When the number of picture elements in a given image is large, the reproduction is clearer (meaning better resolution) at the receiver.

In present day world, watching television is a favourite pastime and the manufacture of television receiver sets occupies a major chunk of modern day electronics industry. The NTSC system (National Television System Committee of the United States) is commonly employed in Japan, Korea, Maxico, Canada, Chile, Taiwan and others. The PAL (Phase Alternate Line) is used in countries like Great Britain, Austria, India, Bangladesh, Switzerland, South Africa, Italy, Israel and others. And the SECAM (Sequential Coding and Memory) system is used in countries like Germany, France, Hungary, Iran, Iraq etc. They differ mainly on the scanning rates.

For good picture reproduction, it is necessary that the following picture qualities are restored

(a) sharpness or pictorial clarity (object edges should be clear and sharp).

(b) contrast between light and dark areas.

(c) motion continuity (illusion of motion to the human eye). The pictures are recreated at 25 pictures per second.

(d) flicker (In television viewing the pictures must be presented at a rate which is higher than that required for continuity of motion).

(e) colour values (it should be acceptable as realistic to the human eye).

(f) freedom from noise, interferences etc.

In the PAL and SECAM system, the basic frame rate is 25 per second and the flash rate is 50 per second. In the NTSC (National Television System Committee) system the frame rate is 30 per second and the flash rate is 60 per second. The higher rate of the NTSC system makes possible substantially brighter flickerless pictures by a factor of about 5 times. The higher picture rate, however, reduces the details of the pictures by about 16·66%.

The PAL system is similar to the NTSC system. In the PAL system, however, the phase of the colour subcarrier is reversed on succeding scanning lines and the inverse variation occurs in the receiver. The PAL system also employs $R-Y$ and $B-Y$ colour difference signals rather than the complex I and Q chrominance signals of the NTSC. The European colour television system employs 625 line scanning at 25 frames per second interlaced at 50 field periods. The video bandwidth employed is 1 to 2 MHz wider compared to that of the NTSC.

For television broadcasting, the frequency assignments from land based stations are set by international agreements and this is depicted in Table 15.1.

Table 15.1 : Frequency Assignments for TV Channels

Broadcast Channel	*Band*	*Frequency Range* (MHz)	*Bandwidth* (MHz)
2	VHF	54–60	6
3	VHF	60–66	6
4	VHF	66–72	6
5	VHF	76–82	6
6	VHF	82–88	6
7	VHF	174–180	6
8	VHF	180–186	6
9	VHF	186–192	6
10	VHF	192–198	6
11	VHF	198–204	6
12	VHF	204–210	6
13	VHF	210–216	6
14	UHF	470–476	6
15	UHF	476–482	6
16	UHF	482–488	6
17	UHF	488–494	6
18	UHF	494–500	6
19	UHF	500–506	6
|	|	|	
82	UHF	878–884	6
83	UHF	884–890	6

FM Radio services are assigned bands between 88–108 MHz and other services are assigned bands between 108 to 300 MHz.

Cable television distribution systems (CATV) employ coaxial cables and have a maximum 300 MHz bandwidth. The high definition television (HDTV) systems started in 1980's and they present a better picture quality.

The frequencies between 806 and 890 MHz, formerly allocated to TV broadcasting, are now allocated to the land mobile services. Operation, on a secondary basis, for TV translators may continue on these frequencies.

In a television system, the picture consists of small picture elements or pixels. Figure 15.1 depicts the picture elements.

Starting from the upper left hand corner with element number 1, the first row of elements is scanned in the sequence 1, 2, 3, 4, ... and so on. For each picture element, an electrical signal proportional to the

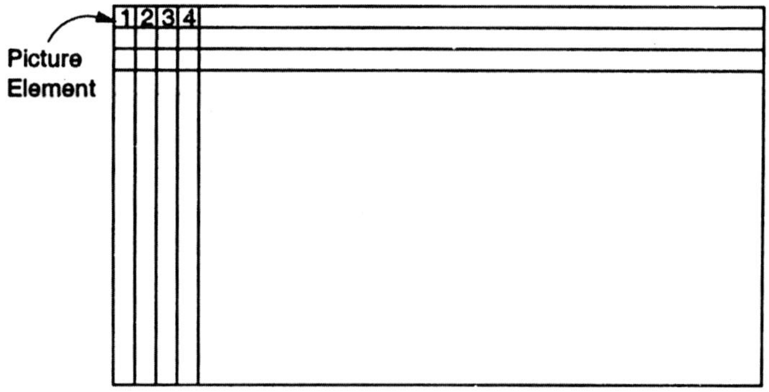

Fig. 15.1 Pixels in a picture (More pixels means better resolution)

brightness of the picture element is transmitted. In this way, electrical signals are generated for each of the picture elements 1, 2, 3, 4, ... and so on for the first row. When the first row scanning is completed, the scanning begins for the second row and the second row picture elements are scanned. When the scanning for the second row is completed, the third row is scanned and in this way all the rows are scanned. We would discuss more about interlacing and vertical synchronization frequency.

15.2 TV CAMERA TUBES

The image is generated by the television camera tube. There are a number of different camera tubes. One example is the **image orthicon.** In the image orthicon tube (Fig. 15.2) a focussed image is formed on the photocathode by the optical system. An electrically charged image is then formed on a surface called the target mosaic. This implies that each point on the target mosaic acquires a positive charge which is proportional to the brightness of the corresponding picture element of the image. Thus an electrically charged image (charge image) is obtained from the visual image. An electron beam from the electron gun now scans the target mosaic surface. The electron beam is controlled by a set of voltages across horizontal and vertical

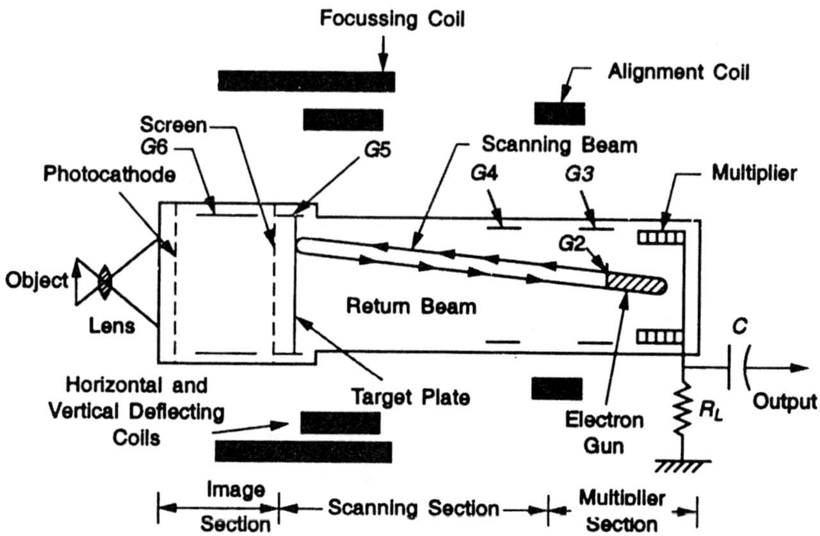

Fig. 15.2 Image orthicon type TV camera

deflection plates. The electron beam takes a time internal of 53·5 microseconds to scan one horizontal line while a retrace time of 1·0 microseconds is required.

In the image orthicon tube, the optical system consists of a photosensitive surface, a target of thin glass and a screen placed very near to the target plate. The number of meshes on the screen is 390 per cm. The photoelectrons are emitted from photocathode in proportion to the image brightness. The screen, which is at a slight positive voltage attracts these electrons, after they pass through the meshes of the screen and strike the target plate. The electric field generated by the ring type grid $G6$ and the axial magnetic field set up by the focussing coils focus electrons emitted from the photocathode to a point on the target plate. The electrons, on striking the target, generate secondary electrons which are attracted to the screen. The target mosaic thus acquires positive charges at various points in proportion to the brightness of the corresponding point of the image.

The other surface of the target mosaic is scanned by the electrons emitted from the electron gun. The cathode of the electron gun is at ground potential. The magnetic field of the alignment coil aligns the electron beam. The focussing of the electron beam is carried out by the magnetic field of the focussing coil and the action of the grid $G4$. The decelaration of the electron beam is done by the electron beam in the horizontal and vertical directions by the sawtooth current in the two sets of deflecting coils.

On the reverse side of the target, an electron beam scans the target. The points at which there are positive charges, the electron beam neutralises the same and the remaining electrons are returned back towards the electron gun. The beam current, thus varies according to the light intensity of the optical image. During the return path, the electrons strike grid G_2 which emits the secondary electrons. The secondary electrons are then deflected into the electron multiplier by grid G_3. The secondary electrons are then multiplied by surfaces kept at increasingly higher voltages. The final output of the multiplier represents an amplified beam current. The output current finally generates an output voltage termed as video voltage. The voltage is developed across the load resistance R_L and then coupled through the capacitor C. The video voltage will depend on the brightness of the pixels. At points where the picture element is dark, the output voltage will be high and at points where the pixel is bright, the output voltage will be low. For intermediate brightness values, the video voltage developed will have intermediate values.

15.3 VIDICON TV CAMERA

The vidicon TV camera is depicted in Fig. 15.3.

In the vidicon TV camera tube, the target is the signal plate. It is a thin transparent conducting film which has a glassplate on one side and a deposit of photoconductive layer on the other side. An optical image is formed on the photoconductive layer. The photoconductive properties of the element depends on the brightness of the image. At points where the image is dark, the photoconductive layer acts as an insulator. At other points where the photoconductive layer is illuminated, the layer becomes partly conducting.

The signal plate is kept at a positive voltage of 30V. The surface on the photoconductive layer on the electron gun side forms a positive potential image of the object to be transmitted.

The electron beam is generated by a cathode which is kept at a potential of – 30V with respect to the signal plate. The electron beam is then controlled by grid G_1 and accelarated by grid G_2. An alignment coil aligns the direction of the beam. The beam is moved horizontally and vertically as needed by necessary deflection coils.

The scanning electron beam deposits electrons on each portion of the potential image till such time that the potential of the corresponding surface acquires the potential of the cathode. Thus a potential difference

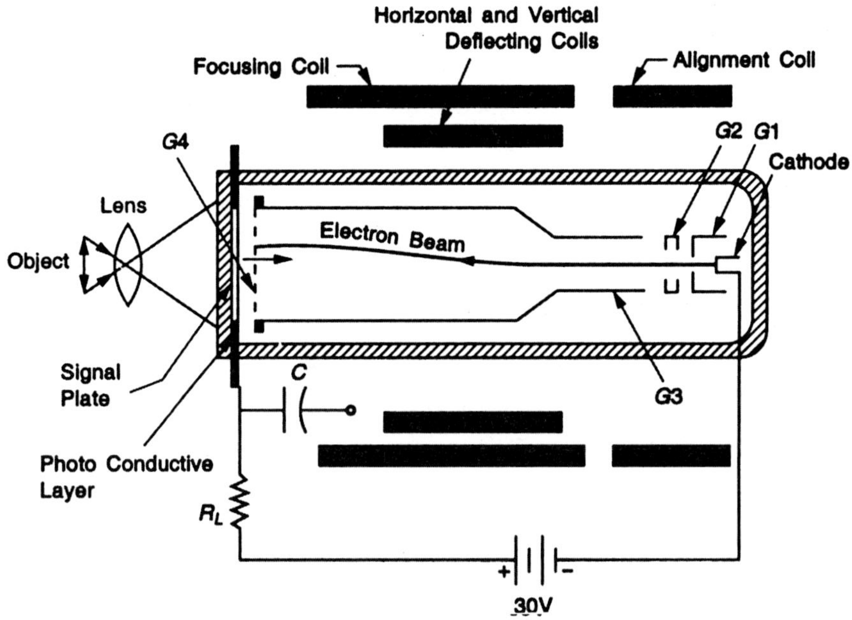

Fig. 15.3 Vidicon TV camera

between the two surfaces of the photoconducting layer being scanned, is produced. This results in a current to flow the load resistance R_L. The current changes with time according to the change in illumination of the successive portions of the photoconductive film. The voltage output developed across R_L is called the output video voltage which is taken through a capacitor *C*. The voltage is then fed through a video amplifier for amplification. The output voltage is a function of the brightness of the optical image. When the picture element is bright, the output voltage is high. If the picture element is dark, the output voltage is low.

The comparison between image orthicon and vidicon camera tubes is depicted in Table in 15.2.

Table 15.2 : Comparison Between Image Orthicon and Vidicon Camera Tubes

	Image Orthicon		*Vidicon*
1.	Operates on the principle of photoemission and electron multiplication	1.	Operates on the property of photoconductivity
2.	Output voltage is low for bright portion of the image	2.	Output voltages is high for bright portion of the image
3.	Output voltage is high for dark portion of the image	3.	Output voltage is low for dark portion of the image
4.	Stable output	4.	Stable output
5.	High sensitivity	5.	High sensitivity
6.	Costly	6.	Cheaper
7.	Linear response characteristics	7.	Non linear response characteristics
8.	Faster	8.	Slower (since photoconductivity changes slowly)
9.	Big size	9.	Small size

15.4 INTERLACED SCANNING

In television, one complete scanning of the picture is necessary for reproducing one complete picture at the television receiver. A large amount of data must be sent during this period. If the picture is scanned at a high rate, the bandwidth required will be very high. If the rate is low, the moving scene would have a stop and jerky effect. Moreover, due to low persistence of TV tube phosphors, the picture intensity decreases between the scans. This produces a flicker at the picture rate. The number of pictures per second should thus be such that at the receiver the flicker will be absent and the picture movement appear to be smooth. The minimum picture rate is thus 35 to 50 pictures per second. In the NTSC system, the frame or picture rate is 30 Hz which is half the supply frequency. The European system employs a 25 Hz frame rate with a 50 Hz a.c. supply.

The 30 Hz frame rate is very small to eliminate flicker. Thus picture frame is divided into two separate fields, each with half the total number of scanning lines. The first picture field consists of lines 1 to 262·5 lines and the second picture field consists of lines 262·5 to 525 lines. They are interlaced. During the first

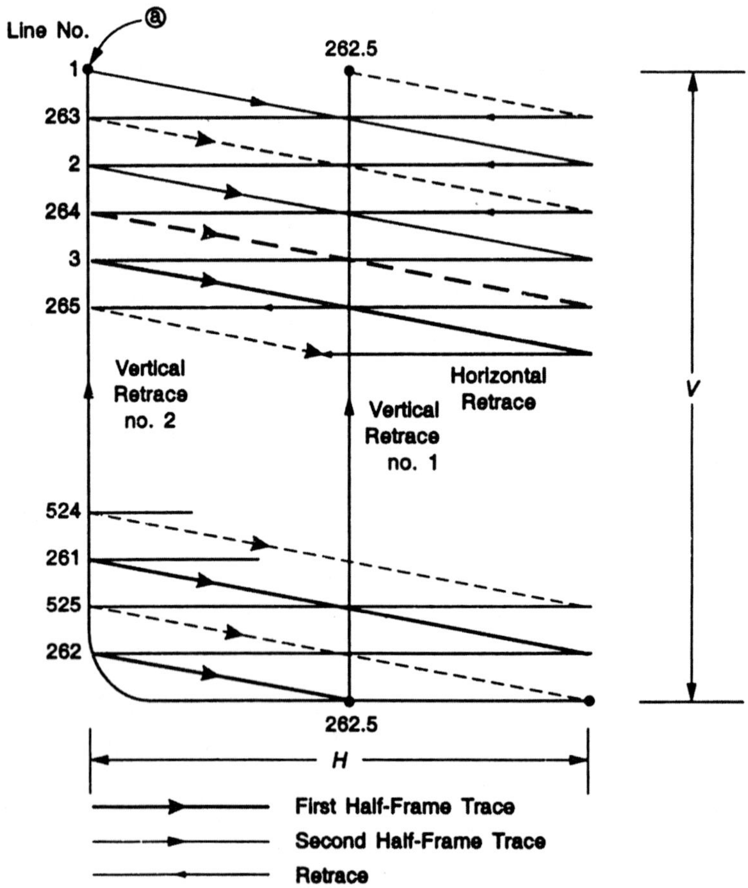

Fig. 15.4 NTSC TV frame and interlaced pattern

field, 262·5 odd numbered lines are scanned from top to bottom of the picture in the order 1, 2, ... upto 262·5. At the 262·5th line scan, the retraces moves the scan spot vertically up in the screen to the top centre and 262·5 lines are scanned during the second field period in the order 263, 264, 265, ... upto 525 from top to bottom. The second field lines are midway between the first field lines to complete the picture. When all the 525 lines are completed, the vertical retrace brings back the spot to the top left (position *a*) in the screen, ready to complete the next frame. Thus the frame repetition rate is 60 Hz which is much higher than the minimum picture rate of 35 to 50 pictures per second.

In each complete picture frame period, the vertical retrace takes place twice. The vertical synchronization frequency (Hz) is thus twice the frame (picture) repetition rate.

In television, the aspect ratio is defined as the ratio of viewing width to viewing height. It has a value of $\frac{4}{3}$.

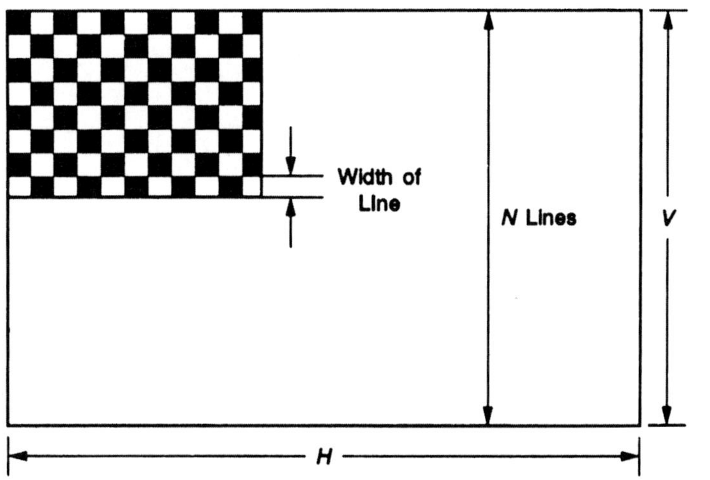

Fig. 15.5 Black and white pixels in the picture

15.5 SYNCHRONIZATION

For the correct reproduction of picture at the TV receiver, the scannings at the transmitter and receiver must be synchronized. This is achieved by synchronization signals which are sent along with the picture from the TV transmitter. The synchronization signals are rectangular type pulses. They are generated periodically at the line and field scanning frequencies. To synchronize the horizontal line scanning, a horizontal sync, pulse is radiated for each horizontal line scanning. There are a total of 525 horizontal lines in each picture frame and the frame scanning rate is 1 / 30 seconds. Thus the total number of horizontal lines scanned per second is 525 × 30 = 15750. The horizontal sync, frequency is thus 15750 Hz. For vertical synchronization, a vertical sync pulse is transmitted for each field. The vertical synchronization frequency is thus 60 Hz. The sync pulses are transmitted during the blanking period so that they do not distrub the picture signal. To carry out the blanking out operation, blanking pulses are also transmitted. When the vertical synchronization pulses are present, the horizontal synchronization is accomplished by serrations that divide the vertical sync pulses into

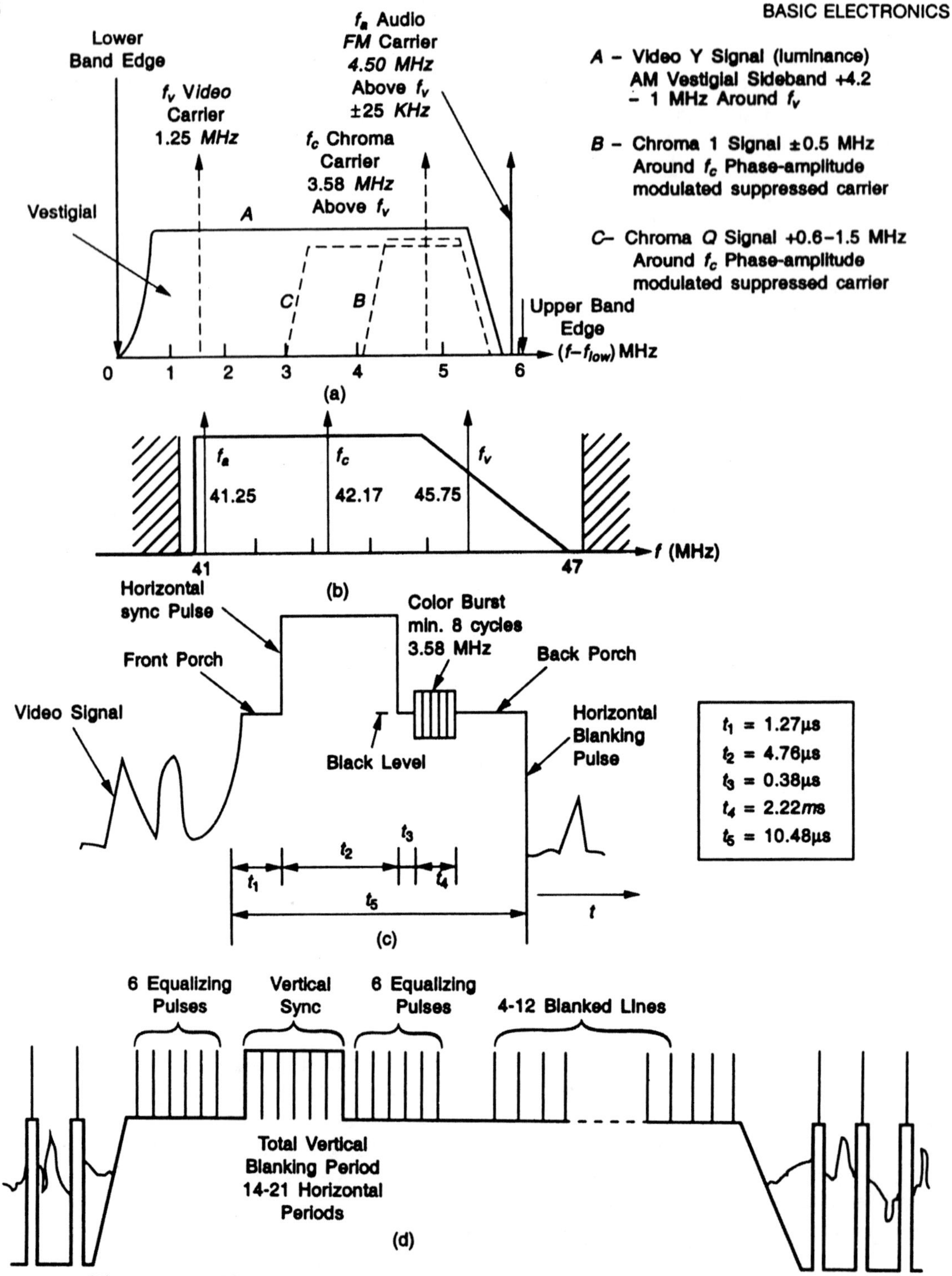

Fig. 15.6 (a) Picture signal with 6 MHz bandwidth (b) Video IF band (c) Horizontal sync pulse shape (d)Vertical sync pulse shape

several blocks. It should be mentioned in this connection that equalizing pulses are also required to reduce the effects of irregularities due to field frequencies.

In a 525 line system, the number of active lines left after deducting lines lost in vertical blanking is $525 - 40 = 485$.

Experiments have shown that a reduction in picture resolution by a factor of 0·7 (known as **Kell factor**) can be tolerated. Thus, vertical resolution = $485 \times 0.7 = 339.5 \cong 340$ lines. Horizontal resolution should also be of the same order. Since aspect ratio is 4 /3, the number of vertical lines for equivalent horizontal resolution is $340 \times 4 / 3 = 453.33 \approx 454$ black and white alternate lines *i.e.* $454 / 2 = 227$ cycles of black and white alternations of elementary areas.

$$\text{Line period} = H = T_H = 63.5 \approx 64\ \mu\text{s}.$$

12 μs are used for blanking of the flyblack retrace.

$$\text{Hence remaining period} = 52\ \mu\text{s}$$

$$\text{Period of square wave} = \frac{52}{227} = 0.2202643 \approx 0.22\ \mu\text{s}$$

$$\text{Hence the bandwidth of the signal is} = \frac{1}{0.22} = 4.54\ \text{MHz}$$

15.6 TELEVISION SIGNAL

The television signal is a complex one. It consists of the following signals :

(i) sound signal

(ii) luminance signal (or picture brightness)

(iii) horizontal synchronization

(iv) vertical synchronization; and

(v) color or chrominance signal

The allocated bandwidth of the TV channel is 6 MHz. The video signal is sent as amplitude modulated vestigial sideband carrier located 1·25 MHz from the lower end of the channel (vide Fig. 15.6(a)). The entire 4·3 MHz upper sideband and 1 MHz lower sideband are sent along with the carrier.

The sound or audio signal is frequency modulated carrier located 4·5 MHz above the video carrier with a bandwidth of 80 kHz about the carrier. Since the sound signal is located above the video, it can be removed easily by filtering action.

The color chrominance signal is double sideband suppressed carrier. They are located at 4·83 MHz above the lower edge of the channel band. The chroma I signal requires a lower sideband of 1·5 MHz and a vestigial upper sideband of 0·6 MHz. The chroma Q signal has a channel bandwidth of $\pm$ 0·5 MHz around 4·83 MHz.

The synchronization (sync) signals are sent during retrace periods in between two lines and in between two fields. The video signal is raised to the black level to blank out during the horizontal retrace time (which is 16·5% of line period). This period is shown as t_5 in Fig. 15.6(c). A horizontal sync pulse (which is $t_2 = 4.76\ \mu\text{s}$) is added to the blanking level. This is also depicted in Fig. 15.6(c). This pulse triggers the horizontal oscillator in the receiver and forces it to synchronize to the signal. In the horizontal blanking period is also sent a burst of 8 cycles of 3·58 MHz chroma carrier. This is used to have a locking of the phase of the receiver chroma oscillator to that of the transmitter for proper demodulation of signals.

During the vertical retrace period, six horizontal sync pulses are sent at twice the rate to have stabilization of synchronization. In the next three horizontal periods, the signal is brought to the synchronizing level and inverted horizontal sync pulses are sent at double the rate to form the vertical sync pulse. Next comes six more equalizing pulses at twice the rate and the normal horizontal synchronizing pulses continue.

15.7 TELEVISION RECEIVERS

The black and white television receivers were developed prior to the development of colour TV receivers. The colour T-V receiver system was designed such that it acted as an additional system without

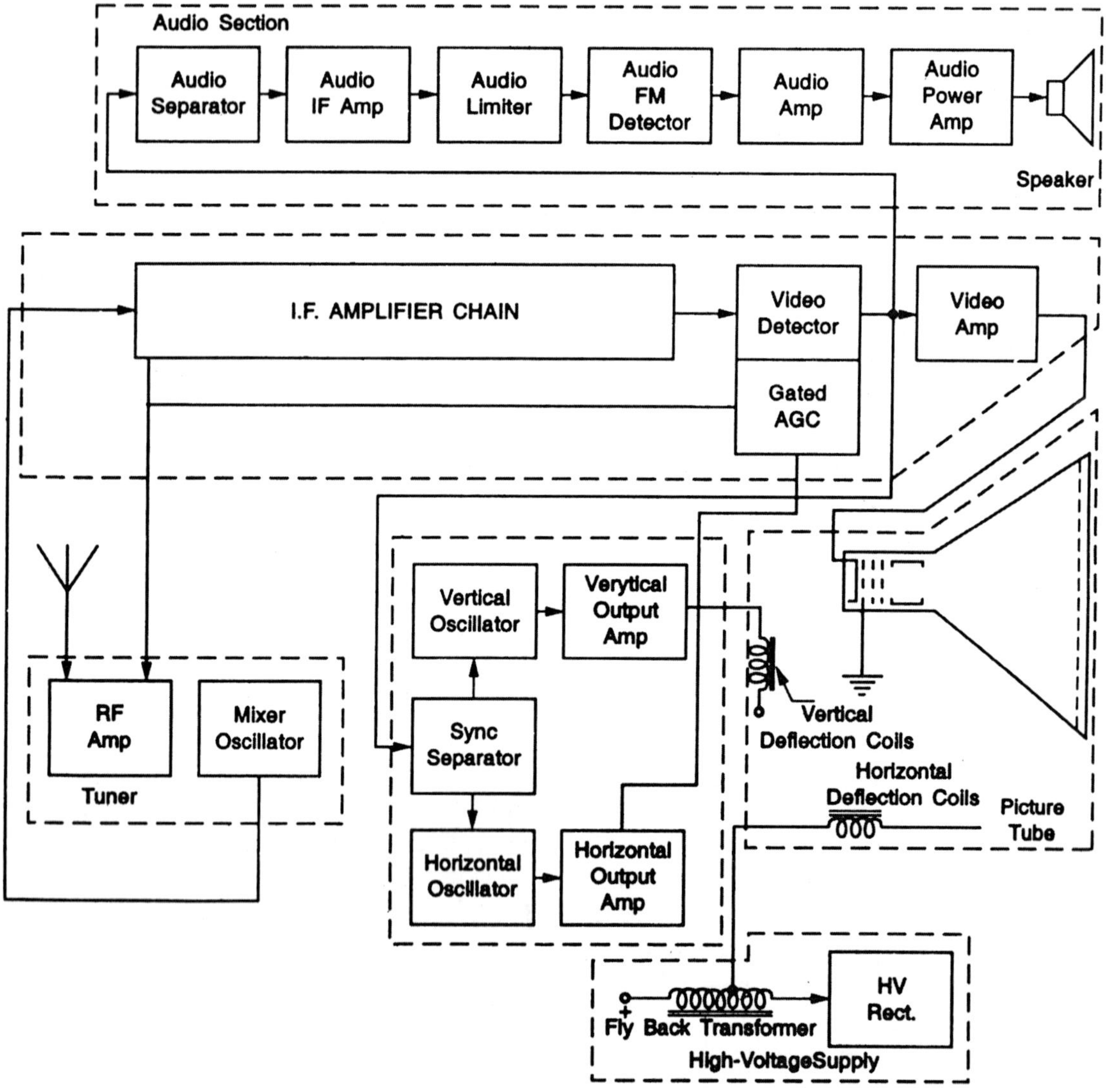

Fig. 15.7 Schematic block diagram of a black and white television receiver set

jeopardizing the performance of black and white TV receivers still in use. Fig. 15.7 depicts the schematic block diagram of a black and white TV receiver set. The TV receiver is fed by either an antenna or a CATV cable. If the outdoor antenna employs a 75Ω coaxial cable, this is to be changed to a new standard DIN 45325 plug (IEC 169–2) and this can be plugged to the antenna jack at the back of the TV set. If the outdoor antenna uses a 300 Ω twin lead flat feeder, this can be changed by a 300 ohms to 75 ohms impedance converter and plugged to the antenna jack at the back of the TV set.

The first block in the TV receiver block diagram is the timer which comprises an RF amplifier and a mixer stage. The RF amplifier is a class *A* broadband small signal amplifier. Normally, two separate tuners are built into the receiver one for the VHF bands (channel 2 to channel 13) and another for the UHF band (channel 14 to channel 83).

Previously, VHF bands were tuned by a turret switch which placed a different coil-capacitor combination for each channel and used a small variable capacitor for fine adjustment. Tuning in UHF bands is accomplished by employing a large variable capacitance for channel selection and a smaller one for fine adjustment. In the present day receiver sets, varactor tuning is very much common with a wafer switch to apply different voltages to the varactor. A potentiometer is used for fine adjustment in timing.

The IF band for the TV receiver occupies the band 41 to 474 MHz with the frequency spectrum inverted (the oscillator frequency being above the channel frequency). The audio carrier is at 41·25 MHz and the video carrier is at 45·75 MHz. Further, the curve is sloped between 44·5 to 47 MHz so as to compensate the effect of the vestigial sideband in the detector. The IF amplifier chain is multistage with 3 to 5 stages. One or more of the amplifiers in the IF stage have AGC applied just as the RF amplifier in the tuner. It should be kept in mind that the AGC maintains the d.c. level of the receiver output constant and smooths out as much of the noise like amplitude fluctuations as possible (it prevents saturation by large signals). A simple envelope detector at the output of the IF amplifier chain is employed for video detection.

The detected video is applied to the input of the audio tuned circuits (separator) that separates the audio carrier at 4·5 MHz alongwith a 80 kHz band around the carrier. The signal then passes through audio IF amplifier, audio limiter, FM detector and audio amplifier. Finally and audio power amplifier applies the signal to the loudspeaker.

The video signal is passed through a video amplifier, which raises the level, filters out the sound carrier and finally produces the output signal. This output is fed as the **control bias for the picture tube electron beam gun.** The light intensity on the face of the CRT is thus modulated.

The video signal is applied to the synchronization (sync.) circuit. In the synchronization circuit, a clamp circuit removes the video portion of the signal to leave only the sync. pulses. **Differentiating the sync. pulses, the horizontal sweep oscillator is triggered.** The linear ramp voltage generated is amplified to drive the horizontal output transformer and horizontal deflection coils. The flyback transformer output is rectified to generate a very high d.c. voltage. This is applied to the picture tube target and this provides the gating signal to control the AGC circuit.

Similarly, integrating the clamped sync. pulses, the generated signals are applied to trigger the vertical oscillator for vertical sweep voltage production. This is then amplified and applied to the vertical deflection coils.

15.8 COLOUR TV RECEIVERS

In a colour TV, it is necessary to decode the colour signals from the video signal before feeding to the picture tube. The following schematic blocks namely the tuners, the IF amplifier chain, the deflection circuits and

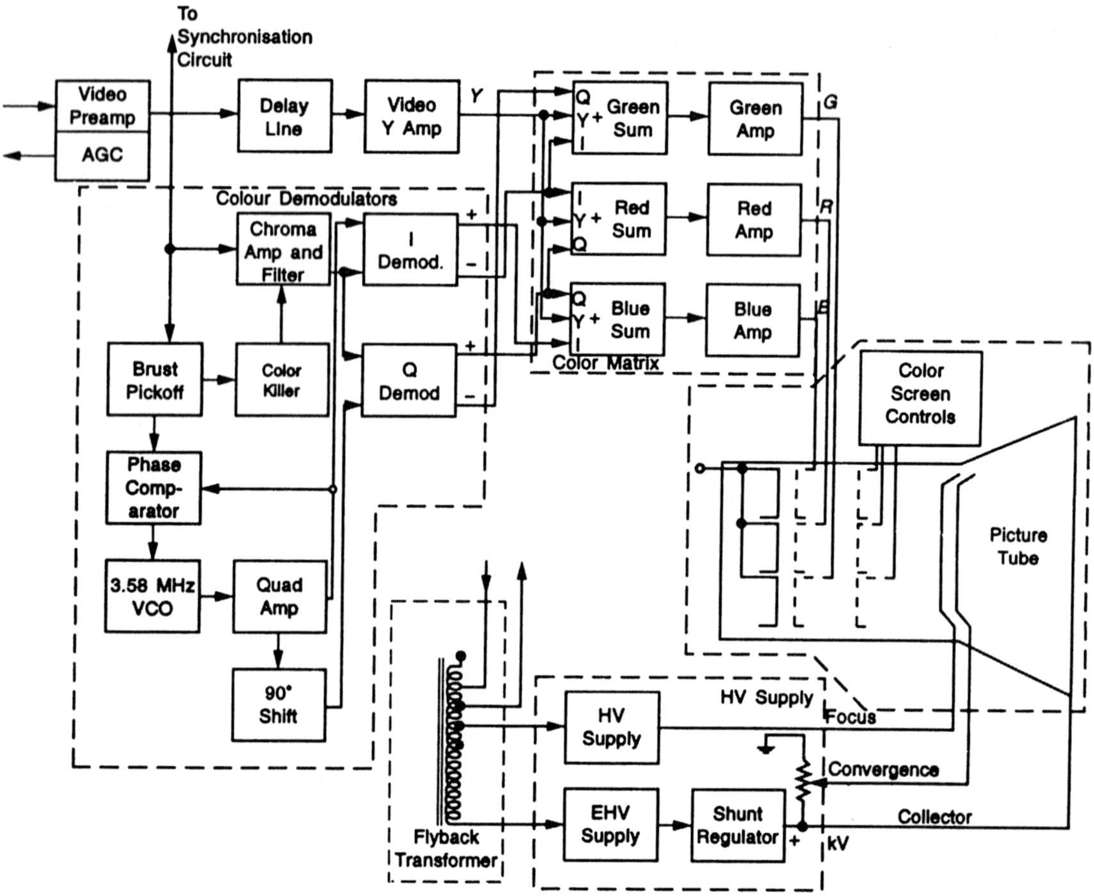

Fig. 15.8 Schematic block diagram of a colour television receiver . Parts of the receiver which are common to a black and white receiver has been shown in this figure.

the audio circuits are the same in a colour TV receivers having similar specifications. The schematic block diagram of a colour TV receiver is depicted in Fig. 15.8.

The receiving antenna output signal is connected to the tuner, (VHF or UHF) and passes through the IF amplifier chain to the video detector. The output of the video detector is connected to the video amplifier. The video preamplifier has a bandwidth of 4·2 MHz. The video amplifier also possesses a notch filter to remove the sound carrier at 4·5 MHz. The output of the video preamplifier is connected to four different paths. In one leg it is connected to a delay line which is then connected to a luminance (*Y*) video amplifier. The luminance video amplifier has a bandwidth of 4·2 MHz. This signal gives the video signal to the picture tube during black and white picture reception. For colour picture reception, it provides much of the minute details of the picture.

In another path, the output of the video preamplifier is connected to synchronization separator and deflection circuits for the vertical and horizontal deflection coils.

In still a third path, the output of the video preamplifier is connected to a chroma amplifier and filter having a + 0·6 MHz/− 1·5 MHz bandwidth around the 3·58 MHz colour carrier frequency. The resulting signal is then connected to colour demodulators.

The fourth connection to which the video preamplifier output in connected is the colour burst pick off circuit. This circuit responds to the 3·58 MHz frequency bursts in the horizontal synchronizing pulses. During black and white reception, I and Q circuits are turned off by the colour killer circuit. The colour burst pick off also gives the phase synchronization signals for locking the 3·58 MHz colour local oscillator.

The functions of the I and Q demodulators is to bring out the two colour signals which are then connected to the colour decode matrix. To the colour decode matrix is also connected the Y signal. The three colours **Green, Blue** and **Red** are then amplified by three amplifiers of the above three colours. The relationship between G, B and R signals with I, Q and Y signals is given by

$$G = -0.28\,I - 0.64\,Q + Y \qquad ...(15.1)$$

$$B = -1.11\,I + 1.73\,Q + Y \qquad ...(15.2)$$

and

$$R = 0.95\,I + 0.62\,Q + Y \qquad ...(15.3)$$

The three colour signals Green (G), Blue (B) and Red (R) excite three different electron beams in the picture tube. The three beams are suitably focussed on the phosphors which are arranged suitably in an alternating dot matrix to build the image.

The high voltage target supply voltages (19 to 24 kilovolts) should have excellent regulation. If the voltage becomes very high, the picture tube will create harmful X-rays. There should be proper shielding arrangements to prevent radiation or flashover.

15.9 COLOUR TV TRANSMITTERS

We are considering colour TV transmitters only because black and white TV transmitters are obsolete now a days. Sometimes black and white moves are shown. But, these signals can be generated very easily by turning off the colour modulators in a colour TV transmitter.

The optical rays from the object passes through the optical lens system which focusses the image and separates it into three beams. The beams pass through colour filters (R, G and B) and camera tubes.

From the output is obtained the three different signals R, G and B. The colour signals are then mixed by the colour matrix circuit such that the signals Y, I and Q are generated. The luminance (Y) signal is

$$Y = +0.30\,R + 0.59\,G + 0.11\,B \qquad ...(15.4)$$

and the In phase (I) and Quadrature (Q) signals are

$$I = +0.60\,R - 0.28\,G - 0.32\,B \qquad ...(15.5)$$

and

$$Q = +0.21\,R - 0.52\,G + 0.31\,B \qquad ...(15.6)$$

The luminance (Y) signal possesses the entire details of the signal. In fact, it contains all the information to construct a black and white TV picture. The Y signal is passed through a Y filter having a bandwidth of 4·2 MHz. The I signal is passed through a I filter having a bandwidth of 1·5 MHz and the Y signal is passed through a Y filter having a bandwidth of 0·5 MHz. The 3·58 MHz chroma carrier oscillator generates the subcarrier for the chroma signals. After necessary phase shifting, the signals are applied to I and Q balanced modulators. The 15750 Hz horizontal, 60 Hz vertical synchronizing signals and the horizontal and vertical blanking pulses are also obtained from the chroma oscillator output after necessary frequency division. These signals determine the scanning of the control system and are also sent to the receiver. The signals are then applied to the summing circuit.

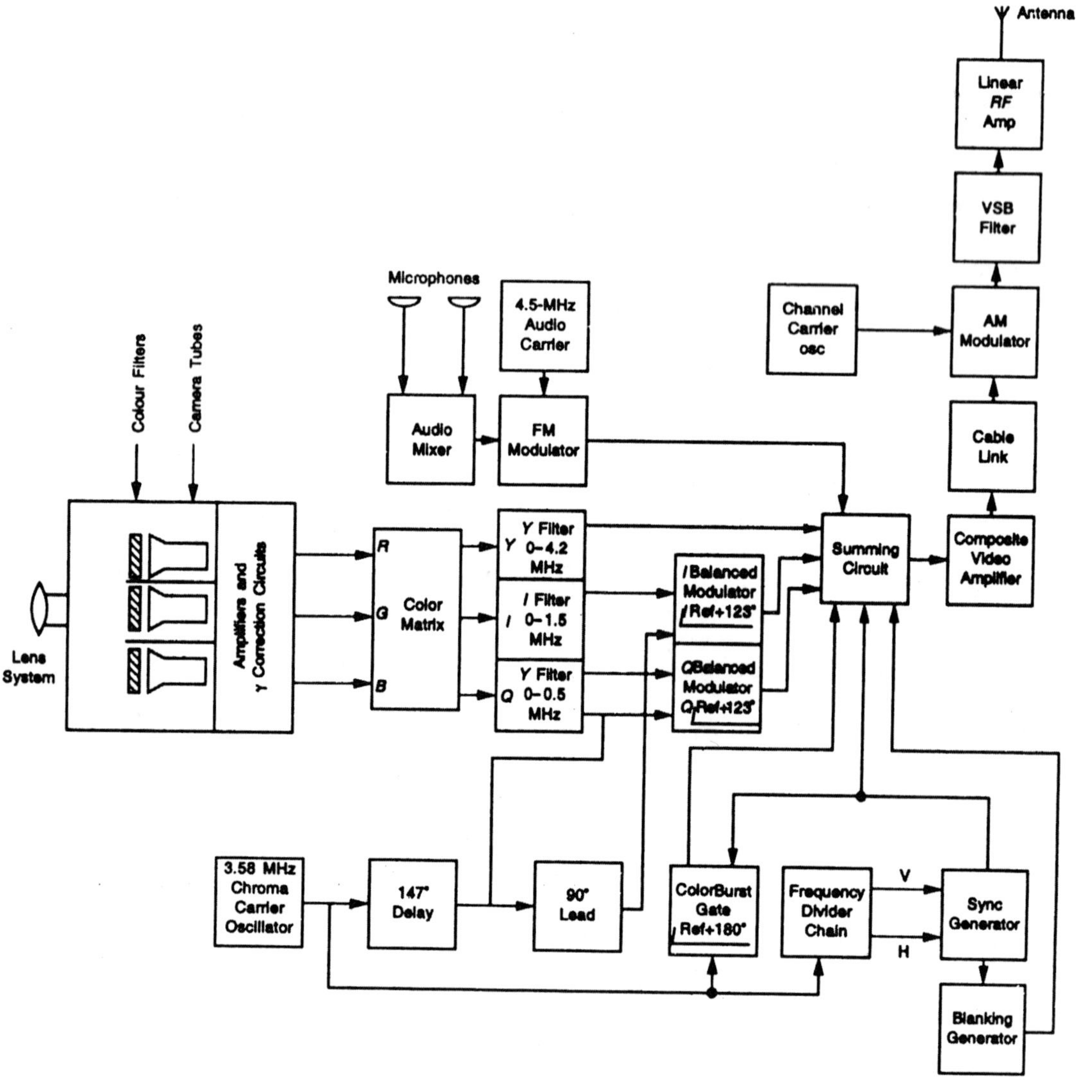

Fig. 15.9 Simplified block schematic of an NTSC colour TV transmitter

The output of the microphones is applied to an audio mixer circuit and is used to frequency modulate the 4·5 MHz audio carrier. The audio carrier is generated by a crystal controlled oscillator.

A summing circuit gives at its output, the composite video signal from the *Y* signal, *I* and *Q* modulated signals, the sound signal, the colour burst, the blanking signals, the horizontal and vertical synchronizing pulses. The video baseband signal has frequency components from d.c. to 4·6 MHz. The video baseband signal is then vestigial sideband modulated at an intermediate frequency such as the channel 2 video carrier at 55·25 MHz. The output signal of the modulator is then passed through a band pass filter to remove all but 1 MHz of the lower sideband and transmit the 1 MHz vestigial lower sideband carrier and full upper sideband. It is then passed through coaxial cable to a point-to-point microwave or satellite link.

Television System Constants

Nominal duration of a horizontal line = 64 μsec

Line blanking period = 11·8 to 12·3 μsec

Front porch = 1·3 to 1·8 μsec

Back porch = 5·5 to 6·1 μsec

Line synchronization pulse = 4·5 to 4·9 μsec

Build uptime for line synchronization pulse = 0·2 to 0·4 μsec

Vertical field period = 20000 μsec

Field Blanking period = 1164 to 1420 μsec

Duration of pre-equalizing pulse sequence = 155 to 165 μsec

Equalizing pulse width = 2·25 to 2·45 μsec

Review Questions

1. What do you understand by the term 'television' ?
2. Distinguish between NISC, PAL and SECAM systems.
3. What are the important qualities of picture ?
4. State the frequency bands, the corresponding frequency range and the corresponding bandwidth allocated to TV.
5. What do you understand by the term 'pixel' ?
6. What is a TV camera tube ? Discuss the image orthicon type TV camera tube.
7. What is a vidicon camera tube. Discuss its details.
8. Make a comparison between image orthicon and vidicon camera tubes.
9. What is meant by interlaced scanning ? Distinguish between picture and frame for NTSC TV transmission.
10. Draw a schematic diagram showing the video *Y* signal.
11. Illustrate with suitable diagrams, the horizontal sync pulse shape.
12. Illustrate with suitable diagrams, the vertical sync pulse shape.
13. What is aspect ratio, Kell factor ?
14. Draw the schematic block diagram of a black and white television receiver set and explain its important functions.
15. Mention the different types of tuners for VHF and UHF bands in a black and white TV receiver.
16. What is a colour TV receiver ? Mention the functioning of a colour TV receiver with suitable block schematic diagram.
17. What are the problems that can arise if the regulation of the high voltage supply is poor in a colour TV receiver ?
18. Mention with suitable block schematics the principle of operation of the NTSC colour TV transmitter.
19. In the composite video signal, state the duration of the following in microseconds
    (a) nominal duration of a horizontal line
    (b) line blanking period
    (c) front porch
    (d) back porch
    (e) vertical field period
    (f) duration of pre-equalizing pulse sequence.

Objective Type Questions

(Tick mark the correct or most appropriate answer)

1. The basic frame rate in PAL (Phase Alternation Lines) system is
(a) 25 (b) 30
(c) 50 (d) 60 per second
2. The allowed BW assigned in TV channel is
(a) 4.75 (b) 5.25 (c) 5.75 (d) 6 Mhz
3. The TV channel frequency range 476–482 MHz falls in
(a) VHF (b) UHF
(c) SHF (d) microwave range
4. A typical cable TV distribution system (CATV) employ coaxial cables and may have a bandwidth of the order of
(a) 6 (b) 50
(c) 300 (d) 2400 MHz
5. In a TV camera tube, the time taken to scan one horizontal line is
(a) 63.5 (b) 53.5
(c) 10.45 (d) 5.5 microseconds
6. A picture field in TV has
(a) 262.5 (b) 525
(c) 625 (d) 50 lines
7. The sound or audio signal is located at
(a) 4 (b) 4.5 (c) 5 (d) 6 MHz
above the video carrier.
8. In a colour TV receiver, the video preamplifier has a bandwidth of
(a) 4 (b) 4.2 (c) 5 (d) 6 MHz
9. The value of the aspect ratio in television is
(a) 0.7 (b) 1 (c) 1.33 (d) 2
10. The value of the Kell factor in a TV picture is
(a) 0.7 (b) 1 (c) 1.33 (d) 2
11. The first block in a TV receiver set is
(a) tuner (b) synch. seperator
(c) video amplifier (d) flyback transformer
12. Present day television receiver sets commonly employ
(a) turret switch with different LC combinations
(b) varactor tuning
(c) large variable capacitors for tuning
(d) RC tuning
13. The sound or audio signal in a television is
(a) FM with carrier at 4.5 MHz above the lower end of the channel
(b) FM with carrier at 4.5 MHz above the video carrier
(c) AM with carrier at 1.25 MHz above the lower end of the channel
(d) AM with carrier frequency at 5.75 MHz above the lower end of the channel
14. In a colour TV, the luminance signal is the
(a) Red signal (b) Green signal
(c) Blue signal (d) a mixture of Red, Green and Blue signals
15. The vertical field period in television is
(a) 1164 (b) 64
(c) 1.8 (d) 20000 microseconds

Appendix-A

Frequency response of BJT Amplifiers

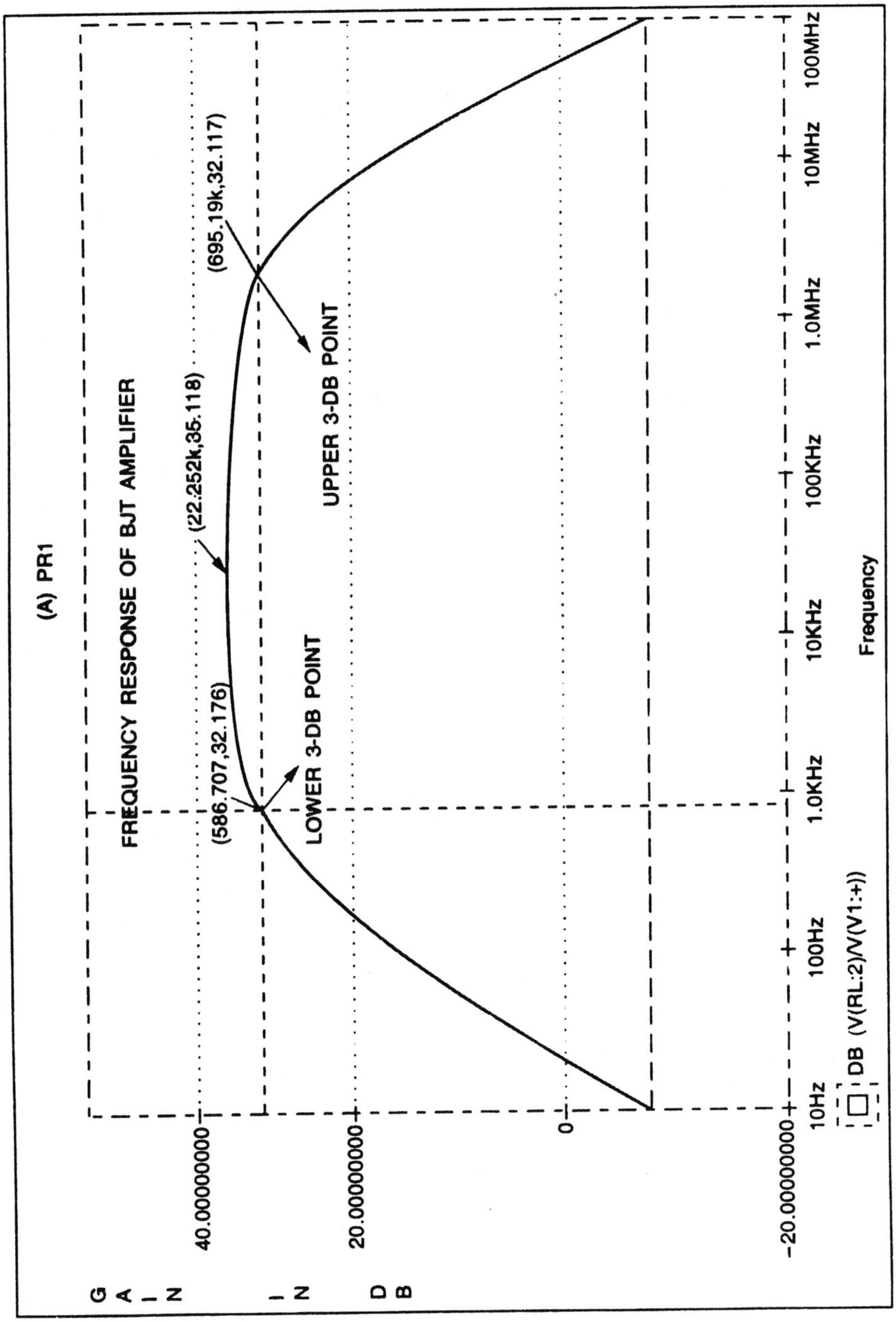

Frequency response of JFET Amplifiers (using PSPICE)

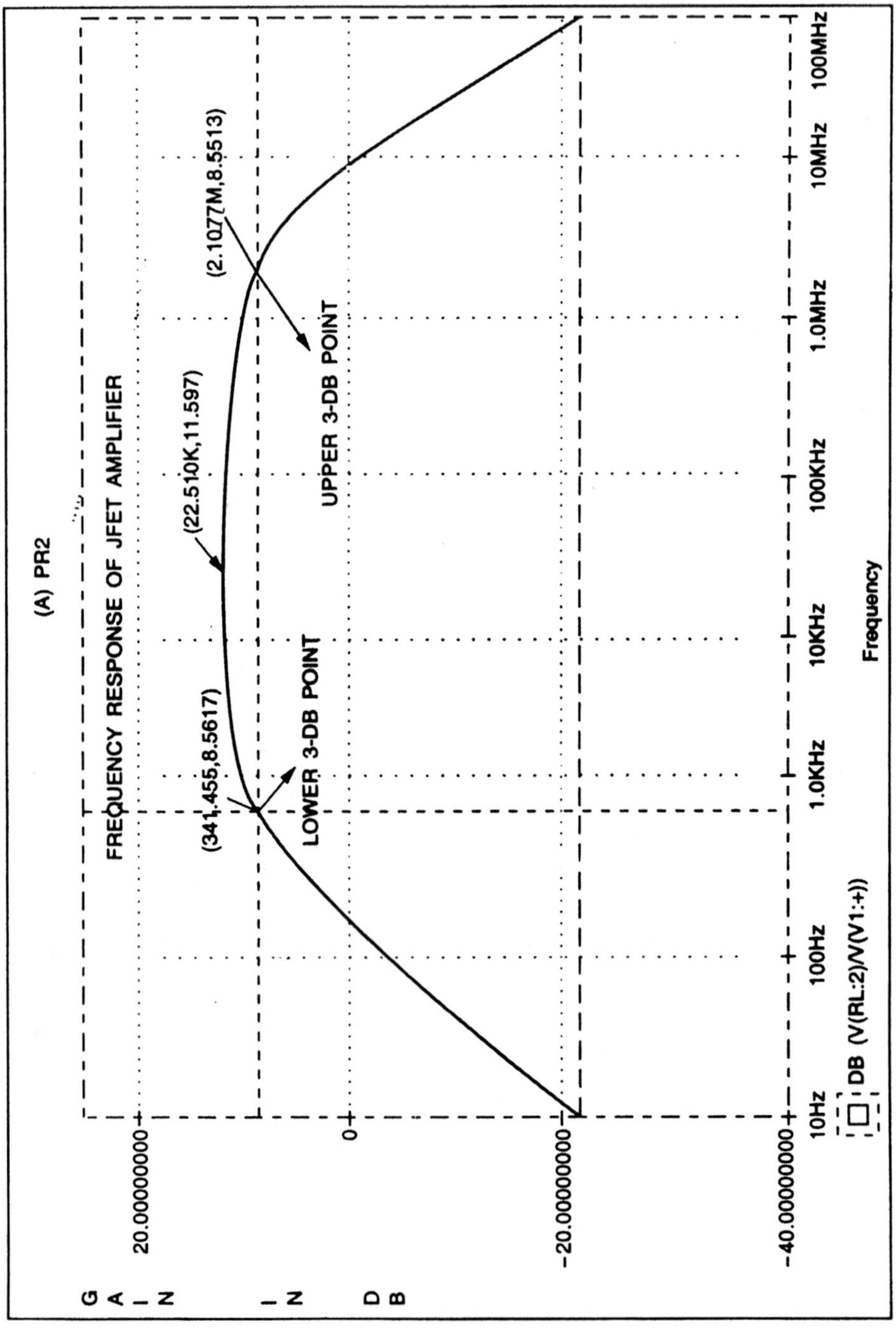

Appendix-B

Case structures of bipolar junction transistors

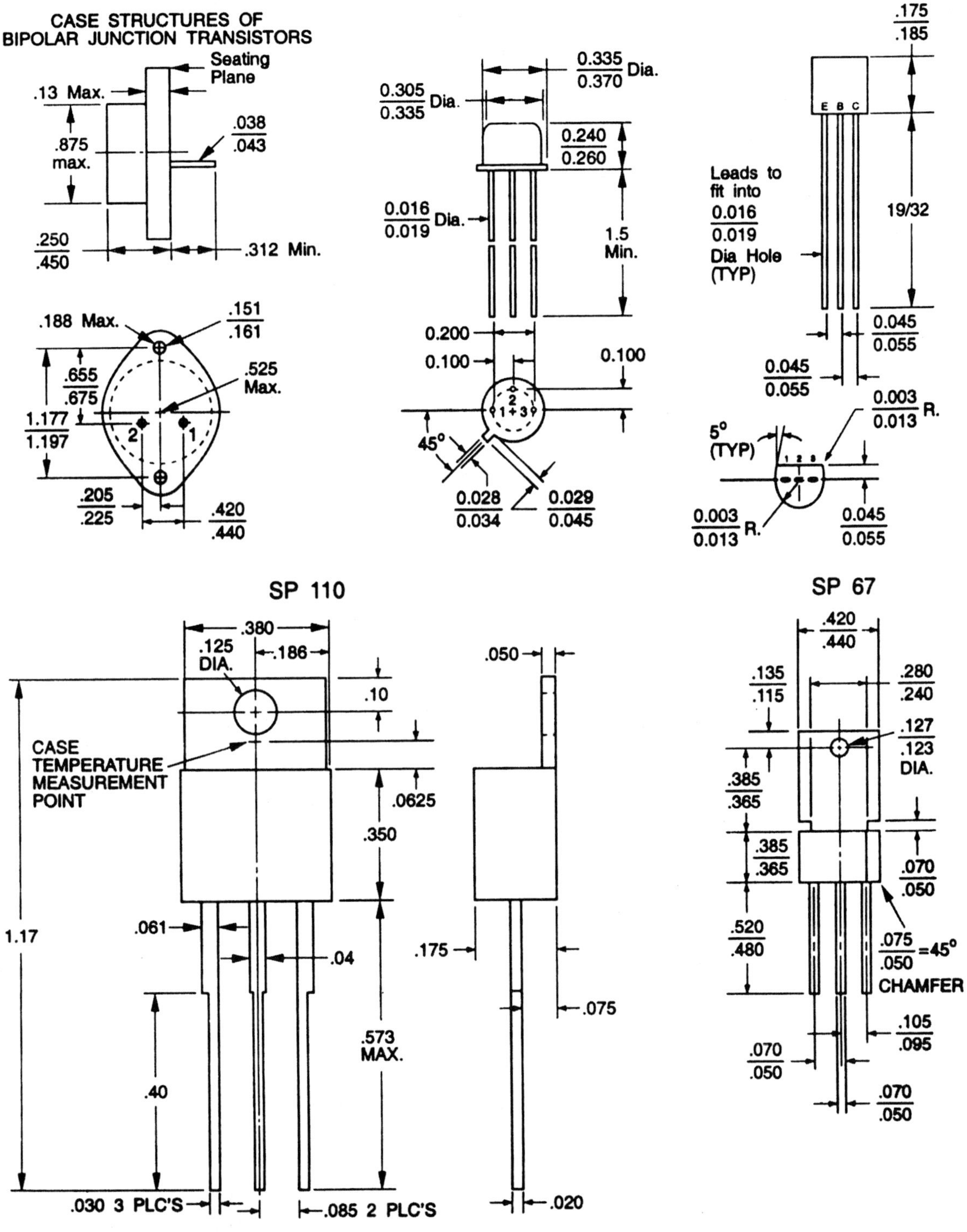

Appendix-C

Derivation of the equation bandwidth, BW = $\frac{0.35}{\text{Risetime}}$

For RC circuit shown in Fig. C.1, let the bandwidth is BW and the risetime is t_r. The charging or discharging of capacitor in the series RC circuit is determined by

$$v = (V - V_0)(1 - e^{-t/RC}) + V_0 \quad \text{...(C.1)}$$

where V_0 is the initial voltage in the capacitor C and v is the instantaneous value. V is the final voltage on the capacitor.

If the capacitor with initial voltage $V_0 = 0$ volts charges up, the time t_1 for the voltage to reach the 10% level is

$$0.1\,V = V(1 - e^{-t_1/RC}) \quad \text{...(C.2)}$$

or, $$0.9\,V = V e^{-t_1/RC}$$

or, $$t_1 = 0.105RC \quad \text{...(C.3)}$$

Similarly, the time t_2 taken to charge upto 90% level is

$$0.9\,V = V(1 - e^{-t_2/RC}) \quad \text{...(C.4)}$$

or, $$0.1 = e^{-t_2/RC}$$

or, $$t_2 = 2.303\,RC \quad \text{...(C.5)}$$

The risetime t_r is the time interval taken to charge up from 10% to 90% level.

Thus t_r is

$$\begin{aligned} t_r &= t_2 - t_1 \quad \text{...(C.6)} \\ &= (2.303 - 0.105)\,RC \\ &= 2.198\,RC \end{aligned}$$

or, $$RC = \frac{t_r}{2.198} \quad \text{...(C.7)}$$

The bandwidth of an RC circuit is determined from the response to a sinusoidal input. Assuming the lower frequency to be zero (or dc), the bandwidth has a value equal to the cut off frequency. Also, at cut off

$$R = \frac{1}{2\pi fC} \quad \text{...(C.8)}$$

or, Bandwidth $$\begin{aligned} \text{BW} &= \frac{1}{2\pi\,RC} \\ &= \frac{1}{2\pi\left(\frac{t_r}{2.198}\right)} \\ &= \frac{0.35}{t_r} \\ &= \frac{0.35}{\text{Risetime}} \quad \text{...(C.9)} \end{aligned}$$

Physical Constants

Charge of electron $e = 1{\cdot}602 \times 10^{-19}$ Coulomb

Rest mass of electron $= m_0 = 9{\cdot}11 \times 10^{-31}$ *Kgs*

Speed of light in free space $= c = 2{\cdot}998 \times 10^{11}$ *cms/sec*

Planck's constant $h = 6{\cdot}62 \times 10^{-27}$ *erg sec*

Boltzmann constant $k_\beta = 1{\cdot}38 \times 10^{-16}$ *erg/k*

$1eV = 1{\cdot}602 \times 10^{-12}$ *ergs* $= 1{\cdot}602 \times 10^{-19}$ Joules

Permittivity of free space $= 8{\cdot}854 \times 10^{-12}$ Farad/*m*

Permeability of free space $= \mu_0 = 4\pi \times 10^{-7}$

V_{on} Klitzing Constant $R_{K-90} = \dfrac{h}{e^2} = 25812{\cdot}807$ *ohms*

Josephoon constant $K_{J-90} = \dfrac{2e}{h} = 483597{\cdot}9$ *GHz/volt*

KEY TO OBJECTIVE QUESTIONS

CHAPTER-1

1d 2c 3d 4a 5c 6c 7d 8a 9a 10c 11d 12c 13c 14d 15b 16b 17c 18a 19c

CHAPTER-2

1a 2b 3a 4b 5d 6c 7d 8b 9c 10b 11d 12d 13c 14c 15b 16a 17c 18b 19d

CHAPTER-3

1c 2c 3a 4b 5d 6b 7d 8a 9a 10b 11d 12d 13c 14d 15b 16c

CHAPTER-4

1d 2c 3b 4a 5c 6b 7b 8a 9d 10b 11b

CHAPTER-5

1c 2b 3b 4d 5a 6c 7d 8a 9b 10c 11a 12c 13a 14c 15a 16b

CHAPTER-6

1b 2b 3a 4c 5c 6d 7b 8a 9b 10c 11b 12d 13b 14a

CHAPTER-7

1e 2b 3a 4c 5d 6b 7b 8a 9c 10a 11a 12c 13a 14c 15c 16d 17b 18a 19b 20a 21c

CHAPTER-8

1b 2d 3d 4a 5c 6d 7a 8a 9a 10b 11c 12b

CHAPTER-9

1c 2c 3a 4b 5b 6c 7a 8d 9a 10b 11c 12c 13a 14b 15b 16b 17a 18a

CHAPTER-10

1b 2d 3d 4b 5d 6c 7d 8a 9c 10b 11a 12d 13b 14c

CHAPTER-11

1a 2b 3b 4c 5d 6a 7b 8b 9c 10b 11d 12c 13d 14d 15d

CHAPTER-12

1b 2a 3a 4d 5d 6c 7a 8b 9d 10b 11b 12b 13c 14a 15d 16a 17d 18b 19d
20d 21d 22c 23d 24a 25c 26c 27c 28b

CHAPTER-13

1d 2a 3a 4b 5c 6d 7c 8c 9a 10c 11a 12a 13a 14d 15c 16d 17c

CHAPTER-14

1a 2d 3d 4d 5b 6a 7d 8a 9c 10c 11b 12c 13b 14a 15c 16b 17c 18d 19c

CHAPTER-15

1a 2d 3b 4c 5d 6a 7b 8b 9c 10a 11a 12b 13b 14d 15d

INDEX